MW01098505

To dear E Be
the Coauthor of our future book
on GMDH
With big respect and love

A. Ivakhnenko 1996

Inductive Learning Algorithms for Complex Systems Modeling

Hema R. Madala

Department of Mathematics and Computer Science
Clarkson University
Potsdam, New York

Alexy G. Ivakhnenko

Ukrainian Academy of Sciences
Institute of Cybernetics
Kiev, Ukraine

CRC Press
Boca Raton Ann Arbor London Tokyo

Madala, Hema Rao.
Inductive learning algorithms for complex systems modeling / Hema Rao Madala and Alexey G. Ivakhnenko.
p. cm.
Includes bibliographical references and index.
ISBN 0-8493-4438-7
1. System analysis. 2. Algorithms. 3. Machine learning.
I. Ivakhnenko, Aleksei Grigo'evich. II. Title.
T57.6.M313 1993 93-24174
003--dc20 CIP

International Standard Book Number 0-8493-4438-7
Library of Congress Card Number 93-24174
Printed in the United States of America 1 2 3 4 5 6 7 8 9 0
Printed on acid-free paper

Preface

One can see the development of automatic control theory from single-cycled to the multicycled systems and to the development of feedback control systems that have brainlike network structures (Stafford Beer). The pattern recognition theory has a history of about fifty years—beginning with single-layered classificators, it developed into multi-layered neural networks and from there to connectionist networks. Analogical developments can be seen in the cognitive system theory starting with the simple classifications of the single-layered perceptrons and further extended to the system of perceptrons with the feedback links. The next step is the stage of "neuronets."

One of the great open frontiers in the study of systems science, cybernetics, and engineering is the understanding of the complex nonlinear phenomena which arise naturally in the world we live in. Historically, most achievements were based on the deductive approach. But with the advent of significant theoretical breakthroughs, layered inductive networks, and associated modern high-speed digital computing facilities, we have witnessed progress in understanding more realistic and complicated underlying nonlinear systems. Recollect, for example, the story of Rosenblatt's perceptron theory. Until recently, the absence of good mathematical description with the demonstration by Minsky and Papert (1969) that only linear descrimination could be represented by two-layered perceptron, led to a waning of interest in multilayered networks. Still Rosenblatt's terminology has not been recovered; for example, we say "hidden units" instead of Rosenblatt's "association units" and so on.

Moving in the direction of unification we consider the inductive learning technique called Group Method of Data Handling (GMDH), the theory originated from the theory of perceptron and is based on the principle of self-organization. It was developed to solve the problems of pattern recognition, modeling, and predictions of the random processes. The new algorithms that are based on the inductive approach are very similar to the processes in our brain. Scientists who took part in the development have accepted "this science" as a unification of pattern recognition theory, cybernetics, informatics, systems science, and various other fields. Inspite of this, "this science" is quickly developing, and everybody feels comfortable in using "this science" for complex problem-solving. This means that this new scientific venture unifies the theories of pattern recognition and automatic control into one metascience. Applications include the studies on environmental systems, economical systems, agricultural systems, and time-series evaluations. The combined Control Systems (CCS) group of the Institute of Cybernetics, Kiev (Ukraine) has been a pioneering leader in many of these developments. Contributions to the field have come from many research areas of different disciplines. This indicates a healthy breadth and depth of interest in the field and a vigor in associated research. Developments could be more effective if we become more attentive to one another.

Since 1968 layered perceptron-like networks have been used in inductive learning algorithms, particularly in the training mode. The algebraic and the finite-difference type of polynomial equations which are linear in coefficients and nonlinear in parameters are used for the process predictions. In the network, many arbitrary links of connection-weights are obtained, several partial equations are generated, and the links are selected by our choice. The approach was originally suggested by Frank Rosenblatt to choose the coefficients of the first layer of links in a random way.

The polynomials of a discrete type of Volterra series (finite-difference and algebraic forms) are used in the inductive approach for several purposes:

First—for the estimation of coefficients by the least-squares method using explicit or implicit patterns of data. When the eigenvalues of characteristic equation are too small, this method leads to very biased estimates and the quality of predictions is decreased. This problem is avoided with the developments of objective systems analysis and cluster analysis.

Second—the polynomial equations are used for the investigation of selection characteristic by using the consistency (Shannon's displacement) criterion of minimum according to Shannon's second-limit theorem (analogical law is known in communication theory). The structure of optimal model is simplified when the noise dispersion in the data is increased. When Shannon's displacement is present, selection of two-dimensional model structures is used. When the displacement is absent, the selection of two one-dimensional model structures are used—first, the optimal set of variables, then the optimal structure of the model are found. The use of objective criteria in canonical form simplifies this procedure further.

Third—the use of polynomial equations are organized "by groups" in the selection procedure to get a smooth characteristic with single minimum. Selection "by groups" allows one to apply the simple stopping rule "by minimum" or "by the left corner rule." In multilevel algorithms, for example, each group includes a model candidate of similar structure of an equal number of members; and

Fourth—the equations are used to prove the convergence of iteration processes in multilayered algorithms. The convergence exists for some criteria in a mean-square sense called internal convergence; for others it is called external convergence. In the latter case, there is a necessity for certain "regularization" means.

This book covers almost last twenty years of research—from basic concepts to the recent developments in inductive learning algorithms conducted by the CCS group.

Chapter 1 is concerned with the basic approach of induction and the principle of self-organization. We also describe the selection criteria and general features of the algorithms.

Chapter 2 considers various inductive learning algorithms: multilayer, single-layered combinatorial, multi-layered aspects of combinatorial, multi-layered with propagating residuals, harmonical algorithms, and some new algorithms like correlational and orthogonalized partial descriptions. We also describe the scope of long-range quantitative predictions and levels of dialogue language generalization with subjective versus multilevel objective analysis.

Chapter 3 covers noise immunity of algorithms in analogy with the information theory. We also describe various selection criteria, their classification and analysis, the aspects of the asymptotic properties of external criteria, and the convergence of algorithms.

Chapter 4 concentrates on the description of physical fields and their representation in the finite-difference schemes, as these are important in complex systems modeling. We also explain the model formulations of cyclic processes.

Chapter 5 coverage is on how unsupervised learning or clustering might be carried out with the inductive type of learning technique. The development of new algorithms like objective computerized clustering (OCC) is presented in detail.

Chapter 6 takes up some of the applications related to complex systems modeling such as weather modeling, ecological, economical, agricultural system studies, and modeling of solar activity. The main emphasis of the chapter is on how to use specific inductive learning algorithms in a practical situation.

Chapter 7 addresses application of inductive learning networks in comparison with the artificial neural networks that work on the basis of averaged output error. The least mean-square (LMS) algorithm (adaline), backpropagation, and self-organization boolean-logic techniques are considered. Various simulation results are presented. One notes that the backpropagation technique which is encouraged by many scientists, is only one of several possible ways to solve the systems of equations to estimate the connection coefficients of a feed-forward network.

Chapter 8 presents the computational aspects of basic inductive learning algorithms. Although an interactive software package for inductive learning algorithms which includes multilayer and combinatorial algorithms was recently released as a commercial package (see *Soviet Journal of Automation and Information Sciences* N6, 1991), the basic source of these algorithms along with the harmonical algorithm are given in chapter 8.

The book should be useful to scientists and engineers who have experience in the scientific aspects of information processing and who wish to be introduced to the field of inductive learning algorithms for complex systems modeling and predictions, clustering, and neural-net computing, especially these applications.

This book should be of interest to researchers in environmental sciences, macro-economical studies, system sciences, and cybernetics in behavioural and biological sciences because it shows how existing knowledge in several interrelated computer science areas intermesh to provide a base for practice and further progress in matters germane to their research.

This book can serve as a text for senior undergraduate or for students in their first year of a graduate course on complex systems modeling. It approaches the matter of information processing with a broad perspective, so the student should learn to understand and follow important developments in several research areas that affect the advanced dynamical systems modeling. Finally, this book can also be used by applied statisticians and computer scientists who are seeking new approaches.

The scope of these algorithms is quite wide. There is a wide perspective in which to use these algorithms; for example, multilayered theory of statistical decisions (particularly in case of short-data samples) and algorithm of rebinarization (continued values recovery of input data). The "neuronet," that is realized as a set of more developed component-perceptrons in the near future, will be similar to the House of Commons, in which decisions are accepted by the voting procedure. Such voting networks solve problems related to pattern recognition, clustering, and automatic control. There are other ideas of binary features applied in the application of "neuronets," especially when every neuron unit is realized by two-layered Rosenblatt's perceptron.

The authors hope that these new ideas will be accepted as tools of investigation and practical use—the start of which took place twenty years ago for original multilayered algorithms. We invite readers to join us in beginning "this science" which has fascinating perspectives.

H. R. Madala and A. G. Ivakhnenko

Acknowledgments

We take this opportunity to express our thanks to many people who have, in various ways, helped us to attain our objective of preparing this manuscript.

Going back into the mists of history, we thank colleagues, graduate students, and co-workers at the Combined Control Systems Division of the Institute of Cybernetics, Kiev (Ukraine). Particularly, one of the authors (HM) started working on these algorithms in 1977 for his doctoral studies under the Government of India fellowship program. He is thankful to Dr. V.S. Stepashko for his guidance in the program. Along the way, he received important understanding and encouragement from a large group of people. They include Drs. B.A. Akishin, V.I. Cheberkus, D. Havel, M.A. Ivakhnenko, N.A. Ivakhnenko, Yu.V. Koppa, Yu.V. Kostenko, P.I. Kovalchuk, S.F. Kozubovskiy, G.I. Krotov, P.Yu. Peka, S.A. Petukhova, B.K. Svetalskiy, V.N. Vysotskiy, N.I. Yunusov and Yu.P. Yurachkovskiy. He is grateful for those enjoyable and strengthening interactions.

We also want to express our gratitude and affection to Prof. A.S. Fokas, and high regard for a small group of people who worked closely with us in preparing the manuscript. Particularly, our heartfelt thanks go to Cindy Smith and Frank Michielsen who mastered the TeX package. We appreciate the help we received from the staff of the library and computing services of the Educational Resources Center, Clarkson University. We thank other colleagues and graduate students at the Department of Mathematics and Computer Science of Clarkson University for their interest in this endeavor.

We are grateful for the help we received from CRC Press, particularly, Dr. Wayne Yuhasz, Executive Editor. We also would like to acknowledge the people who reviewed the drafts of the book—in particular, Prof. N. Bourbakis, Prof. S.J. Farlow and Dr. H. Rice—and also Dr. I. Havel for fruitful discussions during the manuscript preparation.

We thank our families. Without their patience and encouragement, this book would not have come to this form.

Contents

Chapter 1
Introduction

1 SYSTEMS AND CYBERNETICS

Civilization is rapidly becoming very dependent on large-scale systems of men, machines, and environment. Because such systems are often unpredictable, we must rapidly develop a more sophisticated understanding of them to prevent serious consequences. Very often the ability of the system to carry out its function (or alternatively, its catastrophically failing to function) is a property of the system as a whole and not of any particular component. The single most important rule in the management of large scale systems is that one must account for the entire system—the sum of all the parts. This most likely involves the discipline of "differential games." It is reasonable to predict that cybernetic methods will be relevant to the solution of the greatest problems that face man today.

Cybernetics is the science of communication and control in machines and living creatures [133]. Nature employs the best cybernetic systems that can be conceived. In the neurological domain of living beings, the ecological balance involving environmental feedback, the control of planetary movements, or the regulation of the temparature of the human body, the cybernetic systems of nature are fascinating in their accuracy and efficiency. They are cohesive, self-regulating and stable systems; yet they do have the remarkable adaptability to change and the inherent capacity to use experience of feedback to aid the learning process.

Sustained performance of any system requires regulation and control. In complicated machinery the principles of servomechanism and feedback control have long been in effective use. The control principles in cybernetics are the error-actuated feedback and homeostasis. Take the case of a person driving a car. He keeps to the desired position on the road by constantly checking the deviation through visual comparison. He then corrects the error by making compensating movements of the steering wheel. Error sensing and feedback are both achieved by the driver's brain which coordinates his sight and muscular action. Homeostasis is the self-adjusting property that all living organisms possess and that makes use of feedback from the environment to adjust metabolism to changing environmental conditions. Keeping the temperature of the human body constant is a good example of homeostasia.

The application of cybernetics to an environmental situation is much more involved than the servomechanism actuating "feedback correction." The number of variables activating in the system are plentiful. The variables behave in stochastic manner and interactive relationships among them are very complex. Examples of such systems in nature are meteorological and environmental systems, agricultural crops, river flows, demographic systems, pollution, and so on. According to complexity of interactions with various influences in nature, these are called cybernetical systems. Changes take place in a slow and steady manner, and any suddenness of change cannot be easily perceived. If these systems are not studied continuously by using sophisticated techniques and if predictions of changes are not allowed to accumulate, sooner or later the situation is bound to get out of hand.

The tasks of engineering cybernetics (self-organization modeling, identification, optimal control, pattern recognition, etc.) require development of special theories which, although look different, have many things in common. The commonality among theories that form the basis of complex problem-solving has increased, indicating the maturity of cybernetics as a branch of science [37]. This leads to a common theory of self-organization modeling that is a combination of the deductive and inductive methods and allows one to solve complex problems. The mathematical foundations of such a common theory might be the approach that utilizes the black box concept as a study of input and output, the neural approach that utilizes the concept of threshold logic and connectionism, the inductive approach that utilizes the concept of inductive mechanism for maintaining the composite control of the system, the probabilistic approach that utilizes multiplicative functions of the hierarchical theory of statistical decisions, and Gödel's mathematical logic approach (incompleteness theorem) that utilizes the principle of "external complement" as a selection criterion.

The following are definitions of terms that are commonly used in cybernetic literature and the concept of black box.

1.1 Definitions

1. A *system* is a collection of interacting, diverse elements that function (communicate) within a specified environment to process information to achieve one or more desired objectives. Feedback is essential, some of its inputs may be stochastic and a part of its environment may be competitive.
2. The *environment* is the set of variables that affects the system but is not controlled by it.
3. A *complex system* has five or more internal and nonlinear feedback loops.
4. In a *dynamic system* the variables or their interactions are functions of time.
5. An *adaptive system* continues to achieve its objectives in the face of a changing environment or deterioration in the performance of its elements.
6. The rules of behavior of a *self-organizing system* are determined internally but modified by environmental inputs.
7. *Dynamic stability* means that all time derivatives are controlled.
8. A *cybernetic system* is complex, dynamic, and adaptive. Compromise (optimal) control achieves dynamic stability.
9. A *real culture* is a complex, dynamic, adaptive, self-organizing system with human elements and compromise control. Man is in the feedback loop.
10. A *cybernetic culture* is a cybernetic system with internal rules, human elements, man in the feedback loop, and varying, competing values.
11. *Utopia* is a system with human elements and man in the feedback loop.

The characteristics of various systems are summarized in Table 1.1, where 1 represents "always present" and a blank space represents "generally absent." The differences among the characteristics of utopia and cybernetic culture are given in Table 1.2.

Table 1.1. Characteristics of various systems

Characteristics	System	Complex system	Dynamic system	Adaptive system	Self-organizing system	Cybernetic system	Real culture	Cybernetic culture	Utopia
Collection of interacting, diverse elements, process information, specified environment, goals feedback	1	1	1	1	1	1	1	1	1
At least five internal and nonlinear feedback loops		1				1	1	1	
Variables and interactive functions of time			1			1	1	1	
Changing environment deteriorating elements				1		1	1	1	
Internal rules					1		1	1	
Compromise (Optimal) control						1 1	1 1	1 1	
Human elements							1	1	1
Dynamic stability						1		1	
Man in feedback loop						1	1	1	1
Values varying in time and competing						1		1	

Table 1.2. Differences among utopia and cybernetic culture

Characteristic	Utopia	Cybernetic culture
Size	Small	Large
Complex	No	Yes
Environment	Static, imaginery	Changing, real
Elements deteriorate	No	Yes
Rules of behavior	External	Internal
Control	Suboptimized	Compromised
Stability	Static	Dynamic
Values	Fixed	Varying, Competing
Experimentation	None	Evolutionary operation

1.2 Model and simulation

Let us clarify the meaning of the words *model* and *simulation.* At some stage a *model* may have been some sort of small physical system that paralleled the action of a large system; at some later stage, it may have been a verbal description of that system, and at a still later—and hopefully more advanced—stage, it may have consisted of mathematical equations that somehow described the behavior of the system.

A model enables us to study the various functions and the behavioral characteristics of a system and its subsystems as well as how the system responds to given changes in inputs or reacts to changes in parameters or component characteristics. It enables us to study the extent to which outputs are directly related to changes in inputs—whether the system tends to return to the initial conditions of a steady state after it has been disturbed in some way, or whether it continues to oscillate between the control limits. A cybernetic model can help us to understand which behavior is relevant to or to what extent the system is responsible for changes in environmental factors.

Simulation is a numerical technique for conducting experiments with mathematical and logical models that describe the behavior of a system on a computer over extended periods of time with the aim of long-term prediction, planning, or decision-making in systems studies. The most convenient form of description is based on the use of the finite-difference form of equations.

Experts in the field of simulation bear a great responsibility, since many complex problems of modern society can be properly solved only with the aid of simulation. Some of these problems are economic in nature. Let us mention here models of inflation and of the growing disparity between rich and poor countries, demographic models, models for increased food production, and many others. Among the ecological problems, primary place is occupied by problems of environmental pollution, agricultural crops, water reservoirs, fishing, etc. It is well known that mathematical models, with the connected quantities that are amenable to measurement and formalization, play very important roles in describing any process or system. The questions solved and the difficulties encountered during the simulation complex systems modeling are clearly dealt with in this book.

It is possible to distinguish three principal stages of the development of simulation:

(Experts without computers) → (Man-machine Dialogue systems) → (Computer without experts)

We are still at the first stage; man-machine dialogue systems are hardly used at this time. Predictions are realized in the form of two or three volumes of data tables compiled on the basis of the reasoning of "working teams of experts" who basically follow certain rules of thumb. Such an approach can be taken as "something is better than nothing." However, we cannot stay at this stage any longer.

The second stage, involving the use of both experts and computers, is at present the most advanced. The participation of an expert is limited to the supplying of proper algorithms in building up the models and the criteria for choosing the best models with optimal complexity. The decisions for contradictory problems are solved according to the multi-objective criteria.

The third stage, "computers without experts," is also called "artificial intelligence systems." The man-machine dialogue system based on the methods of inductive learning algorithms is the most advanced method of prediction and control. It is important that the artificial intelligence systems operate better than the brain by using these computer-aided algorithms. In contrast to the dialogue systems, the decisions in artificial intelligence systems are made on the basis of general requests (criteria) of the human user expressed in a highly abstract metalanguage. The dialogue is transferred to a level at which contradictions between humans are impossible, and, therefore, the decisions are objective and convincing. For example, man can make the requirement that "the environment be clean as possible," "the prediction very accurate," "the dynamic equation most unbiased," and so on. Nobody would object to such general criteria, and man can almost be eliminated from the dialogue of scientific disputes.

In the dialogue systems, the decisions are made at the level of selection of a point in the "Pareto region" where the contradiction occurs. This is solved by using multi-criterion analysis. In artificial intelligence systems, the discrete points of Pareto region are only inputs for dynamic models constructed on the basis of inductive learning algorithms. Ultimately, the computer will become the arbiter who resolves the controversies between users and will play a very important role in simulations.

1.3 Concept of black box

The black box concept is a useful principle of cybernetics. A black box is a system that is too complex to be easily understood. It would not be worthwhile to probe into the nature of interrelations inside the black box to initiate feedback controls. The cybernetic principle of black box, therefore, ignores the internal mechanics of the system but concentrates on the study of the relationship between input and ouput. In other words, the relationship between input and output is used to learn what input changes are needed to achieve a given change in output, thereby finding a method to control the system.

For example, the human being is virtually a black box. His internal mechanism is beyond comprehension. Yet neurologists have achieved considerable success in the treatment of brain disorders on the basis of observations of a patient's responses to stimuli. Typical cybernetic black box control action is clearly discernible in this example. Several complex situations are tackled using the cybernetic principles. Take the case for instance, of predictions of agricultural crop productions. It would involve considerable time and effort to study the various variables and their effect on each other and to apply quantitative techniques of evaluation. Inputs like meteorological conditions, inflow of fertilizers and so on influence crop production. It would be possible to control the scheduling and quantities of various controllable inputs to optimise output. It is helpful to think of the determinants of any "real culture" as it would be the solution of a set of independent simultaneous equations with many unknowns.

Mathematics can be an extremely good tool in exhausting all the possibilities in that it can get a complete solution of the set of equations (or whatever the case may be). Many mathematicians have predicted that entirely new branches of mathematics would someday have to be invented to help solve problems of society—just as a new mathematics was necessary before significant progress could be made in physics. Scientists have been thinking more and more about interactive algorithms to provide the man-machine dialogue, the intuition, the value judgement, and the decision on how to proceed. Computer-aided self-organization algorithms have given us the scope to the present developments and appear to provide the only means for creating even greater cooperative efforts.

2 SELF-ORGANIZATION MODELING

2.1 Neural approach

Rosenblatt [105], [106] gives us the theoretical concept of "perceptron" based on neural functioning. It is known that single-layered networks are simple and are not capable of solving some problems of pattern recognition (for example, XOR problem) [95]. At least two stages are required: $X \rightarrow H$ transformation, and $H \rightarrow Y$ transformation. Although Rosenblatt insists that $X \rightarrow H$ transformation be realized by random links, $H \rightarrow Y$ transformation is more deterministically only realized by learned links where $X, H,$ and Y are input, hidden, and output vectors. This corresponds to an *a priori* and conditional probabilistic links in Bayes' formulae:

$$p(y_j) = \sum_{1}^{N} \left[p_0 \prod_{i=1}^{m} p(y_j/x_i) \right]; \quad j = 1, 2, \cdots, n, \tag{1.1}$$

where p_0 is an *a priori* link corresponding to the $X \rightarrow H$ transformation, $p(y_j/x_i)$ are conditional links corresponding to the $H \rightarrow Y$ transformation, N is the sample size, m and n are the number of vector components in X and Y, respectively. Consequently, the perceptron structures have two types of versions: probabilistic or nonparametric and parametric. Here our concern is parametric network structures. Connection weights among the $H \rightarrow Y$ links are established using some adaptive techniques. Our main emphasis is on an optimum adjustment of the weights in the links to achieve desired output. Eventually neural nets have become multilayered feedforward network structures of information processing as an approach to various problem-solving.

We understand that information is passed on to the layered network through the input layer, and the result of the network's computation is read out at the output layer. The task of the network is to make a set of associations of the input patterns $\mathbf{x}$ with the output patterns $\mathbf{y}$. When a new input pattern is put in the configuration, its output pattern must be identifiable by its association.

An important characteristic of any neural network like "adaline" or "backpropagation" is that output of each unit passes through a threshold logic unit (TLU). A standard TLU is a threshold linear function that is used for binary categorization of feature patterns. Nonlinear transfer functions such as sigmoid functions are used as a special case for continuous output. When the output of a unit is activated through the TLU, it mimics a biological neuron turning "on" or "off." A state or summation function is used to compute the capacity of the unit. Each unit is analyzed independently of the others. The next level of interaction comes from mutual connections between the units; the collective phenomenon is considered from loops of the network. Due to such connections, each unit depends on the state of many other units. Such an unbounded network structure can be switched over to a self-organizing mode by using a certain statistical learning law that connects specific forms of acquired change through the synaptic weights, one that connects present to past behavior in an adaptive fashion so positive or negative outcomes of events serve as signals for something else. This law could be a mathematical function—either as an energy function which dissipates energy into the network or an error function which measures the output residual error. A learning method follows a procedure that evaluates this function to make pseudo-random changes in the weight values, retaining those changes that result in improvements to obtain optimum output response. The statistical mechanism helps in evaluating the units until the network performs a desired computation to obtain certain accuracy in response to the input signals. It enables the network to adapt itself to the examples of what it should be doing and to organize information within itself and thereby learn.

Connectionist models

Connectionist models describe input-output processes in terms of activation patterns defined over nodes in a highly interconnected network [24], [107]. The nodes themselves are elementary units that do not directly map onto meaningful concepts. Information is passed through the units and an individual unit typically will play a role in the representation of multiple pieces of knowledge. The representation of knowledge is thus parallel and distributed over multiple units. In a connectionist model the role of a unit in the processing is defined by the strength of its connections—both excitatory and inhibitory—to other units. In this sense "the knowledge is in the connections," as connectionist theorists like to put it, rather than in static and monolithic representations of concepts. Learning, viewed within this framework, consists of the revision of connection strengths between units. Back propagation is the technique used in the connectionist networks—revision of strength parameters on the basis of feedback derived from performance and emergence of higher order structures from more elementary components.

2.2 Inductive approach

Inductive approach is similar to neural approach, but it is bounded in nature. Research on induction has been done extensively in philosophy and psychology. There has been much work published on heuristic problem-solving using this approach. Artificial intelligence is the youngest of the fields concerned with this topic. Though there are controversial discussions on the topic, here the scope of induction is limited to the approach of problem-solving which is almost consistent with the systems theory established by various scientists.

Pioneering work was done by Newell and Simon [96] on the computer simulation of human thinking. They devised a computer program called the General Problem Solver (GPS) to simulate human problem-solving behavior. This applies operators to objects to attain targetted goals; its processes are geared toward the types of goals. A number of similarities and differences among the objective steps taken by computer and subjective ways of a human-operator in solving the problem are shown. Newell and Simon [97] and Simon [113] went on to develop the concepts on rule-based objective systems analysis. They discussed computer programs that not only play games but which also prove theorems in geometry, and proposed the detailed and powerful variable iteration technique for solving test problems by computer.

In recent years, Holland, Holyoak, Nisbett and Thagard [25] considered, on similar grounds, the global view of problem-solving as a process of search through a state space; a problem is defined by an initial state, one or more goal states to be reached, a set of operators that can transform one state into another, and constraints that an acceptable solution must meet. Problem-solving techniques are used for selecting an appropriate sequence of operators that will succeed in transforming the initial state into a goal state through a series of steps. A selection approach is taken on classifying the systems. This is based on an attempt to impose rules of "survival of the fittest" on an ensemble of simple productions.

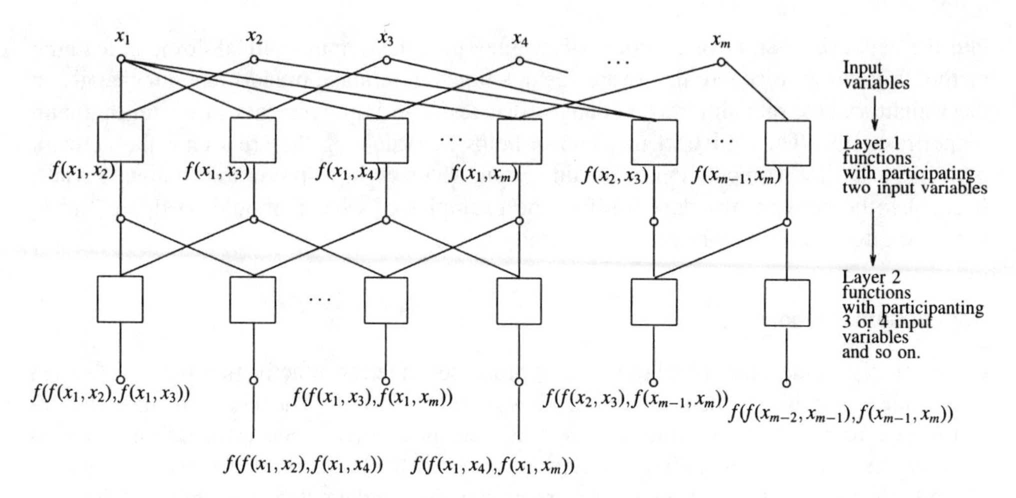

Figure 1.1. Multilayered induction for gradual increase of complexity in functions

This ensemble is further enhanced by criterion rules which implement processes of genetic cross-over and mutation on the productions in the population. Thus, productions that survive a process of selection are not only applied but also used as "parents" in the synthesis of new productions. Here an "external agent" is required to play a role in laying out the basic architecture of those productions upon which both selective and genetic operations are performed. These classification systems do not require any *a priori* knowledge of the categories to be identified; the knowledge is very much implicit in the structure of the productions; i.e., it is assumed as the *a priori* categorical knowledge is embedded in the classifying systems. The concepts of "natural selection" and "genetic evolutions" are viewed as a possible approach to normal levels of implementation of rules and representations in information processing models.

In systems environment there are dependent $(y_1, y_2, \cdots, y_n)$ and independent variables $(x_1, x_2, \cdots, x_m)$. Our task is to know which of the independent variables activate on a particular dependent variable. A sufficient number of general methods are available in mathematical literature. Popular among them is the field of applied regression analysis. However, general methods such as regression analysis are insufficient to account for complex problem-solving skills, but those are backbone for the present day advanced methods. Based on the assumption that composite (control) systems must be based on the use of signals that control the totality of elements of the systems, one can use the principle of induction; this is in the sense that the independent variables are sifted in a random fashion and activated them so that we could ultimately select the best match to the dependent variable.

Figure 1.1 shows a random sifting of formulations that might be related to a specific dependent variable, where $f(\ )$ is a mathematical formulation which represents a relationship among them. This sort of induction leads to a gradual increase of complexity and determines the structure of the model of optimal complexity. Figure 1.2 shows another type of induction that gives formulations with all combinations of input variables; in this approach, model of optimal complexity is never missed. Here the problem must be fully defined. The initial state, goal state, and allowable operators (associated with the differences among current state and goal state) must be fully specified. The search takes place step by step at all the units through alternative categorizations of the entities involved in the set up. This type of processing depends on the parallel activity of multiple pieces of emperical knowledge that compete with and complement each other based on an external knowledge in revising

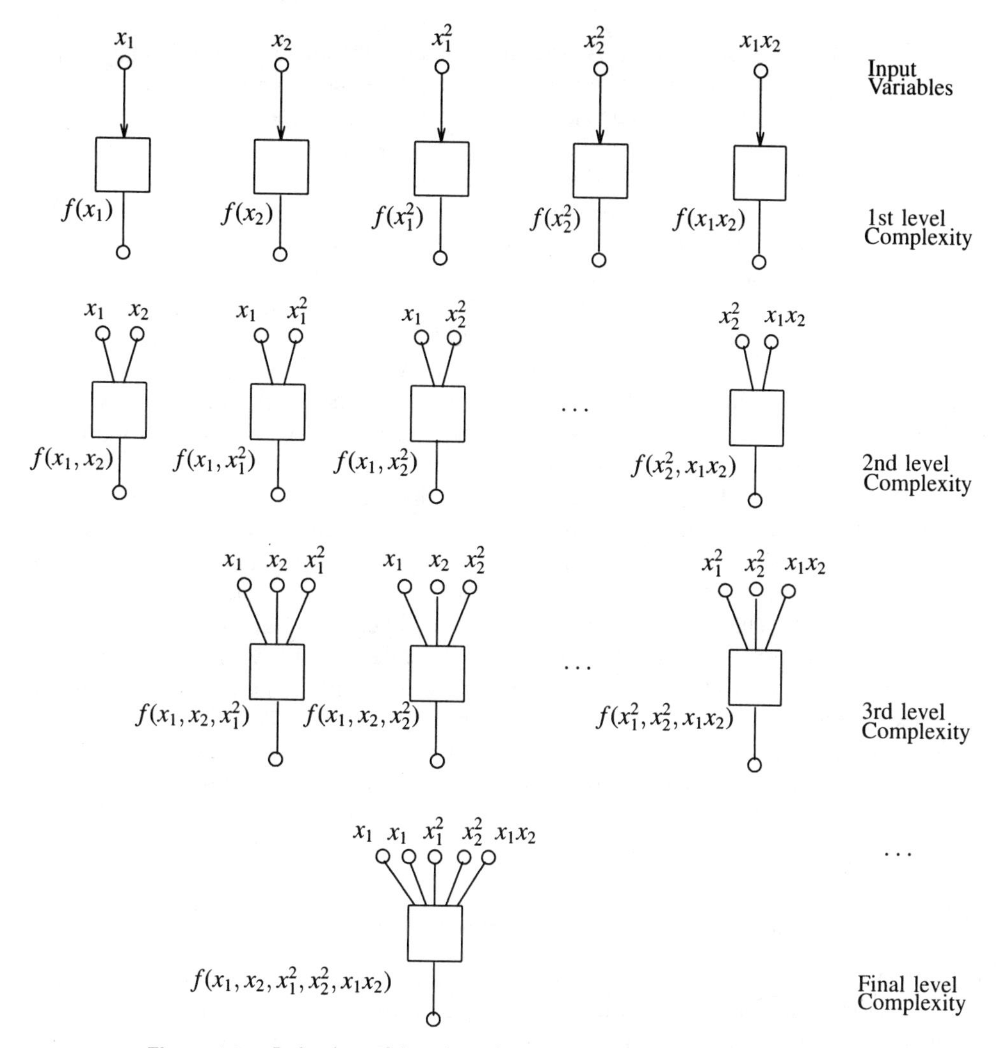

Figure 1.2. Induction of functions for all combinations of input variables

the problem. Such interactive parallelism is a hallmark of the theoretical framework for induction given here.

Simplification of self-organization is regarded as its fundamental problem from the very beginning of its development. The modeling methods created for the last two decades based on the concepts of neural and inductive computing ensure the solution of comprehensive problems of complex systems modeling as applied to cybernetical systems. They constitute an arsenal of means by which—either on the basis of notions concerning system structures and the processes occurring in them, or on the basis of observations of the parameters of these systems—one can construct system models that are accessible for direct analysis and are intended for practical use.

3 INDUCTIVE LEARNING METHODS

Inductive learning methods are also called Group Method of Data Handling (GMDH), Self-organization, sorting out, and heuristic methods. The framework of these methods differs slightly in some important respects. As seen in Chapter 2, the inductive learning algorithms (ILA) have two fundamental processes at their disposal: bounded network connections for generating partial functions and threshold objective functions for establishing competitive learning. The principal result of investigations on inductive learning algorithms (not so much of the examples of computer-designed models presented here), is of a change in view about cybernetics as a science of model construction, in general, and of the role of modern applied mathematics. The deductive approach is based on the analysis of cause-effect relationships. The common opinion is that in the man-machine dialogue, the predominant role is played by the human operator; whereas, the computer has the role of "large calculator." In contrast, in a self-organization algorithm, the role of human operator is passive—he is no longer required to have a profound knowledge of the system under study. He merely gives orders and needs to possess only a minimal amount of *a priori* information such as (i) how to convey to the computer a criterion of model selection that is very general, (ii) how to specify the list of feasible "reference functions" like polynomials or rational functions and harmonic series, and (iii) how to specify the simulation environment; that is, a list of possible variables. The objective character of the models obtained by self-organization is very important for the resolution of many scientific controversies [22]. The man-machine dialogue is raised to the level of a highly abstract language. Man communicates with the machine, not in the difficult language of details, but in a generalized language of integrated signals (selection criteria or objective function). Self-organization restores the belief that a "cybernetic paradise" on earth, governed by a symbiosis between man (the giver of instructions) and machine (an intelligent executer of the instructions) is just around the corner. The self-organization of models can be regarded as a specific algorithm of computer artificial intelligence. Issues like "what features are lacking in traditional techniques" and "how is it compensated in the present theory" are discussed before delving into the basic technique and important features of these methods.

3.1 Principal shortcoming in model development

First of all, let us recollect the important invention of Heisenberg's uncertainty principle from the field of quantum theory which has a direct or indirect influence on later scientific developments. Heisenberg's works became popular between 1925 and 1935 [23], [102]. According to his principle, a simultaneous direct measurement between the coordinate and momentum of a particle with an exactitude surpassing the limits is impossible; furthermore, a similar relationship exists between time and energy. Since his results were published, various scientists have independently worked on Heisenberg's uncertainty principle.

In 1931, Gödel published his works on mathematical logic showing that the axiomatic method itself had inherent limitations and that the principal shortcoming was the so-called inappropriate choice of "external complement." According to his well-known incompleteness theorem [126], it is in principle impossible to find a unique model of an object on the basis of empirical data without using an "external complement" [10]. The regularization method used in solving ill-conditioned problems is also based on this theorem. Hence "external complement" and "regularization" are synonyms expressing the same concept.

In regression analysis, the root mean square (RMS) or least square error determined on the basis of all experimental points monotonically decreases when the model complexity gradually increases. This drops to zero when the number of coefficients n of the model becomes equal to the number of empirical points N. Every equation that possesses n coefficients can be regarded as an absolutely accurate model. It is not possible, in principle, to find a unique model in such a situation. Usually experienced modellers use trial and error techniques to find a unique model without stating that they consciously or unconsciously

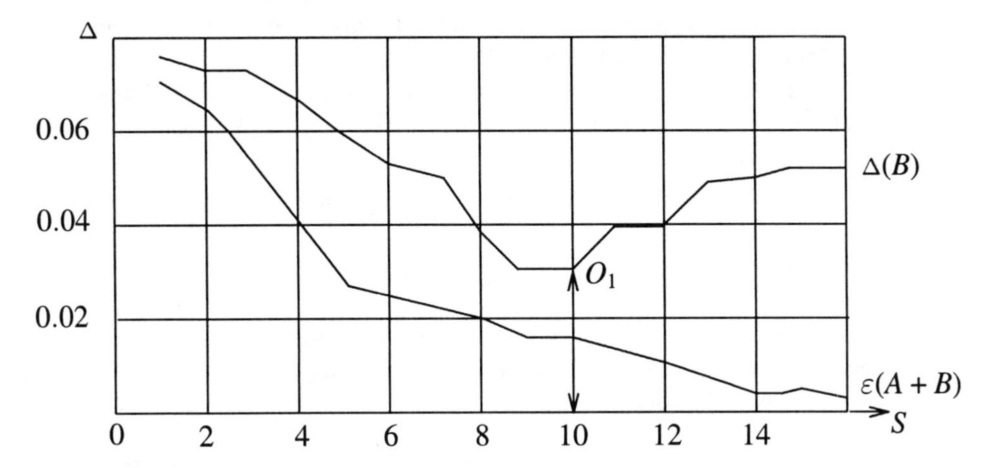

Figure 1.3. Variation in least square error $\varepsilon(A + B)$ and error measure of an "external complement" $\Delta(B)$ for a regression equation of increasing complexity S; O_1 is the model of optimal complexity

use an "external complement," necessary in principle for obtaining a unique model. Hence, none of the investigators appropriately selects the "external complement"—the risk involved in using the trial and error methods.

3.2 Principle of self-organization

In complex systems modeling we cannot use statistical probability distributions, like normal distribution, if we possess only a few empirical points. The important way is to use the inductive approach for sifting various sets of models whose complexity is gradually increased and to test them for their accuracy.

The principle of self-organization can be formulated as follows: When the model complexity gradually increases, certain criteria, which are called selection criteria or objective functions and which have the property of "external complement," pass through a minimum. Achievement of a global minimum indicates the existence of a model of optimum complexity (Figure 1.3).

The notion that there exists a unique model of optimum complexity, determinable by the self-organization principle, forms the basis of the inductive approach. The optimum complexity of the mathematical model of a complex object is found by the minimum of a chosen objective function which possesses properties of external supplementation (by the terminology of Gödel's incompleteness theorem from mathematical logic). The theory of self-organization modeling is based on the methods of complete, incomplete and mathematical induction [4]. This has widened the capabilities of system identification, forecasting, pattern recognition and multicriterial control problems.

3.3 Basic technique

The following are the fundamental steps used in self-organization modeling of inductive algorithms:

1. Data sample of N observations corresponding to the system under study is required; Split them into training set A and testing set B ($N = N_A + N_B$).

2. Build up a "reference function" as a general relationship between dependent (output) and independent (input) variables.
3. Identify problem objectives like regularization or prediction. Choose the objective rule from the standard selection criteria list which is developed as "external complements."
4. Sort out various partial functions based on the "reference function."
5. Estimate the weights of all partial functions by a parameter estimation technique using the training data set A.
6. Compute quality measures of these functions according to the objective rule chosen using the testing data set B.
7. Choose the best measured function as an optimal model. If you are not satisfied, choose F number of partial functions which are better than all (this is called "freedom-of-choice") and do further analysis.

Various algorithms differ in how they sift partial functions. They are grouped into two types: single-layer and multi-layer algorithms. Combinatorial is the main single-layer algorithm. Multi-layer algorithm is the layered feedforward algorithm. Harmonic algorithm uses harmonics with nonmultiple frequencies and at each level the output errors are fed forward to the next level. Other algorithms like multilevel algorithm are comprised of objective system analysis and two-level, multiplicative-additive, and multilayer algorithms with error propagations. We go through them in detail in the second chapter. Modified variants of multilayer algorithms were published by Japanese researchers (usually with suggestions regarding their modifications) [78], [122], [108]. Shankar [110] compared the inductive approach with the regression analysis with respect to accuracy of modeling for a small sample of input data. There were other researchers [6], [7], [12], [84], [94], [109] who solved various identification problems using this approach. Farlow [16] compiled various works of US and Japanese researchers in a compendium form. There are a number of investigators who have contributed to the development of the theory and to applications of this self-organization modeling. The mathematical theory of this approach has shown that regression analysis is a particular case of this method; however, comparison of inductive learning algorithms and regression analysis is meaningless.

3.4 Selection criteria or objective functions

Self-organization modeling embraces both the problems of parameter estimation and the selection of model structure. One type of algorithm generates models of different complexities, estimates their coefficients and selects a model of optimal complexity. The global minimum of the selection criterion, reached by inducting all the feasible models, is a measure of model accuracy. If the global minimum is not satisfied, then the model has not been found. This happens in the following cases: (a) the data are too noisy, (b) there are no essential variables among them, (c) the selection criterion is not suitable for the given task of investigation, and (d) time delays are not sufficiently taken into account. In these cases, it is necessary to extend the domain of sifting until we obtain a minimum. Each algorithm uses at least two criteria: an internal criterion for estimating the parameters and an external one for selecting the optimal structure. The external criterion is the quantitative measure of the degree of correspondence of a specific model to some requirement imposed on it. Since the requirements can be different, in modeling one often uses not one but several external criteria; that is, a multicriterion selection. Successive application of the criteria is used primarily in algorithms of objective systems analysis and multilevel long-range forecasting. Furthermore, several criteria are necessary for increasing the noise immunity of the modeling. Selection criteria are also called objective functions or objective rules as they verify and lead to the obtaining of optimal functions according to specified requirements. We can also say that these functions are used to evaluate the threshold capacity of each unit by the quantitative comparison of models of varying complexity necessary for selecting a subset of the best models from the entire set of model candidates generated in the self-organization process. If one imposes the requirement of uniqueness of choice with respect to one or several criteria, then the application of such a criterion or group of criteria yields a unique model of optimal complexity. We give here the typical criteria, historically the first external criteria and their different forms.

Suppose that the entire set (sample) of the original data points N is partitioned into three disjoint subsets A, B and C (parts of the sample) and denotes the union $A \cup B = W$. All the criteria used in the algorithms can be expressed in terms of the estimates of the model coefficients obtained on A, B and W and in terms of the estimates of the output variables of the models on A, B, C and W.

We assume that the initial data (N points) are given in the form of matrices below:

$$\mathbf{y} = \begin{pmatrix} \mathbf{y}_A \\ -- \\ \mathbf{y}_B \\ -- \\ \mathbf{y}_C \end{pmatrix}, \quad X = \begin{pmatrix} X_A \\ -- \\ X_B \\ -- \\ X_C \end{pmatrix}, \quad \begin{array}{c} \mathbf{y}[N \times 1], \\ \\ X[N \times m], \\ \\ N_A + N_B + N_C = N. \end{array} \tag{1.2}$$

The optimal dependence of the output y on the m input variables $\mathbf{x}$ is sought by the group of data handling in inductive fashion the class of functions that are linear in the coefficients of $\mathbf{y} = X\hat{\mathbf{a}}$. The submatrices of matrices X_A and X_B corresponding to any particular model of complexity s (includes s of the m arguments, $s \leq m$), which is tested in the selection process, are of complete rank.

It is convenient to define criteria by some "elementary" quantities. For example, when partitioning a sample into different parts, we introduce the quantities:

$$\hat{\mathbf{a}}_G = (X_G^T X_G)^{-1} X_G^T \mathbf{y}_G; \quad G = A,\ B, \text{ and } W \tag{1.3}$$

$$\varepsilon_G^2 = \|\mathbf{y}_G - X_G \hat{\mathbf{a}}_G\|^2, \tag{1.4}$$

where ε_G is the least squares error; that is, by the least squares technique the coefficients are estimated using set G; and the error is calculated on the same set.

$$\Delta^2(H) = \Delta^2(H/G) = \|\mathbf{y}_H - X_H \hat{\mathbf{a}}_G\|^2, \tag{1.5}$$

where $H = A, B$; $H \cap G = \emptyset$; the notation $\Delta^2(H/G)$ indicates that the error is calculated on set H of the model—the coefficients of which are estimated on set G.

Regularity criterion

This consists of a squared error calculated on the basis of testing set B.

$$\Delta^2(B) \triangleq \sum_{p \in N_B} (y - \hat{y})_p^2 / \sum_{p \in N_B} y_p^2, \tag{1.6}$$

where $\Delta(B)$ indicates the regularity measure; y and $\hat{y}$ are the desired and estimated outputs, respectively.

Other forms of "regularity criterion" are:

$$\Delta^2(B) \triangleq \sum_{p \in N_B} (y - \hat{y})_p^2 \tag{1.7}$$

$$\Delta^2(B) \triangleq \sum_{p \in N_B} (y - \hat{y})_p^2 / \sum_{p \in N_B} (y_p - \bar{y})^2, \tag{1.8}$$

where $\bar{y}$ is the average value of y. $\Delta^2(B)$ is also renotated as $\Delta^2(B/A)$, which denotes the error of the model calculated on set B using the coefficients obtained on set A. The criterion is given in matrix notation using the Eucledean norm below:

$$\begin{aligned} \Delta^2(B) = \Delta^2(B/A) &= \|\mathbf{y}_B - \hat{\mathbf{y}}_B(A)\|^2 \\ &= (\mathbf{y}_B - X_B\hat{\mathbf{a}}_A)^T\,(\mathbf{y}_B - X_B\hat{\mathbf{a}}_A) \\ &= \|\mathbf{y}_B - X_B\hat{\mathbf{a}}_A\|^2, \end{aligned} \tag{1.9}$$

where $\hat{\mathbf{a}}_A = (X_A^T X_A)^{-1} X_A^T \mathbf{y}_A$, and $\hat{\mathbf{y}}_B(A) = X_B\hat{\mathbf{a}}_A$.

Minimum bias or consistent criterion

This consists of a squared error of difference between the outputs of two models developed on the basis of two distinct sets A and B.

$$\eta_{bs}^2 \triangleq \sum_{p \in W} (\hat{y}^A - \hat{y}^B)_p^2 / \sum_{p \in W} y_p^2, \tag{1.10}$$

where $\hat{y}^A$ is the estimated output of the function, obtained on the basis of set A, and $\hat{y}^B$ is the estimated output of the function based on set B. Usually the data with higher values of variance are included into set A, while those with the smaller variance are put in set B. When the model is consistent with exact weights on both sets of the data, then the outputs are equal, $\hat{y}^A = \hat{y}^B$ and $\eta_{bs} = 0$. Therefore, the comparison of the model equations using this criterion $\eta_{bs} \rightarrow 0$ enables us to obtain consistent models, it is possible to recover an optimal response which represents a physical law of the system hidden in the noisy experimental data. Similar forms in regularity case can be expressed also as

$$\eta_{bs}^2 \triangleq \sum_{p \in W} (\hat{y}^A - \hat{y}^B)_p^2 \tag{1.11}$$

$$\eta_{bs}^2 \triangleq \sum_{p \in W} (\hat{y}^A - \hat{y}^B)_p^2 / \sum_{p \in W} (y_p - \bar{y})^2 \tag{1.12}$$

$$\begin{aligned} \eta_{bs}^2 &\triangleq \|\hat{\mathbf{y}}^A - \hat{\mathbf{y}}^B\|_W^2 \\ &= \|X_W\hat{\mathbf{a}}_A - X_W\hat{\mathbf{a}}_B\|^2 \\ &= (\hat{\mathbf{a}}_A - \hat{\mathbf{a}}_B)^T X_W^T X_W (\hat{\mathbf{a}}_A - \hat{\mathbf{a}}_B). \end{aligned} \tag{1.13}$$

Another form of this criterion expresses somewhat a different requirement.

$$\eta_a^2 \triangleq \|\hat{\mathbf{a}}_A - \hat{\mathbf{a}}_B\|^2, \tag{1.14}$$

where $\hat{a}_A$ and $\hat{a}_B$ are the coefficient vectors estimated on the basis of sets A and B, respectively. The criteria in this group do not take into account the error of the model in explicit form; the criterion of minimum coefficient bias reflects the requirement that the coefficient estimates in the optimal model, calculated on sets A and B, differ only minimally so that they appear to agree. The well-known absolute noise immune criterion is defined as

$$\begin{aligned} V^2 &\triangleq (X_W\hat{\mathbf{a}}_A - X_W\hat{\mathbf{a}}_W)^T(X_W\hat{\mathbf{a}}_W - X_W\hat{\mathbf{a}}_B) \\ &= (\hat{\mathbf{a}}_A - \hat{\mathbf{a}}_W)^T X_W^T X_W(\hat{\mathbf{a}}_W - \hat{\mathbf{a}}_B) \\ &= (\hat{\mathbf{y}}^A - \hat{\mathbf{y}}^W)_W^T(\hat{\mathbf{y}}^W - \hat{\mathbf{y}}^B)_W, \end{aligned} \quad (1.15)$$

where $\hat{a}_W$ is the coefficient vector obtained on the basis of whole set W. It is possible to select a model that is not sensitive to the data on which it is based. The minimum bias criterion requires the model to yield the same results at successive experimental points of N_A and N_B. It is possible to recover a hidden physical law using this criterion from the noisy data.

Prediction criterion

This consists of a squared error calculated on the basis of a separate examin set C, which is not used in estimating the coefficients:

$$\Delta^2(C/W) \triangleq \sum_{p \in N_C} (y - \hat{y})_p^2 / \sum_{p \in N_C} y_p^2. \quad (1.16)$$

In case of finite difference equations, the criterion is also evaluated as

$$i^2(W) \triangleq \sum_{p \in N_W} (y - \hat{y})_p^2 / \sum_{p \in N_W} y_p^2, \quad (1.17)$$

where the estimate $\hat{y}$ is obtained through step by step integration of a difference equation from the given initial conditions. The autoregression form of such model is

$$\hat{y}_p = a_0 + a_1\hat{y}_{p-1} + a_2\hat{y}_{p-2} + \cdots + a_\tau\hat{y}_{p-\tau}. \quad (1.18)$$

This criterion can be also calculated on other parts of the sets $i^2(A)$, and $i^2(B)$.

Combined criteria

Sometimes we choose not only the minimum bias models but also the models with other characteristics. Combined criteria are used as two or more participating criteria in one function. In solving practical problems, it is often necessary to obtain a model that satisfies several requirements simultaneously. Using an objective rule, we then arrive at the familiar problem of multi-criterion selection when some or all of the criteria are contradictory. For example, even the simple problem of selecting a model that will simultaneously be the most regular ("exact") and have the least bias often proves contradictory. Under these conditions, one selects a unique model by using the combined criteria in the form of the sum of the individual criteria with certain weight factors. For example, the combined criteria is given as

$$k^2 \triangleq \alpha k_1^2 + (1 - \alpha)k_2^2, \quad (1.19)$$

where k_1 and k_2 are given criteria and α is the weight factor. One can also use the normalized values of the criteria:

$$k_l^2 \triangleq \bar{k}_{1l}^2 + \bar{k}_{2l}^2 = k_{1l}^2/k_{1\,max}^2 + k_{2l}^2/k_{2\,max}^2, \quad (1.20)$$

where l is the index of the model under consideration and the maximum values of the criteria are determined out of all the F models that participate in the sorting.

$$k_{1\,max}^2 = \max_{l \in F} k_{1l}^2, \quad k_{2\,max}^2 = \max_{l \in F} k_{2l}^2. \tag{1.21}$$

The following are some of the combined criteria used in the algorithms.

One of the combined criteria is "bias plus approximation error." Here the index of the model concerned is not shown.

$$c1^2 \triangleq \bar{\eta}_{bs}^2 + \bar{\varepsilon}^2(W) = \eta_{bs}^2/\eta_{bs\,max}^2 + \varepsilon^2/\varepsilon_{max}^2, \tag{1.22}$$

where $\bar{\varepsilon}^2(W)$ is the normalized approximation error on the data set $W(= A \cup B)$ using the coeficients obtained on W; this is nothing but the least square error. The second form of combined criterion is "bias plus regularity":

$$c2^2 \triangleq \bar{\eta}_{bs}^2 + \bar{\Delta}^2(B). \tag{1.23}$$

Another form of combined criteria which has the best prediction properties ("bias plus error on examination") is

$$c3^2 \triangleq \bar{\eta}_{bs}^2 + \bar{\Delta}^2(C). \tag{1.24}$$

It provides the most unbiased, stable and accurate predicting models, where $\Delta(C) = \Delta(C/W)$ is the mean square error of predictions calculated on data set C using the coefficients obtained on W. The criterion $\Delta(C)$ can also be replaced by $i(W)$; it is appropriate in the case of step-by-step predictions. In calculating criterion $c3$, we can usually divide data in proportions of part A = 40 %, part B = 40 % and part C = 20 %. Sets A and B are used to calculate the minimum bias measure and set C is used for predicting error. In case of criteria $c1$ and $c2$, whole data can be divided into two parts.

There are other forms of combined criteria depending on the combination of various criteria used.

Balance-of-variables criterion

This is used in obtaining a model for long-term predictions in the case of some known *a priori* dependence between variables. For example, if $y = f(x_1, x_2, x_3)$ is the dependent relation, then the balance criterion has the form

$$b \triangleq \sum_{p \in N_C} (y - f(x_1, x_2, x_3))_p^2 / \sum_{p \in N_C} y_p^2, \tag{1.25}$$

where N_c is the set of points in the extrapolation interval and y is the desired output. In the problems, where balance-of-variables is not known, it can be discovered with the help of minimum-of-bias criterion.

Regularity criterion is useful in obtaining an exact approximation of a system as well as of a short-term prediction (for one or two steps ahead) of the processes taking place in it. In the interpolation interval all of the models yield almost the same results (we have the principle of multiplicity of models). In the extrapolation interval the predictions diverge, forming a so called "fan" of predictions.

The minimum-of-bias criterion yields a narrower fan, and hence a longer prediction time than the regularity criterion. This means that prediction is possible for several steps ahead (medium term prediction). However, the theory of self-organization will not solve the problems to which it is applied unless it yielded examples of exact long-term predictions.

The balance-of-variables criterion is proposed for long-range predictions. This requires simultaneous prediction of several interrelated variables. In many examples these variables are constructed artificially. For example, for three variables it is possible to discover the laws:

$$x_1 = f_{11}(x_2, x_3), \quad x_2 = f_{22}(x_1, x_3), \quad x_3 = f_{33}(x_1, x_2), \tag{1.26}$$

where f_{11}, f_{22}, and f_{33} are the functional relations among the variables. The balance-of-variables criterion requires that these relations between pairs of variables be satisfied not only in the interpolation interval, but also in the extrapolation interval. For this purpose, the differences are constructed between "direct" and "inverse" functions. The inverse functions $x_1^* = f_{21}(x_2, x_3)$, $x_2^* = f_{32}(x_1, x_3)$ and $x_3^* = f_{13}(x_1, x_2)$ are computed from the second, third and first laws of the above "direct" equations, respectively. The "inverse" functions can also be obtained as f_{31}, f_{12} and f_{23}. The first subscript is the number of equation and the second is the number of the variable to be determined. If the "direct" and "inverse" functions are exact, the balance criterion requires that

$$b_1 = (f_{11} - f_{21}) \to 0, \quad b_2 = (f_{22} - f_{32}) \to 0, \quad b_3 = (f_{33} - f_{13}) \to 0. \tag{1.27}$$

The balance-of-variables criterion measures the unbalances b_1, b_2, and b_3 in the extrapolation interval as

$$\begin{aligned} B^2 &= \sum_{N_C} [f_{11}(x_2, x_3) - f_{21}(x_2, x_3)]^2 / \sum_{N_C} [f_{11}(x_2, x_3)]^2 \\ &\quad + \sum_{N_C} [f_{22}(x_1, x_3) - f_{32}(x_1, x_3)]^2 / \sum_{N_C} [f_{22}(x_1, x_3)]^2 \\ &\quad + \sum_{N_C} [f_{33}(x_1, x_2) - f_{13}(x_1, x_2)]^2 / \sum_{N_C} [f_{33}(x_1, x_2)]^2 \\ &= (b_1^2 + b_2^2 + b_3^2), \end{aligned} \tag{1.28}$$

where N_C is number ef points in the prediction or examin data set.

This criterion yields reference points in the future; it requires that a law, effective up to the present, continue into the future in the extrapolation interval; the sum of unbalances in the extrapolation interval should be minimal. In cases where exact relations are not known in the interpolation interval, these can be obtained by using minimum bias criterion in one of the inductive learning algorithms.

The correctness of the prediction is checked according to the values of the criterion. By gradually increasing the prediction time, we arrive at a prediction time for which it is no longer possible to find an appropriate trend in the fan of a given "reference function." The value of the minimum function begins to increase; thus appropriate action must be taken. For example, it may be necessary to change the "reference function." For a richer choice of models, it is also recommended that one go from algebraic to finite-difference equations, take other system variables, estimate the coefficients and others.

3.5 Heuristics used in problem-solving

The term *heuristic* is derived from the Greek word *eureka* (to discover). It is defined as "experiential, judgemental knowledge; the knowledge underlying 'expertise'; rules of thumb, rules of good guessing, that usually achieve desired results but do not guarantee them" [17]. Heuristics does not guarantee results as absolute as conventional algorithms do, but it offers efficient results that are specific and useful most of the time. Heuristic programming provides a variety of ways of capturing human knowledge and achieving the results as per the objectives. There is a slight controversy in using heuristics in building up expert and complex systems studies. Knowledge-base and knowledge-inference mechanisms are developed in expert systems. The performance of an expert system depends on the retrieval of the appropriate information from the knowledge base and its inference mechanism in evaluating its importance for a given problem. In other words, it depends on how effective logic programming and the building up of heuristics is in the mechanisms representing experiential knowledge. The main task of heuristics in self-organization modeling is to build up better man-machine information systems in complex systems analysis thereby reducing man's participation in the decision-making process (with higher degree of generalization.)

Basic modeling problems

Modeling is used for solving the problems: (i) systems analysis of the interactions of variables in a complex object, (ii) structural and parametric identification of an object, (iii) long-range qualitative (fuzzy) or quantitative (detailed) prediction of processes, and (iv) decision-making and planning.

Systems analysis of the interactions of variables precedes identification of an object. It enables us not only to find the set of characteristic variables but also to break it into two subsets: the dependent (output) variables and the independent (input or state) variables (arguments or factors).

In identification, the output variables are given and one will need to find the structure and parameters of all elements. Identification leads to a physical model of the object, and hence can be called the determination of laws governing the object. In the case of noisy data, a physical model can only be used for determining the way the object acts and for making short-range predictions. Quantitative prediction of the distant future using such physical model is impossible. Nevertheless, one is often able to organize a fuzzy qualitative long-range prediction of the overall picture of the future with the aid of so-called loss of scenarios according to the "if-then" scheme. There is a basic difference between the two approaches to modeling. The only way to construct a better mathematical model is to use one's experience ("heuristics or rules of thumb"). Experience, however, can be in the form of the author's combined representations of the model of the object or of the empirical data—the results of an active or passive experiment. The first kind of experiment leads to simulation modeling and the second to the experimental method of inductive learning or self-organization modeling. The classical example of simulation modeling is the familiar model of world dynamics [20]. A weak point with simulation method is the fact that the modeller is compelled to exhibit the laws governing all the elements, including those he is uncertain about or which he thinks are simply less susceptible to simulation. In contrast to simulation modeling, the inductive approach chooses the structure of the model of optimal complexity by testing many candidate models according to an objective function.

In mathematical modeling, certain statistical rules are followed to obtain solutions. These rules, based on certain hypothesis, help us in achieving the solutions. If we take the problem of pattern classification, a discriminant function in the form of a mathematical equation is estimated using some empirical data belonging to two or more classes. The mathematical equation is trained up using a training data set and is selected by one of the statistical criterion, like minimum distance rule. The second part of data of discriminant function is tested for its validation. Here our objective is to obtain optimal weights of the function suited for the best classification; this is mainly based on the criterion used in the procedure, the data used for training and testing the function, and the parameter estimation technique used for this purpose. Obtaining a better function depends on all these factors and how these are handled by an experienced modeller. This depends on the experience and on the building up of these features as heuristics into the algorithm. This shows the role of the human element in the feedback loop of systems analysis.

Developing a mathematical description according to the input-output characteristics of a system, and generating partial functions by linear combinations of the input arguments from the description, splitting of data into number of sets and design of "external complement" as a threshold objective function are noted as common features established in learning mechanism of the inductive algorithms. The output response of the network modeling depends highly on how these features are formed in solving a specific problem. Depending on the researcher's experience and knowledge about the system, these features are treated as heuristics in these algorithms.

Mathematical description of the system

A general relationship between output and input variables is built up in the form of a mathematical description which is an overall form of relationship refering to the complex system under study. This is also called "reference function." Usually the description is considered a discrete form of the Volterra functional series which is also called Kolmogorov-Gabor polynomial:

$$y = a_0 + \sum_{i=1}^{m} a_i x_i + \sum_{i=1}^{m}\sum_{j=1}^{m} a_{ij} x_i x_j + \sum_{i=1}^{m}\sum_{j=1}^{m}\sum_{k=1}^{m} a_{ijk} x_i x_j x_k + \cdots, \tag{1.29}$$

where the output is designated as y, the external input vector as $\mathbf{x} = (x_1, x_2, \cdots)$, and $\mathbf{a}$ the vector of coefficients or weights. This is linear in parameters a and nonlinear in x. Components of the input vector $\mathbf{x}$ could be independent variables, functional terms, or finite difference terms. This means that the function could be either an algebraic equation, a finite difference equation, or an equation with mixed terms. This polynomial represents the full form of mathematical description. This can be replaced with a system of partial polynomials of the form

$$y_i = a_0 + a_1 x_i + a_2 x_j + a_3 x_i x_j + a_4 x_i^2 + a_5 x_j^2, \tag{1.30}$$

where $i, j = 1, 2, \cdots, m;\ i \neq j$.

Mathematical descriptions can be grouped into three forms as single input-single output forms (trend equations), multi-input-single output forms (multivariate equations), and multi-input-multi-output forms (system of equations). Specific terms like moving averages, logarithmic terms, time function, time, harmonic trends, and so on, can be considered under these descriptions.

(i) When we think about rationalized descriptions according to our understanding of the system, we have to consider interaction of independent variables in the "reference functions." There are various hypotheses regarding the interaction of these variables. For example, there are four variables (x_1, x_2, x_3, x_4).

The first hypothesis is that these variables do not interact with each other; then the description is considered as

$$y = a_0 + f_1(x_1) + f_2(x_2) + f_3(x_3) + f_4(x_4). \tag{1.31}$$

The second hypothesis is that the first variable x_1 does not interact with the others, but that the others interact among themselves.

$$y = a_0 + f_5(x_1) + f_6(x_2, x_3, x_4). \tag{1.32}$$

The third hypothesis is that the first variable interacts with the second, and the third interacts with the fourth.

$$y = a_0 + f_7(x_1, x_2) + f_8(x_3, x_4);\ \text{etc.,} \tag{1.33}$$

where f_1, f_2, f_3, f_4, and f_5 are first-degree polynomials; f_7 and f_8 are second-degree polynomials; and f_6 is a third-degree polynomial.

The examples are easily continued. Also it is clear from physical considerations that one of the hypotheses is true and that the number of hypotheses is small. Thus, the purpose of optimization of the mathematical form is achieved. Possible combinations of all variables can also be regarded as a sorting of a number of hypotheses—one of which is true. Realization of such combinations cannot lose the optimal model because it ensures complete sorting of all possible models for a given support function.

(ii) One needs to investigate the convergence to trends of the process using multilayered inductive approach. The level of trends can be done using algebraic equations or finite difference analogues of differential equations. The most general systems analysis is based on equations of the form:

$$y^t = f_1(u, t) + f_2(y^{t-1}, y^{t-2}, \cdots, y^{t-\tau}), \tag{1.34}$$

where f_1 is a source function as a linear trend with variables, time t, and a control vartiable u. To simplify the overall investigation, the analysis is broken into two parts:

1. First analysis is of the trends, $y_1^t = f_1(u, t)$; for example, $f_1(u, t) = a_0 + a_1 t + a_2 u$; and
2. Second analysis is of the dynamics, $y_2^t = f_2(y^{t-1}, y^{t-2}, y^{t-3}, \cdots, y^{t-\tau})$.

Although $y^t \neq y_1^t + y_2^t$, to obtain $y^t = y_1^t + y_2^t$ it is necessary to use points of deviations from the first analysis of trends for the second analysis of dynamics.

(iii) If the physical law of the system is considered a "reference function," this would mean that the scope of search for an optimal model in the self-organization modeling is reduced. If there is noise in the empirical data, the physical models cannot guarantee long-range predictions. Our studies show that physical models cannot be used for long-range predictions because of noise in the data. The physical models are suitable only for identification and short-range predictions.

(iv) Sometimes the modeller cannot ascertain which are the output variables in the system. It is very important to find the "leading" variable in the set of output variables of the complex object. The "leading" variable is the variable that is predicted better on-more accurately than the other variables. To identify the "leading" variables certain algorithms are recommended.

(v) Mathematical descriptions with variable coefficients have been used widely as "reference functions" in case of ecological modeling. For example, if we have three control variables (u_1, u_2, and u_3), and four other variables (y_1, y_2, y_3, and y_4), we can write the complete polynomial as an algebraic equation.

$$\begin{aligned} y_1 = {} & (a_0 + a_1 u_1 + a_2 u_2 + a_3 u_3) + (b_0 + b_1 u_1 + b_2 u_2 + b_3 u_3) y_2 \\ & + (c_0 + c_1 u_1 + c_2 u_2 + c_3 u_3) y_3 + (d_0 + d_1 u_1 + d_2 u_2 + d_3 u_3) y_4. \end{aligned} \tag{1.35}$$

One can include time as a variable along with the control variables in the above form. In the same way, complete polynomials for y_2, y_3, and y_4 can be written. Finite difference form with two delayed arguments is written as

$$\begin{aligned} y_i^{t+1} = {} & (a_0 + a_1 u_1 + a_2 u_2 + a_3 u_3) + (b_0 + b_1 u_1 + b_2 u_2 + b_3 u_3) y_i^t \\ & + (c_0 + c_1 u_1 + c_2 u_2 + c_3 u_3) y_i^{t-1} + (d_0 + d_1 u_1 + d_2 u_2 + d_3 u_3) y_i^{t-2} \end{aligned} \tag{1.36}$$

for $i = 1, 2, 3, 4$. These types of polynomials are also used in studies of inflation stability.

(vi) One must take necessary care when the mathematical description is described. The following are four features to improve, in a decisive manner, the existing models of complex objects and to give them an objective character.

1. Descriptions that are limited to a certain class of equations and to a certain form of support functions lead to poor informative models with respect to their performance on predictions. For example, a difference equation with a single delayed argument with constant coefficients is considered a "reference function":

$$x_i^t = a_0 + a_1 x_i^{t-1} + a_2 t + a_3 t^2. \tag{1.37}$$

   The continuous analogous of such equation is first-order differential equation; the solution of such equation is an exponential function. If many variants are included in the description, the algorithm sorts out the class of equations and support functions according to the choice criteria.
2. If the descriptions are designed with arbitrary output or dependent variables, then output variables are unknown. Those types of descriptions lead to biased equations. Inductive learning algorithms with special features are used to choose the leading variables.
3. There is a wrong notion that physical models are better for long-range predictions. The third feature of the algorithms is that nonphysical models are better for long-range predictions of complex systems. Physical models (that is, models isomorphic to an object which carry over the mechanism of its action), in the case of inexact data are unsuitable for quantitative long-range prediction.
4. The variables which hinder the object of the problem must be recognized. The fourth feature of the algorithms is that predictions of all variables of interest are found as functions of "leading" variables.

Splitting data into training and testing sets

Most of the selection criteria require the division of the data into two or more sets. In inductive learning algorithms, it is important to efficiently partition the data into parts (the efficiency of the selection criteria depends to a large extent on this). This is called "purposeful regularization." Various ways of "purposeful regularization" are as below:

1. The data points are grouped into a training and a checking sequence. The last point of the data belongs to the checking sequence.
2. The data point are grouped into training and checking sequences. The last point belongs to the training sequence.
3. The data points are arranged according to the variance and are grouped into training and testing parts. This is the usual method of splitting data. Half of the data with the higher values is used as the training set and another half is used as the testing set.
4. The data points represent the last year. Points correspond to the past data for all years that differ from the last by a multiple of prediction interval T_{pre}. For example, the last year in the data table corresponds to the year 1990; prediction interval is made for the year 1994 (ie., $T_{pre} = 4$ years). The checking sequence comprises the data for the years 1990, 1986, 1982, 1978, etc. and the other data belong to the training sequence.

5. The checking sequence consists of only one data point. For example, if we have data of N years and the prediction interval is T_{pre}, then the points from 1 to $N - T_{pre} - 1$ belong to the training sequence and Nth point belongs to the checking sequence. This is used in the algorithm for the first prediction.

   The second prediction is obtained based on the same algorithm, with another checking point which consists of $N - 1$ point; the training sequence contains from 1 to $N - T_{pre} - 2$ points.

   The third prediction is based on the $(N - 2)$nd point for checking sequence and 1 to $N - T_{pre} - 3$ points for training sequence.

   The predictions are repeated ten to twenty times and one obtains prediction polynomials. All the polynomials are summed up and taken average of it. Each prediction is made for an interval length of T_{pre}, and the series of prediction equations is averaged.
6. The data points are grouped into two sequences: the last points in time form the training sequence; and the checking sequence is moved backward l years, where l depends on the prediction time and on the number of years for which the prediction is calculated; i.e., it indicates the length of the checking sequence.

Although each method has its unique characteristics of obtaining the model in optimal complexity, only under special conditions are they used. The most usual method is the third method which has to do with the variance and helps minimize the selection layers in case of multi-layer inductive approach.

The following are some examples to show the effect of partitioning of data.

1. It is the method of optimization of allocation of data sample to training and testing sets. There were 14 points in the data sample. Experiments were conducted with different proportions of training and testing sets to obtain the optimal model using the regularity criterion. Figure 1.4 illustrates that a choice of proportionality 9:5 is optimal from the point of view of the number of selection layers in the multilayer iterative algorithm. The simplest and most adequate model was obtained with such an allocation of points. It was noted that the regularity criterion could be taken as the reciprocal of the mean square error in the testing set.
2. Here is another example of the effect of partitions on the global minimum achieved by using the combined criterion $c3$ that is defined as

$$c3^2 \triangleq \alpha\, \eta_{bs}^2 + (1 - \alpha)\, \Delta^2(C), \tag{1.38}$$

   where

$$\eta_{bs}^2 \triangleq \sum_N (\hat{y}^A - \hat{y}^B)^2 / \sum_N y^2,$$
$$\Delta^2(C/W) \triangleq \sum_{N_C} (y - \hat{y})^2 / \sum_{N_C} y^2.$$

   A random data of 100 points is arranged as per its variance and is divided into proportions $A : B : C$, as shown in the Table 1.3. The combined criterion measure at each layer is given for different values of α. Global minimum for each experiment is indicated with "$*$". When $\alpha = 1$, only minimum bias criterion is participated. As the value of α decreases, the participation of $\Delta^2(C)$ increases in selecting the optimal model. From the global values of the criteria, one can note that the optimum splitting of data is 45:45:10.
3. One of the experiments was done by finding the required partition of empirical data points using the extremal values of the minimum bias selection criterion on the set of

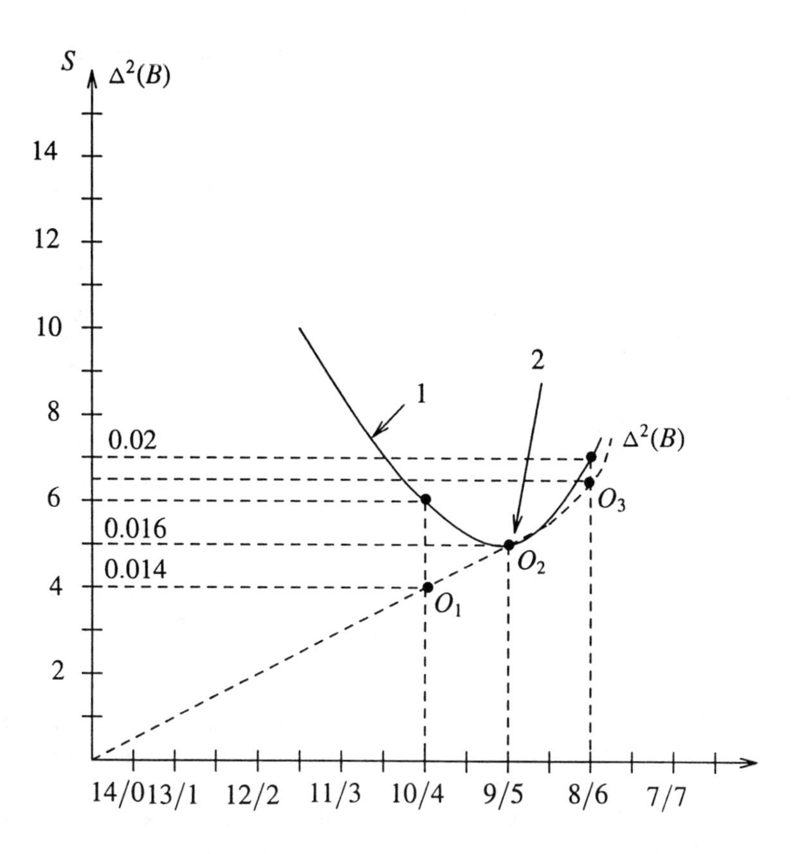

Figure 1.4. Optimum allocation of data to training and testing sets, where S is the number of selection layers, $\Delta^2(B)$ is the error measure using regularity criterion. 1. plot of number of selection layers and 2. chosen optimum allocation

all possible versions of data partition in a prescribed relationship [128]. It was shown that the different possible partitions effect the global minimum.

Objective functions

Thinking of objectives in mathematical form is one of the difficult tasks in these algorithms. Extensive has been work done in this direction and enormous contributions have been made to the field in recent years. Most of the objective functions are related to the standard mathematical modeling objectives such as regularization, prediction, unbiasedness and so on. There are standard statistical criteria used by various researchers according to statistical importance. One can also design his own set of criteria with regard to specific objectives. The following is a brief sketch of the development of these functions.

(i) In the beginning stages of self-organization modeling (1968 to 1971), it was applied to pattern recognition, identification, and short-range prediction problems. These problems were solved by regularity criterion only.

$$\Delta^2(B) \triangleq \sum_{p \in N_B} (y - \hat{y})_p^2 / \sum_{p \in N_B} y_p^2, \tag{1.39}$$

Table 1.3. c3 values for different values of α with different partitions

A:B:C	Layer: 1	2	3	4	5	6	7
$\alpha = 1$:							
45:45:10	0.152	0.053	0.073	0.007*	0.120	0.048	0.034
40:40:20	0.176	0.052*	0.146	0.099	0.126	0.158	0.149
35:35:30	0.181	0.151	0.109	0.059*	0.193	0.097	0.159
$\alpha = 0.75$:							
45:45:10	0.323	0.262*	0.360	0.362	0.440	0.440	0.439
40:40:20	0.293	0.249	0.233*	0.306	0.242	0.263	0.265
35:35:30	0.307	0.300	0.313	0.281*	0.452	0.374	0.368
$\alpha = 0.5$:							
45:45:10	0.416	0.423	0.390	0.332*	0.409	0.400	0.373
40:40:20	0.376	0.351	0.346*	0.389	0.407	0.408	0.462
35:35:30	0.389	0.347*	0.362	0.405	0.370	0.370	0.359
$\alpha = 0.25$:							
45:45:10	0.489	0.420	0.384	0.385	0.335*	0.369	0.436
40:40:20	0.443	0.423	0.380*	0.469	0.468	0.467	0.428
35:35:30	0.455	0.427	0.420	0.417*	0.471	0.427	0.453

where y is the desired output variable, $\hat{y}$ is the estimated output based on the model obtained on training set A (about 70% of data), and N_B, is the number of points in the testing set (about 30% of data) used for computing regularity error.

Sometimes this criterion was used in the form of a correlation coefficient between y and $\hat{y}$ variables or in the form of a correlation index (for nonlinear models).

(ii) Later, during 1972 to 1975, the ideas of multicriteria choice of models were developed in pattern recognition theory, minimum bias, balance of variables, and combined criteria. Minimum bias criterion is recommended to obtain a physical model; balance-of-variables criterion is preferred to identify a model for long-range predictions. Various criteria like prediction criterion and criteria for probabilistic stability were also proposed during this period. We were convinced that the wide use of the minimum bias and balance of variables criteria, together with the solution of the noise resistance problem, were the major ways of improving the quality of the models.

(iii) During the eighties, there was fruitful research in the direction of developing noise immune criteria which lead to the successful development of various algorithms such as objective system analysis and multilevel algorithms. The noise stability of self-organization modeling algorithms and noise immune external criteria will be discussed in Chapter 3.

There is confusion with the notations used for the selection criteria as developments progressed through the years. Here we try to give various forms of criteria with standard notations.

All the individual criteria, which are of quadratic form, are divided into two basic groups:

(i) accuracy criteria, which express the error in the model being tested on various parts of the sample (example, regularity),

(ii) matching (consistent) criteria, which are a measure of the closeness of the estimates obtained on different parts of the sample (example, minimum bias).

By adding other two groups, such as balance and dynamics (step-by-step integral) criteria, all external criteria are classified into four groups, as given in the Table 1.4, where β is the parameter used in averaging the term and $\hat{\mathbf{y}}_W(y_0, \hat{a}_W)$ is the step-by-step integrated

Table 1.4. External criteria

Criteria group	Group symbol	Criteria	Notation Old	Notation New	Computational formula
Accuracy criteria	A	(i) Regularity	$\Delta^2(B)$	AB	$\|\|\mathbf{y}_B - X_B\hat{\mathbf{a}}_A\|\|^2$
		(ii) Symmetric (dual) regularity	d^2	AD	$\|\|\mathbf{y}_A - X_A\hat{\mathbf{a}}_B\|\|^2 + \|\|\mathbf{y}_B - X_B\hat{\mathbf{a}}_A\|\|^2$
		(iii) Stability	S^2	AS	$\|\|\mathbf{y}_W - X_W\hat{\mathbf{a}}_A\|\|^2 + \|\|\mathbf{y}_W - X_W\hat{\mathbf{a}}_B\|\|^2$
Balance criteria	B	(i) Predictions balance (linear)	B^2	BL	$\|\|\hat{Q} - (\hat{q}_1\| \cdots \|\hat{q}_L)\beta\|\|^2$
		(ii) Variables balance	B^2	BV	$\|\|\hat{\Phi} - \psi(\hat{y})\|\|^2$
Matching criteria	C	(i) Consistency (mini mum- bias)	η_{bs}^2	CB	$\|\|X_W\hat{\mathbf{a}}_A - X_W\hat{\mathbf{a}}_B\|\|^2$
		(ii) Unbiasedness in coefficients	η_a^2	CC	$\|\|\hat{\mathbf{a}}_A - \hat{\mathbf{a}}_B\|\|^2$
		(iii) Variability (absolute noise immune)	V^2	CV	$(X_W\hat{\mathbf{a}}_A - X_W\hat{\mathbf{a}}_W)^T (X_W\hat{\mathbf{a}}_W - X_W\hat{\mathbf{a}}_B)$
Dynamic criteria	D	(i) Integral	i^2	DI	$\|\|\mathbf{y}_W - \hat{\mathbf{y}}_W(y_0, \hat{\mathbf{a}}_W)\|\|^2$

Table 1.5. Classification of criteria

No.	Internal	External
i	*Accuracy type*: (a) Mean square error $\varepsilon^2 = \sum_{i\in N_W}(\hat{y} - y)_i^2$ (b) Correlational $\rho_W = \| \text{Cov}\ (y\hat{y})\|_1^{N_W}$ (c) Distance $d = \sigma(x_i, x_j)$	(a) Ideal $J = \sum_{i\in N_W}(\hat{y} - \overset{\circ}{y})^2$ (b) Correlational $\rho_B = \| \text{Cov}\ (y\hat{y})\|_1^{N_B}$ (c) Regularity $\Delta^2(B) = \sum_{i\in N_B}(y - \hat{y}^A)_i^2$
ii	*Integral type(Dynamic)*: Stepwise prediction $i^2(W) = \sum_{t=1}^{N_W-1}(\hat{y}_{t+1} - ay_t)^2$	Stepwise prediction $i^2(C) = \sum_{t=1}^{N_C}(\hat{y}_{t+1} - a\hat{y}_t)^2$
iii	*Differential type (Balance and Matching)*: - -	(a) Balance-of-predictions $B^2 = \sum_{i\in N}[\hat{Q} - \frac{1}{4}(\hat{q}_1 + \hat{q}_2 + \hat{q}_3 + \hat{q}_4)]_i^2$ (b) Minimum- bias or consistency $\eta_{bs} = \sum_{i\in N_W}(\hat{y}^A - \hat{y}^B)_i^2$

output value which is initialized with the first value y_0 using the estimated coefficients $\hat{a}_W$. "Symmetric" and "nonsymmetric" forms of certain criteria are shown. "Symmetric" criterion means one in which the data information in parts A and B of the sample are used equally; when it is not, the criterion is "nonsymmetric." These are further discussed in later chapters. Here we have given old and new notations of these criteria; the old notation is followed throughout the book. The new notation will be helpful in following the literature as it varies from author to author.

As it is clear that the internal criteria are the criteria that participate in the interpolation region in estimating or evaluating the parameters of the models; on the other hand, the external criteria are the criteria that use the information from the extrapolation region (partially or fully) in evaluating the models. Table 1.5 demonstrates some of these criteria, where $\overset{\circ}{y}$ is the ideal output value (without noise).

The inductive approach proposes a more satisfactory way to find optimum decisions in self-organization models for identification and for short- and long-range predictions. This is particularly useful with noisy data. Communication theory and inductive theory differ from one another by the number of dimensions used in self-organization modeling, but they have common analogy according to the principle of self-organization. The internal criteria currently used in the traditional theories does not allow one to distinguish the model of optimal complexity from the more complex overfitted ones.

Chapter 2
Inductive Learning Algorithms

1 SELF-ORGANIZATION METHOD

The existence of man on earth began about two million years ago. The twentieth century is the turning point for changes, the likes of which have never before occurred in human history. The change in the way of living, modern production processes, scientific-technical revolution, and important changes in and rapid development of computers for information processing and control of complex objects, including the hereditary characteristics of living organisms, constitute but a short list of new phenomena that characterize our age. The expanding possibilities are accompanied by growing anxieties such as how to cope with increasing pollution problems and how to find new food and energy sources in time to feed and warm an increasing population. These and other important problems will be solved in the near twenty-first century—the century in which our children and grandchildren will develop. It is clear why our eyes are turned to the future, to the twenty-first century. We cannot say that we are completely ignorant, because this is not true. For example, we know the fundamental trends of development of scientific-technical progress. We know that this leads to the qualitative mathematical predictions of the future that are possible only with the methods of quantitative mathematical modeling which answer questions about the time periods of the predictions (when?), the forms (how, in which way, and under what conditions?), and the place (where?). Futurology is a young mathematical science whose purpose is to develop methods for the prediction of the future. This branch of science is slowly coming to a mature stage when some of its unavoidable meanderings and errors can be executed. However, we will see here some of its successes and achievements.

A long time ago scientists knew how to calculate exactly the motions of the planets, how to predict eclipses of the moon and of the sun because these are described by exact deterministic equations. Even these were achieved in successful prediction of variable quantities, averaged over a long period of time. Thus, it is not difficult to predict the average temperature of the earth for the entire twenty-first century or to calculate the amount of precipitation which will fall on the mainland of Asia during the same long period of time. The question is, however, how much can the averaging time intervals and area sectors be decreased? How does one obtain predictions for each month and season or year when the prediction interval extends over scores of years? Can scientists actually predict the distant future in general? Such a question usually comes into our minds because of lack of success in many detailed (nonaveraged) quantitative forecasts. The first fiasco was suffered by the so-called probabilistic methods of long-term predictions. Probability is also a type of averaging, and is useful in predictions of the future. Probabilistic methods have their own drawbacks because of their lack of inductive analysis.

Simulation modeling is another area used by Forrester [20] in studying world dynamics. The fact is that simulation modeling does not require any test data. Equations that describe separate parts of the object of prediction are invented by modelers on the basis of their subjective ideas of the object. These models apparently do not take into account the variability of the object characteristics. The common feature of probabilistic and simulation methods is that "the more complex the model, the more accurate" the interval of empirical observations used for estimating the parameters. The modeler cannot tell whether the model is accurate or not in the interval of the object that he lacks the knowledge about. The self-organization modeling that is described here plays an important role in such conditions.

The self-organization method uses a very general meta-language, rather than a language of detailed instructions. The quantitative model built up from the observations should be the same as the model built up from other observations taken at different times and places. This is the prerequisite in obtaining a predictive model through the inductive approach. Let us first go through the basic iterative algorithm based on the inductive approach and the basic network structures that have been in use since the beginning of the usage of these algorithms. Later we will study basic directions, principal characteristics of the algorithms and advanced multilevel achievements in complex problem solving.

1.1 Basic iterative algorithm

In problem solving, the main strategy is to specify a set of proper input-output associations and the main goal is to design an efficient learning algorithm, which is regarded as a search procedure that correctly solves the problem. Learning in the networks takes a variety of forms; mainly it discovers statistical features for detecting regularity. Self-organization is considered while building the connections among the units through a learning mechanism to represent discrete items. The self-organizing process that is established using various heuristics in the network structure helps in obtaining the optimum output response. In the network each unit is represented independently as a black box to generate the input-output relationship as a state function and group of units are treated as a layer of certain thresholding hierarchical stage. The relationships are established through the connecting weights that are estimated by adapting a parameter estimation technique. The measure of the objective function as an "external complement" is used as a threshold value on a competitive basis to make the unit "on" or "off" and if the unit is "on" its output is fed forward to the next layer as inputs. The measure is also considered as an objectivity measure of the unit. Overall, this works as a search in the domain of solution space through a sort of competitive learning. Relevance of local minimum depends on the complexity of the task on which the system is trained by building up heuristics like design of objective function, design of input-output relationships or summation functions at the units, and usage of empirical data for training and testing the network. When one of the units in a particular layer achieves the global minimum of the measure, the processing is stopped and information about the optimal response is recollected from the associated units in the preceding layers. The global minimum is guaranteed because of steepest descent in the output error with respect to the connection weights in the solution space in which it is searched as per a specific objective through cross validating the weights.

Suppose we have a sample of N observations, a set of input-output pairs (I_1, o_1), (I_2, o_2), $\cdots$, $(I_N, o_N) \in N$, where N is a domain of observations corresponding to the empirical data, and we have to train the network using these input-output pairs to solve two types of problems: (1) identification problem—the given input $I_j (1 \leq j \leq N)$ of variables x corrupted by some noise expected to reproduce the output o_j and to identify the physical laws, if any, embedded in the system; (2) prediction problem—the given input I_{N+1} expected to predict exactly the output o_{N+1} from a model of the domain studied during the training.

In solving these, a general relationship between output and input variables is built in the form of a mathematical description, which is also called a reference function. Usually the description is considered as a discrete form of the Volterra functional series or Kolmogorov-Gabor polynomial:

$$y = a_0 + \sum_{i=1}^{m} a_i x_i + \sum_{i=1}^{m}\sum_{j=1}^{m} a_{ij} x_i x_j + \sum_{i=1}^{m}\sum_{j=1}^{m}\sum_{k=1}^{m} a_{ijk} x_i x_j x_k + \cdots, \tag{2.1}$$

where the output is designated as y, the input vector as $\mathbf{x} = (x_1, x_2, \cdots)$ and $\mathbf{a}$ is the vector of coefficients or weights. This is linear in parameters a and nonlinear in x. Components of the input vector $\mathbf{x}$ could be independent variables or functional terms or finite difference terms. This means that the function could be either an algebraic equation or a finite difference equation, or an equation with mixed terms. The partial form of this function as a state or summation function is developed at each simulated unit and is activated in parallel to build up the complexity.

Unit level

Each simulated unit k receives input variables—for example, $(x_i, x_j) \subset \mathbf{x}$; $i \neq j$—and generates a function $f()$ which is a partial form of the reference function.

$$f(x_i, x_j) = \nu_0^{(k)} + \nu_1^{(k)} x_i + \nu_2^{(k)} x_j + \nu_3^{(k)} x_i x_j + \nu_4^{(k)} x_i^2 + \nu_5^{(k)} x_j^2, \tag{2.2}$$

where $\nu^{(k)}$ are the connecting weights. If we denote o as the desired values and y as the estimated values of the outputs for the function being considered, the output errors would be given by

$$e_p = y_p - o_p; \quad p \in N_A. \tag{2.3}$$

The total squared error for that input vector is

$$E = \sum_{p \in N_A} e_p^2. \tag{2.4}$$

This corresponds to the minimization of the average error E in estimating the weights $\nu^{(k)}$; this is the least squares technique. The weights are computed using a specific training sample N_A which is a part of the whole data points N specified for this purpose.

Layer level

Each layer contains a group of units that are interconnected to the units in the next layer. The weights at each unit are estimated by minimizing the error E. The measure of an objective function is used as the threshold value to make the unit "on" or "off" in comparison with the testing data N_B which is another part of N and, at the same time, it is considered to obtain the optimum output response; i.e., this is used as threshold as well as objectivity measures simultaneously. The outputs of the units which are "on" as per the threshold values are connected as inputs to the units in the next layer; that means that the output of kth unit, if it is in the domain of local threshold measure, would become input to some other units in the next level. The process continues layer after layer. The estimated weights of the connected units are memorized in the local memory.

Functional flow of the algorithm

The flow of the algorithm can be described as follows. We feed the input data of m input variables of x randomly; for example, if they are fed in pairs at each unit, then a total of C_m^2 ($= m(m-1)/2$) partial functions of the form below are generated at the first layer:

$$y = f(x_i, x_j); \quad i,j = 1,2,\cdots,m; \quad i \neq j, \tag{2.5}$$

where $f()$ is the partial function as in Equation 2.2 and y is its estimated output. Then outputs of F_1 ($\leq C_m^2$) functions ("freedom-of-choice") are selected as per the threshold measure to pass on to the second layer as inputs in pairs. In the second layer we check the functions of the form

$$z = f(y_i, y_j); \quad i,j = 1,2,\cdots,F_1; \quad i \neq j, \tag{2.6}$$

where $f()$ is the same form of the partial function as in Equation 2.2 and z is its estimated output. The number of such functions is $C_{F_1}^2$. Outputs of F_2 functions are selected to pass on to the third layer. In the third layer we estimate the functions of the form:

$$v = f(z_i, z_j); \quad i,j = 1,2,\cdots,F_2; \ i \neq j, \tag{2.7}$$

where v is the estimated output of the type of function as in Equation 2.2. The number of such functions is $C_{F_2}^2$. The process continues and is stopped when the threshold value begins to increase. The parameters of the optimal function are retrieved through the path of the connecting units from the preceding layers.

2 NETWORK STRUCTURES

The network structures differ as per the interconnections among the units and their hierarchical levels. There are three main inductive learning networks: multilayer, combinatorial, and harmonic. There are other networks based on these three using the concept of self-organization modeling. Multilayer algorithm uses a multilayered network structure with linearized input arguments and generates simple partial functionals. Combinatorial algorithm uses a single-layered structure with all combinations of input arguments including the full description. This could be realized in different ways at each layer of multilayer structure by restricting the number of selected nodes at each layer. Harmonic algorithm follows the multilayered structure in obtaining the optimal harmonic trend with nonmultiple frequencies for oscillatory processes.

2.1 Multilayer algorithm

Multilayer network is a parallel bounded structure that is built up based on the connectionistic approach given in the basic iterative algorithm with linearized input variables and information in the network flows forward only. Each layer has a number of simulated units depending upon the number of input variables. Two input variables are passed on through each unit. For example, x_i and x_j are passed on through kth unit and build a summation function. Weights are estimated using the training set A. At the threshold level, error criterion is used to evaluate this function using the test set B. If there are m input variables, the first layer generates M_1 ($= C_m^2$) functions. $F_1(\leq M_1)$ units as per the threshold values are made "on" to the next layer. Outputs of these functions become inputs to the second layer and the same procedure takes place in the second layer. It is further repeated in successive layers until a global minimum on the error criterion is achieved. If it is not achieved, it means the heuristic specifications must be considered for alteration. The partial function that achieves the global minimum is treated as an optimal model under the given specifications.

The mathematical description of a system can be considered as a nonlinear function in its arguments which may include higher ordered terms and delayed values of the input variables.

$$y = f(x_1, x_2, \cdots, x_1^2, x_2^2, \cdots, x_1x_2, x_1x_3, \cdots, x_{1(-1)}, \cdots, x_{1(-2)}, \cdots), \tag{2.8}$$

where $f()$ is a function of higher degree and y is its estimated output. This can be renotated as a linearized function by calculating all arguments of x in the following form of full description.

$$\begin{aligned} y &= f(u_1, u_2, \cdots, u_m) \\ &= a_0 + a_1u_1 + a_2u_2 + \cdots + a_mu_m, \end{aligned} \tag{2.9}$$

where u_i, $i = 1, 2, \cdots, m$ are the renotated terms of x; a_k, $k = 0, 1, \cdots, m$ are the coefficients and m is total number of arguments. These m input variables become inputs to the first layer. The partial functions generated at this layer are

$$\begin{aligned} y_1 &= \nu_{01}^{(1)} + \nu_{11}^{(1)}u_1 + \nu_{21}^{(1)}u_2, \\ y_2 &= \nu_{01}^{(2)} + \nu_{11}^{(2)}u_1 + \nu_{21}^{(2)}u_3, \\ &\cdots \\ y_{M_1} &= \nu_{01}^{(M_1)} + \nu_{11}^{(M_1)}u_{m-1} + \nu_{21}^{(M_1)}u_m, \end{aligned} \tag{2.10}$$

where M_1 $(= C_m^2)$ is the number of partial functions generated at the first layer, y_j and $\nu_{i1}^{(j)}$, $j = 1, 2, \cdots, M_1$; $i = 0, 1, 2$ are the estimated outputs and corresponding weights of the functions. Let us assume that F_1 functions are selected for the second layer and that there are M_2 $(= C_{F_1}^2)$ partial functions generated at the second layer.

$$\begin{aligned} z_1 &= \nu_{02}^{(1)} + \nu_{12}^{(1)}y_1 + \nu_{22}^{(1)}y_2, \\ z_2 &= \nu_{02}^{(2)} + \nu_{12}^{(2)}y_1 + \nu_{22}^{(2)}y_3, \\ &\cdots \\ z_{M_2} &= \nu_{02}^{(M_2)} + \nu_{12}^{(M_2)}y_{F_1-1} + \nu_{22}^{(M_2)}y_{F_1}, \end{aligned} \tag{2.11}$$

where z_j and $\nu_{i2}^{(j)}$, $j = 1, 2, \cdots, M_2$; $i = 0, 1, 2$ are the estimated outputs and corresponding coefficients of the functions. In the same way, assume that F_2 functions are passed on to the third layer; this means that there are M_3 $(= C_{F_2}^2)$ partial functions generated in this layer.

$$\begin{aligned} v_1 &= \nu_{03}^{(1)} + \nu_{13}^{(1)}z_1 + \nu_{23}^{(1)}z_2, \\ v_2 &= \nu_{03}^{(2)} + \nu_{13}^{(2)}z_1 + \nu_{23}^{(2)}z_3, \\ &\cdots \\ v_{M_3} &= \nu_{03}^{(M_3)} + \nu_{13}^{(M_3)}z_{F_2-1} + \nu_{23}^{(M_3)}z_{F_2}, \end{aligned} \tag{2.12}$$

where v_j and $\nu_{i3}^{(j)}$, $j = 1, 2, \cdots, M_3$; $i = 0, 1, 2$ are the estimated outputs and corresponding coefficients of the functions. The process is repeated by imposing threshold levels of $m \geq F_1 \geq F_2 \geq F_3 \geq \cdots \geq F_l$ so that finally an unique function is selected at one of the layers. The multilayer network structure with five input arguments and five selected nodes

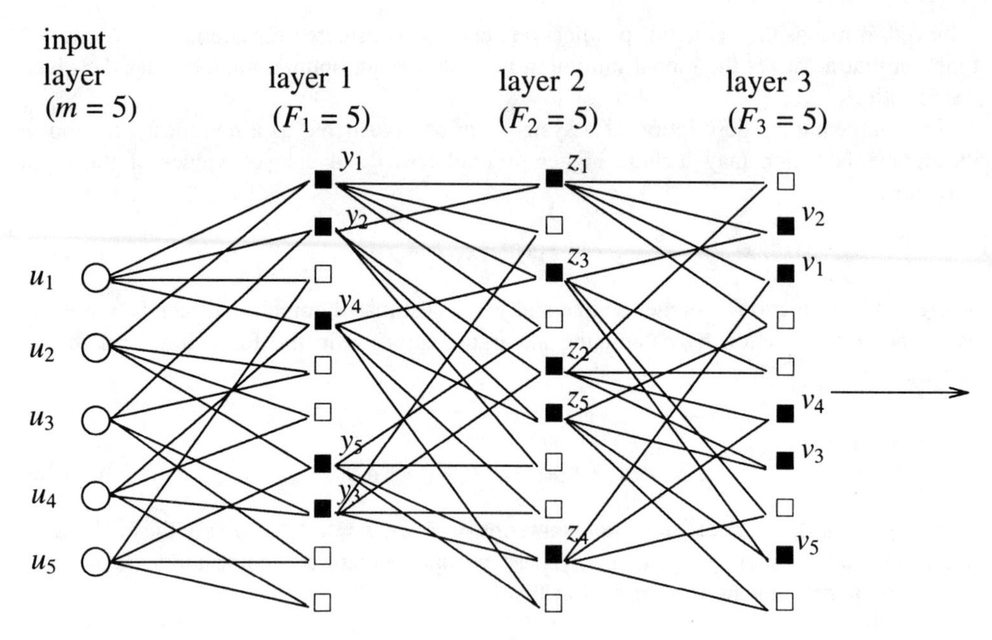

Figure 2.1. Multilayer network structure with five input arguments and selected nodes

at each layer is exhibited in Figure 2.1. For example, if the function $v_2 = \nu_{03}^{(2)} + \nu_{13}^{(2)} z_1 + \nu_{23}^{(2)} z_3$ in Equation 2.12 achieves the global minimum, then it traces back to the preceding layers to recollect the functional values of all connecting weights from the associated units. Finally, we get the optimal function in terms of the input arguments as shown below:

$$\begin{aligned} v_2 &= f(z_1, z_3) \\ &\equiv f(\, f(y_1, y_2), f(y_1, y_4)\,) \\ &\equiv f(u_1, u_2, u_3, u_5) = f(X). \end{aligned} \tag{2.13}$$

This is demonstrated in Figure 2.2. One could obtain more functions which are nearer to global measure for further evluation.

The computational aspects that are considered as the multilayer network procedure is more repetitive in nature. It is important to consider the algorithm in modules and facilitate repetitive characteristics. This is given in the last chapter of the book.

2.2 Combinatorial algorithm

This uses a single-layered structure. Summation functions are generated for all combinations of input variables; i.e., this is like "all types of regressions" in the regression analysis. Let us describe the mathematical description of a system as shown below with three input arguments.

$$y = a_0 + a_1 u_1 + a_2 u_2 + a_3 u_3, \tag{2.14}$$

where y is the estimated output, u_1, u_2, and u_3 are the input arguments, and a_i are the weights. The algorithm uses a single-layered structure because of its complexity in model building. This is given below.

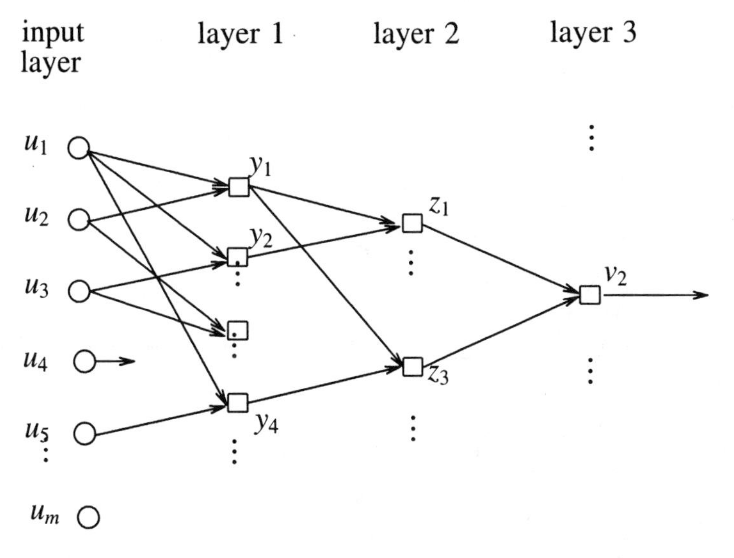

Figure 2.2. Schematic output flow to unit 2 of layer 3 in the multilayer structure

Layer level

1. Summation functions for all combinations of the input arguments u_1, u_2, u_3 (in this case there are seven units, Figure 2.3) are generated:

$$\begin{aligned} y_1 &= a_0^{(1)} + a_1^{(1)} u_1, \\ y_2 &= a_0^{(2)} + a_2^{(2)} u_2, \\ y_3 &= a_0^{(3)} + a_1^{(3)} u_1 + a_2^{(3)} u_2, \\ y_4 &= a_0^{(4)} + a_3^{(4)} u_3, \\ y_5 &= a_0^{(5)} + a_1^{(5)} u_1 + a_3^{(5)} u_3, \\ y_6 &= a_0^{(6)} + a_2^{(6)} u_2 + a_3^{(6)} u_3, \\ y_7 &= a_0^{(7)} + a_1^{(7)} u_1 + a_2^{(7)} u_2 + a_3^{(7)} u_3, \end{aligned} \tag{2.15}$$

   where y_k is the estimated output of kth unit, $k = 1, 2, \cdots, 7$; and $a_i^{(k)}$, $i = 0, 1, 2, 3$ are their connecting weights.
2. The weights are estimated by using the least squares technique with a training set at each unit.
3. Then the unit errors are compared as per the threshold objective function using a testing set, and
4. Units with selected output responses are made "on" and evaluated further.

The schematic flow of the algorithm is given in Figure 2.4.

Usage of "structure of functions." If there is increase in the input arguments, there is corresponding increase in the combinations of them. Suppose there are m variables, then the total combinations are $M_1 = (2^m - 1)$. This is the main difference of this single-layer algorithm in comparison with the multilayered algorithm described in the previous section. There is restriction on the number of input variables with which to use this algorithm as per the capacity of the computer. Efforts are given to build up the algorithm in a more

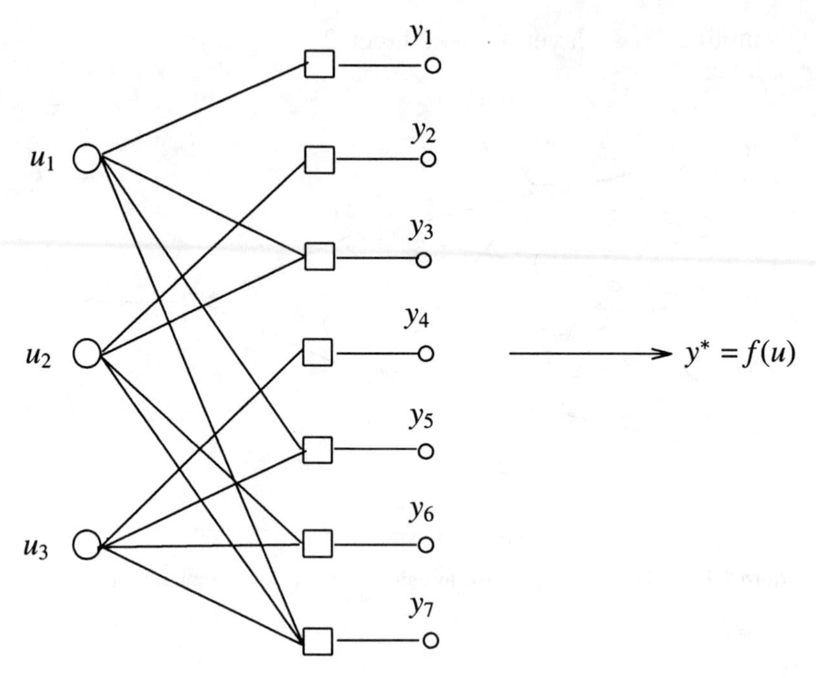

Figure 2.3. Single-layer layout of combinatorial structure

economical way in generating the functions with all combinations. It uses a binary matrix structure of "0"s and "1"s, where each row indicates a partial function with its parameters represented by "1", number of rows indicates total number of units in the layer, and number of columns indicates total number of parameters in the full description. Terms in the matrix are made equal to "0" if the parameters are not present in the function. This is called "structure of functions" (2.16), and includes the full description and the function with all arguments. Usually the "structure of functions" contains the constant term a_0 which is present in all functions:

$$\begin{array}{ccccc} i & a_0^{(i)} & a_3^{(i)} & a_2^{(i)} & a_1^{(i)} \\ 1 & 1 & 0 & 0 & 1 \\ 2 & 1 & 0 & 1 & 0 \\ 3 & 1 & 0 & 1 & 1 \\ 4 & 1 & 1 & 0 & 0 \\ 5 & 1 & 1 & 0 & 1 \\ 6 & 1 & 1 & 1 & 0 \\ 7 & 1 & 1 & 1 & 1. \end{array} \tag{2.16}$$

This is referred further in forming the normal equations for each function. Connection weights of each unit are estimated using a training set and evaluated for its threshold measure in comparison with a testing set. Finally it gives out the selected output responses as per the threshold values.

Gradual increase of complexity. As we have seen above, the combinatorial algorithm is based on an inductive approach; this is done by sorting of all possible models from a given basis of a reference function with fixed input variables. The best of them are chosen according to the external criteria. The complexity of the models is increased by sorting, which is done by gradual increase of arguments in the polynomials or partial functions. The

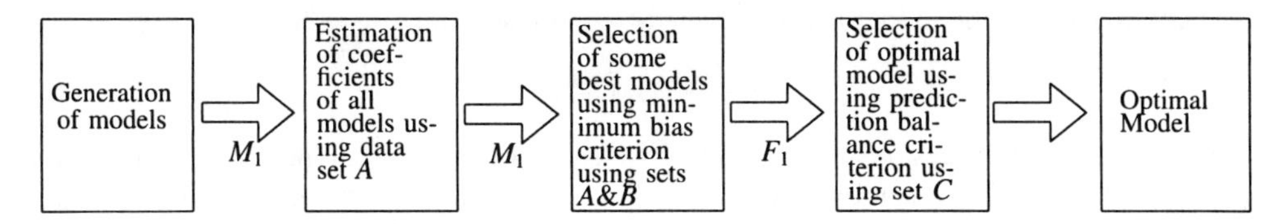

Figure 2.4. Schematic flow of single-layer combinatorial algorithm

important thing here is that no possible variants of the model that appears with the complete set are missed. Let us see how this is realized with three variables of the complete quadratic polynomial that has the form

$$\begin{aligned} y &= a_0 + a_1x_1 + a_2x_2 + a_3x_3 + a_4x_1^2 + a_5x_2^2 + a_6x_3^2 \\ &+ a_7x_1x_2 + a_8x_1x_3 + a_9x_2x_3. \end{aligned} \tag{2.17}$$

There are ten terms in the full polynomial that includes the constant term a_0 ($m = 10$). Sometimes arguments like $1/x_1, 1/x_2$, and $1/x_3$ and other higher order nonlinear terms are to be included based on the global minimum attained on the external criterion. The partial polynomials are linear in coefficients a, and the least squares technique is used to estimate the coefficients by reindexing the nonlinear terms in linear form. Here we give the scheme for gradual increase of complexity in the partial functions. The scheme is as follows.

In the first step, all models with single arguments are determined.

$$y_1 = a_0, \quad y_2 = a_1x_1, \quad y_3 = a_2x_2, \cdots, \quad y_{10} = a_9x_2x_3. \tag{2.18}$$

That means there are $C_{10}^1 = 10$ partial models. Then in the second step, all models with two arguments are determined.

$$\begin{aligned} y_{11} &= a_0 + a_1x_1, \quad y_{12} = a_0 + a_2x_2, \cdots, y_i = a_0 + a_9x_2x_3 \\ y_j &= a_1x_1 + a_2x_2, \quad y_k = a_1x_1 + a_3x_3, \cdots, y_{45} = a_8x_1x_3 + a_9x_2x_3. \end{aligned} \tag{2.19}$$

There are $C_{10}^2 = 45$ partial models. Similarly, in the third step models with three arguments are built up, in the fourth step with four arguments, and so on until $C_{10}^{10} = 1$ model, which is the complete polynomial. The total number of all possible models constructed for m arguments is

$$M_1 = \sum_{s=1}^{m} C_m^s = 2^m - 1. \tag{2.20}$$

The value of M_1 increases with the increase of m; for example, if $m = 10$, then $M_1 = 1,023$ and if $m = 15$, then $M_1 = 32,767$. This algorithm with the given program at the end, where the complexity of the partial models is not changed sequentially but rather according to the binary matrix of m-digit counter, sorts all possible models for $m \leq 18$ in an acceptable time. However, the inclusion of an additional argument in the input set doubles the computational time. We give here one of the optimal sorting schemes [115] that enables us to increase the input set to $m = 23$.

2.3 Recursive scheme for faster combinatorial sorting

Suppose we are given N measurement points of output y and input variables $\mathbf{x}$ of a system. For the given output y we set up the measurement function

$$\mathbf{y} = X\mathbf{a}, \quad \text{where } \mathbf{y}[N \times 1], \; X[N \times m], \; \text{and } \mathbf{a}[m \times 1]. \tag{2.21}$$

The coefficients of the model are determined by the least-squares method

$$\hat{\mathbf{a}} = (X^T X)^{-1} X^T \mathbf{y}. \tag{2.22}$$

Let us consider the question of obtaining inverse matrices of the type $(X^T X)^{-1}$ for each partial polynomial in the combinatorial algorithm. First of all we consider for one individual output; this means that at a certain stage of the algorithm we have obtained the estimates of the parameters of a partial model containing k arguments $x_1, x_2, \cdots, x_k$. Then we know that

$$\hat{\mathbf{a}}_{\mathbf{k}} = H_k^{-1} \mathbf{g}_k, \tag{2.23}$$

where

$$H_k = X_k^T X_k, \qquad \text{and} \qquad \mathbf{g}_k = X_k^T \mathbf{y} \tag{2.24}$$

denote the elements of the matrix of the normal system

$$[H_k \mid \mathbf{g}_k] \triangleq [X_k^T X_k \mid X_k^T \mathbf{y}]. \tag{2.25}$$

In estimating the coefficients $\mathbf{a}_k$, the inverse of the matrix H_k is computed by using the Gauss method. As a further step, for example, the system considers another partial model with an additional argument x_{k+1}, and the matrix of the normal system takes the form

$$[H_{k+1} \mid \mathbf{g}_{k+1}] = \left[\begin{array}{c|c|c} H_k & \mathbf{h}_{k+1} & \mathbf{g}_k \\ \hline \mathbf{h}_{k+1}^T & \vartheta_{k+1} & \gamma_{k+1} \end{array}\right], \tag{2.26}$$

where $\mathbf{h}_{k+1}[k \times 1]$ is a k dimensional vector; ϑ_{k+1} and γ_{k+1} are scalars whose values are computed using the measurements of the $(k+1)$st argument. In the combinatorial algorithm, the estimate of the vector $\hat{\mathbf{a}}_{k+1}$ is found by the least squares method; that means that one performs operations analogous to a new inversion of the matrix H_{k+1} for finding the parameter estimates $\hat{\mathbf{a}}_{k+1} = H_{k+1}^{-1} \mathbf{g}_{k+1}$. This estimate can be obtained in terms of the known H_k^{-1} and other elements $\mathbf{h}_{k+1}$ and ϑ_{k+1} as below.

$$\hat{\mathbf{a}}_{k+1} = \left[\begin{array}{c} \hat{\mathbf{a}}_k^* \\ \hline \hat{\alpha}_{k+1} \end{array}\right] = \left[\begin{array}{c|c} H_k & \mathbf{h}_{k+1} \\ \hline \mathbf{h}_{k+1}^T & \vartheta_{k+1} \end{array}\right]^{-1} \left[\begin{array}{c} \mathbf{g}_k \\ \hline \gamma_{k+1} \end{array}\right]. \tag{2.27}$$

Specifically, if we write the inverse of the matrix as

$$H_{k+1}^{-1} \triangleq B_{k+1} = \left[\begin{array}{c|c} B_k & \mathbf{b}_{k+1} \\ \hline \mathbf{b}_{k+1}^T & \beta_{k+1} \end{array}\right] \tag{2.28}$$

and solve the equation $B_{k+1} H_{k+1} = I_{k+1}$, where I_{k+1} is the identity matrix, we obtain the following formulae by inverting the above block matrices.

$$\begin{aligned} \beta_{k+1} &= 1/(\vartheta_{k+1} - \mathbf{h}_{k+1}^T \mathbf{c}_{k+1}), \quad \mathbf{c}_{k+1} \triangleq H_k^{-1} \mathbf{h}_{k+1}, \quad H_0^{-1} \triangleq 0. \\ \mathbf{b}_{k+1} &= -\beta_{k+1} \mathbf{c}_{k+1}, \quad B_k = H_k^{-1} + \beta_{k+1} \mathbf{c}_{k+1} \mathbf{c}_{k+1}^T. \end{aligned} \tag{2.29}$$

Substituting the results above, we get

$$\begin{aligned} \hat{\mathbf{a}}_k^* &= H_k^{-1} \mathbf{g}_k + \beta_{k+1} \mathbf{c}_{k+1} (\mathbf{c}_{k+1}^T \mathbf{g}_k - \gamma_{k+1}) \\ \hat{\alpha}_{k+1} &= -\beta_{k+1} (\mathbf{c}_{k+1}^T \mathbf{g}_k - \gamma_{k+1}). \end{aligned} \tag{2.30}$$

Thus, the new estimate $\hat{\mathbf{a}}_{k+1}$ can be expressed explicitly in terms of the estimate already calculated.

$$\hat{\mathbf{a}}_{k+1} = \begin{bmatrix} \hat{\mathbf{a}}_k^* \\ --- \\ \hat{\alpha}_{k+1} \end{bmatrix} = \begin{bmatrix} \hat{\mathbf{a}}_k - \hat{\alpha}_{k+1}\mathbf{c}_{k+1} \\ ------- \\ \hat{\alpha}_{k+1} \end{bmatrix}, \quad \hat{\mathbf{a}}_0 \triangleq 0. \tag{2.31}$$

It is obvious that the above operation considerably decreases the number of computations needed for estimating $\mathbf{a}_{k+1}$ in comparison with the direct determination of the matrix H_{k+1}^{-1}. If the next term x_{k+2} is adjoined to the model as another partial polynomial, then the new inverse matrix is obtained directly as

$$H_{k+1}^{-1} = \left[\begin{array}{c|c} H_k^{-1} + \beta_{k+1}\mathbf{c}_{k+1}\mathbf{c}_{k+1}^T & -\beta_{k+1}\mathbf{c}_{k+1} \\ \hline -\beta_{k+1}\mathbf{c}_{k+1}^T & \beta_{k+1} \end{array}\right]. \tag{2.32}$$

The above recursive technique is convenient to use in constructing models of the partial polynomials of gradually increasing complexity that begin with a single argument. This type of approach is called "method of bordering."

The above table of "structure of functions" (Equation 2.16) shows the sequential change in the states of the binary counter that corresponds to the changes in structures of partial polynomials. In the shift from 1 $\rightarrow$ 3 or 2 $\rightarrow$ 3, 4 $\rightarrow$ 5 and 6, and 5 $\rightarrow$ 7 or 6 $\rightarrow$ 7 models, one can notice that a new term is added; the inverse matrices in these cases can simply be computed using the above recursive algorithm. This type of combinatorial scanning of models accelerates the calculation of the model coefficients and reduces the computational time; one has to think about the optimum way of utilizing the computer memory to store these inverse matrices as the number of arguments increases. The solution is to have an optimum way of sequencing the rows of binary matrix in a specific way. For example, let us consider the sequencing as follows:

i	$a_0^{(i)}$	$a_3^{(i)}$	$a_2^{(i)}$	$a_1^{(i)}$
1	1	0	0	1
2	1	0	1	1
3	1	0	1	0
4	1	1	1	0
5	1	1	1	1
6	1	1	0	1
7	1	1	0	0.

(2.33)

In this sequencing, one can notice that from 1 $\rightarrow$ 2 one argument is added, from 2 $\rightarrow$ 3 one argument is eliminated, and so on. When an argument is added, the above recursive procedure can be used for obtaining the new inverse matrix. When an argument is eliminated, a new procedure which works in reverse is needed; one can introduce the inverse operation of it to calculate H_k^{-1} and $\mathbf{a}_k$ from the known H_{k+1}^{-1} and $\hat{\mathbf{a}}_{k+1}$ as

$$H_{k+1}^{-1} = \left[\begin{array}{c|c} B_k & \mathbf{b}_{k+1} \\ \hline \mathbf{b}_{k+1}^T & \beta_{k+1} \end{array}\right], \quad \hat{\mathbf{a}}_{k+1} = \begin{bmatrix} \hat{\mathbf{a}}_k^* \\ --- \\ \hat{\alpha}_{k+1} \end{bmatrix}, \tag{2.34}$$

and eliminate the $(k+1)$st argument by

$$\begin{aligned} H_k^{-1} &= B_k - \frac{1}{\beta_{k+1}}\mathbf{b}_{k+1}\mathbf{b}_{k+1}^T, \\ \hat{\mathbf{a}}_k &= \hat{\mathbf{a}}_k^* - \frac{\hat{\alpha}_{k+1}}{\beta_{k+1}}\mathbf{b}_{k+1}. \end{aligned} \tag{2.35}$$

This type of refinement of the recursive algorithm yields maximum decrease in the time of total scanning of the models at the cost of increased memory utilization. In the above "structure of functions" (Equation 2.33), one argument is added to the models in the shift from $1 \to 2$, $3 \to 4$, and $4 \to 5$; and in the shift from $2 \to 3$, $5 \to 6$, and $6 \to 7$, one argument is eliminated. The above procedure with the alternate use of recursive and inverse recursive routines can be used as required for computing H_k^{-1} and H_{k+1}^{-1} matrices alternately from an initial matrix H_k^{-1}. The least square error ε_k^2 can be calculated after estimating the coefficients $\hat{\mathbf{a}}_k$.

$$\varepsilon_k^2 = (\mathbf{y}_A - \hat{\mathbf{y}}_A)^T(\mathbf{y}_A - \hat{\mathbf{y}}_A) \equiv \mathbf{y}_A^T\mathbf{y}_A - \hat{\mathbf{a}}_k^T\mathbf{g}_A, \tag{2.36}$$

where $\mathbf{g}_A = X_A^T\mathbf{y}_A$; here the subscript A corresponds to the training set. The recurrent algorithm enables us to compute the least squares error ε_{k+1}^2 recursively when a new argument is added.

$$\varepsilon_{k+1}^2 = \varepsilon_k^2 + \hat{\alpha}_{k+1}(\gamma_{k+1} - \mathbf{c}_{k+1}^T\mathbf{g}_k) \equiv \varepsilon_k^2 + \frac{\hat{\alpha}_{k+1}^2}{\beta_{k+1}}. \tag{2.37}$$

2.4 Multilayered structures using combinatorial setup

One version of a multilayered structure is that the combinatorial algorithm could be realized at each layer of the multilayer network structure by keeping the limit on the "freedom of choice" at each layer. The unit outputs are fed forward layer by layer as per the threshold measure to obtain the global output response for optimal complexity. This structure is exhibited in Figure 2.5 with three input arguments and three selected nodes at each layer.

2.5 Selectional-combinatorial multilayer algorithm

Here is another version of a multilayered structure that is called a selectional-combinatorial algorithm that realizes the above recursive procedure [116] [49]. The general outline of the algorithm is as follows.

In the first layer, all models containing single arguments are estimated, and some of the best are selected as per certain external criteria and passed on to the next layer. In the second layer, different arguments are selected and added to these models, which improved the response as per the external criteria. This continues until it deteriorates. In contrast to the original multilayer set up, it does not pass on the outputs of the units. The multilayer error is not passed on because of its retainment of the original basis functions; their number of arguments coincides with the layer number, and the total number of layers cannot exceed m.

The important aspect of this algorithm is its realization of the recursive procedure for successive estimation of coefficients of the partial models according to the least squares method. The matrix of normal equations for a model is represented with $l+1$ arguments that is obtained from the complete normal matrix $[m \times (m+1)]$ matrix $[H \mid \mathbf{g}] = [X^TX \mid X^T\mathbf{y}]$ using the form

$$H_{l+1} = \left[\begin{array}{c|c} H_l & \mathbf{h}_{l+1} \\ --- & --- \\ \mathbf{h}_{l+1}^T & \vartheta_{l+1} \end{array}\right], \quad \mathbf{g}_{l+1} = \left[\begin{array}{c} \mathbf{g}_l \\ --- \\ \gamma_{l+1} \end{array}\right]. \tag{2.38}$$

The coefficients $\hat{\mathbf{a}}_{l+1}$ are estimated using the following:

$$\beta_{l+1} = 1/(\vartheta_{l+1} - \mathbf{h}_{l+1}^T\mathbf{c}_{l+1}), \quad \mathbf{c}_{l+1} = H_l^{-1}\mathbf{h}_{l+1},$$

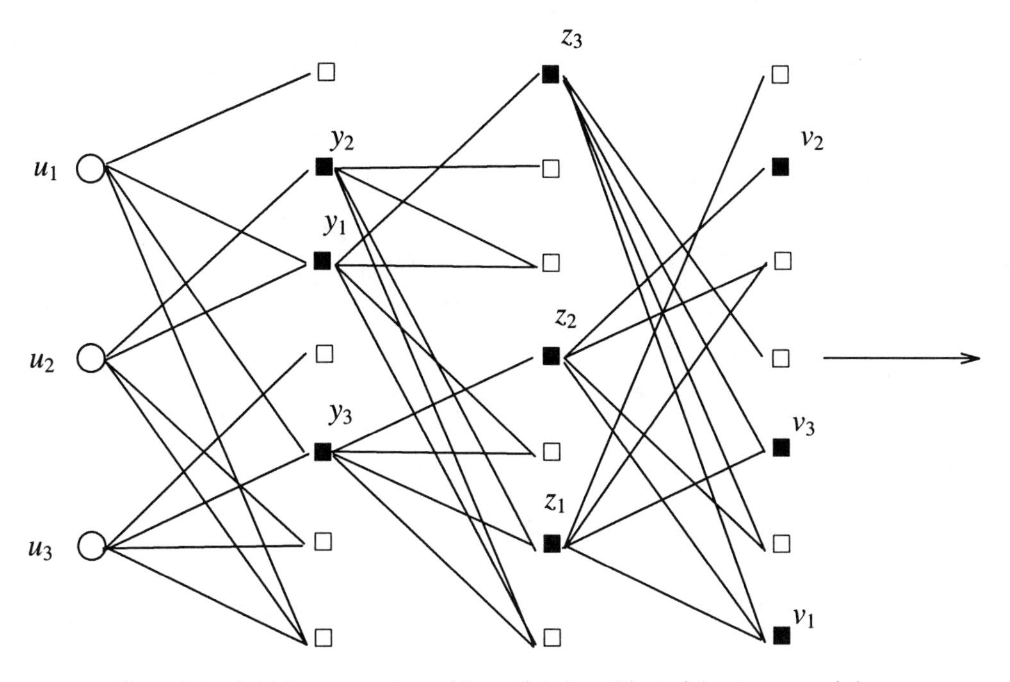

Figure 2.5. Multilayer structure with restricted combinatorial set up at each layer

$$\hat{\mathbf{a}}_{l+1} = \begin{bmatrix} \hat{\mathbf{a}}_l \\ --- \\ \hat{\alpha}_{l+1} \end{bmatrix} = \begin{bmatrix} \hat{\mathbf{a}}_l - \hat{\alpha}_{l+1}\mathbf{c}_{l+1} \\ ------ \\ \beta_{l+1}(\gamma_{l+1} - \mathbf{g}_l\mathbf{c}_{l+1}) \end{bmatrix}. \tag{2.39}$$

Let us explain the sequence of calculations of the $(l+1)$st layer from the lth layer.

1. For each ith model ($i = 1, \cdots, M_1$) of the first layer ($l = 1$), the inverse matrices $H_1^{-1}(i)$ are calculated. These matrices will be updated at all the consecutive layers using the recursive procedure. Let us assume that $H_l^{-1}(i)$ are the inverse matrices at the lth layer and F_l models are selected as per the external criteria.
2. Partial models of the $(l+1)$st layer are generated by an ordered addition to each selected ith model of one of the arguments absent from it that correspond to the zeroth elements of the binary structure of the vector for ith model. Here the partial models M_{l+1} are generated with the complexity $l+1$ arguments. Each new model is uniquely defined from the preceding F_l models of lth layer, and the x_j is the new added argument.
3. Estimates $\hat{\mathbf{a}}_{l+1}(i,j)$ are calculated for all M_{l+1} models using the recursive procedure for each i on the basis of the matrix $H_l^{-1}(i)$. From the denominator terms, one can easily sort out even the ill-conditioned normal matrix.
4. The values of the external criteria are computed for each model.
5. The best F_{l+1} models are chosen for the next $(l+2)$nd layer from the condition of improvement of the minimal value of the external criterion; for example, $\Delta(B)$ is

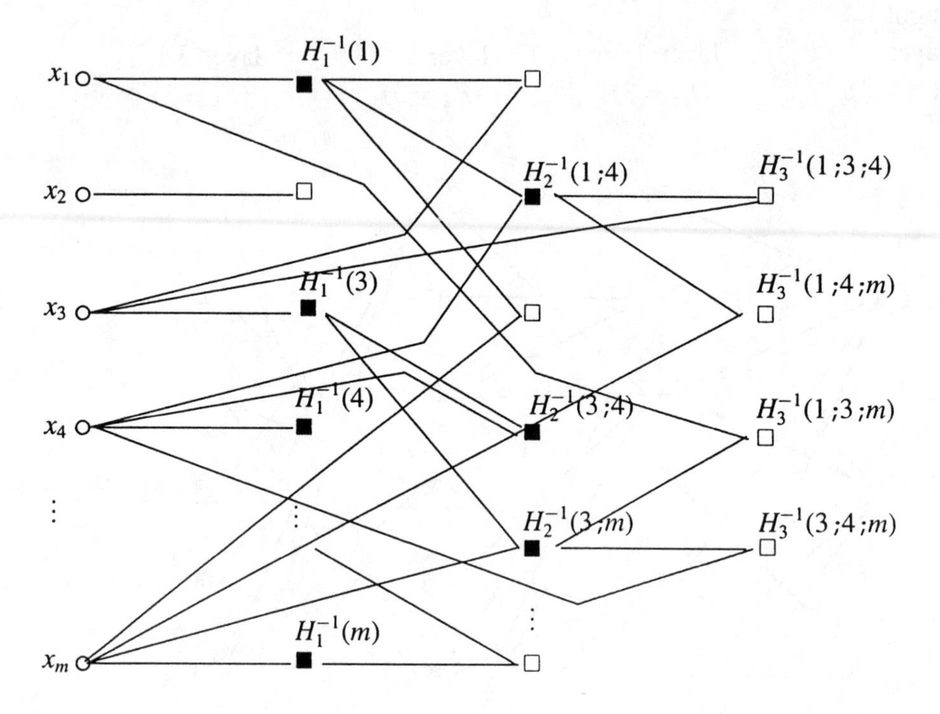

Figure 2.6. Multilayered version of selectional-combinatorial algorithm; the inverse matrices $H_l^{-1}(i)$ are updated layer after layer using the recursive technique

obtained at the preceding layer l.

$$F_{l+1} = \sum_{p=1}^{M_{l+1}} \delta_p, \qquad \delta_p = \begin{cases} 1 & \text{if } \Delta_p(B) < \theta_l \\ 0 & \text{if } \Delta_p(B) \geq \theta_l, \end{cases} \tag{2.40}$$

$$\text{where} \quad \theta_l = \min_{i \in F_l} \Delta_i(B).$$

This procedure is applicable in a strict sense when the external criterion actually behaves like an ideal one in selecting a unimodel.

In general, one can choose selection of models on a competitive basis as in the usual multilayer algorithm.

To overcome the possible local minimum when $l+1 \ll m$ and $F_{l+1} = 0$, it is better to fix the lower boundary (for example, $F_{l+1_{min}} = m - (l+1)$) and an upper boundary F_{max}. The freedom of choice at the $(l+1)$st layer is determined with the constraint

$$F_{l+1_{min}} \leq F_{l+1} \leq F_{max}. \tag{2.41}$$

6. When $F_{l+1} = 0$, the procedure is stopped automatically which would indicate the minimum is achieved at the previous layer.

The schematic diagram of this algorithm is shown in Figure 2.6 with the passage of inverse matrix H and an additional term from layer to layer.

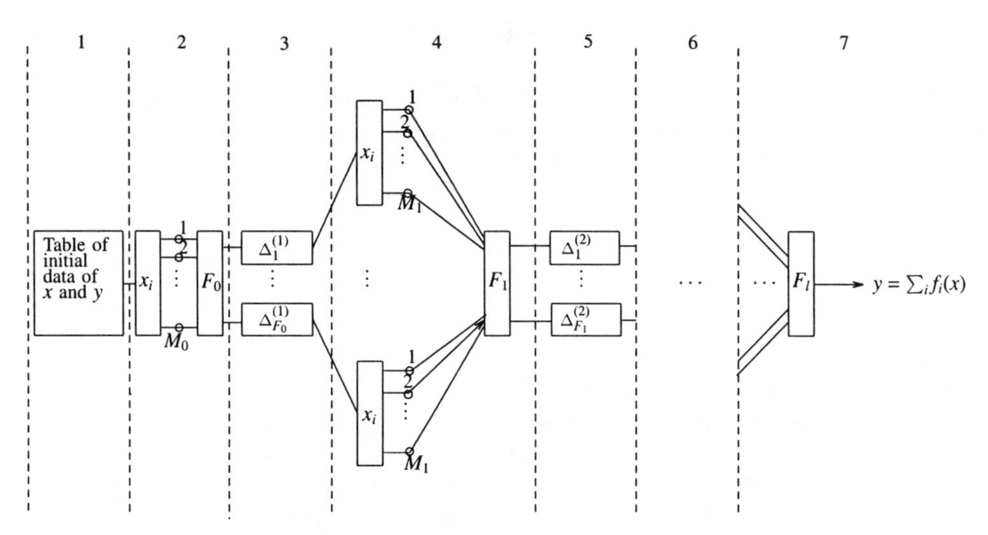

Figure 2.7. Schematic illustration of front propagation algorithm with the calculation of output error residuals, where $M_0, M_1, \cdots$ denote the total partial models; $F_0, F_1, \cdots, F_l$ are the number of best models (freedom of choice) at zeroth, first and last layers correspondingly

2.6 Multilayer algorithm with propagating residuals (front propagation algorithm)

This algorithm is built up based on forwarding the output errors to the next layers as outputs and using the combinatorial induction on original input variables at each layer. The schematic flow of such algorithm is given in Figure 2.7.

Each block of the figure is explained below.

1. Table of initial empirical data points of N, m input variables of x_i and output variable y.
2. The initial layer which is called the zeroth layer uses the combinatorial algorithm in choosing the best F_0 models.
3. The first differences (error residuals for each model) which are denoted as $\Delta_i^{(1)}, i = 1, \cdots, F_0$ are computed. Each vector, $\Delta_i^{(1)}$ is of $[N \times 1]$; $\Delta_i^{(1)} = \mathbf{y} - \hat{\mathbf{y}}_{(i)}$, where $\hat{\mathbf{y}}_{(i)}$ is the vector of estimated output corresponding to ith model.
4. This is the first selection layer in which the F_0 vectors of first differences are used as output variables independently at each combinatorial sorting procedure. The best F_1 models are chosen.
5. The second differences $\Delta_j^{(2)}, j = 1, \cdots, F_1$ are computed; $\Delta_j^{(2)} = \Delta^{(1)} - \hat{\Delta}^{(1)}$, where $\Delta^{(1)}$ is the vector of actual values of first differences from the first layer and $\hat{\Delta}^{(1)}$ is the vector of estimated output corresponding to each jth model.
6. This block denotes the similar follow-up of layers and calculation of further residuals.
7. Last layer with the model of optimal complexity.

Some features of the algorithm

One can see below the overall effect of the forward propagation of residuals.

First-differences (residuals) at layer "0" are computed as

$$\Delta^{(1)} = \mathbf{y} - \hat{\mathbf{y}}$$

$$\equiv \mathbf{y} - X\hat{\mathbf{a}}^{(0)}, \tag{2.42}$$

where $\mathbf{y}, \hat{\mathbf{y}}$ and $\Delta^{(1)}$ are the actual, estimated and first difference vectors with the dimension of $[N \times 1]$, $\hat{\mathbf{a}}^{(0)}$ is the vector of least squares estimates of coefficients with the dimension $[m \times 1]$, and matrix X is the initial data of $[N \times m]$.

Second-differences (residuals) at layer "1" are computed as

$$\begin{aligned}\Delta^{(2)} &= \Delta^{(1)} - \hat{\Delta}^{(1)} \\ &\equiv \mathbf{y} - X\hat{\mathbf{a}}^{(0)} - X\hat{\mathbf{a}}^{(1)} \\ &= \mathbf{y} - X(\hat{\mathbf{a}}^{(0)} + \hat{\mathbf{a}}^{(1)}),\end{aligned} \tag{2.43}$$

where $\hat{\mathbf{a}}^{(1)}$ is the vector of estimated coefficients at the first layer.

The propagating residuals help in the finer adjustment of the coefficients as the process proceeds layer by layer and ultimately an optimal model is obtained.

The external criterion $c3^2 \triangleq \bar{\eta}_{bs}^2 + \bar{\Delta}^2(C)$ decreases monotonically with the prolifeiration of the selection layers and a relatively small decrease in it serves as a signal to stop the procedure. In comparison with the original multilayer algorithm, the objective nature of the choice of model is not fully conserved because the error decreases monotonically.

This algorithm is worked out for finite-difference reference functions [54]. In the absence of constraints on the coefficients, this algorithm with forward error propagations resembles the exponential-harmonic algorithm.

2.7 Harmonic Algorithm

The principal aim of this algorithm is to extend the use of the inductive self-organization principle to identify the oscillatory processes. It is assumed that the effective reference functions of such processes are in the form of the sum of harmonics with nonmultiple frequencies. The harmonic function is composed of several sinusoids with arbitrary frequencies which are not necessarily related. This type of function produces a multifrequency resultant and exhibits similar spectral characteristics. A balance relation plays an important role in obtaining the frequencies of the process and as an objective function in selecting the optimal trend in the multilayered structure. The algorithm is explained below with the derivation of the balance relation.

Suppose a function $f(t)$ is a process having the sum of m harmonic components with pairwise discrete frequencies $w_1, w_2, \cdots, w_m$.

$$\begin{aligned}f(t) &= \sum_{k=1}^{m} [\mathcal{A}_k \, sin\,(w_k t) + \mathcal{B}_k \, cos\,(w_k t)] \\ &\equiv \sum_{k=1}^{m} \Phi(w_k, t),\end{aligned} \tag{2.44}$$

where $\mathcal{A}_k$ and $\mathcal{B}_k$ are the coefficients; and $w_i \neq w_j$, $i \neq j$, $0 < w_i < \pi$, $i = 1, 2, \cdots, m$. The process has a total interval length of N $(1 \leq t \leq N)$ and points at discrete intervals of time δ_t.

For a fixed point i and any p of Equation 2.44, one can obtain the formula:

$$f(i+p) + f(i-p) = 2\sum_{k=1}^{m} cos\,(pw_k)\; \Phi(w_k, i). \tag{2.45}$$

Summing Equation 2.45 for p from 0 to $m-1$, with weighing coefficients $\mu_0, \mu_1, \cdots, \mu_{m-1}$, we derive a relation:

$$\sum_{p=0}^{m-1} \mu_p \left[f(i+p)+f(i-p)\right] = 2\sum_{k=1}^{m} \Phi(w_k, i)[\mu_0 + \sum_{p=1}^{m-1} \mu_p cos\,(pw_k)]$$
$$= 2\sum_{k=1}^{m} \Phi(w_k, i)\ cos\,(mw_k)$$
$$= f(i+m)+f(i-m). \quad (2.46)$$

This is considered as a balance relation of the process.

$$b_i = [f(i+m)+f(i-m)] - \sum_{p=0}^{m-1} \mu_p[f(i+p)+f(i-p)]. \quad (2.47)$$

If the process is expressed exactly in terms of a given sum of harmonic components, then $b_i = 0$; i.e., the discrete values of $f(t)$, which are symmetric with respect to a point $i\,(m+1 \leq i \leq N-m)$, satisfy the balance relation. The coefficients μ_p are independent of i.

The trigonometric relation used in the calculations of Equation 2.46 is:

$$\mu_0 + \sum_{k=1}^{m-1} \mu_p\, cos\,(pw_k) = cos\,(mw_k). \quad (2.48)$$

This could be formed as mth degree algebraic equation in $cos\, w$ by using the recursive trigonometric relations

$$\mathcal{D}_m(cos\, w)^m + \mathcal{D}_{m-1}(cos\, w)^{m-1} + \cdots + \mathcal{D}_1(cos\, w) + \mathcal{D}_0 = 0, \quad (2.49)$$

where $\mathcal{D}_i,\ i = 0, 1, \cdots, m$ are the functions of μ_p.

It is possible to determine uniquely the coefficients $\mu_p,\ p = 0, 1, \cdots, m-1$ from the relation (Equation 2.46) for $i = m+1, \cdots, N-m$. $(N-m)-(m+1) \geq m-1$; i.e., $N \geq 3m$. Substituting the values of μ_p in Equation 2.49, it can be solved for m frequencies w_k of harmonics by using standard numerical techniques. These frequencies are passed through a multilayered network to form various combinations of the harmonic components. The parameters of the harmonic components $\mathcal{A}_k$ and $\mathcal{B}_k$ are estimated at each unit by using the least squares technique. The optimal trend is obtained at one of the units by considering the balance relation as a threshold objective function.

Suppose $y(t)$ is the given discrete time series data and is to be identified for its harmonic trend $f(t)$. The data are to be separated into two data sets as training set points N_A and testing set points N_B. We can allot some more points N_C as a checking set for checking the trend, i.e., $N = N_A + N_B + N_C$. The maximum number of harmonics is chosen as M_{max} $(< N/3)$. The coefficients μ_p are estimated by using the least squares technique by forming the balance equations with the training set. The system of equations has the form:

$$\sum_{p=0}^{m-1} \mu_p[y(i+p) + y(i-p)] = y(i+m) + y(i-m);$$
$$i = m+1, \cdots, N_A - m. \quad (2.50)$$

By substituting the values of μ_p, the frequencies of the harmonics are determined by solving the mth degree Equation 2.49. This has m roots that uniquely determine m frequencies w_k.

These frequencies are fed through the input layer of a multilayered structure as components of the harmonic terms. The procedure of complete sifting of trends would take place by multilayer selection of trends using the inductive principle. The linear normal equations are constructed in the first layer for any $1 \leq m \leq M_{max}$ number of harmonics, and the coefficients $\mathcal{A}_k$ and $\mathcal{B}_k$ are estimated for all combinations of M_{max} harmonics based on the training set using the least squares technique. All harmonic trends are evaluated for their threshold values in comparison with the testing set, and the output errors of the best trends from F_1 units are fed forward as inputs to the second layer. This procedure is repeated in all subsequent layers. The complexity of the model increases layer by layer as long as the value of the "imbalance" decreases on the testing set points N_B. The balance criterion B (Equation 2.51) is used as the objective function which takes into account the balance relation, b_i (Equation 2.47).

$$B \triangleq \sum_{i=m+1}^{N_B-m} b_i^2 \quad \rightarrow \quad min. \tag{2.51}$$

This unique solution is guaranteed as it is in the multilayer algorithm; the performance of the optimal trend can be tested further using the checking set N_C.

2.8 New algorithms

According to the form of the reference functions, the inductive learning algorithms can be divided into several main classes that could be constructed based on the addition (additive algorithms) or multiplication (multiplicative algorithms); they could be extended further as additive-multiplicative and multiplicative-additive with the factors considered as unit, integer, or noninteger powers. In addition to these, there are other algorithms like correlational and orthogonalized (generalized) algorithms. Let us first give some of the types of polynomials and later study some algorithms.

(i) *Additive polynomials with unit powers of the factors*:

$$y = a_0 + \sum_{j=1}^{m} a_j x_j, \tag{2.52}$$

where m is the number of independent variables; a are the coefficients; y and x are the output and input variables correspondingly.

(ii) *Multiplicative polynomials with unit powers of the factors*:

$$y = a. \prod_{j=1}^{m} x_j, \tag{2.53}$$

where a is the single coefficient; y and x are the output and input variables; and m is the number of independent variables.

(iii) *Multiplicative-additive polynomial with unit powers of factors*: This can have different forms as per the complexity of the terms. One of the forms is given as

$$y = \left\{ \sum_{k=1}^{m} a_k . \prod_{j=1}^{k} x_j \right\}, \tag{2.54}$$

and another form can have factors in integer or noninteger powers

$$y = \left\{ \sum_{k=1}^{s} a_k . \prod_{j=1}^{m} x_j^{p_{kj}} \right\}, \quad p_{kj} \in \{0, \Delta h, \cdots\}, \tag{2.55}$$

where s denotes the complexity of the equation, and Δh is the least fractional power (a certain minimum fractional power of the original factors is indicated; for example, $x_j^{0.5}$ and $x_j^{-0.5}$). The model of optimal complexity in this case can be represented in the form of linear and multiplicative terms with the powers not higher than the powers specified in advance.

Multiplicative-additive (generalized) algorithm

The algorithm described below allows one to obtain models with the multiplicative-additive terms [51].

When the basic inductive algorithms, where the variables have integer powers, do not lead to unbiased and accurate predictions, it is necessary to shift the solution space to another region of functional space; for example, to the region of polynomials with other than integer powers of generalized arguments. This is possible with the following multiplicative-additive algorithm. First of all, one has to choose certain multiplicative models with optimal complexity on the basis of the external criteria. An original model is represented in the form of a product of given arguments with unknown powers;

$$y = a_0 x_1^{k_1} x_2^{k_2} x_3^{k_3} \cdots x_m^{km}. \tag{2.56}$$

This can be rewritten in the following form by taking logarithms on both sides:

$$\ln y = \ln a_0 + k_1 \ln x_1 + k_2 \ln x_2 + \cdots + k_m \ln x_m. \tag{2.57}$$

Using the original data table of the quantities y, x_1, $\cdots, x_m$, a new data table for the variables with the logarithmic values can be set up. Data is separated into training, testing, and examining sets. Several partial, but best, models can be chosen by using one of the inductive learning algorithms (combinatorial or multilayer) with the combined criteria of "minimum-bias plus prediction." By inverting the logarithms of these optimal models one can obtain the best multiplicative models of the given process.

At the second level, to obtain the generalized multiplicative-additive model, we combine the selected multiplicative models into a single complete polynomial as

$$y = b_0 + b_1 \hat{y}_1 + b_2 \hat{y}_2 + \cdots + b_l \hat{y}_l + \cdots, \tag{2.58}$$

where y is the desired output of the process; $\hat{y}_j$, $j = 1, 2, \cdots, l, \cdots$ are the estimated outputs of the selected multiplicative models; and b are the coefficients.

The combinatorial algorithm enables us to obtain a unique optimal model. This model can be rewritten in terms of the original input variables; this is in the form of the multiplicative-additive model with the sum of covariance terms with noninteger powers of the factors and their products which could be used for further analysis.

Sometimes, at the first level, when the conditional equations are formed, it is necessary to take the logarithms of negative quantities. This situation can be avoided either by discarding that particular conditional equation or by reformulating the data in advance. The constant term is chosen in such a way that the error caused by applying the least squares technique to the logarithms of the variables, rather than to the variables themselves, is compensated.

Algorithm for correlation models

Rosenblatt [106] asserts that an infinite perceptron can execute a classification of images according to classes without any *a priori* information. Analogously, one might assert that a model of optimal complexity can be found by sorting the points of the entire function space without any *a priori* specification (or "prompting"). Actually, infinitely large sortings are impossible either in a perceptron or by using inductive algorithms. That's why the class of functions that can be used to choose the complete description must be chosen on the basis of certain *a priori* "prompting." The majority of the inductive algorithms are based on the following "prompting": the complete description is designed as a polynomial which is in its linear form. This means that, as per this constraint, the object is described by a linear algebraic or finite-difference equation that remains unchanged at the interpolation or prediction intervals; the stability condition for these types of equations is unchanged.

In the regions of other types of functional spaces (for example, in the correlational functional space) it can be prompted by computer with the condition of correlational stability. This condition provides a way of extending the function manifold that can be used by applying correlation models of standard form [53], [68]. Here "prompting" is set up by a complete set of support functions of the correlational type,

$$K_y(\tau) = \frac{1}{T - \tau - 1} \sum_{i=\tau}^{T} y_i y_{i-\tau}; \quad 1 \leq i \leq T, \tag{2.59}$$

where τ is the displacement along the axis.

The sum of the discrepancies of these equations from the values of the correlation function for $\tau = 1, 2, \cdots, n$ is

$$\Delta_\Sigma^2 = \Delta_1^2 + \Delta_2^2 + \cdots + \Delta_\tau^2, \tag{2.60}$$

where $\Delta_1, \Delta_2, \cdots, \Delta_n$ are the mismatchings and $\Delta_j = K_y(\tau) - K_y(\tau + j)$.

One can obtain a correlational model of standard structure for a fixed value of τ. The combinatorial algorithm is used to choose the optimal number of mismatchings of the correlation equations. For each sum of the discrepancies, a system of normal equations are formed as

$$\frac{\partial \Delta_\Sigma^2}{\partial y_1} = 0, \quad \frac{\partial \Delta_\Sigma^2}{\partial y_2} = 0, \quad \cdots. \tag{2.61}$$

The model of optimal complexity contains the optimal set of mismatchings. Differentiating this sum with respect to the first sought prediction y_{pr} at the point T, the inverse transformation of the correlation function into a prediction that best minimizes the sum of discrepancies Δ_Σ is obtained. The obtained equation $y_{pr}^3 + a y_{pr} + b = 0$ is the correlational predicting model and the prediction is the real root of this equation. The procedure is repeated for calculating the prediction at the point $T + 1$ by replacing the interval for determination of the estimates of the correlation function.

In the *case of two-dimensional models*, we have

$$\begin{aligned}
K_y(\tau_i, \tau_j) &= \frac{1}{(N_i - \tau_i - 1)(N_j - \tau_j - 1)} \sum_{i=\tau_i+1}^{N_i} \sum_{j=\tau_j+1}^{N_j} y_{i,j} y_{i-\tau_i, j-\tau_j}; \\
K_y(\tau_i + 1, \tau_j) &= \frac{1}{(N_i - \tau_i)(N_j - \tau_j - 1)} \sum_{i=\tau_i+2}^{N_i} \sum_{j=\tau_j+1}^{N_j} y_{i,j} y_{i-\tau_i-1, j-\tau_j}, \\
&\quad (1 \leq i \leq N_i, \ 1 \leq j \leq N_j); \\
\Delta_{10} &= K_y(\tau_i, \tau_j) - K_y(\tau_i + 1, \tau_j),
\end{aligned} \tag{2.62}$$

where τ_i is the displacement along the horizontal line corresponding to i; τ_j is the displacement along the vertical line corresponding to j; N_i and N_j are the number of discrete values of the function along the horizontal and vertical lines respectively; and Δ_{10} is the discrepancy.

One has to compute all discrepancies for the specified pattern area of the two-dimensional discrete field and obtain a correlational model of standard structure from the system of normal equations. Differentiating this sum with y_{pr}, which denotes the predicted value of the variable, the inverse transformation of the correlation function is obtained as a cubic model; the real root of the function represents the value of the prediction. The usage of combinatorial algorithm is similar to the above case.

Case of multivariable (multifactor) fields. An advantage of the algorithm with correlational models over the harmonic algorithm is the possibility of solving multifactor problems. For example, if two variables (the output y and its factor x) are indicated in each cell of the pattern in the two-dimensional field, the functions similar to the above must be formed corresponding to two autocorrelation functions ($K_{(y_i,y_j)}$ and $K_{(x_i,x_j)}$) and one cross-correlation ($K_{(y_i,x_i)}$).

Differentiating the sum of squares of the discrepancies and setting $\frac{\partial \Delta_\Sigma^2}{\partial y_{pr}} = 0$, one obtains an estimate of the extrapolation that is averaged in the mean-square sense. A nonlinear equation is obtained if the terms include a discrepancy of the expression for correlation coefficient without displacement; otherwise, linear equations are obtained. The variety of multi-dimensional autocorrelation and cross-correlation functions and the set of their ordinates make it possible to obtain a corresponding variety of correlational models of standard structure. The optimal model is selected by using the combinatorial or multilayer algorithm with the help of external criteria. This is subjected to the sorting on the basis of the criteria.

Multilayer algorithm using the correlation models of standard structure. Let us assume that the initial data sample of N points is supplied for the output variable y and the input arguments (factors) $x_1, x_2, \cdots, x_m$; and that the number of data points $N(= A \cup B)$ are comparatively small, where A and B are the training and testing sets correspondingly.

(i) At the first layer, partial descriptions in the form of systems of equations are formed using the output variable y with a pair of input variables x_i and x_j;

$$
\begin{aligned}
K_{yx_i}(\tau) &= \frac{1}{N_\tau} \sum_k y_k x_{i,k-\tau}, \\
K_{yx_j}(\tau) &= \frac{1}{N_\tau} \sum_k y_k x_{j,k-\tau}, \\
K_{x_i y}(\tau) &= \frac{1}{N_\tau} \sum_k x_{i,k} y_{k-\tau}, \\
K_{x_j y}(\tau) &= \frac{1}{N_\tau} \sum_k x_{j,k} y_{k-\tau}, \\
K_{yy}(\tau) &= \frac{1}{N_\tau} \sum_k y_k y_{k-\tau}.
\end{aligned}
\tag{2.63}
$$

It is possible to make C_m^2 such partial descriptions for each value of the displacement τ.

(ii) The values of the functions $K_{x_i y}(\tau)$ and $K_{yy}(\tau)$ are found by averaging the sequence of observations of length N_τ over several intervals. The values of y are taken as actual values from the data table.

(iii) The estimated values of the output variable $\hat{y}$ are computed from the correlation models. The best system of equations of the ordinates of the correlation function is chosen by minimizing the sum of squares of the discrepancies of that system as the relation $\Delta_{\Sigma}^2 = \sum_i \Delta_i^2$ is satisfied. The optimal vector $\hat{y}$ is obtained by solving the normal equations of the form $\frac{\partial \Delta_{\Sigma}^2}{\partial \hat{y}_i} = 0$. Calculation of $\hat{y}$ enables us to expand the initial data table; the number of columns of the table may be increased by F_1, where F_1 is the freedom-of-choice, $(F_1 < C_m^2)$.

The algorithm described above can be used for identification and prediction of the sequence of observations. This can be realized in a multilayered structure.

Correlational models provide an inverse transition from correlation functions to the original field or process for predicting it. Success with the correlation models is based on the following features.

1. Correlational models are to be constructed for stationary fields or processes, or for their remainders (obtained after removing their regular trend), where the condition for correlation stability holds.
2. The coefficients of the correlation models are to be estimated by minimizing the condition of the sum of their squared deviations.
3. The noncontradictory correlation models are to be chosen by using an inductive learning algorithm with the minimum-bias criterion.
4. Correlational models are nonphysical; i.e., they do not easily lend themselves to interpretation; that's why, when they are sorted out, one has to consider a great variety of candidate models so that the criterion indicated must be ensured a definite confidence level.

Generalized algorithm with orthogonal partial descriptions

Modeling of complex systems is frequently hindered by possible selection of initial independent variables; sometimes this might result in the loss of stability of the model structure. When one changes either the selection criterion or the composition of the data sequences of training and testing, the composition of the original arguments in a model and its structure begin to change strongly. There usually appear many models of similar quality. The stability is lost in estimating the coefficients because of the mutual dependence of the arguments, causing a mutual dependence on the corresponding coefficients. These disadvantages are especially evident in standard regression analysis where a single sequence of experimental points are used; better results are obtained by using the inductive learning algorithms. The major advantage of inductive learning algorithms is that one can avoid such biased results in identification of the object by using the minimum-bias criterion on the specific selection of the experimental points. However, the adaptation of coefficients and estimation of their confidence intervals over the entire data sample become impossible if the final form of the model contains dependent initial variables. The orthogonalized inductive algorithms allow one to improve stability in determination of coefficients. The algorithm with orthogonalized partial descriptions [102] [123] enables one to improve the stability in determining the coefficients by facilitating and (i) to use dependent variables in the set of experimental data, (ii) to obtain independent estimates of the coefficients, and (iii) to perform adaptation of the optimal model coefficients by refining their estimates over the entire sequence of the experimental points.

Let us consider that the complete polynomial of the object with dependent variables is in the form of a Kolmogorov-Gabor polynomial:

$$y = a_0 + \sum_{j=1}^{m} a_j \prod_{i=1}^{k} x_i^{p_{ji}},$$
$$p_{ji} = 0, 1, 2, \cdots; \; m \leq l; \; \sum_{i=1}^{k} p_{mi} \leq s, \tag{2.64}$$

where y and x are the output and input variables correspondingly; s denotes the maximum allowed degree of the polynomial; l specifies the maximum allowed terms in the model; and k indicates the number of initial input variables.

Let us denote the vector forms of the data matrices;

$$Y\ [N \times 1]; \;\; X = [x_{ij}], \; 1 \leq i \leq k, \; 1 \leq j \leq N, \tag{2.65}$$

where N is the number of experimental points.

Data is divided into training A and testing B sequences; $N = A \cup B$.

At the *first layer*, the actual output quantity y is projected orthogonally onto each argument from the generalized arguments $\hat{x}_{1m}$ separately for the training and testing sequences. The partial descriptions have the form

$$\hat{y}_{1m}^{A} = a_{1m}^{A} \hat{x}_{1m}, \;\; a_{1m}^{A} = \frac{E(y.\hat{x}_{1m})_A}{E(\hat{x}_{1m}^2)_A},$$
$$\hat{y}_{1m}^{B} = a_{1m}^{B} \hat{x}_{1m}, \;\; a_{1m}^{B} = \frac{E(y.\hat{x}_{1m})_B}{(\hat{x}_{1m}^2)_B}, \tag{2.66}$$

where E is the notation for mathematical expectation.

Each element of the set of generalized arguments is estimated from the initial variables by using the formulas

$$\hat{x}_{1m} = \prod_{i=1}^{k} x_i^{p_{mi}} - E\left(\prod_{i=1}^{k} x_i^{p_{mi}}\right),$$
$$p_{mi} = 0, 1, 2, \cdots; \; \sum_{i=1}^{k} p_{mi} \leq s. \tag{2.67}$$

The partial descriptions obtained from above are compared to each other by using the minimum-bias criterion of coefficients;

$$\eta_a^2 = \frac{(a_{1m}^A - a_{1m}^B)^2}{(a_{1m}^A)^2 + (a_{1m}^B)^2}. \tag{2.68}$$

From this analysis, the best generalized arguments, which are denoted as $\hat{x}_1$, are selected and introduced into the model to obtain the value of a_1 that is refined (adapted) by using the entire data sample. Thus, the adaptation of coefficients is combined with the selection of variables. Then the remainder of the output quantity Δ_1 is computed and used in the subsequent layers.

$$\Delta_1 = y - a_1 \hat{x}_1 \tag{2.69}$$

Subsequent selection layers are analogous to the first one. The computed remainder output quantity after the selection layer r is Δ_r;

$$\Delta_r = \Delta_{r-1} - a_r \hat{x}_r. \tag{2.70}$$

The calculations for the elements in the set of generalized arguments are also changed as follows:

$$\hat{x}_{rm} = \prod_{i=1}^{k} x_i^{p_{mi}} - E\left(\prod_{i=1}^{k} x_i^{p_{mi}}\right) - \sum_{j=1}^{r-1} b_{rm_j} \hat{x}_j,$$
$$p_{mi} = 0, 1, 2, \cdots; \; \sum_{i=1}^{k} p_{mi} \leq s. \tag{2.71}$$

The coefficients b_{rm_j} are obtained from the orthogonalized condition for $\hat{x}_{rm}$ and for each of the generalized arguments obtained during the preceding layers as

$$b_{rm_j} = \frac{E\left[\left\{\prod_{i=1}^{k} x_i^{p_{mi}} - E\left(\prod_{i=1}^{k} x_i^{p_{mi}}\right)\right\}\hat{x}_j\right]}{E\left(\hat{x}_j^2\right)}. \tag{2.72}$$

This shows that a partial description obtained from each layer is orthogonal to all previous descriptions. That is why the estimates of all coefficients a_r are independent from one another, which allows us to adapt them separately after each selection. One has to keep in mind that the orthogonalization and centering of the generalized variables are performed separately as shown above using the training and testing sets during the selection of variables, and that the refining of the model coefficients is performed on the entire sample.

In view of the linearity of the orthogonalization transformation with respect to the initial variables, one can use an inverse transformation on the final model obtained. Such an inverse transformation is necessary for interpretation of the results of the modeling.

Stopping rule. The independence of the model coefficients enables one to obtain independent estimates of their confidence intervals. The confidence interval of each coefficient is obtained as

$$d(a_r) = \sqrt{\frac{\sigma(\hat{x}_r)}{\sigma(\Delta_r)}}.t_{1-\frac{\alpha(n-2)}{2}}, \tag{2.73}$$

where $d(a_r)$ is the confidence interval of coefficient a_r; $\sigma(\hat{x}_r)$ and $\sigma(\Delta_r)$ are the estimated variances of the generalized argument and the remainder quantity, correspondingly; and $t_{1-\frac{\alpha(n-2)}{2}}$ is the quantile of the student t-distribution for the probability $(1-\alpha)$ and $(n-2)$ degrees of freedom.

The student criterion can be used as a second external criterion in order to determine the optimum complexity of the model. When $\frac{|a_r|}{d(a_r)} > 1$, then the coefficient a_r is significant with a probability of $(1-\alpha)$ and is included in the model. When this condition is not satisfied, the selection is stopped. The coefficients of the final model are statistically significant, making the model highly reliable. Thus, the criterion used for stopping the selection could also be the principal criterion used in the algorithm.

It is concluded that (i) the algorithm with orthogonal partial descriptions ensures stability of the model structure and of the estimates of its coefficients for complex system modeling with changing dependent variables, and (ii) the criterion proposed for measuring the significance of coefficients enables us to obtain a statistically reliable model with optimal complexity.

The structures of inductive learning algorithms are analogous to each other for different reference functions. The three basic forms of structures, single-layer combinatorial, multilayer, and harmonic algorithms are given here. Obtaining finite-difference models with lagging terms using the former type of polynomial algorithms is analogous to obtaining a harmonic model using the later type of harmonic algorithm [49]. The different forms of multilayer structures, multilayer algorithms with forward error propagations, selectional-combinatorial with the realization of recursive procedure and with restrictions on the freedom of choice are covered. There is another way of looking at the algorithms according to the types of polynomials; various recently developed algorithms are briefly presented.

However, the properties of the algorithms with different types of reference functions depend only on the structure. Note that (a) combinatorial algorithm does not produce errors due to multilayeredness and does not admit "loss" of optimal model; (b) multilayer algorithm without propagating errors has an explicitly expressed minimum which defines the model complexity because of an external criterion; (c) multilayer algorithm with propagation of errors has the error of monotonic nature, and the choice of model is made according to the "left rule."

Multilayer algorithms enable us to obtain polynomials with more terms than number of points in the data sample and with number of harmonics exceeding the number of harmonics of the first layer. By appropriately choosing the combined criterion (particularly choosing the minimum bias criterion as one of them), one can arrange for all terms of the original function to be reproduced using a very small number of data points. The results of self-organizing models on the basis of multilayer algorithms with and without front propagation of errors can coincide only for a rather large sample of initial data and in case of an identification problem using the minimum-bias criterion.

Multiplicative-additive algorithms of correlation models and algorithms with orthogonalized partial functions extend the region of functional space used with the inductive approach, and thus increase the possibility of solving complex problems.

3 LONG-TERM QUANTITATIVE PREDICTIONS

The subjective character of the models and the inaccuracy of long-term predictions obtained by various authors who used probabilistic and simulation methods have somewhat undermined the authority of cybernetics. The self-organization method should be able to change such a situation drastically. A computer can become an arbiter of controversies between various scientists only when objective methods are placed at its disposal. This means that we are approaching the creation of a collective man-machine superintellect that will be capable of solving the most complicated problems in prediction and control. The domain of activity includes the problems of nature that require more knowledge and skill than possessed by human experts. A computer that works on the basis of inductive learning algorithms is able to participate in the creative process as an equal partner with a human being. Let us see how it is achieved considering the following facts and characteristics.

3.1 Autocorrelation functions

Statistical prediction of random processes uses empirical data of the process (its previous history) to estimate its future values by applying the probabilistic characteristics of the process and corresponding algorithms. From the prediction point of view, one of the most important characteristics which indicates the statistical connection between the values of the process separated by some interval of time τ is the correlation function A_y:

$$A_y(t_1, t_2) = A_y(\tau) = E[\overset{\circ}{y}(t_1)\,\overset{\circ}{y}(t_1 + \tau)], \tag{2.74}$$

where $\tau = t_2 - t_1$; E denotes the expected value, and $\overset{\circ}{y}(t) = y(t)$ is an m_y-centered process (m_y is the mathematical expectation of the process).

Usually normalized correlation function is used as

$$\rho_y(\tau) = \frac{A_y(\tau)}{A_y(0)}, \tag{2.75}$$

where $A_y(0)$ is the variance of the process. One of the properties of the correlation function is that it is an even function: $A_y(\tau) = A_y(-\tau)$. In practice, when dealing with ergodic stationary processes, averaging over the set of realizations is replaced with averaging over time, and in place of the correlation function its time estimate is used

$$A_y(\tau) = \frac{1}{T-\tau}\int_0^{T-\tau} \overset{\circ}{y}(t)\,\overset{\circ}{y}(t+\tau)dt, \tag{2.76}$$

where T is the length of realization.

There is one-to-one correspondence between the correlation function and the power spectrum of the process; specifically, the power spectrum is the Fourier transform of the correlation function.

$$S_y(w) = \frac{1}{2\pi}\int_{-t}^{+t} A_y(\tau)e^{-jwt}d\tau. \tag{2.77}$$

In turn, the correlation function is defined in terms of the inverse Fourier transform,

$$A_y(\tau) = \int_{-\infty}^{+\infty} S_y(w)e^{jwt}dw; \tag{2.78}$$

i.e., the form of the correlation function depends essentially on the frequency spectrum of the original signal. The higher the frequency of the harmonics contained in that signal, the faster the correlation function decreases; a narrow spectrum corresponds to a broad correlation function and vice versa. In the limiting case, the correlation function of white noise is a *delta*-function with its singular point at the coordinate origin. Thus, the correlation function is a measure of the smoothness of the process being analyzed, and it can serve as a measure of the accuracy of prediction of its future values.

A relay autocorrelation function is called the sign-changing function $A'_y(\tau)$;

$$A'_y(\tau) = \lim_{T\to\infty}\frac{1}{2T}\int_{-T}^{+T} \overset{\circ}{y}(t)A\ \mathrm{sign}\ [\overset{\circ}{y}(t+\tau)]dt. \tag{2.79}$$

Analogously, a relay cross-correlation function is given below.

$$K'_{xy}(\tau) = \lim_{T\to\infty}\frac{1}{2T}\int_{-T}^{+T} \overset{\circ}{y}(t)A\ \mathrm{sign}\ [\overset{\circ}{x}(t+\tau)]dt. \tag{2.80}$$

Relay autocorrelation functions reflect only the sign and not the magnitude of $x(t)$. They have properties analogous to those of ordinary correlation functions, and in particular they coincide with them in sign. The advantage of relay functions (auto- and cross-correlations) is in the simplicity of the apparatus used for obtaining them. When the phase of the function $y(t)$ changes by 180°, the sign of the correlation function reverses. This means that in extremal regulation systems the correlation functions (ordinary or relay) can be used for determining which side of an extremum the system is on.

In practical computations associated with the random processes, one frequently estimates the so-called correlation interval, which is the time τ_c, over which the statistical connection between sections of the process is kept—in that the correlation moment between these sections exceeds some given level; for example, $|A(\tau)| > 0.05$ (Figure 2.8a).

Sometimes the meaning of the correlation interval is taken as the rectangular height $A(0)$ with area equal to the area under the correlation function (Figure 2.8b).

$$\tau_c = \frac{1}{A(0)}\int_{-\infty}^{+\infty} A(\tau)d\tau. \tag{2.81}$$

This is a convenient definition in case of a nonnegative correlation function.

The correlation time or interval is also defined as half the base of a rectangle of unit height whose area is equal to the area under the absolute value of the correlation function (Figure 2.8c).

$$\tau_c = \frac{1}{2} \int_{-\infty}^{+\infty} \mid A(\tau) \mid d\tau. \tag{2.82}$$

Among these three definitions we shall use the first one because of its simplicity.

3.2 Correlation interval as a measure of predictability

Various types of mathematical details (language) of modeling can be used. The influence of the degree of detailedness (sharpness) of the modeling language on the modeling accuracy—or in case of prediction, the limits of predictability of the process—is of great interest. One of the simplest devices for changing the diffuseness of description of a time series is to change the intervals of averaging (smoothing) of the data (for example, mean monthly, mean seasonal, mean annual, mean 11 years, etc.). The spectrum of the process in question then narrows down to the original and its correlation function broadens; that is, the correlation interval increases. This in turn extends the scope of predicting the process.

The problem encountered now is how to estimate, at least approximately, the achievable prediction time. The maximum achievable prediction time $T_{p_{max}}$ of a one-step forecast is determined by the correlation interval time called coherence time τ_c of the autocorrelation function A_y. This time is equal to the shift that reduces the autocorrelation function (or its envelope) to a value determined by the allowed prediction error $\delta\%$ following this level which it no longer exceeds.

The maximum allowed prediction time of a multiple (step-by-step) forecast is equal to the coherence time multiplied by the number of steps; i.e., $T_{p_{max}} = n\tau_c$. The prediction error increases with each integration step, which imposes a definite limit on the step-by-step forecast. We give here a brief view on the maximum capabilities of multiple step-by-step prediction, assuming that they are determined by the coherence time in the same way as those for one-step prediction.

Because of one-to-one dependence between the correlation and spectral characteristics of a random process, one can use some limiting correlation frequency as a measure of process predictability instead of correlation interval. The spectrum amplitude for the limiting correlation frequency is less than some threshold $S(w) \leq \theta$. Obviously these measures of diffuseness of the modeling language are not universal and are suitable only for evaluating certain mathematical modeling languages—primarily languages differing as regards the interval of averaging of the variables.

Example 1. Let us look at the influence of the interval of averaging on the form of its correlation function, its interval, and hence on the limit of its predictability; the example given here is an analysis on outflow $q(t)$ of a river over a period of one hundred years [44]. The autocorrelation functions for different averaging times are constructed.

$$A_q(\tau) = \frac{1}{N-\tau} \sum_{i=1}^{N-\tau} q_i q_{i-\tau}, \tag{2.83}$$

where q is the mean monthly outflow, N is the number of data points, and τ is the step in computation of the correlation function. It shows that averaging of variables in time increases the coherence time, in the same way as averaging time interval of variables over the surface of the earth, as shown in Figure 2.10.

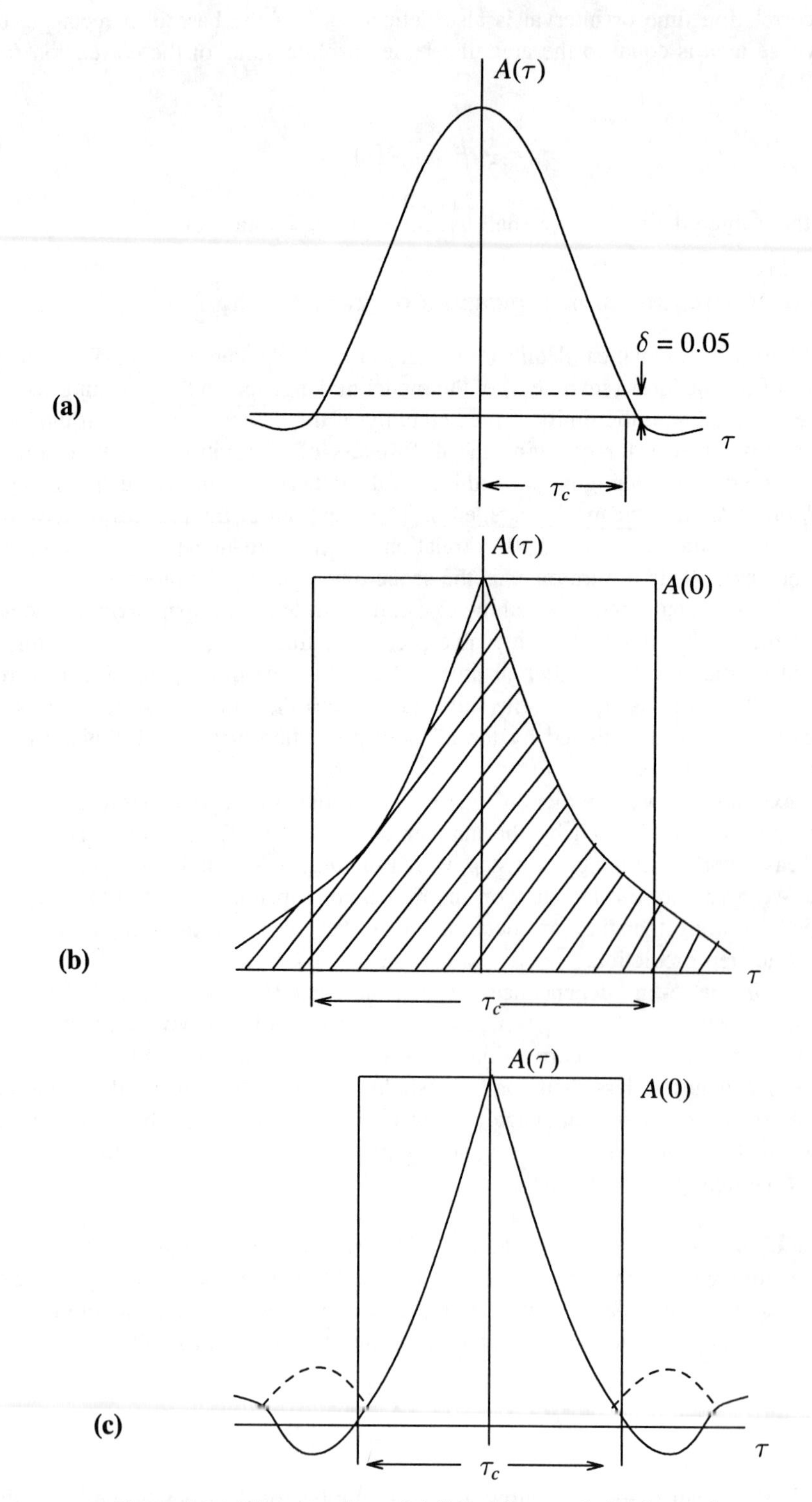

Figure 2.8. Three versions of defining the correlation interval

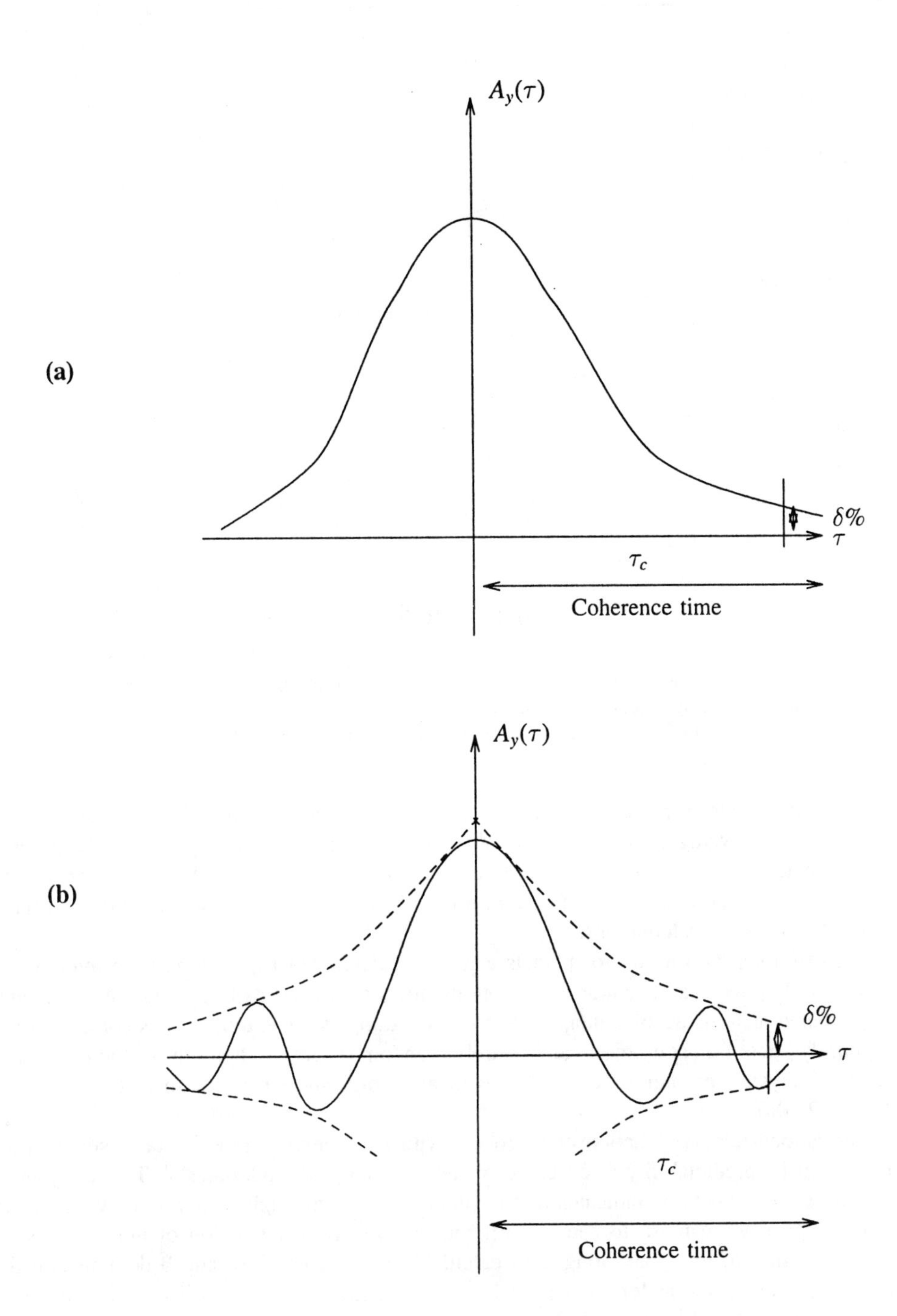

Figure 2.9. Autocorrelation functions; (a) monotonically decreasing and (b) oscillating

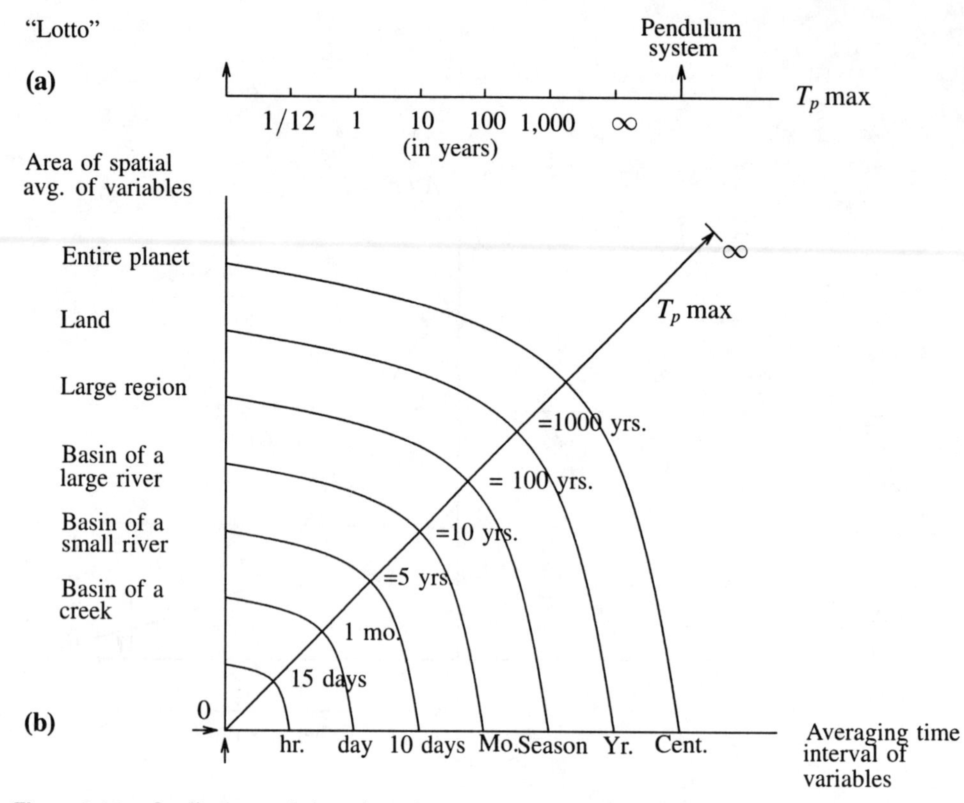

Figure 2.10. Qualitative variation of maximum prediction validity time T_{pmax} as a function of object properties and averaging interval of variables; (a) axis of maximum prediction time with constant averaging, (b) location of axis (a) in the plane of time and space averages

It is appropriate to remember that the achievable prediction time of a forecast depends not only on the averaging interval of variables, but also on physical properties of the process being predicted, as well as on the quality and characteristics of the mathematical prediction apparatus. If an exact deterministic description of the process is known, then prediction is reduced to detailed calculations.

For example, the motions of planets can be predicted exactly for long time intervals in advance. Outputs of a generator of random numbers or the results of a "lotto" game cannot be predicted as a matter of principle. These two examples are extreme cases corresponding to "purely" deterministic objects and "purely" random objects with equiprobable outcomes. In actual physical problems we are always located somewhere between these two extremes (Figure 2.10a).

The autocorrelation function of a process with its coherence time contains some information on its predictability (the degree of determinancy or randomness). The analysis of autocorrelation functions indicates that by increasing the averaging interval of variables in time or space we can, so to speak, shift the process from the region of unpredictability into the region of exact and long-term calculability. Figures 2.11a and b demonstrate the autocorrelation functions for one with calendar averaging and another with moving averages on the empirical data of river outflow.

One can see that with the increase in the interval of averaging of the data, the correlation function for a single time scale becomes ever more sloping, and the correlation interval increases. In the moving average case, a smaller step of sampling the initial data enables

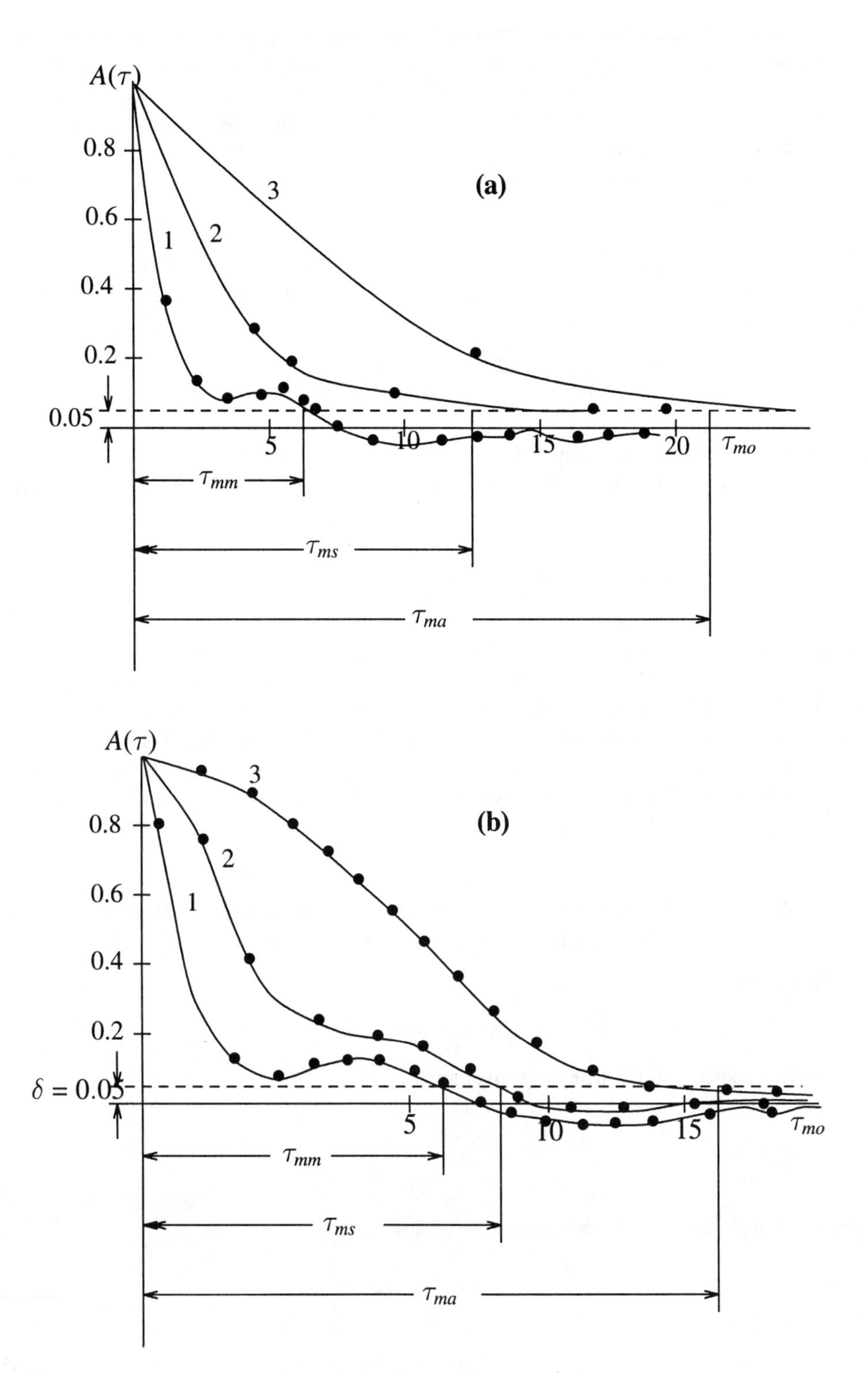

Figure 2.11. Autocorrelation functions of a river outflow; (a) with calendar averages and (b) moving averages on (1) monthly data, (2) seasonal data, and (3) annual data

us to keep unchanged the number of sample data (all monthly values), which leads to a broadening of the spectrum of the original signal and to a corresponding narrowing of its correlation function. The correlation function obtained in the case of moving averages occupies an intermediate position between the correlation functions of unsmoothed data and the data of calendar smoothing. Thus, the correlation time can serve not only as a measure of the limit of predictability of the process, but also as a measure of detailedness of a number of modeling languages.

Example 2. In the harmonic algorithm the trend is represented as a sum of a finite number of harmonic components (usually the optimal number of components does not exceed $m =$ 20).

$$q_{mo}(t) = \sum_{i=1}^{m}(A_i \cos w_i t + B_i \sin w_i t), \tag{2.84}$$

where $q_{mo}(t)$ is the mean monthly data.

Running moving average is an approximation of the operation of integration over a given interval of time.

$$q_{run}(t) = \frac{1}{3}\int_0^3 q_{mo}(t)dt \;\cong\; \frac{1}{3}\sum_{j=0}^{3}\sum_{i=1}^{m}(A_i \cos w_i t + B_i \sin w_i t), \tag{2.85}$$

where $q_{run}(t)$ is the running average of three on the mean monthly data.

Integration does not change the number of harmonics to be added or their frequencies, but it does decrease the amplitudes by a factor of $1/w_i$. As a result, the components with comparatively high frequencies decrease more than the others, and the curve $q_{mo}(t)$ becomes much smoother than the original curve. This also explains the smoothing effect shown in the figures above. The same reasoning holds true for the curves of correlation functions for seasonal and yearly data.

Example 3. This is demonstrated using the same data by constructing algebraic, differential (difference) and integral type models on an interval of 20 years.

algebraic model:

$$q = a_0 + a_1 t + a_2 t^2;$$

differential model and its difference analogue:

$$\frac{dq}{dt} = a_1 + a_2 t,$$
$$\Delta q = q_{+1} - q_0 = b_0 + b_1 t;$$

integral model and its discrete summation analogue:

$$\int_0^t q\,dt = a_0 t + a_1\frac{t^2}{2} + a_2\frac{t^3}{3} + C,$$
$$\sum_0^n q = b_0 t + b_1\frac{t^2}{2} + b_2\frac{t^3}{3} + C_1. \tag{2.86}$$

The autocorrelation functions for these three types of models over the 80-year period are shown in Figure 2.12.

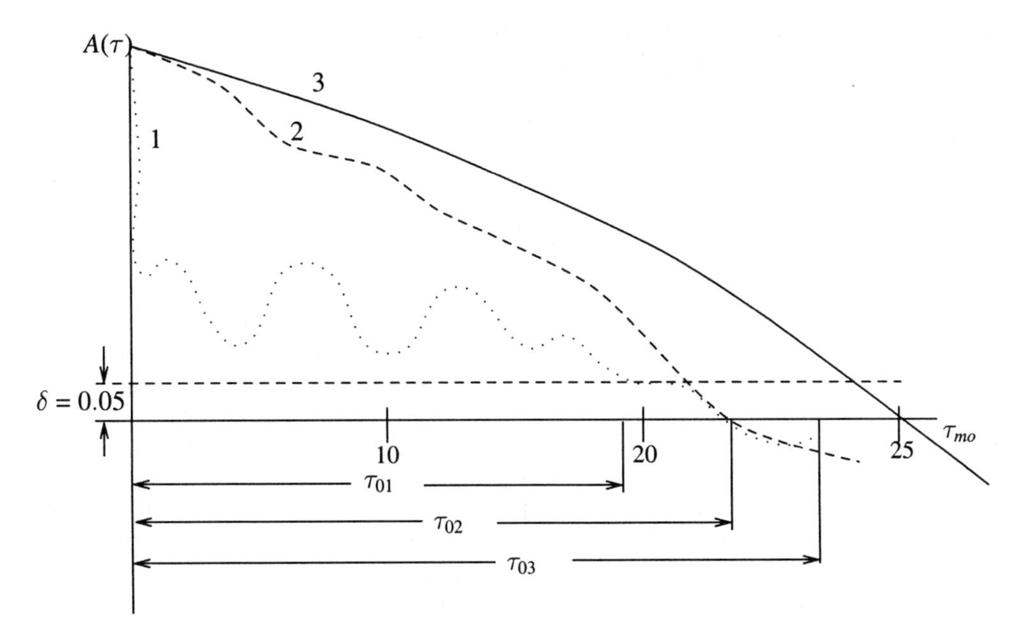

Figure 2.12. Autocorrelation functions for languages of (1) integral, (2) algebraic and (3) differential equations

One can see that the language of differential equations is the most diffuse of the three modeling languages; it is more suitable for long-range predictions. This explains the widespread use of differential equations in the equivalent analogue of finite-difference equations in modeling as compared with algebraic and integral models.

Let us take the problem of *weather forecasting*. Weather forecasters use data gathered by satellite in order to predict the weather quite successfully over an extended period of time, but this prediction is only possible in terms of a very general language. They convey the future weather picture qualitatively ("it will be warmer," "precipitation," "cold," etc.). More quantitative predictions require the use of mathematical models. As per various studies it is indicated that the daily prediction interval cannot exceed 15 days and practical predictions have even shown for a much shorter interval of time (not more than 3 to 4 days). The mean monthly values of variables are less correlated than the average daily variables; the maximum length of the prediction interval of mean monthly values does not exceed 3 to 4 months. Average yearly values of variables have an intermediate degree of correlation, and the maximum achievable prediction interval of average yearly values is 8 to 10 years. It is important to point out that the limit imposed on the interval of prediction, measured in the same units of time, increases together with the interval over which the variables are averaged. In other words, the interval span for average daily values is 15 days, the span for average monthly values is $4 \times 30 = 120$ days, and the interval for average yearly values is $10 \times 365 = 3650$ days, etc.

Reliable long term predictions of weather are frequently related to the idea of analogues. This idea is simple and interesting: one must find an interval in the prehistoric measured data whose meteorological characteristics are identical to the currently observed data. The future of this interval (observed in the past) will be the best forecast at present. Nevertheless, attempts to apply the idea of analogues always produced results that were not very convincing. The fact is that for such a large number of observed variables (and also many unobserved ones) it is impossible to find exact analogues in the past. Resorting to group analogues, introduction of weighing coefficients for each measurement, and other measures first bring us to regression analysis and then, after further improvements, to the inductive approach algorithms. Therefore, inductive learning can be interpreted as an improved method of group analogues in which the analogues of the present state of the atmosphere are selected by using special criteria and summed up with specific weighing coefficients to produce the most probable forecast. Weather forecasting is an object whose structure switches when a new type of circulation is established randomly at the time of equilibrium. Nevertheless, it is possible to investigate an optimum method for overcoming the predictability limit applicable to some weather variables (temperature and pressure at surface layer, etc.). This will be discussed in later chapters. Further research is needed on this subject.

It seems that insurmountable barriers have been established for quantitative predictions. However, the self-organization method enables one to overcome these limitations and to solve the problem of long-term predictions, because the limit of predictability depends on the time interval of averaging. Self-organization uses two or three averaging intervals for correcting the variable under study; for example, the daily prediction is corrected according to a 10-day prediction, the 10-day prediction is corrected according to the mean monthly prediction, and the mean monthly prediction is corrected in accordance with the average yearly prediction. In this way we can achieve a breakthrough in methods of long-term and very long-term prediction which has heretofore not been achievable by any other method.

3.3 Principal characteristics for predictions

The principle characteristic of achieving an objective goal is for detailed (sharp) predictions in a low-level language which contain the greatest amount of detail while maintaining the prediction lead time that is typically obtained by using the most general high-level language. The more general the language, the longer the achievable prediction lead time (Figure 2.10).

Let us give here some examples indicating the levels of languages:

(i) *Prediction of processes in economic and ecological systems.*

A language which preserves probabilistic moments of the process is used at the upper level to select quantitative predictions by using the mean annual values of variables and the mean seasonal or monthly values. The middle-level language consists of modeling mean annual values and the lower level (detailed) consists of modeling average seasonal or monthly values.

(ii) *Prediction of river flows.*

The upper level uses the language which preserves the nature of probability of distributions, the middle level consists of predictions of average annual run-off, and the lower level involves predictions of average seasonal or monthly values. The conversion from statistical to quantitative predictions should be performed by taking into account the principle—that is, by using rationalized (multilevel) scanning of quantitative predictions.

(iii) *Long-term weather forecasting.*

The upper level can be a language which preserves the weather forecast for a large region (or a long averaging time). The middle level will then consist of predictions for small parts of the region (or medium averaging time), and finally the lower level will give predictions for a specific point and specific time.

The examples given above contain three levels of detailedness of the modeling language, which is obviously not required for all problem-solving tasks.

As we know, the principle of self-organization is realized in single-layer (combinatorial) and multilayer inductive learning algorithms. Using the basic structures of these algorithms, multilevel prediction algorithms are operated in several different languages simultaneously, within which the predictions expressed in a more general language are used for selection of an optimum quantitative prediction in the more detailed language. Several levels are needed to overcome the "limit of predictability" of detailed predictions, and also to eliminate the multivalued choice of a prediction on the basis of general criteria. Let us go through different cases of self-organization modeling for clarity in multicriterion analysis.

Case of exact data

In case of exact data, exact computation takes place for prediction (for example, motion of heavenly bodies, prediction of eclipses, etc.) from the solution of a system equations as mathematical models of the cosmic system of bodies.

Under the conditions of exact empirical data, self-organization modeling can only have as its purpose the discovery of laws hidden in the data. It is sufficient to use any one internal or external criterion like regularity or minimum bias criterion in sorting out the models. It is important to note that we do not require multicriterion choice of a model. More complex problems arise within the field of noisy data.

Case of noisy data

It is sufficient to impose on one of the variables (usually the output) a very small additive or multiplicative noise so that the position of the variable is changed cardinally. If we try to obtain an optimal model using only internal criteria, we always end up with a more complex model, that will be more accurate in the least squares sense; only external criteria provide a model with optimal complexity. Let us consider various systems of equations describing an object; they are not equally valuable since they are connected with measurement of different variables. The optimal system with the fewest excessively noisy variables can be sorted out among variants of the system of equations using the system criterion of minimum bias:

$$\eta_{s_{(bs)}} = \frac{1}{s}(\eta_{bs_1} + \eta_{bs_2} + \cdots + \eta_{bs_s}), \tag{2.87}$$

where $\eta_{s_{(bs)}}$ is the system criterion for the system of equations and the η_{bs_i}, $i = 1, 2, \cdots, s$ are the criteria for each equation in the system of s equations.

As we know from the information theory point of view, increasing the noise stability decreases the transmission capacity; this means that with an increase in the noise level, a model simpler than a physical model becomes optimal. (Here physical model means a model corresponding to the governing law hidden in the noisy data.) It is expedient to distinguish two kinds of models: (i) a physical or identification model which is suitable for analysis of interrelations and for short-range predictions, (ii) a nonphysical or descriptive model for long-range predictions. One can discover a physical model with various concepts of modeling, but detailed long-range predictions are impossible without the help of inductive learning.

If the data are noisy, even to obtain a physical model requires one to organize rational sorting of physical models by self-organization using several criteria which have definite physical meanings. Usually one needs a model which is not only physical but also easy to interpret instantaneous unaveraged values of the variables; that means the model is chosen based on the simultaneous selection of minimum bias criterion and short-range prediction criterion.

$$\eta_{bs}^2 = \sum_{i=1}^{N} \frac{(\hat{y}_A - \hat{y}_B)_i^2}{y_i^2}, \quad \Delta^2(C/W) = \sum_{i=1}^{N_C} \frac{(y - \hat{y})_i^2}{(y_i - \bar{y})^2}, \tag{2.88}$$

where y is the output variable, $\hat{y}_A$ and $\hat{y}_B$ are the estimates of the models obtained based on the sets A and B, respectively, $\hat{y}$ is the estimated prediction, and $\bar{y}$ is the average value of y.

In the plane of two criteria, each model corresponds to its own characteristic point; the point corresponding to the model of optimal complexity lies closer to the coordinate origin than do the points of other models participating in the sorting. Here we can say that one can find a physical model using both deductive reasoning of man and self-organization of machine with respect to choice of many criteria.

In obtaining nonphysical models for long-range detailed predictions, the role of man, as he remains the author of the model, consists of supplying the most efficient set of criteria for sorting the models. The dialogue between man and machine is in the language of criteria and not in the language of exact instructions. In addition, to use the minimum bias criterion on two sets of data A and B, the step-by-step prediction criterion is to be included for calculating the prediction error on entire interval ($W = A + B$) of data. The above short-range prediction criterion $\Delta(C/W)$ is used as long-range prediction criterion $i(W)$ as per notation by replacing N_C with N_W for the entire range of data points. This criterion is desirable to use not only for choosing the structure of the model but also for removing the bias of the estimates of the coefficients in the model. In addition to these criteria, in multi-criteria choice of an optimal nonphysical model for long-range predictions, stability criteria of moments (upper and lower) and probabilistic characteristics of correlation functions are used; these will be explained later in the chapter. This means that multicriterion choice is one of the basic methods of increasing noise stability of inductive learning algorithms.

The physical and nonphysical models differ not only in their purpose but also in their informational basis because of reasoning of the objective criteria. The arguments of physical model can be all input variables and their lagged values (for dynamic models). The arguments of nonphysical predicting models can only include different intervals of averaging and the time variables which are known on the entire interval of long-range prediction. Physical models that are obtained are usually linear and nonphysical models are nonlinear with respect to time.

Case of time series data

If an algorithm is used for obtaining a single "optimum" prediction (according to any criteria) using pre-history data, then such algorithm is meant for only short-range or average-term prediction (for one to two or three to five time intervals in advance respectively). If the algorithm envisions the use of empirical data in order to obtain a single prediction over a large averaging interval (for example, one year), and several predictions (in accordance to multicriteria) over a small averaging interval of variables (for example, seasonal) in order to use the balance criterion over the interval of predictions (ten to 20 years in advance), then the choice of seasonal models on the basis of yearly model is done on the basis of balance-of-predictions criterion [58], [65].

$$B^2_{season} = \sum_{i=1}^{N_C} b_i^2$$

$$b_i = \hat{Q}_{yr} - \frac{1}{4}(\hat{q}_w + \hat{q}_{sp} + \hat{q}_{su} + \hat{q}_f)_i, \tag{2.89}$$

where N_c is the number of prediction points, $\hat{Q}_{yr}$ is the prediction based on the yearly model (a single prediction), $\hat{q}_w, \hat{q}_{sp}, \hat{q}_{su}$, and $\hat{q}_f$ are predictions based on different variants of the set of seasonal models for winter, spring, summer, and fall correspondingly, and N_c is at the range of prediction interval of ten years.

In the same fashion one can build an algorithm which envisions over a very long averaging interval (for example, 11 years) and at the same time several predictions over shorter averaging intervals (for example, one year or one season); if the algorithm uses a two-level balance-of-predictions criterion, then that would be successful for very long-range predictions (40 or more years in advance) [58]. The choice of the yearly models and the model which uses the averaging interval of 11 years is based on the following balance-of-predictions criterion:

$$B^2_{11yrs} = \sum_{i=1}^{N_C} b_i^2$$

$$b_i = \hat{Q}_{11yrs} - \frac{1}{11}(\hat{q}_1 + \hat{q}_2 + \hat{q}_3 + \cdots + \hat{q}_{11})_i, \tag{2.90}$$

where $\hat{Q}_{11yrs}$ is the prediction based on the model which uses the averaging interval of 11 years (a single prediction); $\hat{q}_1, \hat{q}_2, \hat{q}_3, \cdots \hat{q}_{11}$ are predictions based on various versions of the set of yearly models.

The rules for building up such algorithms realize the principle of "freedom of choice of decisions" formulated by Gabor [22]. The basic long-term prediction is harmonic or polynomial prediction of variables when the averaging interval is of maximum length. The criterion of prediction balance "pulls up" the accuracy and the averaging time of predictions for small averaging intervals to the accuracy and prediction time obtained when the averaging interval is long.

Another issue where the self-organization stands firm is when a decision is to be made in case of two or more contradictory requirements, which is called "Pareto problem." The "Pareto region" is the region where the solutions contradict each other and which requires the use of experts. This is achieved by the self-organization method yielding a new problem formulation of multicriterion control selection done heuristically on the basis of physical properties of the system to be predicted. The lead time of prediction interval usually reaches the time of interval used for validity of the criterion. In order to eliminate multivalued selection, scanning of forecasts for different intervals is replaced by multilevel algorithm development as scanning of algorithms and models, generating a variety of predictions on the basis of their external criteria.

4 DIALOGUE LANGUAGE GENERALIZATION

Complex systems analysis is based on modeling of a system with interactive elements in order to identify the system structure and parameters, to perform various tasks like short- and long-term predictions of processes, and to optimize the control task. Usually during algorithm development, the computer has a passive role; that is, it is unable to participate in creative modeling. Interpolation problems are multi-solution problems; additional data set or *a priori* testing set is necessary to obtain a unique solution. Commonly used simulation methods are based on a large volume of *a priori* information that is difficult to obtain.

Self-organization modeling is directed to reduce *a priori* information as much as possible. The purpose of self-organization is not to eliminate human participation (it is impossible unless a complete intelligence model is developed), but to make this participation less laborious, reduce some specific problems, and avoid expert participation. This can be achieved in ergatic information systems by using more generalized "man-machine" meta-language, which uses general criteria given by man—the learning is done by the computer. In addition to the generalized criteria, man provides the empirical data. In some cases man may be involved in final model corrections. Here it is shown that many things still can

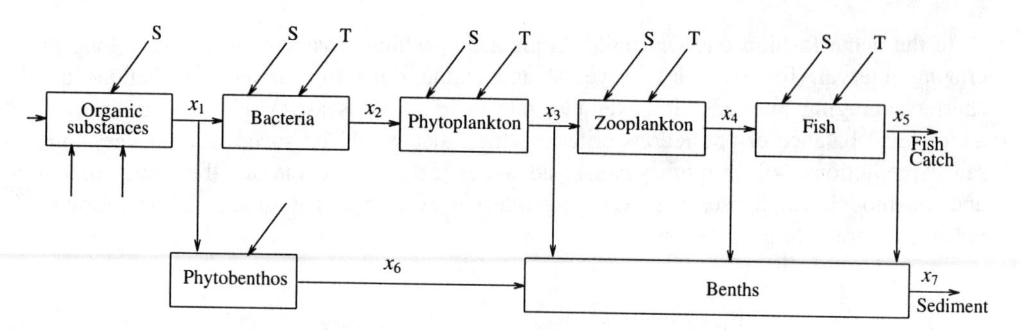

Figure 2.13. Usual (subjective) system analysis (example)

be done to reduce human involvement in the creative modeling process and make it much easier.

4.1 Regular (subjective) system analysis

The regular (subjective) system analysis shown in Figure 2.13 is a system imitation model for the northwest region of the Black Sea [33] (details of the model are not shown to simplify description).

An organic substance is formed (variable x_1 is substance production) from biogene substances P and N that were exposed to the sun. Bacteria (biomass x_2) eat the organic substance, and the phytoplankton (x_3) eat the bacteria. Zooplankton (x_4) eat the phytoplankton and the fish (ichthyomass x_5) eat the zooplankton. If we are interested in analyzing fish catch, the following equations (for the surface layer of water) are used.

$$\begin{aligned}
\tau_1 \frac{dx_1}{dt} + x_1 &= a_0 + a_1 S + a_2 P + a_3 N, \\
\tau_2 \frac{dx_2}{dt} + x_2 &= b_0 + b_1 S + b_2 T + b_3 x_1, \\
\tau_3 \frac{dx_3}{dt} + x_3 &= c_0 + c_1 S + c_2 T + c_3 x_2, \\
\tau_4 \frac{dx_4}{dt} + x_4 &= d_0 + d_1 S + d_2 T + d_3 x_3, \\
\tau_5 \frac{dx_5}{dt} + x_5 &= l_0 + l_1 S + l_2 T + l_3 x_4. \qquad (2.91)
\end{aligned}$$

By excluding the intermediate variables, we derive a linear differential equation of the fifth order for the output variable x_5 (fish catch) and analyze its solutions for given initial conditions. If the equations are nonlinear, then we can substitute the derivatives by finite-differences and find the results using simultaneous step-by-step integration of the system of nonlinear equations.

This example shows the basic characteristics of imitation modeling and commonly used system analysis: (i) this model requires in-depth knowledge of the subject; this knowledge is based on a large volume of information that is entered in the computer by the modeler; (ii) empirical data are not needed but may be used for scaling the coefficients by using the least squares method. One may design and analyze this model using a calculator; (iii) the results of such knowledge are subjective because the model is based on the author's subjective understanding (there may be as many different models as many modelers). The model does not resolve scientific disputes between experts on the subject; and (iv) only physical models can be obtained, but these are not suitable for long-term predictions.

4.2 Multilevel (objective) analysis

The idea of sorting many variants using some set of external criteria in the form of an objective function in order to find a mathematical model of a given complex subject seems unreal. Self-organization method tries to rationalize such sorting so that an optimal model is achieved. Multilevel algorithgms of inductive learning serve just this purpose. They allow changes of large number of variables to be considered. The model structure, which is characterized by the number of polynomial elements and its order, is found by sorting a large number of variants and by estimating the variants according to specific first level selection criteria (regularity, minimum bias, balance of variables and others). If the objectivity of the model is not achieved, then the high level criteria are used.

Here we give the concept of multilevel objective analysis under various conditions of multicriteria. The single-level analysis using one of the basic network structures like combinatorial, multilayer or harmonic is sometimes not sufficient for detailed analysis and we go for multistage analysis which is described as a multilevel algorithm. These prediction algorithms operate in separate different languages simultaneously as the predictions at a general language are used for obtaining a more detailed model at the next detailed language. Several levels are very essential, as one is to overcome the limit of predictability of detailed predictions and another is to avoid the multivalued choice of a model using the general criteria. Thus, in the stages of these algorithms, three basic directions of dialogue language are preserved; (i) the self-organization principle, which asserts that with gradual increase in the complexity of model, the external criteria pass through their minima, enabling us to choose a model of optimum complexity, (ii) an algorithm for multilevel detailed long-range predictions, and (iii) an algorithm for narrowing the "Pareto region" in case of multi-criterion choice of decisions.

4.3 Multilevel algorithm

The multilevel system is subjected to all the general laws governing the behavior of multilevel decision-making systems which realize the principle of incomplete induction. As in multilayer algorithm, here there is possibility of losing the best predictive model; an increase of the "freedom of choice" decreases the possibility of such loss. Various principles related to selection and optimization of "freedom of choice" in multilayer algorithm also apply to the multilevel system of languages having different levels of details.

If we had a computer with large capacity, then the problem of selecting detailed models could be solved by simply scanning all versions of partial models using combinatorial algorithm with a large ensemble of criteria. Since the capacity is limited, it is necessary to expose the basic properties of the models step by step.

In order to reduce the volume of scanning and to achieve uniqueness of choice, the principle discussed above is realized in several levels whose schematic structure for one version is shown in the Figure 2.14.

Let us explain the operations performed during these levels.

Objective systems analysis

The purpose of this level is to divide the system variables into output, input variables and variables which have no substantial effect on the outputs. Here structure of and number of equations is to be chosen in such a way that the overall model is consistent. The structure as well as number of equations must not be changed significantly when a new data set is added. The estimation of coefficients should not be changed. This type of sifting for

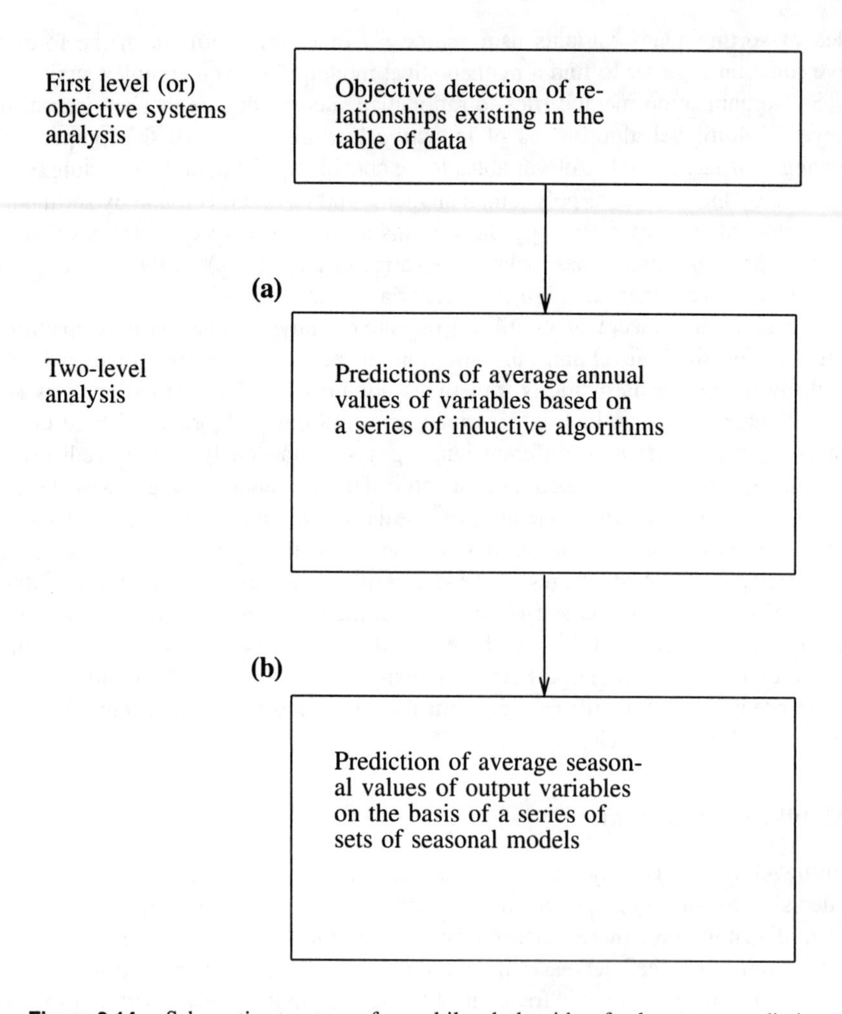

Figure 2.14. Schematic structure of a multilevel algorithm for longterm prediction

systems of equations is done systematically; that's why this level is called an objective systems analysis (OSA). An objective model is identified as a set of output variables and the connections between the system components as a result of learning. The analysis involves testing of several hypotheses about the model structure below:

First layer: The following hypothesis is tested for a single equation using empirical data. There are $M_1 = n$ equations formed, one for each variable.

$$x_i^t = a_0 + a_1 x_i^{t-1} + a_2 u, \tag{2.92}$$

where $x_i, \quad i = 1, \cdots, n$ are the state variables, u is an external influence, and superscripts t and $t-1$ indicate arguments with no delay and one-step delay, correspondingly. There are two methods to determine the variables of external influences u. In the first one, experts specify the disturbances *a priori* before execution of the program and in the second, the suitable control disturbances are chosen from the variables, which are already sorted out in the program. However, the role of these variables is considered as less important.

F_1 best models are obtained using the minimum-bias criterion.

$$\eta_{bs_i}^2 = \sum_{p \in N} (\hat{x}_i^A - \hat{x}_i^B)_p^2, \quad i \in n, \tag{2.93}$$

where N is number of observations of empirical data, $\hat{x}_i^A$ is the estimated output of the model based on the training set A, and $\hat{x}_i^B$ is the estimated output of the model based on the testing set B.

Second layer: The hypothesis is tested for the structure of models described by two equations including the delayed and nondelayed arguments (one can use more delayed arguments).

$$\begin{aligned} x_i^t &= a_0 + a_1 x_i^{t-1} + a_2 x_j^t + a_3 x_j^{t-1} + a_4 u, \\ x_j^t &= b_0 + b_1 x_i^t + b_2 x_i^{t-1} + b_3 x_j^{t-1} + b_4 u, \end{aligned} \tag{2.94}$$

where $i, j = 1, 2, \cdots, n$; $i \neq j$. There are $C_{n-1}^1 = n - 1$ equations for each variable, and overall there are a total of $n. C_{n-1}^1 = n(n-1)$ equations. The system obtains F_2 best models of optimal complexity from among all two-set models (M_2) of state variables i, j using the system criterion of minimum-bias

$$\eta_{s_{(bs)}} = \frac{1}{2}(\eta_{bs_i} + \eta_{bs_j}), \tag{2.95}$$

where η_{bs_i} and η_{bs_j} are minimum-bias estimates of ith and jth equations.

Third layer: The system models consisting of three equations is found at this layer.

$$\begin{aligned} x_i^t &= a_0 + a_1 x_i^{t-1} + a_2 x_j^t + a_3 x_j^{t-1} + a_4 x_k^t + a_5 x_k^{t-1} + a_6 u, \\ x_j^t &= b_0 + b_1 x_j^{t-1} + b_2 x_i^t + b_3 x_i^{t-1} + b_4 x_k^t + b_5 x_k^{t-1} + b_6 u, \\ x_k^t &= c_0 + c_1 x_k^{t-1} + c_2 x_i^t + c_3 x_i^{t-1} + c_4 x_j^t + c_5 x_j^{t-1} + c_6 u, \end{aligned} \tag{2.96}$$

where $i, j, k = 1, 2, \cdots, n$; $i \neq j \neq k$. There are C_{n-1}^2 equations for each variable and there are a total of $n. C_{n-1}^2$ equations. All three-set models (M_3) of variables (i, j, k) are evaluated using the system criterion of minimum-bias.

$$\eta_{s_{(bs)}} = \frac{1}{3}(\eta_{bs_i} + \eta_{bs_j} + \eta_{bs_k}). \tag{2.97}$$

Better sets of models (F_3) are obtained from this layer based on the criterion measure.

It proceeds further and tests for four-, five-set models, and so on until the system criterion of minimum bias starts increasing. Ultimately, the overall process determines the set of variables for the complex object and its linearized structure. Usually the system consists of three to five equations. The variables in the selected set of equations are called "characteristic" variables. Figure 2.15 shows how the minimum bias error of system criterion is reduced as the number of equations increases; each column of the points in the figure corresponds to group of models having similar structure. The approximate limit for successful analysis of modeling is established as $\eta_{s_{(bs)}} \leq 10^{-5}$ on the practical use of the objective system analysis.

If one of the equations has high minimum bias value, then such an equation is considered inconsistent and is excluded from the analysis. If none of the equations is good, then the analysis fails. This can happen if the state variables are too noisy or if the given state variables do not contain any characteristic variables. Noise immunity can be improved by designing specific criteria; the noise immunity depends on the mathematical form of the criterion and on the method of convolution of the criteria into general form. The second level of such criteria are given below; the multicriteria analysis, symmetrical, and combined criteria significantly improve the noise immunity of the algorithm.

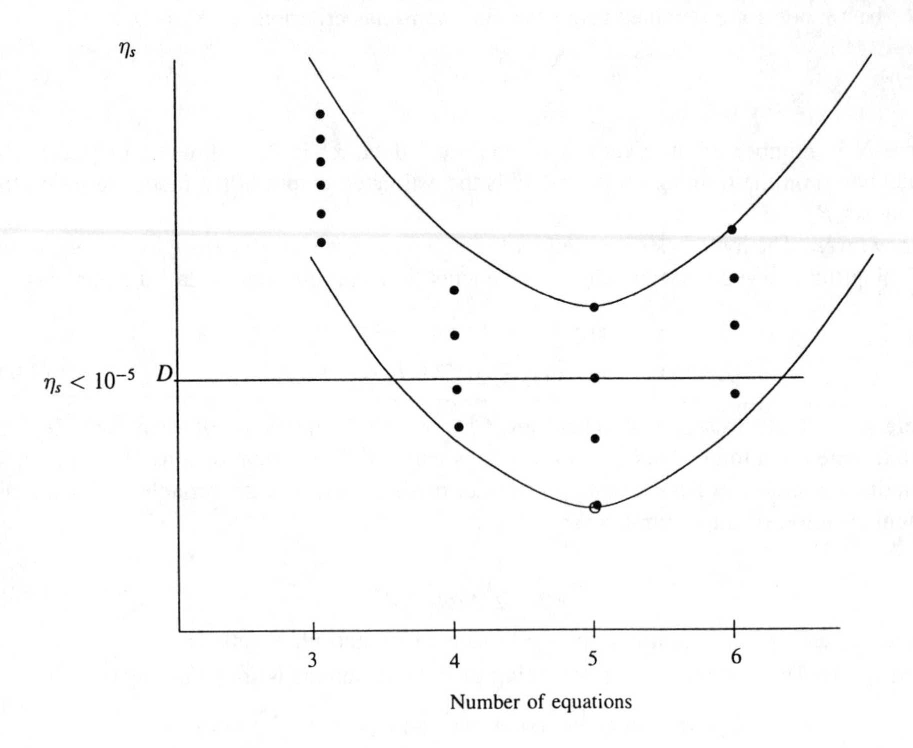

Figure 2.15. Objective system analysis where D is the depth of the minimum and "." is the characteristic point of the model

Another form of minimum-bias criterion is

$$\eta_{bs}^2 = \sum_{p=1}^{N_W} \frac{(\hat{x}_A - \hat{x}_B)_p^2}{x_p^2}. \tag{2.98}$$

Symmetric regularity criterion:

$$\Delta^2(AB) = \sum_{p=1}^{N_B} (\hat{x}_A - x)_p^2 + \sum_{p=1}^{N_A} (\hat{x}_B - x)_p^2 = \Delta^2(B/A) + \Delta^2(A/B). \tag{2.99}$$

It is equal to the sum of two regularity criteria, which represent the usual case, when N_A and N_B are used as data points in training and testing sets, alternatively.

Another form of regularity criterion is

$$\Delta^2(AB) = \sum_{p=1}^{N_W} (\hat{x}_A - x)_p^2 + \sum_{p=1}^{N_W} (\hat{x}_B - x)_p^2 = \Delta^2(W/A) + \Delta^2(W/B), \tag{2.100}$$

where $N_W = N_A + N_B$.

Combined criterion:

$$\Delta^2(AB) = \eta_{bs}^2 + 2\sum_{p=1}^{N} (\hat{x}_A - x)_p(\hat{x}_B - x)_p. \tag{2.101}$$

Sometimes this type of convolution may lead to additional problems in selecting the final optimal set of equations, but one must use them with care. All equations with the characteristic points below a certain confidence level D are considered equivalent. The final set of equations determines the input-output variables and connection diagram for such set of system components as shown in Figure 2.13.

In this algorithm, the result of calculations on each consecutive layer (with increased number of equations with increased complexity) does not cancel variables of the previous layer, but only adds new "characteristic" variables. Overall, the total number of equations generated are $M = n. \sum_{s=1}^{n} C_{n-1}^{s-1}$ in maximum of n layers; sets of models are formed in each layer and evaluated among them. It is also possible to reduce significantly the scope of calculations at each layer as follows: the first layer uses the computer capacity in full, the next layer uses the set of output variables determined on the basis of first layer, and so on. The system of equations which corresponds to the minimum of the system criterion is chosen as the optimal system. Variables which do not appear in this system are excluded from further consideration. The results from the OSA are passed on to the next levels and used to solve two types of problems: (a) identification of the physical model which is suitable for short-term predictions, (b) identification of nonphysical models for long-term predictions (two-level analysis).

Physical model for short-term predictions

From the above objective system analysis, the set of characteristic (output) variables are identified. Based on the set, nonlinear physical models are developed for the system and its components. The multilayer algorithm with redenoted variables are used for obtaining the optimal model even with very short data samples. We call this model a physical model because of its characteristic variables and its evaluation from a single-level analysis. The physical models obtained this way are not suitable for long-term predictions even though the noise level is relatively low.

Another fact is that not all characteristic variables resulting from OSA can be predicted with the same success. One can use one of the following accuracy criteria for evaluating short-term predictions of the variables either short-range prediction criterion

$$\Delta_i^2(C) = \sum_{p \in N_C} \frac{(x_i - \hat{x}_i)_p^2}{(x_i - \bar{x}_i)_p^2}, \tag{2.102}$$

or the criterion of step-by-step prediction

$$\Delta_i^2(W) = \sum_{p \in W} (x_i - \hat{x}_i)_p^2. \tag{2.103}$$

The variable that has the least convolution value for these criteria is called the "leading" variable. Considering the prediction for the "leading" variable, we find predictions for all other variables which are not even characteristic variables.

Nonphysical model for long-term predictions (two-level analysis)

The *first stage* of two-level analysis is to divide the set of predictions of the average annual values of variables (those not discarded during the objective system analysis) into "good", "satisfactory" and "unsatisfactory" predictions, and to select the best predictions (one for each variable). Input variables that are not satisfactory are excluded for further cosideration.

Output variables are retained regardless of the quality of their annual predictions because the ultimate goal of the entire algorithm is to predict the output quantities. Predictions of models which are obtained because of various algorithms like multilayer, combinatorial, and harmonic algorithms are subjected to comparison as they use different reference functions. The choice of prediction models in all algorithms is made with reference to two criteria; the minimum bias η_{bs} and the prediction criterion $\Delta(C)$, or in the case of small number of data points, the regularity $\Delta(B)$ and the prediction criterion $\Delta(C)$. The models which are more predictive as per these criteria (one prediction for each algorithm) are evaluated further with reference to two other criteria—prediction criterion $\Delta(C)$ and criterion of preservation of first two moments $\rho(m)$. The criterion $\Delta(C)$ is used on examin set C; predictions are assumed to be "good" for $0 < \Delta(C) < 0.5$, "satisfactory" for $0.5 < \Delta(C) < 0.8$, and "unsatisfactory" for $0.8 < \Delta(C)$. Input variables whose predicted annual values are below some threshold are excluded from further consideration.

$$\rho(m) = [(\frac{x_{iav} - \hat{x}_{iav}}{x_{iav} + \hat{x}_{iav}})^2 + (\frac{\sigma_i - \hat{\sigma}_i}{\sigma_i + \hat{\sigma}_i})^2], \tag{2.104}$$

where x_{iav} and σ_i are the mean value and the variance of the variable x_i according to the test set B—i.e., on the interpolation interval N_B—and $\hat{x}_{iav}$ and $\hat{\sigma}_i$ are the mean value and the variance of estimated predictions of $\hat{x}_i$ on the interpolation and prediction intervals $N_B + N_C$. These are computed as below:

$$x_{iav} = \frac{1}{N_B} \sum_{p \in N_B} x_{ip}; \quad \sigma_i = \frac{1}{N_B} \sum_{p \in N_B} (x_i - x_{iav})_p^2 \tag{2.105}$$

$$\hat{x}_{iav} = \frac{1}{(N_B + N_C)} \sum_{p \in (N_B + N_C)} \hat{x}_{ip}; \quad \hat{\sigma}_i = \frac{1}{(N_B + N_C)} \sum_{p \in (N_B + N_C)} (\hat{x}_i - \hat{x}_{iav})_p^2 \tag{2.106}$$

The criteria $\Delta(C)$ and $\rho(m)$ are used in sequence. Algorithms under consideration are first examined on the basis of $\Delta(C)$ and in the next scanning based on $\rho(m)$ they are identified as "good" and "satisfactory." One or more better algorithms are selected for each variable for small values of $\rho(m)$ and for $\Delta(C) < 0.8$. The reliability of annual prediction estimated according to the criterion $\rho(m)$ normally improves if the average prediction is better as per $\Delta(C)$. If none of the algorithms provides satisfactory predictions, then it is necessary to introduce one more level of detailedness—for example, the averaging interval is longer than one year. The output variables that have performed good predictions of annual values ($\Delta(C) \leq 0.8$ and $\rho(m) \leq 0.01$) are hereafter called the "leading" output variables.

The *second stage* is to identify the system of seasonal models using the long-term predictions of average annual and average seasonal values of variables. This means that, the levels of detail contained in various predictions are analyzed such that the average seasonal values of variables are corrected on the basis of average annual values, which are evaluated as per the first stage. The main purpose of this stage is to obtain long-term predictions of the average seasonal values of the output variables.

It was indicated that optimum seasonal (detailed) predictions are not obtained by scanning a large number of competing random predictions, but rather by scanning a relatively small number of models, each of which generates one prediction according to its own criteria. In case of cyclic models, the scanning must include all sets of seasonal models which preserve their natural sequence. Here, usually, cyclic means we consider the seasonal models in the sequence of seasonal changes; i.e., winter, spring, summer and autumn, but in a number of cases the cyclic behavior can be created artificially. Using harmonic algorithm we find

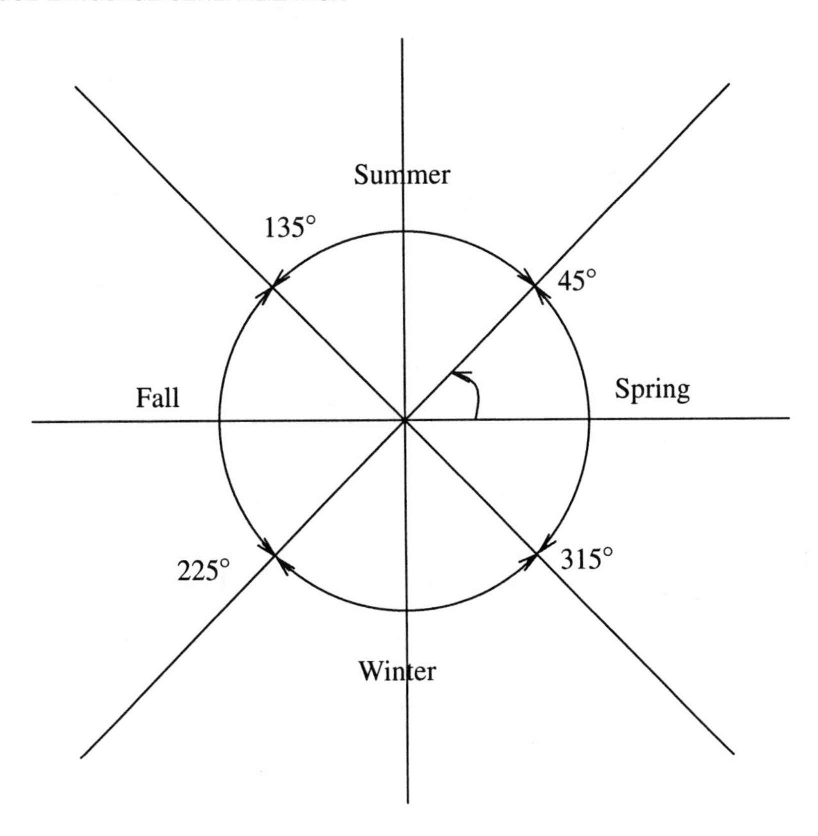

Figure 2.16. Cyclic pattern for four seasons

harmonic series containing only harmonic components which approximate a given process. We select the fundamental term of the process with the largest amplitude and we divide its period into four seasons as shown in Figure 2.16: "summer" 45°–135°, "winter" 225°–315°, "spring" 315°–45°, and "fall" 135°–225°. Before this, the polynomial trend is estimated and subtracted from the data to leave only the cyclic oscillatory part.

For noncyclic processes, the balance criterion is expressed by the sum of squares of the differences with the system of algebraic equations obtained from the previous stage.

Seasonal models are obtained using combinatorial or multilayer algorithms by scanning through a large number of competing models using the minimum bias criterion η_{bs} and prediction criterion $\Delta(C)$. When there is only one "leading" variable, then select up to ten models with different structures for this particular variable and select only one model for every other variable. The scanning of the sets of seasonal models is organized to find the optimum set. Here it is necessary to use different data bases for yearly and seasonal data in the algorithm. The yearly predictions are performed based on the one-dimensional pattern, and the seasonal predictions use the Γ-shape pattern with two-dimensional time count (refer chapter 4 for details). Best models are selected from both the levels. The balance-of-predictions criterion is used to determine the optimal model.

$$B^2 = \sum_{p \in N_C} b_p^2 \rightarrow \min; \quad b_p = \hat{Q}_{yr} - \frac{1}{4}(\hat{q}_w + \hat{q}_{sp} + \hat{q}_{su} + \hat{q}_f), \tag{2.107}$$

where $\hat{Q}_{yr}$ is the yearly prediction value of the leading variable x_i on an examin set N_C, and

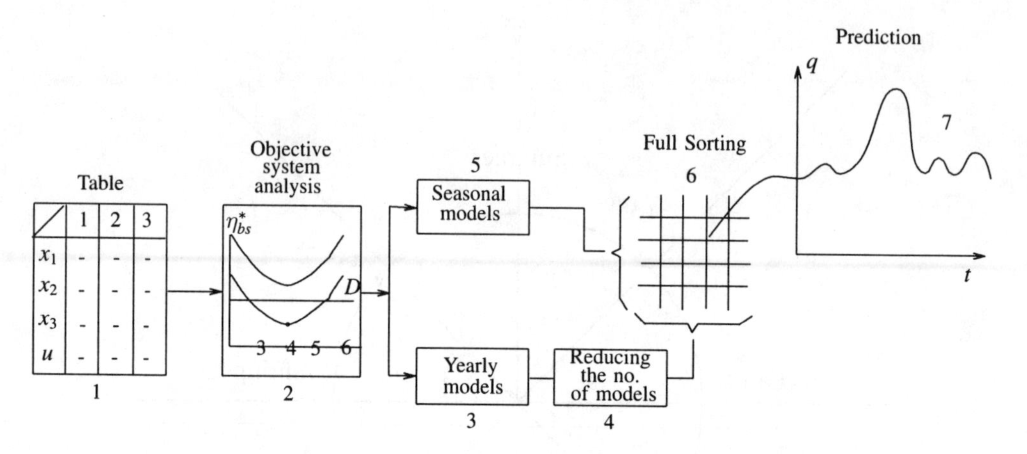

Figure 2.17. Use of OSA for long-range predictions

$\hat{q}_w, \hat{q}_{sp}, \hat{q}_{su}, \hat{q}_f$ are seasonal predicted values of same variable for winter, spring, summer, and fall respectively.

Step-by-step integration of optimum system equations gives the desired long-term predictions simultaneously for all output variables. When there are several "leading" output variables, the better set of models is selected on the basis of system criterion of balance of predictions:

$$B^* = \sqrt{\frac{1}{s}\sum_{i=1}^{s} B_i^2}, \tag{2.108}$$

where s is the number of leading variables that have good and satisfactory annual predictions.

Some practical examples are presented in later chapters. The general scheme of the multilevel algorithm is given in Figure 2.17: the first block indicates the supply of initial data table, the second block denotes first-level analysis which is called an objective system analysis (output variables are determined here), then onwards to two-level analysis; the third and fourth blocks show the first stage of the two-level analysis, and fifth and sixth blocks show the second stage of the analysis. In the first stage of two-level analysis, the third block denotes the selection of F_1 systems of equations for mean annual values of the output variables. The fourth block denotes the choice of $F_2(< F_1)$ systems of equations according to an external criterion. In the second stage of two-level analysis the fifth block denotes the selection of F_3 systems of equations for mean quarterly or seasonal values of the output variables. The sixth block denotes the sorting of the variants of the predictions in the space of system structures according to the criterion of balance of predictions, and the seventh block indicates the long-range predictions of a specific output variable.

The models used for two-level prediction with two-dimensional time count are considered as nonphysical; for example, they include both yearly and seasonal values of the variables simultaneously. The parameters of two-dimensional time coordinates (t and T) can also be considered into the systems of equations for mean annual and mean seasonal data.

The reliability of choice of a better set of models will increase when the number of scanned predictions is increased. Let p be the number of intervals of the detailed prediction within a year (months, seasons, etc.), let s be the number of leading output variables, and k be the number of models selected for each leading variable in accordance with the combinatorial algorithm. Then the number of compared model sets will be $C = (k^p)^s$.

The freedom of choice can be increased by four to five times in the same length of computer time by changing the averaging intervals to "season-year"; i.e., one can scan through eight model versions for each season. The number of compared predictions (for a single "leading" variable) will be $C_{season-year} = k^8 = 8^4 = 4096$. Therefore seasonal prediction models are preferred over monthly prediction models whenever they are adequate.

The improvement of ergatic or man-machine systems is based on the gradual reduction of human participation in the modeling process. The human element involves errors, instability, and undesired decisions. One approach to this problem is to specify the objectives, or—using technical language—determine the set of criteria. Based on such objective criteria, inductive learning algorithms are able to learn the complexities of the complex system. In self-organization processing the experts must agree on the set of criteria of lower level (regularity, minimum bias, balance of variables, and prediction criteria). If for some reason they cannot come to an agreement, then the solution is to use second-level criteria based on improvement of noise immunity. However, the important problems of sequential decision making, (such as the set of criteria determining their sequence, level of "free choice" and so on), are solved during this decade. Man still participates in the process but his task is made easier. The second area is multicriteria decision making in the domain of more "efficient solutions," where the criteria contradict each other. The solution is to use a number of random process realizations for each probability characteristic like transition graph, correlation function, probability distributions, etc. Additional *a priori* information is needed in order to choose one realization. One may have to balance the realizations of two processes that have two different averaging intervals for the variables (balance of seasonal and yearly, etc).

We conclude this section by saying that the ergatic information systems do not have any "bottle-neck" areas in which the participation of man, needed in principle, cannot be reduced or practically eliminated by moving the decision-making process on the level with a higher degree of generalization, where the solutions are obvious.

Chapter 3
Noise Immunity and Convergence

According to the principle of self-organization, the depth of the minimum of the principal selection criterion (i.e., regularity, minimum bias, balance of variables) is taken as an indicator of the successful synthesis of a model. Suppose we have m input variables of x and an output variable y with N observations. In the combinatorial inductive setup, we make all possible partial structures from the reference function $y = f(x)$. The choice of the optimal model depends on the given external criterion and on the given partition of data sets. An unbiased equation can be obtained with the help of the minimum bias criterion η_{bs} as the principal selection criterion. The same result can also be obtained, for low noisy data, using the regularity criterion $\Delta(B)$. The deeper the minimum of the unbiasedness $(0 \leq \eta_{bs} \leq 0.05)$ or regularity, the more reliable the prediction of the changing character of the process. Nevertheless, biased equations can be useful for approximating a process in the interpolation interval. If the global minimum is not achieved according to our expectations, it signifies that the problem is not solved. Then it is necessary to take measures like (i) reformulating the problem, (ii) changing the list of feasible variables, (iii) introducing new reference functions, (iv) increasing the freedom of choice for further evaluation, and so on.

Noisy data is characterized by its noise level α as a measure of noise-to-signal ratio. Noise intensity in the data plays an important role in obtaining the deep minimum. If a sufficiently deep minimum of the principal selection criterion is reached, it is possible to assume that the problem is solved. The results of potential noise stability indicate the exact limit of satisfactory modeling from the noisy data using an inductive algorithm that can be attained by using actual external selection criteria or multicriterion analysis. The degree of noise stability of the selection criterion can be determined by gradually increasing the noise level of data and finding its critical value α^*, above which the criterion fails. Before going into experimental studies, we give an analogy with the well-established information theory.

1 ANALOGY WITH INFORMATION THEORY

The concept of a signal and its noise stability are well studied and established in the field of information theory [111]. The importance of the studies in information theory exerts a favorable influence on other branches of science and technology—in particular, with the self-organization theory. The information theory assumes that input signal is frequency-band limited and that an additive noise is superimposed on it (even if the noise level is very high). According to the self-organization theory, usually only a small sample of data represents the system. It takes into account the fact that additive noise is superimposed on the output variable. Comparison of the properties of different systems in modulating a

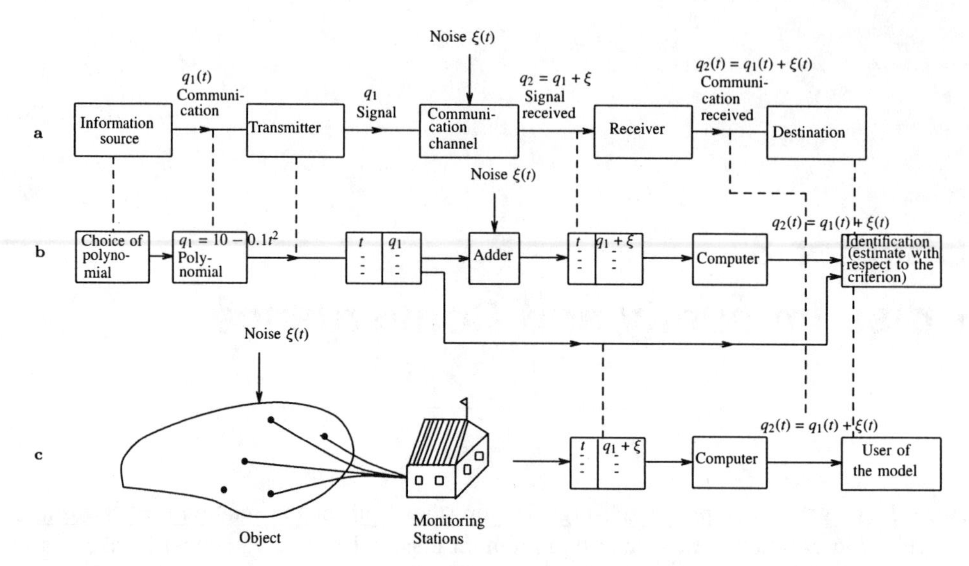

Figure 3.1. Schematic diagram of (a) a communication system, (b) a computational experimental setup, and (c) a self-organization modeling system

signal, which include Shannon's coding theory, constitutes an important part of information theory.

We give an analogy between the basic concepts of information theory and self-organization theory in identifying the processes. The main purpose of this analogy is to show the possibility of the exchange of basic ideas between these theories. We restrict our assumptions such that we are dealing only with simple amplitude modulation used in the communications and with the simplest polynomial (linear in weights) models of the form $q = a_0+a_1x_1+\cdots+a_mx_m$, where q is the dependent variable and x is the relabeled independent argument of nonlinear nature (for example, $q = 10 - 0.1t^2$).

In systems modeling, one usually considers the identification of a model only, and not the self-organization of predicting models, although communication theory does include a prediction method that is used for decreasing the redundancy of a signal. This does not restrict our study of drawing meaningful analogues between communication systems and self-organization modeling systems.

Let us put our analogy in the form of block diagrams as shown in Figure 3.1, where (a) is a communication system, (b) is a computational experimental setup, and (c) is a self-organization modeling system for obtaining an objective model (omitting the functions of specific elements).

In the communication system the information source chooses the particular form of communication from a set of possible communications. In the computational experimental scheme, we choose a polynomial (for example, $q = a_0 + a_1x = 10 - 0.1t^2$). In the self-organization system, the information source is the object of investigation (for example, ecological system) that "transmits communication" within a period of time.

In the communication system, the transmitter maps the space of communications into the space of non-noisy signals as $q_1(t) \rightarrow q_1$. In the computational experimental scheme, the polynomial $q_1(t)$ is represented in a data table with the columns of t and q_1. In the self-organizing system, the actual data is hidden in the system itself.

The communication channel in the communication system is the link at which noise intrudes. At its output, we obtain a copy of the signal; namely, the table of noisy data with t and $q_2 = q_1 + \xi$. The noisy signal is received by the receiver and is mapped into the space of received communications $q_2 \rightarrow q_2(t)$. In all the systems, the data table with t and q_2 is transformed into a polynomial for $q_2(t)$, called the physical model. The receiver corresponds to the algorithm in self-organization modeling. The destination is the place where the communication (model) is expected to go.

Information theory studies the signal at the output of the communication channel; self-organization theory studies the experimental data sample at the output of the object of investigation. Overall, one can see that the most important parts of the systems from the communication channel to the destination or user is the same for all three systems.

Analogy between the approaches in information and self-organization theories. Both theories focus on the quasistatic part of the processes (known as the signal or trend) that consider noise as a dynamic component. Both of them assume that the data being processed contain information of true input signal that conceal the governing laws acting on the object. The objective goals concentrate on a receiving device for restoring as accurately as possible the original signal (governing laws); here the receiver corresponds to the modeling algorithm of self-organization modeling.

The information theory assumes that the signal at the input of a communication channel is frequency-band limited and that an additive noise is superimposed on it. The self-organization theory also takes into account that additive noise is superimposed on the output.

The communication theory pragmatically defines the "true input signal" $q_1(t)$ and the concept of noise $\xi(t)$; for example, a portion of the output voltage permitting transmission of communication appears in the signal. Similarly, in systems modeling, the useful part of the data is the part that is utilized for identification or prediction depending on the problem; everything else is noise. The noise hinders performance of modeling and lowers the minimum of criterion for selecting a model.

Information theory assumes that noise is independent of signal and additive with normal distribution. Self-organization theory asserts that if noise is independent, then the information theory is directly applicable; but if noise is dependent on the signal, it is applicable only to orthogonalized inductive algorithms.

1.1 Basic concepts of information and self-organization theories

Signal transmission time versus interval of data points. In the information theory, the signal at the input of a communication channel is characterized by the quantities: amplitude $q_1(t)$, power $P_1(t) = q_1^2$, frequency band ω_1, maximum transmitter frequency ω_{max}, signal-to-noise ratio as $\log_2(P_1/\xi^2)$, and volume $V_1 = \omega_1 T_1 \log_2(P_1/\xi^2)$.

The signal at the channel output is determined by the quantities: amplitude $q_2(t) = q_1(t) + \xi(t)$, power $P_2 = P_1 + \xi^2$, frequency band ω_2, signal-to-noise ratio as $\log_2(P_2/\xi^2)$, and channel volume $V_2 = \omega_2 T_2 \log_2(P_2/\xi^2)$. The signal duration T is analogous to the period of observations (length of experiment) of the modeling object; i.e., the total time interval of data observations from first observation to the last one. The divisions of data must be no wider than $1/(2\omega_1)$, where ω_1 is the frequency band. Consequently, the signal transmission time corresponding to the minimum length of the measurements is as follows:
when there is no noise,

$$T_1 = \frac{N_1}{2\omega_1} \text{ sec},$$

when there is noise,

$$T_2 = \frac{N_2}{2\omega_2} \text{ sec; here}$$

$$T_2 \leq T_1.$$

N_1 and N_2 are the algebraic minima of points required in self-organization modeling with and without noises, respectively. For polynomial models, the number of points is equal to the number of terms in the individual polynomials. For harmonic models, it is three times the number of harmonic components of the model. Here N specifies the number of terms in the polynomial. At the same time, it is also the minimum number of data points required to estimate the coefficients using the least-squares technique.

Transmission capacity versus minimization of external criterion. The transmission capacity C_t of a communication system in the sense of Hartley is logarithmic to the base two of the number of communications that can be transmitted per unit time with a given accuracy. The optimal admissibility of a communication system is given in terms of its transmission capacity (speed of transmission) as

$$C_t = \omega_2 \log_2 \left(\frac{P_1 + \xi^2}{\xi^2} \right) \text{ bits/sec.} \tag{3.1}$$

In time T, it is possible to transmit $J = C_t T$ bits of information through the communication system. The formula shows that for equal information, that is for J = constant, signal power P_1 can be traded off for bandwidth ω or for transmission time T, and so on.

In self-organization modeling, the problem is solved in a much more modest way. If we confine ourselves to stationary models with constant coefficients, we need to transmit only one communication; i.e., to construct a single model. The optimal system for obtaining a self-organizing linear model in the absence of noise requires a number of measurements equal to the algebraic minimum of the N_1 points.

We can treat the reciprocal of the minimum of the selection criterion as the analogue of the transmission capacity of the communication system ($C_t = k/\Delta(B)_{min}$), where k is an arbitrary constant. As noise increases, the minimum depth of the criterion decreases; i.e., the transmission capacity drops (Figure 3.2).

Transmission capacity versus noise stability. The noise stability of a communication system is determined by the minimum limiting admissible value of the signal-to-noise ratio for which it is still possible to receive the signal.

In self-organization modeling, one uses two limits. One of the limits is determined by the confidence level of the the external criterion through a computational experiment and the other by the polynomial structural changes.

The efficiency E of a communication system is directly proportional to the transmission capacity C_t and the maximum noise stability, and is inversely proportional to the signal observation time T.

The efficiency of an inductive learning algorithm is directly proportional to the ratio of the algebraic minimum number of points necessary for constructing the model to the number of points in the data table.

$$E = k\frac{N_1}{N_{max}} = k\frac{V_1}{V_{max}}. \tag{3.2}$$

The greater the ratio of the volume of the communication channel to that of the signal, the greater the noise stability, but the lower the efficiency of use of the given communication channel (or the efforts made to obtain the experimental data).

The efficiency of communication characterizes the possibility of transmission along channels with narrow-band with low energy expenditure. The efficiency of modeling character-

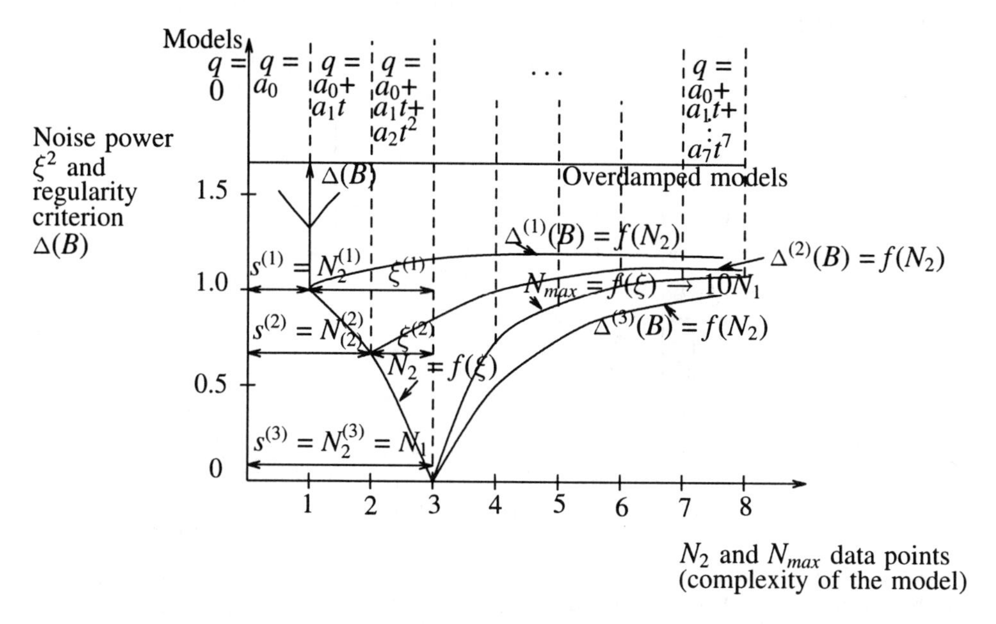

Figure 3.2. Decrease in the discrete values of the optimal complexity of model corresponding to N_2 and increase in the optimal length of the data sample N_{max} with increase in the noise power ξ^2 for a specified model complexity as at the point (3,0) as $q = 10 - 0.1t^2$

izes the possibility of constructing a sufficiently accurate model from a small number of points with a small expenditure of time on measurement, collection, and processing of data.

This can be applied directly to one-dimensional modeling problems, though two dimensional models require the introduction of two frequency bands as in two dimensional cases of communications [35].

1.2 Shannon's second theorem

The theorem is formulated as follows: Let P denote the signal power—supposing that the noise is independent—and white is the variance of ξ^2 in a frequency band ω. The optimal transmission speed attained is

$$C_{t\max} = \omega_2 \log_2\left(\frac{P+\xi^2}{\xi^2}\right) = \omega_2 \log_2\left(\frac{\omega_1}{\omega_1 - \omega_2}\right). \tag{3.3}$$

The greater the signal power in comparison with the noise variance, the greater will be the attainable transmission capacity. Thus, the theorem establishes a bound for the transmission capacity of the communication system that is attainable for optimal choice of the coding method and channel band ω_2 (the signal band ω_1 is assumed to be given) (Figure 3.3).

In self-organization modeling, the theorem enables us to choose the model with optimal complexity N_2 (complexity of the modeling object N_1 is given). The greater the noise, the lesser the depth of the minimum of the selection criterion, and the simpler the model (Figure 3.2). The theorem indicates the optimal (limiting attainable) values of the signal band (and the complexity of the models), and thus makes it clear why it is necessary, in the presence of noise, to use nonphysical models. The physical models correspond to the point (3,0) indicated in Figures 3.2 and 3.3.

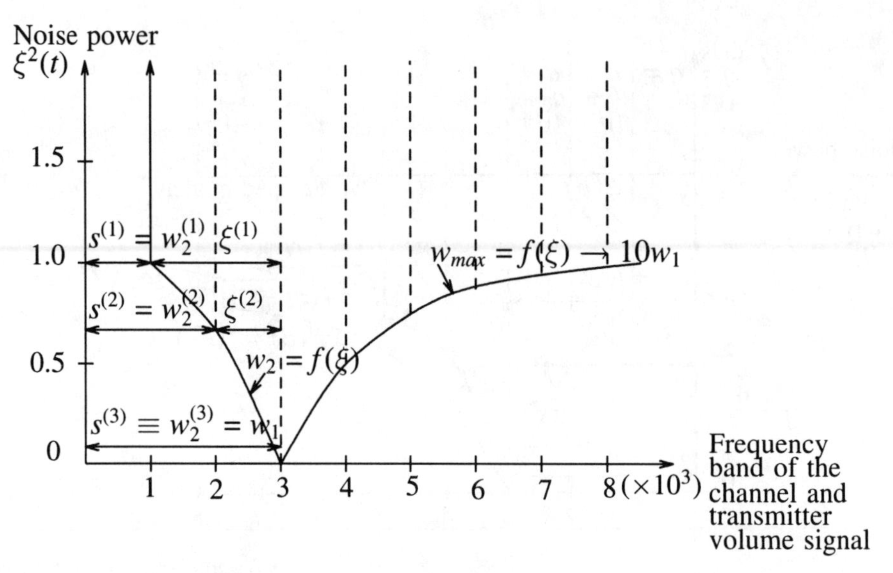

Figure 3.3. Decrease in the discrete values of the optimal band of the channel corresponding to ω_2 and increase in the optimal band of transmitter ω_{max} with increase in the noise power $\xi^2(t)$ for a given signal band of $\omega_1 = 3 \times 10^3$ = constant

Shannon's geometrical construction of the theorem

Shannon's geometrical construction is an interpretation of Shannon's second theorem (noisy coding theorem) about the limiting transmission capacity of a communication system. The input signal (like the table of input data in modeling) is at all times filled with new points with a discrete interval step of $\Delta \leq \frac{1}{2\omega_1}$. With the appearance of each new point, the dimension of the hyperspace increases by one. However, the mean value of the signal is stable. This is represented by a hypersphere with unit radius r_1 and with a center at zero. The noise is equal to the variance of the deviations of the signal from its average value. It is represented by a hypersphere with radius δ_{x_1} corresponding to noise and with the center at point A on the outer hypersphere with radius $r_2\,(= 1 + \delta_{N_1}^2)$ and with the center at zero.

In the absence of noise, the number of models is infinite as they lie on the inner hypersphere of unit radius. With reference to the Figures 3.2 and 3.3, all of them correspond to the point (3,0) and are often combined into a single "physical model."

In the presence of noises, the number of models, called nonphysical, is finite and lies on the outer hypersphere of radius r_2, which satisfies the relationships of Shannon's limit theorem. In Figures 3.2 and 3.3 they correspond to the points (1,1) and (2,0.75).

- for Figure 3.2: $N_1 \equiv P + \xi^2$; $N_2 \equiv P$; $N_1 - N_2 \equiv \xi^2$
- for Figure 3.3: $\omega_1 \equiv P + \xi^2$; $\omega_2 \equiv P$; $\omega_1 - \omega_2 \equiv \xi^2$

so that

$$\frac{N_2}{N_1} = \frac{P}{P+\xi^2} \quad \text{or} \quad \frac{\omega_2}{\omega_1} = \frac{P}{P+\xi^2}. \tag{3.4}$$

If the noise power ξ^2 is given and the physical model corresponding to N_1 is known, then the theorem enables us to find a model of optimal complexity with N_2 deductively; i.e., without sorting the partial models. The theorems applies only to self-organization modeling on the basis of external criteria characterizing its accuracy (regularity, prediction, etc.).

Criterion of convolution stability

The basic purpose of geometric construction of the theorem is to find the area or chord lengths of the surface in the hypersphere (corresponding to noise) intersecting the inner hypershere of unit radius. Specifically, this is formulated as

$$h = \sqrt{2T\omega \frac{P\xi^2}{P+\xi^2}}, \tag{3.5}$$

where T is the signal duration, ω is the frequency band, P is the signal power, and ξ^2 is the noise variance.

This enables us to find a criterion for stability of convolution of chords that is convenient for self-organization modeling and that can be used for solving various problems such as pattern recognition and long-range predictions.

For example, in selecting the best predicting model, this is represented from analytical formulas of form:

$$h_{t_i} = \frac{2\sigma^2_{(t_i)}}{1+\sigma_i^2}, \tag{3.6}$$

where $\sigma_{(t_i)}$ and σ_i are the variances of the prediction models and the variable features correspondingly, and are calculated by averaging deviations. t is the prediction model numbers and i, the variable numbers $i = 1, 2, \cdots, m$.

h_t is computed as the mean chord length (convolution)

$$h_t = \sqrt{\frac{1}{m}(h^2_{t_1} + h^2_{t_2} + \cdots + h^2_{t_m})}. \tag{3.7}$$

For example, if there are ten prediction models, h_t is computed for each model. It is chosen so that the model for the convolution of the chords is least. If $h_{10} < h_t$, $t = \overline{1,9}$, then the optimal model according to the criterion of stability of chord length of Shannon's construction is the tenth.

Similar formulations are used to solve problems of pattern recognition and vector optimization. This criterion is also called Shannon's displacement criterion.

1.3 Law of conservation of redundancy

The properties of a communication system are determined by the value of its redundance. The properties are different for wide-band and narrow-band communication systems. In wide-band systems, the redundancy exceeds zero and the channel volume exceeds the signal volume $V_{max} > V_1$ or $\omega_{max} > \omega_1$. In wide-band systems for self-organizing models, the candidate models (from the very simplest models to the models whose complexity considerably exceeds the complexity of the actual or physical model) are put up for sorting according to a set of criteria. Algebraic models can serve as the analogue of a wide-band communication system in modeling. For them, increase or decrease of data points (with subsequent operations with the data table) is useless.

In narrow-band communication systems, the channel volume is less than the signal volume, and there is no redundancy; $V_{max} < V_1$ or $\omega_{max} < \omega_1$. The optimal relationships of Shannon's limit theorem (shown in Figures 3.2 and 3.3) are violated. In this case reduction of the signal proves feasible.

In narrow-band self-organization systems, we choose models whose complexity is no greater than the complexity of the actual model $N_{max} \leq N_1$. Finite-difference models of complexity lower than the optimal model can serve as an example of a narrow-band link. Difference models are the analogues of differential equations only for small steps in the data sample.

Here the sequence of the following operations (using seasonal and annual values of a system)—collecting average seasonal data points of the variables, expanding the data table with the average annual values, self-organization modeling for obtaining a model with optimal complexity, detailed identification (seasonal data), and smoothed identification (annual data)—can be extremely efficient. Without expanding the data table, the model with optimal complexity cannot be attained because of the insufficient number of points of initial data.

1.4 Model complexity versus transmission band

In self-organization modeling, one often uses the term "complexity of a model." The complexity of the models is gradually increased until the minimum of the selection criterion is found. In linear polynomials, the complexity of the model is determined by the number of terms on the right-hand side of the equations.

The complexity of models obtained from the inductive algorithms varies from zero to N_{max} and passes through the value N_2 sought. In connection with this, in self-organization modeling, it is convenient to look at the quantities $N_1 = 2\omega_1 T_1$ (the algebraic minimum of points necessary for obtaining the true physical model), $N_2 = 2\omega_2 T_2$ (the algebraic minimum of points necessary for obtaining the optimal model using the inductive learning algorithm), and $N_{max} = 2\omega_{max} T_{max}$ (the algebraic minimum of points necessary for the most complex model that can be obtained as a result of self-organization, or the number of data points actually represented in the data table).

The following laws (Figure 3.2) come into effect in self-organization modeling:

1. In the absence of noise, beginning with some complexity equal to the complexity of the actual model N_1, further increase in complexity is not required; for $\xi^2(t) = 0$, we have $N_2 = N_1$ and $N_{max} \geq N_1$.
2. In the presence of noise, the model with optimal complexity appears earlier. The algebraic minimum of points (the complexity of the optimal model) decreases; for $\xi^2 > 0$, we have $N_2 < N_1$ and $N_{max} \geq N_2$.

The analogous laws are known in information theory (Figure 3.3). Since the bandwidths ω_1 and ω_2 can only be approximated, every communication channel gives distortion, just as every data sample, even when $V_2 = V_1$:

1. For exact transmission of a communication, it is necessary for the channel volume to be at least equal to the signal volume; for $\xi^2(t) = 0$, we have $V_2 = V_1$ and $V_{max} \geq V_1$.
2. When there are noises, the optimal channel volume is somewhat less than the signal volume; for $\xi^2 > 0$, we have $V_2 < V_1$ and $V_{max} \geq V_2$.

This means that the transmission band of special receivers designed for operation under noise conditions is narrower than wide-band receivers intended for the case of small noise. Thus, the communication channel band is analogous to the model complexity estimated according to the algebraic minimum of points

$$V_1 = \omega_1 T_1 \log_2(P_1/\xi^2) = \frac{N_1}{2} \log_2(P_1/\xi^2),$$

$$V_2 = \omega_2 T_2 \log_2(P_2/\xi^2) = \frac{N_2}{2}\log_2(P_2/\xi^2). \tag{3.8}$$

The influence of noise on the model accuracy can be overcome to some extent by increasing the number of measurements. However, when the number of data points becomes excessive, the accuracy and noise stability of the model decrease. Thus, there exists an optimal number of data points for stationary and non-stationary signals. Because of the necessity of decreasing the influence of noise, one chooses the table length about 10 times as great as the algebraic minimum of the points $T_\xi = 10.T_1$ (Figure 3.3). During this interval, the system will collect $J = C_t T_\xi = \omega.\log_2(\frac{P_1+\xi^2}{\xi^2}).T_\xi$ bits of information.

An analogy between the optimal complexity of models for the inductive algorithm and the transmission band for a communication system is shown in Figures 3.2 and 3.3, where N_1 is the complexity of the physical model, N_2 is the complexity of the non-physical model of optimal complexity, N_{max} is the optimal range of complexity of model candidates, ω_1 is the band of the true signal, ω_2 is the optimal band of the receiver, and ω_{max} is the optimal volume of the transmitter signal.

The law of compromization. The important result of investigations arrived at through the information theory is the establishment of a connection between transmission capacity and noise stability. Increase in noise stability decreases transmission capacity. Here one varies the parameters of the communication system; for example, by varying the frequency band ω_2.

An analogous law holds for self-organization modeling using the selection criterion such as regularity; an increase in the power (amplitude) of the noise leads to the choice of simpler noise-stable models for which the algebraic minimum of points is less than that of the model of the object obtained under conditions of absence of noise. Here one varies the parameters of the model; for example, by varying the algebraic minimum of points N_2.

Here we conclude that the noisy coding theorem (Shannon's second theorem) plays a central role in this analogy between information theory and self-organization theory. In fact, the theorem states that it is possible to transmit information through the channel with as small a probability of error as desired if it is transmitted at any rate less than the channel capacity. In other words, it guarantees the existence of a code that may be transmitted at any rate close to but less than that of channel capacity and still be received and decoded with arbitrarily small probability of error. It proves that channel capacity is a fundamental property of a communication channel. This is conceptualized analogously to the theory of self-organization modeling. In particular, it shows that, in the presence of noise, non-physical models obtained by self-organization modeling are optimal.

2 CLASSIFICATION AND ANALYSIS OF CRITERIA

Let us assume that the initial data is given in the form of the matrices

$$y = \begin{pmatrix} y_A \\ --- \\ y_B \\ --- \\ y_C \end{pmatrix}, \quad X = \begin{pmatrix} X_A \\ --- \\ X_B \\ --- \\ X_C \end{pmatrix}, \quad \begin{matrix} y\,[N \times 1] \\ X\,[N \times m] \\ N_A + N_B + N_C = N \\ N_A + N_B = N_W \end{matrix}. \tag{3.9}$$

The entire data sample is partitioned into three disjoint subsets A, B, and C. The set W is the union of A and B. The optimal dependence relation between output y and input variables X is sought by the inductive learning that are linear in the coefficients of $y = X\hat{a}$.

It is assumed that the submatrices X_A and X_B, which are used in the selection process to any particular model of complexity $s\,(\leq n)$, are of complete rank.

The external criteria used in the inductive algorithms can be expressed in terms of the estimates of the output variables of the models and their coefficients obtained on $A, B, W,$ and C. Here the basic quadrature, and combined and correlational criteria are described.

All the external criteria that have the quadratic form can be grouped into two basic groups: (i) accuracy criteria, which express the error in the model being tested on various parts of the model and (ii) matching (consistent) criteria, which are a measure of the consistency of the estimates obtained on different sets. There are symmetric and nonsymmetric forms of the criteria in both the groups, where symmetric means one in which the information in sets A and B is used equally; otherwise, it is nonsymmetric.

2.1 Accuracy criteria

Regularity criterion (nonsymmetric)

This is the typical quadratic criterion and historically the first one.

$$\begin{aligned}\Delta^2(B) = \Delta^2(B|A) &= \|y_B - \hat{y}_B^A\|^2 = (y_B - X_B\hat{a}_A)^T(y_B - X_B\hat{a}_A) \\ &= \|y_B - X_B\hat{a}_A\|^2, \end{aligned} \tag{3.10}$$

where $\hat{a}_A = (X_A^T X_A)^{-1} X_A^T y_A$, and $\hat{y}_B^A = X_B\hat{a}_A$.

We can obtain another nonsymmetric regularity criterion by replacing A by B and, vice versa, $\Delta^2(A) = \Delta^2(A/B)$.

Regularity criterion (symmetric)

This can be built up using the both the nonsymmetric versions of the regularity criterion.

$$\begin{aligned}d^2 = d^2(A,B|B,A) &= \Delta^2(B|A) + \Delta^2(A|B) \\ &= \|y_B - X_B\hat{a}_A\|^2 + \|y_A - X_A\hat{a}_B\|^2, \end{aligned} \tag{3.11}$$

where sets A and B are used equally. It smooths out the influence of the noise that acts on both parts of the data sample.

Stability criterion (nonsymmetric)

If we require an optimal model, which must be sufficiently accurate on both the sets—training set A and testing set B for the coefficients estimated on the set A—then this compromise can be obtained by the criterion

$$\begin{aligned}\kappa^2 = \kappa^2(W|A) &= \|y_W - X_W\hat{a}_A\|^2 \\ &= \Delta^2(A|A) + \Delta^2(B|A) = \varepsilon^2(A) + \Delta^2(B), \end{aligned} \tag{3.12}$$

where $\varepsilon^2(A)$ is the least squares error or residual sum of squares.

Stability criterion (symmetric)

$$\begin{aligned}S^2 = S^2(W|A,B) &= \kappa^2(W|A) + \kappa^2(W|B) \\ &= \|y_W - X_W\hat{a}_A\|^2 + \|y_W - X_W\hat{a}_B\|^2. \end{aligned} \tag{3.13}$$

It is expedient to use this criterion if the finiteness of the data is considered. The sensitivity to the separation of data is lowered and the influence of noise is averaged (a kind of filtering takes place). In other words, this has higher noise immunity.

Averaged regularity criterion [122]

According to this criterion the mean value is calculated on N_W for each particular model being tested under the condition that each point in the set W is, in its turn, the testing sample and the remaining $N_W - 1$ points constitute the training sample.

$$\Delta_{av}^2(W) = \frac{1}{N_W} \|y_j - \hat{y}_j^{W_j}\|_{j \in W}^2, \tag{3.14}$$

where $\hat{y}_j^{W_j} = x_j^T \hat{a}_{W_j}$, x_j are the argument measures at the jth point, W_j is the training sample without jth point, and $\hat{a}_{W_j}$ is the estimate of the coefficients on W_j. It is expedient to use this criterion for a small number of points.

Step-by-step prediction criterion

In case of finite-difference equations, it is expedient to use this external integral criterion.

$$i^2(W) = i^2(W|W) = \|y_W - \hat{y}_W^W\|^2, \tag{3.15}$$

where the estimate $\hat{y}_W^W$ is obtained by step-by-step integration of the difference equation from the given initial conditions. This criterion can also have the forms of $i^2(A)$ and $i^2(B)$.

The above accuracy criteria, like all other types of criteria, are used in modeling of both static and dynamic objects.

2.2 Consistent criteria

The criteria in this group do not take into account the error of the model in explicit form, but measures the consistency of the model on two different data sets.

Criterion of minimum coefficient bias

This reflects the requirement that the coefficient estimates in the optimal model estimated on sets A and B, differ minimally; i.e. they are in agreement.

$$\eta_a^2 = \eta_a^2(A, B) = \|\hat{a}_A - \hat{a}_B\|^2. \tag{3.16}$$

Minimum bias criterion

This is the most widely used form of the criterion.

$$\begin{aligned} \eta_{bs}^2 = \eta_{bs}^2(W|A, B) &= \|\hat{y}_W^A - \hat{y}_W^B\|^2 \\ &= \|X_W \hat{a}_A - X_W \hat{a}_B\|^2 \\ &= (\hat{a}_A - \hat{a}_B)^T X_W^T X_W (\hat{a}_A - \hat{a}_B), \end{aligned} \tag{3.17}$$

which differs from η_a by the presence of the weight matrix $X_W^T X_W$ and expresses a different minimum requirement of consistency on the set W from the estimates of the model outputs that obtained coefficients from the sets A and B.

Absolute noise immune criterion

$$V^2 = V^2(W|A,B,W) = (\hat{a}_W - \hat{a}_A)^T X_W^T X_W (\hat{a}_B - \hat{a}_W)$$
$$= (\hat{y}_W^W - \hat{y}_W^A)^T (\hat{y}_W^B - \hat{y}_W^W). \tag{3.18}$$

This uses the estimates of the model output for the coefficients obtained on three sets—A, B, and W. It got its name because it satisfies the most important condition of noise immunity. It rejects excessively complex models under noise conditions [67].

The above minimum bias criteria are symmetric. It is easy to write nonsymmetric forms of η_{bs} and V^2. For example, on the set B [129]

$$\eta_{bs}(B) = \|\hat{y}_B^A - \hat{y}_B^B\|^2 = \|X_B\hat{a}_A - X_B\hat{a}_B\|^2, \tag{3.19}$$

$$V^2(B) = (\hat{a}_W - \hat{a}_A)^T X_B^T X_B (\hat{a}_B - \hat{a}_W). \tag{3.20}$$

One useful way is to clarify the connections among certain external criteria. One can easily show that $\eta_{bs}^2(W) = \eta_{bs}^2(A) + \eta_{bs}^2(B)$ and, in the same way, $V^2(W) = V^2(A) + V^2(B)$ because of the relation $X_W^T X_W = X_A^T X_A + X_B^T X_B$.

2.3 Combined criteria

In addition to the criteria $c1, c2$, and $c3$ introduced in chapter 1, here is another form of the combined criterion $c4$.

Minimum bias plus symmetric regularity

$$c4^2 = \bar{\eta}_{bs}^2 + \bar{d}^2. \tag{3.21}$$

It is recommended that the sequential use of two-criterion selection is preferred in the combined criteria. F number of models are selected using the consistent criterion like η_{bs}^2, then the best model is selected using a accuracy criterion like $\Delta^2(C)$. Such sequential application of the criteria increases the efficiency of the modeling, including noise immunity.

2.4 Correlational criteria

These criteria impose definite requirements on the relationship of correlation characteristics of the output variables of model and the object. Unlike the quadratic criteria, they can be both positive and negative. This is one of the reasons for separating them as a special group of criteria. Their applicability for model selection is ensured by the fact that coefficients of the model are estimated on the set A and the correlation relationship is computed with respect to the set B.

Correlational regularity criterion

$$K(B) = \frac{(y_B - \bar{y}_B)^T(\hat{y}_B^A - \bar{\hat{y}}_B^A)}{\|y_B - \bar{y}_B\| \cdot \|\hat{y}_B^A - \bar{\hat{y}}_B^A\|}, \tag{3.22}$$

where y_B is the actual output; $\hat{y}_B^A$ is the model output, the coefficients of which are estimated on set A; $\bar{y}_B$ and $\bar{\hat{y}}_B^A$ are the mean values of the actual and model outputs, respectively. The best model is based on the condition $K(B) \to 1$.

Table 3.1. Classification of external criteria

Type	Criterion	Criterion form	
		nonsymmetric	symmetric
Accuracy	regularity	$\Delta^2(B), \Delta_2(A)$	$d^2(W)$
	stability	$\kappa^2(B), \kappa^2(A)$	$S^2(W)$
	averaged regularity	-	$\Delta^2_{av}(W)$
	prediction	$i^2(B), i^2(A)$	$i^2(W)$
Consistent	minimum bias	$\eta^2_{bs}(A), \eta^2_{bs}(B)$	$\eta^2_{bs}(W)$
	abs. noise immune	$V^2(A), V^2(B)$	$V^2(W)$
Correlational	regularity	$K(B), K(A)$	$K(A) + K(B)$
	NL agreement	$J_s(B), J_s(A)$	$J_s(A) + J_s(B)$
Combined	bias + regularity	$\eta^2_{bs} + \Delta^2(B)$	$\eta^2_{bs} + d^2$
	bias + MSE	$\eta^2_{bs} + \varepsilon^2(A)$	$\eta^2_{bs} + \varepsilon^2(W)$
	bias + prediction	$\eta^2_{bs} + \Delta^2(C)$	-

Correlational criterion with nonlinear agreement [129]

This has three different components; one is equivalent to the correlational regularity, the other is the agreement criterion for the degree of nonlinearity, and the third is the agreement criterion for the mean values of the actual and estimated outputs. These components are based on the mean-squared error as follows:

$$\begin{aligned}\varepsilon^2 &= \frac{1}{N}(y - \hat{y})^T(y - \hat{y}) \\ &= (1 - J_c^2 + J_s^2 + J_m^2)z_v^2,\end{aligned} \tag{3.23}$$

where y_i and $\hat{y}_i$, $i \in N$ are the actual and estimated outputs of N data points. The quantities J_c, J_s, and J_m are expressed in terms of the centered vectors $v = y - \bar{y}$ and $\hat{v} = \hat{y} - \bar{\hat{y}}$ and the estimates of the variances as $z_v = \sqrt{(v^T v/N)}$ and $z_{\hat{v}} = \sqrt{(\hat{v}^T \hat{v}/N)}$.

$$J_c = r(\hat{v}, v) = \hat{v}^T v/N z_{\hat{v}} z_v \tag{3.24}$$

$$J_s = z_{\hat{v}}/z_v - r(\hat{v}, v) \tag{3.25}$$

$$J_m = (\bar{y} - \bar{\hat{y}})/z_v. \tag{3.26}$$

It was proposed in [129] that the components J_c, J_s, and J_m of the error vector can be used as independent selection criteria, calculated on the set B with the estimates $\hat{a}_A$ obtained on the set A. The component $J_c(B)$ coincides exactly with the criterion $K(B)$. The component $J_s(B)$ is called the agreement criterion for the degree of nonlinearity; this should satisfy the condition $J_s(B) \to 0$. The component $J_m(B)$ is also called the agreement criterion for the mean values, but it does not seem to have any independent significance. One can convert the criterion $J_c(B) \equiv K(B)$ into a minimization form $|1 - K(B)| \to \min$.

The above mentioned correlational criteria are nonsymmetric; to make them symmetric, an expression must be added to each in which the sample parts A and B swap roles. Various groups of criteria are given in Table 3.1.

2.5 Relationships among the criteria

In this section we derive the number of relationships that express the connection among different external criteria.

Let us consider the quadratic criteria of symmetric type. We can write the relationship of S^2 in terms of d^2.

$$S^2 = d^2 + \varepsilon^2(A) + \varepsilon^2(B), \tag{3.27}$$

where $\varepsilon^2(A) = \Delta^2(A|A)$ and $\varepsilon^2(B) = \Delta^2(B|B)$ are the mean square errors on the sets A and B, respectively. We can write the minimum bias criterion as

$$\begin{aligned}\eta_{bs}^2 &= \|(y_W - \hat{y}_W^B) - (y_W - \hat{y}_W^A)\|^2 \\ &= \|y_W - \hat{y}_W^A\|^2 + \|y_W - \hat{y}_W^B\|^2 \\ &\quad -2(y_W - \hat{y}_W^A)^T(y_W - \hat{y}_W^B),\end{aligned} \tag{3.28}$$

since $X_W^T X_W = X_A^T X_A + X_B^T X_B$, $X_W^T y_W = X_A^T y_A + X_B^T y_B$, $\hat{a}_B = (X_B^T X_B)^{-1} X_B^T y_B$, $\hat{a}_A = (X_A^T X_A)^{-1} X_A^T y_A$. The term from the above expansion can be evaluated further as

$$\begin{aligned}(y_W - \hat{y}_W^A)^T(y_W - \hat{y}_W^B) &= y_A^T y_A - y_A^T X_A \hat{a}_A + y_B^T y_B - y_B^T X_B \hat{a}_B \\ &= \varepsilon^2(A) + \varepsilon^2(B).\end{aligned} \tag{3.29}$$

Knowing this, we can obtain a relationship between S^2 and η_{bs}^2:

$$S^2 = \eta_{bs}^2 + 2(\varepsilon^2(A) + \varepsilon^2(B)); \tag{3.30}$$

between d^2 and η_{bs}^2

$$d^2 = \eta_{bs}^2 + \varepsilon^2(A) + \varepsilon^2(B). \tag{3.31}$$

Now let us consider the criterion V^2:

$$\begin{aligned}V^2 &= [(y_W - \hat{y}_W^A) - (y_W - \hat{y}_W^W)]^T[(y_W - \hat{y}_W^W) - (y_W - \hat{y}_W^B)] \\ &= (y_W - \hat{y}_W^A)^T(y_W - \hat{y}_W^W) + (y_W - \hat{y}_W^B)^T(y_W - \hat{y}_W^W) \\ &\quad -(y_W - \hat{y}_W^A)^T(y_W - \hat{y}_W^B) - \|(y_W - \hat{y}_W^W)\|^2.\end{aligned} \tag{3.32}$$

The term $(y_W - \hat{y}_W^A)^T(y_W - \hat{y}_W^B)$ is given above as $\varepsilon^2(A) + \varepsilon^2(B)$. Since $\hat{a}_W = (X_W^T X_W)^{-1} X_W^T y_W$, one can obtain the relation as

$$\begin{aligned}(y_W - \hat{y}_W^A)^T(y_W - \hat{y}_W^W) &\equiv (y_W - \hat{y}_W^B)^T(y_W - \hat{y}_W^W) \\ &\equiv \|y_W - \hat{y}_W^W\|^2 = \varepsilon^2(W)\end{aligned} \tag{3.33}$$

by establishing that $y_W^T \hat{y}_W^A \equiv \hat{y}_W^{W^T} \hat{y}_W^A$, and $y_W^T \hat{y}_W^B \equiv \hat{y}_W^{W^T} \hat{y}_W^B$. Ultimately, we can obtain the formula [35]

$$V^2 + \varepsilon^2(A) + \varepsilon^2(B) = \varepsilon^2(W). \tag{3.34}$$

Using the above examination, one can easily write the relationships

$$V^2 + S^2 = d^2 + \varepsilon^2(W) \tag{3.35}$$

$$V^2 + d^2 = \eta_{bs}^2 + \varepsilon^2(W). \tag{3.36}$$

One can show that the absolute noise immune criterion V^2 is a quadratic, not a nonnegative one, by the relation

$$\begin{aligned}V^2 &= (\hat{a}_W - \hat{a}_A)^T X_W^T X_W (\hat{a}_B - \hat{a}_W) \\ &= \hat{a}_A^T X_A^T X_A \hat{a}_A + \hat{a}_B^T X_B^T X_B \hat{a}_B - \hat{a}_W^T X_A^T X_A \hat{a}_W - \hat{a}_W^T X_B^T X_B \hat{a}_W.\end{aligned} \tag{3.37}$$

This can be expressed into the sum of two quadratic forms:

$$\begin{aligned} V^2 &= \|\hat{y}_A^A - \hat{y}_A^W\|^2 + \|\hat{y}_B^B - \hat{y}_B^W\|^2 \\ &= (\hat{a}_A - \hat{a}_W)^T X_A^T X_A (\hat{a}_A - \hat{a}_W) + (\hat{a}_B - \hat{a}_W)^T X_B^T X_B (\hat{a}_B - \hat{a}_W), \end{aligned} \tag{3.38}$$

so that we can always have $V^2 \geq 0$.

The relationships established between the criteria S^2, d^2, and V^2 interconnect all the symmetric quadratic criteria. In addition to this, the formulas reexpressed for the criteria S^2 and d^2 in terms of the minimum bias and mean-square errors allow one to group these criteria into the group of the combined criteria. They are, however, fundamentally different from the combined criteria because of the components included in them and there is no need for normalization.

Similarly, one can obtain the relationships connecting the nonsymmetric criteria. For example, the regularity criterion $\Delta^2(B)$ can be represented [129] as

$$\Delta^2(B) = \varepsilon^2(B) + \eta_{bs}^2(B), \tag{3.39}$$

where $\eta_{bs}^2(B)$ is the nonsymmetric form of the minimum bias criterion on the set B.

The connection can be established among the regularity criterion and the correlational criteria directly from the relationship $\varepsilon^2 = (1 - J_c^2 + J_s^2 + J_m^2) z_v^2$ as

$$\Delta^2(B) = (1 - K^2(B) + J_s^2(B) + J_m^2(B)) z_v^2 N. \tag{3.40}$$

The representations of some criteria in terms of other criteria enable us to determine the characteristics of unique models derived from the original ones. For example, after calculating the squared errors $\varepsilon^2(A), \varepsilon^2(B), \Delta^2(A), \Delta^2(B)$, and $\Delta^2(C)$, one can also determine d^2, S^2, and η_{bs}^2 directly; after estimating $\varepsilon^2(W)$ one can calculate V^2.

The reader can find the usage of canonical forms in analyzing the noise immunity of quadratic criteria in the works [135], [119]. Here the expected value of the criterion is considered the sum of two components: one takes into account the non-noisy data and decreases (possibly nonmonotonically) with the increase of complexity of models; the second reflects the presence of noise that is directly proportional to its variance and increases monotonically with the increase of complexity of the models. With an increase in the noise level, the minimum of the external criteria (V^2 and d^2) moves into the region of simpler structures, which is analogous to the behavior of the ideal external criterion.

3 IMPROVEMENT OF NOISE IMMUNITY

We assume that noise can be additive, multiplicative, or a combination of these two types and that it does not contain a regular component. When the noise intensity (amplitude) is very high, the external criterion used might select a model that does not correspond to the system under study. The criterion is called noise-immune if it selects the true model even at a significant level of noise immunity. The analytical properties of selection procedures based on certain selection criteria are given here. Emphasis is made on improving the noise stability of the criteria in extracting the optimal model with true structure in the presence of noise. This identifies the true structure by comparing different structures that determine the maximum allowed noise level.

Let us assume that y is an output variable with a normally distributed noise. Its unit variance is represented as

$$y = \overset{\circ}{y} + \xi, \quad E[\xi] = 0, \quad \sigma^2 = E[\xi^T \xi] = 1, \tag{3.41}$$

where $\overset{\circ}{y}$ is the noise-free output connecting a set of m arguments (input variables) as $\overset{\circ}{y} = \phi(z)$. Let us assume that we have some noise realization of ξ_0 with N values. We obtain a series of output data for varying intensity of this realization of the noise, yielding N values of the function

$$y_\xi = \overset{\circ}{y} + \xi_0. \tag{3.42}$$

The sample of noise-free data obtained from the function $\overset{\circ}{y} = \phi(z)$ can be called the signal and output samples for different variances of ξ can be called the signal with noise. Each sample of noisy data is characterized by a value of the noise-to-signal ratio or by the noise level as

$$\alpha^2 = \frac{\sigma^2}{s^2} = \sigma^2 / \sum_{j=1}^{N} (\overset{\circ}{y}_j - \bar{y}^\circ)^2, \tag{3.43}$$

where s^2 and σ^2 are the signal and noise variance or power, correspondingly; $\bar{y}^\circ$ is the average value of the signal. For a fixed signal the variance and noise level are connected by a one-to-one relationship $\sigma^2 = \alpha^2.s^2$ or $\sigma = \alpha.s$.

Suppose the function $\phi(z)$ is a linear (in coefficients) convolution of some number of functions (for example, a set of polynomial functions $f_1(z), f_2(z), \cdots, f_m(z)$), equivalent to the vector of arguments $x = f(z)$. Then for each noise level α, the exact model is restored by optimizing the structure and estimating the coefficients a of the model.

$$y = a^T f(z) = a^T x \tag{3.44}$$

for the given sample of input and output values.

Here two types of study results are presented to show the efforts in improving the noise immunity of various external criteria. The first part consists of the initial studies [129] conducted on the minimum-bias criterion. This reveals the importance of the extension of the time interval to the extrapolation region of the data and shows that the largest noise immunity was possessed by special forms of the criterion with some specified general properties. The second part is concerned with the finding of noise stability of various criteria (single- as well as two-criterion analysis) by increasing noise levels for different data divisions. This gives some comparative results on several most commonly applied criteria for obtaining single- and two-criterion choices of models.

3.1 Minimum-bias criterion as a special case

The original form of the minimum bias criterion is

$$\eta_{bs}^2 = \sum_{p=1}^{N_W} \frac{(\hat{y}^A - \hat{y}^B)_p^2}{y_p^2}, \tag{3.45}$$

where $\hat{y}^A$ is the estimated output of the model, the coefficients of which are obtained using the set A; $\hat{y}^B$ is the estimated output of the model, the coefficients of which are obtained using the set B; and y is the actual output.

Geometric interpretation of the minimum bias

Suppose in an N-dimensional space R^N (N is the length of the data sample), $\hat{y}_{LS}$ is an orthogonal projection of the vector $y^T = (y_1, y_2, \cdots, y_N)$ from the output of the linear model

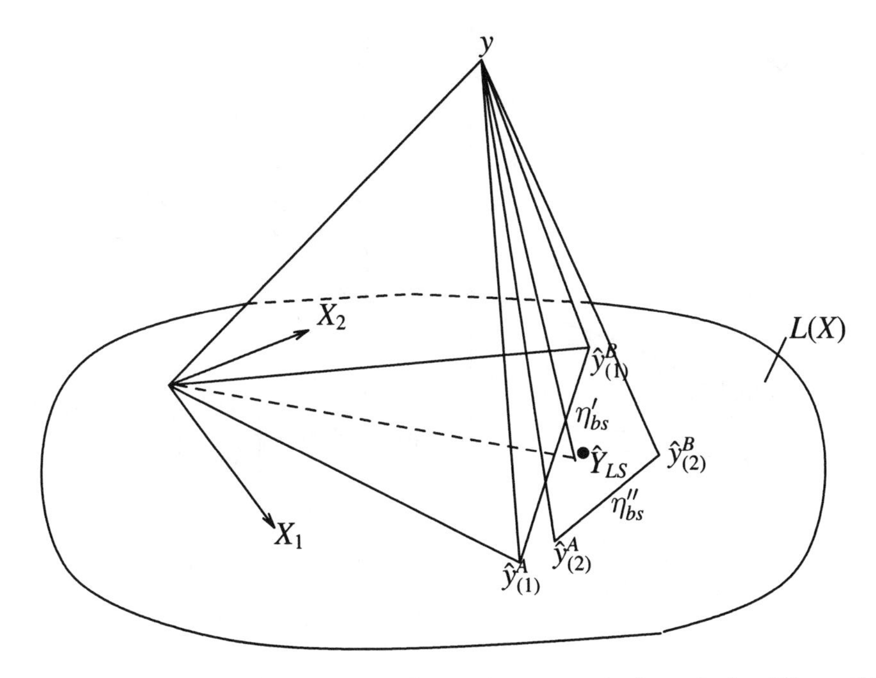

Figure 3.4. Minimum bias of solutions as a distance between projections of y for different divisions of data sample

$y = x^T a$, which is estimated by the least squares method onto a linear subspace $L(X)$ (Figure 3.4), formed by the vectors of m arguments $X_i^T = (x_1, x_2, \cdots, x_N)$, $i = 1, 2, \cdots, m$, ie., $a, x \in R^m$, and $x_i \in R^N$; and $X[N \times m]$ consists of the sets of matrices with real elements.

The data sets A and B are used as training sequence to estimate the coefficients of two models of similar structure and to the approximations of y as a total sample. Projections $\hat{y}^A$ and $\hat{y}^B$ of the same vector y on to the $L(X)$ are formed and these are usually non-orthogonal. Each ith version of the division of the data has corresponding vectors $\hat{y}^A_{(i)}$ and $\hat{y}^B_{(i)}$ belonging to $L(X)$. The ensemble of such vectors forms a "cluster of projections"; i.e., a set of points in $L(X)$; all points of the cluster are grouped around $\hat{y}_{LS}$. Models with false structures are more sensitive to the variations of the training sequence and, as a result, become significantly displaced from $\hat{y}_{LS}$—causing the cluster to widen. Different forms of the minimum bias criterion provide us with the possibility of estimating the dimensions of this cluster; i.e., an ability to compare different models.

(i) *Increasing the time interval of data in the criterion by introducing a noise immunity coefficient* δ_T. The minimum bias criterion has a relatively low noise-immunity because the approximating properties of the models are usually identical on the interval of interpolation. The squared errors are small for models of any structure except for the simplest linear models. The performance of models diverge in the extrapolation interval in which the differences between the model outputs become significant and, consequently, more immune to noise. Figure 3.5 shows an example of bias estimation for two polynomial models. The shaded areas indicate the differences of two sets of models (area between the integral curves $\hat{y}^A_{(1)}$ and $\hat{y}^B_{(1)}$ and area between the curves $\hat{y}^A_{(2)}$ and $\hat{y}^B_{(2)}$). Consequently, the bias estimate of the second polynomial is significantly smaller than the first one; i.e., $\eta^2_{bs_2} \ll \eta^2_{bs_1}$.

Here it is recommended that the minimum bias criterion with additional length of time

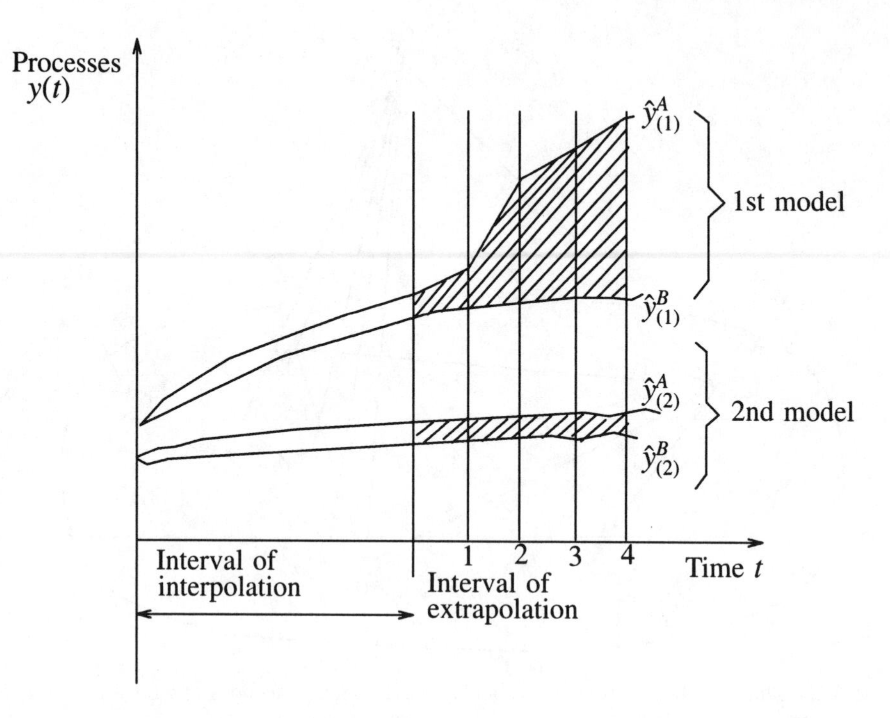

Figure 3.5. Explanation of estimation of bias of two models

interval should be

$$\eta_{bs}^2 = \sum_{p=1}^{\delta_T N_W} \frac{(\hat{y}^A - \hat{y}^B)_p^2}{y_p^2}. \tag{3.46}$$

One can notice that it is applicable only to the nonlinear functions that have quadratic or higher order arguments.

(ii) *Extraction of first harmonic of the output variable.* The output variable is approximated with the harmonic equation using the sets A and B as

$$\begin{aligned} y^A &= a_0 + a_1 \sin(\omega_1 t + Q_1), \\ y^B &= b_0 + b_1 \sin(\omega_2 t + Q_2), \end{aligned} \tag{3.47}$$

where ω_1 and ω_2 are the fundamental frequencies, Q_1 and Q_2 are the phase shifts, and a's and b's are the estimated coeficients. It is assumed that the frequency expansion of the useful signal without noise occupies a portion of the spectrum which is different from the signal with noise. If this is justified, then the noise immunity of the minimum bias criterion can be increased because of the first harmonic. The first harmonic of set A should coincide as nearly as possible with the first harmonic of set B. The minimum bias criterion is recommended as

$$\eta_{bs}^2 = \sum_{p=1}^{\delta_T N_W} \frac{(\hat{y}^A - \hat{y}^B)_p^2}{(\hat{y}^A + \hat{y}^B)_p^2}, \tag{3.48}$$

where $\hat{y}_A$ and $\hat{y}_B$ are the estimated outputs of the first harmonics. In practical situations, the fundamental frequencies ω_1 and ω_2 should be closer within the limits of the specified set of structures. If the spectral content of the signal and noise are identical, then it is difficult to filter out the noise.

(iii) *Extraction of the linear trend of output variable.* If the outputs of the models are smooth and can be approximated by polynomials, extracting linear parts instead of first harmonics is recommended.

$$y^A_{(\text{lin})} = a_0 + a_1 t,$$
$$y^B_{(\text{lin})} = b_0 + b_1 t. \tag{3.49}$$

The two models to be identified have identical structures. The model $y^A_{(\text{lin})}$ should coincide as nearly as possible with the model $y^B_{(\text{lin})}$; i.e., the structure of the model is estimated in accordance to the minimum bias criterion

$$\eta^2_{bs} = \sum_{p=1}^{\delta_T N_W} \frac{(\hat{y}^A_{(\text{lin})} - \hat{y}^B_{(\text{lin})})^2_p}{(\hat{y}^A_{(\text{lin})} + \hat{y}^B_{(\text{lin})})^2_p}. \tag{3.50}$$

This provides justification to calculate the minimum bias criterion based on the linear parts. When the linear parts of the models are slightly dependent on the noise, such a criterion will have an increased noise immunity.

Example 1. An experiment is conducted to show the effect of the data interval on the noise immunity of the criterion. The true model is taken as $\overset{\circ}{y} = 2 - 0.1t^2$. Twelve values of the output variable $y(t)$ are taken, corresponding to $t = 1, 2, \cdots, 12$. The noise intensity is increased step by step and the optimal models are extracted for each set of data. The combinatorial algorithm is used with a reference function of the third-degree polynomial in t.

$$y = a_0 + a_1 t + a_2 t^2 + a_3 t^3. \tag{3.51}$$

The first minimum-bias criterion most immune to the noise is found by extracting the linear part with the noise immunity coefficient value $\delta_T \approx 2.0$. The second one most immune to noise has data points arranged according to variance. The lowest one has data points arranged as even and odd points.

In all the cases, preliminary extraction of linear parts or trends and widening of data interval with δ_T have significant effect; the noise immune coefficient δ_T is found in the range of 1.5 to 3.0.

3.2 Single and multicriterion analysis

Several qualitative estimates of the degree of noise stability can be obtained analytically by considering just one fixed structure of the model. Suppose the equation $y = X\hat{a}$ is written for the chosen structure. Consider the prediction problem using the prediction criterion

$$\Delta^2(C) = \sum_{i \in C} \frac{(\hat{y} - y)^2_i}{(y - \bar{y})^2_i} \leq 1.0, \tag{3.52}$$

where $\hat{y}$ is the estimated output, $\bar{y}$ is the average value of the output, y is the actual output, and C is the prediction data set. We obtain the estimates of the coefficients $\hat{a}$ using the data set $W = A \cup B$.

$$\hat{a} = (X^T X)^{-1} X(\overset{\circ}{y} + \xi_0) = a^0 + \sigma^2 a_\xi. \tag{3.53}$$

This is the sum of the exact value of the coefficient vector and the added quantity which depends linearly on the noise level (the value of a_ξ is independent of noise variance σ^2). The predictions based on this model have the form of

$$\hat{y}_i = \overset{\circ}{y}_i + \sigma^2 a_\xi^T x_i = \overset{\circ}{y}_i + \alpha^2 s^2 a_\xi^T x_i, \quad i \in C, \tag{3.54}$$

where σ is substituted as $\alpha.s$. The prediction accuracy is obtained as

$$\Delta^2(C) = \sigma^2 \sum_{i \in C} (a_\xi^T x_i)^2 / s^2 = \alpha^2 \sum_{i \in C} (a_\xi^T x_i)^2. \tag{3.55}$$

We obtain the critical noise level α^* on the basis of the condition $\Delta(C) = 1$ as

$$\alpha_p^* = 1 / \sqrt{\sum_{i \in C} (a_\xi^T x_i)^2}. \tag{3.56}$$

Thus, the critical noise level α_p^* depends on both the volume and grouping of the data, and on the realization of the noise. However, this estimate does not coincide with the limiting noise stability of any criterion since, with increase in the noise level, the inductive algorithm chooses another simpler model, which can predict a noise-free signal more accurately. Even this is true with the identification and filtering problems. The analytical study of critical noise levels for identification (α_i^*) and filtering (α_f^*) can be developed.

The combinatorial algorithm is used to obtain the optimal model of complexity by sorting all possible polynomials from the complete basis according to a given external criterion or set of criteria for the given partition of sets. The degree of noise stability of the selection criterion is determined by gradually increasing the noise level and finding the critical value of α in each case.

Example 2. An experiment on estimation of the noise stability of various selection criteria is made with $y^0 = 10 - 0.1t^2$, $t = 1, 2, \cdots, 22$—and a normally distributed white noise with unit variance is obtained for 22 values. This is realized for the output variable y for different noise levels of α with percentage values of 3, 5, 10, 20, 40, 60, 80, 100, 130, 160, 200, 230, 260, 300, 330, 360, 400. Four variants of partitioning of data are used: (i) $N_A + N_B + N_C = 7 + 7 + 8$, (ii) $N_A + N_B + N_C = 8 + 8 + 6$, (iii) $N_A + N_B + N_C = 9 + 9 + 4$, and (iv) $N_A + N_B + N_C = 4 + 4 + 3$ (in all the cases, the points are chosen successively).

The reference function considered here is the third-degree polynomial in t. Combinatorial algorithm is used for sorting all combinations of the structures (15 polynomials of varying structure). The following criteria are tested for their noise stability.

Regularity

$$\Delta^2(B) = \Delta^2(B/A) = \sum_{i \in B} \frac{(\hat{y} - y)_i^2}{(y - \bar{y})_i^2}, \tag{3.57}$$

where $\Delta^2(B/A)$ denotes the model calculated on the set B using the coefficients obtained on A.

Minimum bias

$$\eta_{bs}^2 = \sum_{i \in W} \frac{(\hat{y}^A - \hat{y}^B)_i^2}{(y - \bar{y})_i^2}, \tag{3.58}$$

where the estimates $\hat{y}^A$ and $\hat{y}^B$ correspond to the same model structure but with coefficients calculated on sets A and B, respectively.

Symmetric regularity

$$\begin{aligned} d^2 &= \Delta^2(A/B) + \Delta^2(B/A) \\ &= \|y_A - X_A\hat{a}_B\|^2 + \|y_B - X_B\hat{a}_A\|^2, \end{aligned} \tag{3.59}$$

where parts A and B are used equally.

Here is another form of symmetric criterion:

$$\begin{aligned} S^2 &= \Delta^2(W/A) + \Delta^2(W/B) \\ &= \|y_W - X_W\hat{a}_A\|^2 + \|y_W - X_W\hat{a}_B\|^2, \end{aligned} \tag{3.60}$$

which is an overall estimate on W for the same structure, but with coefficients estimated on different sequences (just as in the criteria η_{bs} and d).

The combined type: ("minimum bias plus prediction")

The noise immunity can be increased significantly by using the following criterion.

$$c3^2 = \eta_{bs}^2 + \Delta^2(C), \tag{3.61}$$

where η_{bs} is one of the realizations of the minimum bias criteria and $\Delta(C)$ is the prediction criterion that computes the sum of square errors using the set C. This criterion requires that a model be unbiased and is also the best predictive method.

A common difficulty with direct application of the combined criteria is the incommensurability of their input quantities. They evaluate different characteristics of the model, such as minimum bias and regularization or extrapolation. Therefore, using them requires choosing and applying weights for each problem.

$$c^2 = \gamma\eta_{bs}^2 + (1-\gamma)k^2, \tag{3.62}$$

where k^2 indicates a stabilizing term of the form $\Delta^2(C)$ or d^2. To obviate selection of weights, one uses a normalized form as

$$c^2 = \frac{\eta_{bs}^2}{\eta_{max}^2} + \frac{k^2}{k_{max}^2} = \bar{\eta}_{bs}^2 + \bar{k}^2, \tag{3.63}$$

where $\bar{\eta}_{bs}$ and $\bar{k}$ are the normalized values, and η_{max} and k_{max} are the maximum values of the criteria of all the models being compared.

All the criteria given above can be used individually as a single criterion choice; at the same time, the combined criteria can be used as two criterion choices. One can also use a stepwise choice; first choose F number of models with the minimum bias criterion, then choose the best model among them using the prediction criterion.

Noise stability of single-criterion selection

The combined criterion $c3$ with its normalized form exhibits the lowest noise stability. Individual criteria operate efficiently with successive application. The regularity criterion is

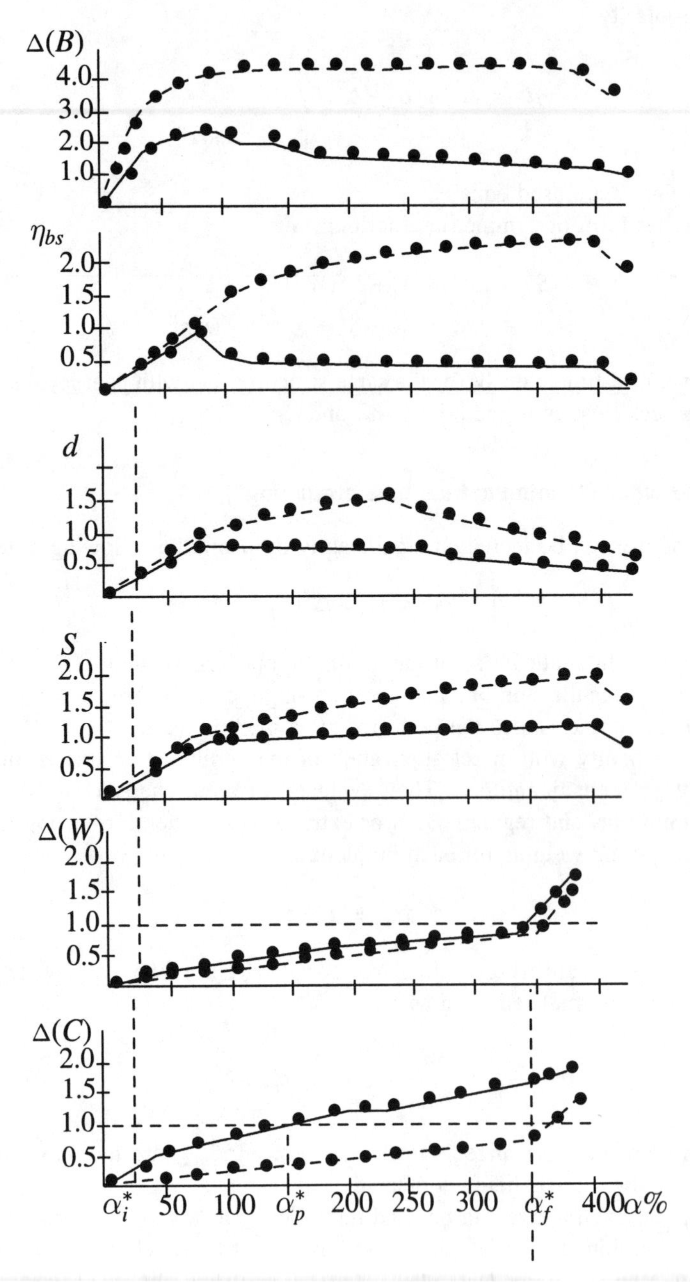

Figure 3.6. Relationship between selection criteria and percentage of the noise level α; the solid line is for models chosen with respect to the minimum of the criterion and the dashed line is for the model $y = 10 - 0.1t^2$

Table 3.2. Values of α_i^* for different selection criteria

Criterion	Different partitions of data			
	4+4+3	7+7+8	8+8+6	9+9+4
$\Delta(B)$	20	360	0	40
η_{bs}	20	60	20	80
$c3$	0	10	10	0
d	20	80	20	80
S	20	130	20	80

the most sensitive criterion to the partitioning of the data; care must be taken in using this criterion for noisy data.

The symmetric criteria ($\eta_{bs}, d,$ and S) proved to be stable with respect to the partitioning of the data; they virtually have the same noise stability in the case of individual application.

The results of determining the limiting noise stability α_i^* (identification case) for which the original model structure was still acurately reproduced is shown in Table 3.2. Figure 3.6 shows the values of the criteria for the structures obtained on the division $N_A + N_B + N_C =$ 8+8+6. The solid curve shows the optimal structures based on the minimum of the criterion and the dashed curve shows the actual structure $y = 10 - 0.1t^2$. The limiting noise levels α_f^* and α_p^* (filtering and prediction cases correspondingly) are considerably higher than the level of the structure for α_i^*.

Noise stability of two-criterion selection

Two-criterion selection is a widely used device in inductive self-organization modeling. Here we use external criteria of a different nature (for example, η_{bs} and $\Delta(B)$) that are related to different parts of data sets (for example, η_{bs} and $\Delta(C)$). There are two types of two-criterion analysis—one is in the form of convolution and another is successive in nature. Sometimes the former may lead to difficulties because of normalization of the criteria. It often turns out that η_{max} exceeds $\Delta(C)$ or d by higher magnitudes so that the bias becomes insignificant and the model is incorrectly chosen by the second criterion. This difficulty is avoided by the successive use of the criteria. The first criterion is used to select F number of models—the best one is chosen using the second criterion. The basic results of successive application of different combinations of two criteria based on the above example are discussed below.

(i) The combination $\Delta(B) \rightarrow \Delta(C)$. The noise stability increases to $\alpha_i^* = 60\%$ (for the criterion $\Delta(B)$, it is 0%).

(ii) The combination $\eta_{bs} \rightarrow \Delta(C)$. The noise stability is very significant; the correct structure of the model is better to a level of $\alpha_i^* = 260\%$ (just for η_{bs} separately, it is 20%).

(iii) The combination $d \rightarrow \Delta(C)$ *and* $S \rightarrow \Delta(C)$. They yield the same results as the preceding pair of criteria.

The use of two-criterion selection of models also increases the level of noise stability in predicting and filtering problems; in case of the combination $\eta_{bs} \rightarrow \Delta(C)$, the noise stability of filtering α_f^* is preserved at the level of 360%, and the noise stability of prediction α_p^* increases from 130% to 400%.

Usually it is impossible to determine the noise stability levels α^* for actual problems on the basis of noisy data because the information regarding the exact structure of the model and the characteristics of the noise is unknown. However, it can be controlled in the course of calculations. The values of the errors $\Delta(A + B)$ and $\Delta(C)$ are noticeably correlated with the ideal estimates of $R(A + B)$ and $R(C)$ (new notations). The difference between $\Delta(A + B)$ and $R(A + B)$ is the denominator term $\sum_{i \in W}(y_i - \bar{y})^2$, that represents the signal variance, similarly between $\Delta(C)$ and $R(C)$. In almost all cases, $\Delta(A + B) > R(A + B)$, and $\Delta(C) > R(C)$. This makes it possible to determine the prediction and filtering satisfactory with the additional conditions $\Delta(C) < 1$ and $\Delta(A + B) < 1$.

There are three ways of testing the operability of an inductive algorithm—with exact data, with a given noise distribution, and with noise distribution peculiar to the class of modeling objects. To verify the results, one has to perform a large number of tests in all the cases. Apparently, it is the only way to solve the problem of definitive verification of the algorithms, and this is done before it is recommended for practical use. Insufficient study of verification might lead to certain difficulties in the practical use of these algorithms. Nonetheless, they can be recommended for solving problems for which other algorithms are unsuited; for example, problems of detailed long-range predictions.

Correct choice of criteria and of the application sequence ensure achievement of qualitative noise stable modeling. Further increase in the noise stability is achieved through the use of multilevel schemes which are described in Chapter 2.

4 ASYMPTOTIC PROPERTIES OF CRITERIA

In this section we present the recent work of Stepashko [120] on asymptotic properties of external criteria for model selection.

The structural identification problem consists of choosing an estimate of the model $\overset{\circ}{y} = a_0^T \overset{\circ}{x}$, $\overset{\circ}{x} = (\overset{\circ}{x}_1, \cdots, \overset{\circ}{x}_{s^0})$, where $\overset{\circ}{y}$ and $\overset{\circ}{x}$ are the output and input vector actions, correspondingly, and a_0 is the actual parameter vector, which is optimal according to a specified combined criterion of minimum-bias plus regularity from a set of various models which contain all possible combinations of $m (\geq s^0)$ input variables. The best regression model is obtained according to the combined criterion from the $2^m - 1$ possible models under the conditions of noisy output $y_i = \overset{\circ}{y}_i + \sigma\xi_i$, $E[\xi_i] = 0$, $E[\xi_i\xi_j] = \sigma^2\delta_{ij}$, where E is the mathematical expectation, δ_{ij} is the kronecker delta, and σ^2 is an unknown finite variance. N_W is the number of points in the given data set.

A simplified version of this problem, which does not restrict the generality of the obtained conclusions about the asymptotic properties of the external criteria, is investigated here.

It is considered as searching an optimal model by successive inclusion of regressors x_s. The set of compared models consists of m various models of the form

$$\hat{y}_s = X_s\hat{a}_s, \quad s = 1, 2, \cdots, m, \tag{3.64}$$

where $X_s = (x_1, \cdots, x_s) = (X_{s-1}, x_s)$, s is the complexity of the model, and where the parameters are estimated by the least squares method as $\hat{a}_s = (X_s^T X_s)^{-1} X_s^T y$.

The structural identification problem is reduced to the determination of the optimal complexity of the model as

$$s^* = \arg \min_{s=\overline{1,m}} c2(y, \hat{y}_s), \tag{3.65}$$

where $c2$ is the combined criterion evaluated by using the actual and estimated values of the output. The whole data set W is partitioned into two subsets A and B such that

$$X = \begin{pmatrix} X_A \\ X_B \end{pmatrix}, \; y = \begin{pmatrix} y_A \\ y_B \end{pmatrix}, \quad \text{rank }\{X_A\} = \text{ rank }\{X_B\} = m, \quad N = A \cup B. \tag{3.66}$$

The ideal (theoretical) criterion of minimum variance of the forecast J (and of its estimate, the combined criterion) are examined. Two variants of the theoretical criterion that are averaged over the number of points in the sample for which they are calculated, are given below:

$$J(s, N) = \frac{1}{N} E\| \overset{\circ}{y} - X_s \hat{a}_s \|^2, \tag{3.67}$$

$$J_B(s, N_A, N_B) = \frac{1}{N_B} E\| \overset{\circ}{y}_B - X_{Bs} \hat{a}_{As} \|^2, \tag{3.68}$$

and the external criteria; regularity and minimum-bias

$$\Delta_B(s, N_A, N_B) = \frac{1}{N_B} \| y_B - X_{Bs} \hat{a}_{As} \|^2, \tag{3.69}$$

$$\eta_{bs}(s, N) = \frac{1}{N} \| X_s \hat{a}_{As} - X_s \hat{a}_{Bs} \|^2. \tag{3.70}$$

An optimal smoothing model corresponds to the solution of the problem with respect to the minimum of $J(s, N)$, and an optimal forecasting model corresponds to the minimum of $J_B(s, N_A, N_B)$. The external criteria Δ_B and η_{bs} are their estimates. To investigate the behavior of the theoretical as well as external criteria as $N \to \infty$, it is assumed that the matrix X satisfies the strong regularity condition;

$$\lim_{N \to \infty} \frac{1}{N} X_N^T X_N = \bar{H}, \tag{3.71}$$

where H is a nonsingular finite $m \times m$ matrix.

The characteristic results of the solution of the structural problem according to the given criteria for $\sigma^2 = \text{ var}, N = \text{ const}$, and $\sigma^2 = \text{ const}, N \to \infty$ are compared below for the adopted assumptions.

4.1 Noise immunity of modeling on a finite sample

When solving the above structural identification problem, one has to estimate the parameters for each set of regressors as $s = 1, 2, \cdots, m$. This can be done conveniently by the recursive algorithm presented in the preceding chapter for constructing the partial models of gradually increasing complexity, beginning with a single argument ("method of bordering"). (Refer to the section on "Recursive scheme for faster combinatorial sorting" in chapter 2.)

For quick reference, we briefly give the algorithm here. $X^T X$ and $X^T y$ are denoted as H and g, correspondingly, and H_s, g_s, and $\hat{a}_s$ are represented in the form of

$$H_s = \begin{bmatrix} H_{s-1} & h_s \\ h_s^T & \vartheta_s \end{bmatrix}, \; g_s = \begin{bmatrix} g_{s-1} \\ \gamma_s \end{bmatrix}, \; \hat{a}_s = \begin{bmatrix} \hat{a}_{s-1}^* \\ \hat{\alpha}_s \end{bmatrix}, \tag{3.72}$$

where $h_s = X_{s-1}^T x_s$, $\vartheta_s = x_s^T x_s$, and $\gamma_s = x_s^T y$.

The following recursive algorithm is valid for the calculation of H_s^{-1} and $\hat{\alpha}_s$;

$$\beta_s = 1/(\vartheta_s - h_s^T c_s), \; c_s = H_{s-1}^{-1} h_s, \; H_0^{-1} \Delta{=}0, \tag{3.73}$$

$$H_s^{-1} = \begin{bmatrix} H_{s-1} + \beta_s c_s c_s^T & | & -\beta_s c_s \\ ------- & | & ---- \\ -\beta_s c_s^T & | & \beta_s \end{bmatrix}, \tag{3.74}$$

$$\hat{a}_s = \begin{bmatrix} \hat{a}_{s-1} - \hat{\alpha}_s c_s \\ ------- \\ \beta_s(\gamma_s - g_{s-1}^T c_s) \end{bmatrix} = \begin{bmatrix} \hat{a}_{s-1}^* \\ \hat{\alpha}_s \end{bmatrix}, \quad \hat{a}_0 \Delta{=}0. \tag{3.75}$$

This algorithm can be used directly for the criterion $J(s,N)$, while for $J_B(s,N_A,N_B)$ and Δ_B, it is applied using the subset A (using the index A). For η_{bs}, the quantities are computed on both the subsets A and B.

Properties of the criteria $J(s,N)$ and $J_B(s,N_A,N_B)$

These criteria are reduced to the form;

$$J(s,N) = J^0(s) + J^*(s) = \frac{1}{N}\| \overset{\circ}{y} - X_s a_s\|^2 + \frac{\sigma^2}{N} s, \tag{3.76}$$

$$J_B(s,N_A,N_B) = J_B^0(s) + J_B^*(s) = \frac{1}{N}\| \overset{\circ}{y}_B - X_{Bs} a_{As}\|^2 + \frac{\sigma^2}{N_B} \text{ tr } (H_{As}^{-1} H_{Bs}). \tag{3.77}$$

The parameters a_s and a_{As} are estimated either by using the least-squares technique or by using the above recursive algorithm and substituting $\overset{\circ}{y}$ for y; i.e., $a_s = E[\hat{a}_s]$, $a_{As} = E[\hat{a}_{As}]$. Here, $J^0(s)$ and $J_B^0(s)$ characterize the structural bias, while $J^*(s)$ and $J_B^*(s)$ reflect the effect of noise. Obviously, $a_{s^0} = a_{As^0} = a_0$, so that $J^0(s^0) = J_B^0(s^0) = 0$.

Let us examine $J(s)$; one can obtain

$$J^0(s) = J^0(s-1) - \frac{1}{N}\alpha_s^2/\beta_s = J^0(s-1) - \frac{1}{N}\beta_s(\gamma_s - g_{s-1}^T c_s)^2, \tag{3.78}$$

where $\beta_s = x_s^T(I - X_{s-1}(X_{s-1}^T X_{s-1})^{-1} X_{s-1})x_s$ is positive and equal to the ratio of the determinants of the matrices H_{s-1} and H_s. Thus, $J^0(s)$ is a monotonically decreasing function of the complexity s so that in view of $J^*(s) = J^0(s)$ the function $J(s,n)$ for $\sigma^2 > 0$ always has a single minimum at the point $s^* \leq s^0$. As σ^2 increases, the complexity s^* decreases. A simpler model becomes J-optimal. This property is named "noise immunity"; i.e., the error in reconstructing the noise-free vector $\overset{\circ}{y}$ decreases due to the simplification of the model. This means that the model of s^0 loses its J-optimality for the variance $\sigma^2 > \sigma_{cr}^2(s^0) = \alpha_{s^0}^2/\beta_{s^0}$. In general, for arbitrary complexity s, the transition from $s^* = s$ to $s^* = s-1$ occurs for $\sigma^2 > \sigma_{cr}^2(s)$, where

$$\sigma_{cr}^2 = N(J^0(s-1) - J^0(s)) = \alpha_s^2/\beta_s. \tag{3.79}$$

α_s is the coefficient of the sth regressor and 'cr' indicates the critical value.

Let us examine $J_B(s)$; one can obtain $J_B^0(s)$ as

$$\begin{aligned} J_B^0(s) &= J_B^0(s-1) - 2\alpha_{As}(\overset{\circ}{y}_B - X_{B,s-1}a_{A,s-1})^T(x_{Bs} - X_{B,s-1}c_{As}) \\ &\quad + \alpha_{As}^2\|x_{Bs} - X_{B,s-1}c_{As}\|^2. \end{aligned} \tag{3.80}$$

Here, one cannot guarantee that the increment will be negative for every s (except when the regressors are orthogonal), so that in the general case the decrease from $J_B^0(1)$ to $J_B^0(s^0) = 0$

is not monotonic. The monotonicity of the dependence on s is preserved for the noise component $J_B^*(s)$. This fact is proven below on the basis of the recursive algorithm, with $h_{Bs} = X_{B,s-1}^T x_{Bs}$ and $\vartheta_{Bs} = x_{Bs}^T x_{Bs}$ taken into account.

$$J_B^*(s) = \frac{\sigma^2}{N_B} \text{ tr } (H_{As}^{-1} H_{Bs}) = J_B^*(s-1) + \frac{\sigma^2}{N_B} \beta_{As} (h_{As}^T H_{A,s-1}^{-1} H_{B,s-1} H_{A,s-1}^{-1} h_{As} -$$
$$-2h_{As}^{-1} H_{A,s-1} h_{Bs} + \vartheta_{Bs} = J_B^*(s-1) + \frac{\sigma^2}{N_B} \beta_{As} \| x_{Bs} - X_{B,s-1} H_{A,s-1}^{-1} h_{As} \|^2 \quad (3.81)$$

The increment change in this equation is estimated, having determined the extremum of the vector argument $\varphi(z) = z^T H_{B,s-1} z - 2z^T h_{Bs} + \vartheta_{Bs}$, where $z \triangle = H_{A,s-1}^T h_{As}$. If we differentiate φ with respect to z and equate it to zero, we obtain $z_0 = H_{B,s-1}^{-1} h_{Bs}$. Since $H_{B,s-1}$ is a positive-definite matrix, $\varphi(z)$ for $z = z_0$ has a positive minimum; $\varphi(z_0) = \vartheta_{Bs} - h_{Bs}^T H_{B,s-1}^{-1} h_{Bs} = 1/\beta_{Bs}$. Thus, the minimal increment change of the trace in the equation of $J_B(s, N_A, N_B)$ equals β_{As}/β_{Bs}, and is attained when the relation

$$H_{A,s-1}^{-1} h_{As} = H_{B,s-1}^{-1} h_{Bs} \quad (3.82)$$

is satisfied. In particular, this relation is satisfied when $H_{Bs} = \lambda^2 H_{As}$ or $X_B = \lambda X_A$, where λ is an arbitrary constant. Here $\text{tr}(H_{As}^{-1} H_{Bs}) = \text{tr } (H_{A,s-1}^{-1} H_{B,s-1}) + \lambda^2$, so that the rate of growth of the trace is proportional to the value of λ^2 as s is to one; i.e., even if the relation is satisfied, the examined increment may be arbitrary. Thus, as the component $J_B^0(s)$ in the equation of $J_B(s, N_A, N_B)$ decreases, and $J_B^*(s)$ increases monotonically as the complexity of the model s increases, the minimum of $J_B(s, N_A, N_B)$ is possible only for $s^* \leq s^0$. Qualitatively the behavior of the criteria $J(s, N)$, $J_B(s, N_A, N_B)$ is the same. Moreover, for a model of complexity s^0, one can establish the threshold of the J-optimality loss. For this, it is considered that $\overset{\circ}{y}_B = X_{Bs^0} a_0 = X_{B,s-1} a_{s^0-1}^0 + x_{Bs^0} \alpha_0$, where $a_0 = (a_{s^0-1}^{0^T} \alpha_0)^T$. Furthermore, according to the recursive relation, $a_{s^0-1}^0 = a_{A,s^0-1} - \alpha_{As^0} c_{As^0}$, $\alpha_{As^0} \equiv \alpha_0$, and, consequently,

$$\overset{\circ}{y}_B - X_{B,s^0-1} a_{A,s^0-1} = \alpha_{As^0} (X_{Bs^0} - X_{B,s^0-1} c_{As^0}),$$
$$J_B^0(s^0) = J_B^0(s^0 - 1) - \alpha_{As^0} \| X_{Bs^0} - X_{B,s^0-1} c_{As^0} \|^2. \quad (3.83)$$

From the conditions $J_B(s^0) = J_B(s^0 - 1)$, $J_B^0(s^0) = 0$, considering the equations for $J_B^0(s^0)$ and $J_B^*(s)$, we obtain

$$\sigma_{cr}^2(s^0) = \alpha_{As^0}^2 / \beta_{As^0}. \quad (3.84)$$

Thus, the condition for losing the J-optimality for a model of actual complexity s^0 (with an unbiased structure) turns out to be completely identical in problems of search for optimal smoothing and prediction models. This property is determined solely by the properties of subset A. This result can also be obtained by using different transformations. It is noted that $\sigma_{cr}^2(s^0)$ does not depend on the number of points N_B but depends implicitly on N_A.

Properties of the external criteria

The mathematical expectations of the regularity ($\Delta_B(s, N_A, N_B)$) and minimum-bias ($\eta_{bs}(s, N)$) criteria are equal to

$$\bar{\Delta}_B(s) = \frac{1}{N_B} \| \overset{\circ}{y}_B - X_{Bs} a_{As} \|^2 + \frac{\sigma^2}{N_B} (N_B + \text{ tr } (H_{As}^{-1} H_{Bs})), \quad (3.85)$$

$$\bar{\eta}_{bs}(s) = \frac{1}{N}\|X_s a_{As} - X_s a_{Bs}\|^2 + \frac{\sigma^2}{N}(2s + \text{ tr } (H_{As}^{-1}H_{Bs} + H_{Bs}^{-1}H_{Bs})). \tag{3.86}$$

Comparison of the equation Δ_B with its expectation $\bar{\Delta}_B$ yields

$$\bar{\Delta}_B(s) = J_B(s, N_A, N_B) + \sigma^2. \tag{3.87}$$

This means that the minimum of the regularity criterion Δ_B gives an unbiased J-optimal model, since the minima of $J_B(s)$ and $\bar{\Delta}_B(s)$ always correspond to the same optimal complexity s^*. Hence, the regularity criterion has the necessary property of noise immunity and other properties of the criterion $J_B(s, N_A, N_B)$; for example, the actual structure is optimal for $\sigma^2 < \sigma_{cr}^2(s^0)$.

The criterion $\bar{\eta}_{bs}(s)$ was worked out in detail in the work [118]; it was shown that, if the condition $(\tilde{X}_A^T\tilde{X}_A)^{-1}\tilde{X}_A^T \overset{\circ}{X}_A \neq (\tilde{X}_B^T\tilde{X}_B)^{-1}\tilde{X}_B^T \overset{\circ}{X}_B$ is satisfied, then it has a single global minimum. Biased values of its model structures decrease from $\eta_{bs}(1)$ to $\eta_{bs}(s^0) = 0$ (possibly nonmonotonically) while the noise component increases monotonically. Consequently, the minimum-bias criterion $\eta_{bs}(s)$ has the noise immune property.

4.2 Asymptotic properties of the external criteria

As $N = N_A \cup N_B$, one has to examine the case of $N \to \infty$ as well as its variants: $N_A \to \infty$, $N_B \to \infty$, and $N_A, N_B \to \infty$. In addition to the assumption that $\lim_{N\to\infty} \frac{1}{N}X_N^T X_N = \bar{H}$, let us assume that the matrices X_A and X_B are regular and are formed independently.

$$\lim_{N_A\to\infty} \frac{1}{N_A} X_{AN}^T X_{AN} = \bar{H}_A, \quad \lim_{N_B\to\infty} \frac{1}{N_B} X_{BN}^T X_{BN} = \bar{H}_B, \tag{3.88}$$

where $\bar{H}_A$, $\bar{H}_B$ are finite nonsingular matrices. The limits in the above equations exist for individual element of the matrices and for each of their blocks. Thus,

$$\lim_{N\to\infty} \frac{1}{N} x_{iN}^T x_{jN} = \bar{h}_{ij}, \; i,j = 1, 2, \cdots, m,$$

$$\lim_{N\to\infty} \frac{1}{N} X_{sN}^T X_{sN} = \bar{H}_s = \begin{bmatrix} \bar{h}_{11} & \cdots & \bar{h}_{1s} \\ & \cdots & \\ \bar{h}_{s1} & \cdots & \bar{h}_{ss} \end{bmatrix}, \tag{3.89}$$

where $\bar{h}_{ij}$ and $\bar{H}_s$ are the finite numbers and matrices, respectively.

Taking the actual model ($\overset{\circ}{y} = a_0^T \overset{\circ}{x}$, $\overset{\circ}{x} = (\overset{\circ}{x}_1, \cdots, \overset{\circ}{x}_{s^0})^T$) of the object into account, one can write the following relation in matrix form as

$$\overset{\circ}{y}_N = \overset{\circ}{X}_N a_0 = \overset{\circ}{X}_N a_0 + \tilde{X}_N \emptyset \Delta = X_N a^*, \tag{3.90}$$

where $\emptyset$ is the zero or empty vector of dimension $(m - s^0)$, and $a^* = (a_0^T, \emptyset^T)^T$ is the finite vector of the actual parameters.

The assumption of limiting transition as $N \to \infty$ implies the existence of the finite limits as

$$\lim_{N\to\infty} \frac{1}{N} \overset{\circ}{y}_N^T \overset{\circ}{y}_N = a^{*T} \bar{H} a^* \Delta = \bar{\mu}, \tag{3.91}$$

$$\lim_{N\to\infty} X_N^T \overset{\circ}{y}_N = \bar{H} a^* \Delta = \bar{g}. \tag{3.92}$$

Similarly, the existence of $\bar{\mu}_A, \bar{\mu}_B, \bar{g}_A,$ and $\bar{g}_B$ for other assumptions is $N_A \to \infty$ and $N_B \to \infty$.

In asymptotic problems of control theory, the condition of mean-square summation (integrability) of functions is usually a common assumption. From the well-known relation between the elements of the normal matrix $X^T X$: $x_i^T x_j < (x_i^T x_i + x_j^T x_j)/2$, the mean square summation of observations of all the individually taken regressors is written as

$$\lim_{N\to\infty} \frac{1}{N} x_{Ns}^T x_{Ns} = \lim_{N\to\infty} \frac{1}{N} \sum_{i=1}^{N} x_{is}^2 = \bar{\vartheta}_s, \quad s = 1, 2, \cdots, m, \tag{3.93}$$

based on the convergent sequences for all $\frac{1}{N} x_{Ni}^T x_{Nj}$, $i \neq j$; $i, j = 1, 2, \cdots, m$.

Properties of the criteria $J(s, N)$ and $J_B(s, N_A, N_B)$

The structural components of these criteria are represented in the form

$$J^0(s, N) = \frac{1}{N} \left(\overset{\circ}{y}{}_N^T \overset{\circ}{y}_N - a_{sN}^T X_{sN} \overset{\circ}{y}_N \right), \tag{3.94}$$

$$J_B^0(s, N_A, N_B) = \frac{1}{N_B} (\overset{\circ}{y}{}_{N_B}^T \overset{\circ}{y}_{N_B} - 2a_{sN_A}^T X_{sN_B}^T \overset{\circ}{y}_{N_B} + a_{sN_A}^T X_{sN_B}^T X_{sN_B} a_{sN_A}). \tag{3.95}$$

The convergence of the parameters in these equations is established on the basis of the limiting transitions given as

$$\lim_{N\to\infty} a_{sN} = \bar{H}_s^{-1} \bar{g}_s \triangle = \bar{a}_s, \quad \lim_{N_A\to\infty} a_{sN_A} = \bar{H}_{As}^{-1} \bar{g}_{As} \triangle = \bar{a}_{As}. \tag{3.96}$$

Taking $J^*(s, N) = \sigma^2 s/N$ into account, we obtain

$$\tilde{J}(s) = \lim_{N\to\infty} J(s, N) = \lim_{N\to\infty} J^0(s, N) = \bar{\mu} - \bar{g}_s^T \bar{H}_s^{-1} \bar{g}_s. \tag{3.97}$$

It is obvious that for $s = 1, 2, \cdots, s^0$, $\tilde{J}(s)$ decreases monotonically, while for $s \geq s^0$ the quantity $a_s = a^*$, so that $J(s) = 0$. Thus, if the quantity $J(s, N)$ for $\sigma^2 > 0$, and $N > \infty$ has a minimum of $s^* \leq s^0$, then there is a compromise between its structural $J^0(s, N)$ and noise $J^*(s, N)$ components. As $N \to \infty$, the component $J^*(s, N)$ disappears (an increase in the amount of information removes the uncertainty) and the minimum of $\tilde{J}(s)$ corresponds to the actual unbiased model structure.

In the case of $J_B(s, N_A, N_B) = J_B^0(s, N_A, N_B) + J_B^*(s, N_A, N_B)$, one has to consider the limiting transitions for $N_A \to \infty$ and $N_B \to \infty$. Finite values are obtained for the structural component $J_B^0(s, N_A, N_B)$ by taking into account its convergence property of parameters.

$$\tilde{J}_B^0(s, N_B) = \lim_{N_A\to\infty} J_B^0(s, N_A, N_B) = \frac{1}{N_B} (\mu_B - 2\bar{a}_{As}^T g_{Bs} + \bar{a}_{As}^T H_{Bs} \bar{a}_{As}), \tag{3.98}$$

$$J_B^0(s, N_A) = \lim_{N_B\to\infty} J_B^0(s, N_A, N_B) = \bar{\mu}_B - 2a_{sN_A}^T \bar{g}_{Bs} + a_{sN_A}^T \bar{H}_{Bs} a_{sN_A}, \tag{3.99}$$

where $\mu_B = \overset{\circ}{y}{}_{N_B}^T \overset{\circ}{y}_{N_B}$, $\bar{\mu}_B = \lim_{N_B\to\infty} \mu_B$. Convergence of the noise component $J_B^*(s, N_A, N_B)$ is determined by the asymptotics of the trace $\mathrm{tr}(H_{As}^{-1} H_{Bs})$ in the equation given for $J_B(s, N_A, N_B)$.

$$\begin{aligned} \tilde{J}_B^*(s, N_B) &= \lim_{N_A\to\infty} J_B^*(s, N_A, N_B) \\ &= \frac{\sigma^2}{N_B} \lim_{N_A\to\infty} \frac{1}{N_A} \mathrm{tr} \left(\frac{1}{N_A} H_{As} \right)^{-1} H_{Bs} = 0, \end{aligned} \tag{3.100}$$

$$J_B^*(s, N_A) = \lim_{N_B \to \infty} J_B^*(s, N_A, N_B) = \sigma^2 \text{ tr } (H_{N_A s}^{-1} \bar{H}_{Bs}) < \infty. \tag{3.101}$$

Thus, as $N_A \to \infty$, the noise component disappears and the properties of $\tilde{J}_B(s, N_B)$ become analogous to the properties of $\tilde{J}(s)$. However, as $N_B \to \infty$, the uncertainty caused by the parameter estimates a_{As} from a finite sample is not removed, and the criterion $\tilde{J}_B(s, N_A) = \tilde{J}_B^0(s, N_A) + \tilde{J}_B^*(s, N_A)$, where N_A is finite has the same properties as the equation given for $J_B(s, N_A, N_B)$. This means that if the parameters of the model obtained on a finite sample A and the model is applied on a infinite sample B, then the minimal variance of the forecast, in general, is achieved by a model according to noise immunity (J-optimal), rather than to an unbiased model which depends on σ^2.

If $N_A \to \infty$ and $N_B \to \infty$, we obtain from the above equations

$$\tilde{J}_B^*(s) = \lim_{N_B \to \infty} \lim_{N_A \to \infty} J_B^*(s, N_A, N_B) = 0$$

$$\begin{aligned}\tilde{J}_B(s) = \tilde{J}_B^0(s) &= \lim_{N_B \to \infty} \lim_{N_A \to \infty} J_B^0(s, N_A, N_B) \\ &= \bar{\mu}_B - 2\bar{a}_{As}\bar{g}_{sB} + \bar{a}_{As}\bar{H}_{Bs}\bar{a}_{As} < \infty,\end{aligned} \tag{3.102}$$

where $\tilde{J}_B(s|s \geq s^0) = 0$, as it is for $\tilde{J}_B(s, N_B)$.

This follows that the criteria $\tilde{J}(s)$ for $N \to \infty$, $\tilde{J}(s, N_B)$ for $N_A \to \infty$, and $\tilde{J}_B(s)$ for $N_A, N_B \to \infty$ converge for any s. Their minima equal to zero which corresponds to the actual model; $s^* = s^0$. This result is because of the consistency of the least squares estimates of the parameters of unbiased structures $s \geq s^0$ and the convergence of these estimates for the biased structures $s < s^0$. This established regularity of the asymptotic behavior of the criteria $J(s, N)$ and $J_B(s, N_A, N_B)$ is called "consistency property." As the sample length increases, the actual model which corresponds to the minimum or zero value of the criterion, becomes the limit of the optimal smoothing and forecasting model. Because there is no appearance of the expression concerning σ^2, the indicated property is valid for any variance. This means that the critical values of the variances $\sigma_{cr}^2(s)$ and $\sigma_{cr}^2(s^0)$ (the expressions given above) should approach infinity as $N \to \infty$ and $N_A \to \infty$, accordingly. In view of the established convergence, the concerned parameters turn out to be finite: $\bar{\alpha}_s = \lim_{N \to \infty} \alpha_{sN} < \infty$, as well as $\bar{\alpha}_{As^0} < \infty$. At the same time, for any s, the given equation for β in the recursive algorithm is obtained as

$$\begin{aligned}\bar{\beta}_s &= \lim_{N \to \infty} \beta_{sN} \\ &= \lim_{N \to \infty} \left[1/ \left\{ N \left(\frac{1}{N}\vartheta_{sN} - \frac{1}{N}h_{sN}^T \left(\frac{1}{N}H_{s-1,N} \right)^{-1} \frac{1}{N}h_{sN} \right) \right\} \right] = 0.\end{aligned} \tag{3.103}$$

Analogously, the relation $\beta_{As^0} = 0$ can be established by virtue of the limiting transitions that proves the assertion that $\sigma_{cr}^2(s, N) \to \infty$ for $\sigma_{cr}^2(s^0, N_A) \to \infty$ and $N_A \to \infty$.

It is obvious that any estimates of the criteria $J(s, N)$ and $J_B(s, N_A, N_B)$ used in practice must have the "consistency property."

Properties of the external criteria

The convergence of the regularity criterion $\bar{\Delta}_B(s) = \bar{\Delta}_B(s, N_A, N_B)$ for the cases $N_A \to \infty$, $N_B \to \infty$ and $N_A, N_B \to \infty$, and any s follows from the relation $\bar{\Delta}_B(s) = J_B(s, N_A, N_B) + \sigma^2$. The first and third cases are of interest with regard to "consistency property." One can obtain any s by taking into account the obtained finite values on $\tilde{J}_B^0(s, N_B)$ and $\tilde{J}_B^0(s)$.

$$\tilde{\Delta}_B(s, N_B) = \lim_{N_A \to \infty} \bar{\Delta}_B(s, N_A, N_B) = \tilde{J}_B^0(s, N_B) + \sigma^2, \tag{3.104}$$

$$\tilde{\Delta}_B(s) = \lim_{N_B \to \infty} \lim_{N_A \to \infty} \bar{\Delta}_B(s, N_A, N_B) = \tilde{J}_B^0(s) + \sigma^2. \tag{3.105}$$

For the limiting transitions considered, taking into account the properties of the quantities $\tilde{J}_B^0(s, N_B), \tilde{J}_B^0(s)$, the minimum of the criterion $\bar{\Delta}_B(s, N_A, N_B)$ is

$$\min \tilde{\Delta}_B(s, N_B) = \min \tilde{\Delta}_B(s) = \sigma^2. \tag{3.106}$$

Thus, the minimum of the mathematical expectation of the criterion $\Delta_B(s, N_A, N_B)$ is an asymptotically unbiased estimate with the unknown noise variance and corresponds in the limit to the actual model, which has the "consistency property."

The asymptotic properties of the consistency criterion $\eta_{bs}(s) = \eta_{bs}(s, N_A, N_B)$ are to be determined by performing on the established relation $\bar{\eta}_{bs}(s)$ a double limiting transition such as $N_A \to \infty$ and $N_B \to \infty$. It is convenient to adopt the commonly applied condition $N_A = N_B$; then $N = 2N_A = 2N_B$. First, the deterministic component is considered and represented as

$$\eta_{bs}^0(s, N_A, N_B) = \frac{1}{N}(a_{As} - a_{Bs})^T H_s (a_{As} - a_{Bs}). \tag{3.107}$$

Taking the limits, we have

$$\tilde{\eta}_{bs}^0(s) = \lim_{N_A \to \infty} \lim_{N_B \to \infty} \eta_{bs}^0(s, N_A, N_B) = (\bar{a}_{As} - \bar{a}_{Bs})^T \bar{H}_s (\bar{a}_{As} - \bar{a}_{Bs}) < \infty, \tag{3.108}$$

where $\tilde{\eta}_{bs}^0(s|s \geq s^0) = 0$.

Second, the noise component (from the equation $\bar{\eta}_{bs}(s)$) is considered for performing the limiting transformations. This is analogous to the relation we got for $\tilde{J}_B^*(s, N_B)$;

$$\tilde{\eta}_{bs}^*(s) = \lim_{N_B \to \infty} \lim_{N_A \to \infty} \eta_{bs}^*(s, N_A, N_B) = 0. \tag{3.109}$$

This means that the mathematical expectation of the minimum-bias criterion also converges for any s and has the "consistency property." Moreover, this criterion can be viewed as the asymptotically unbiased estimate of the values of $\tilde{J}(s)$. Hence, using the minimum-bias criterion is preferred in searching for the optimal smoothing model, while it is better to use the regularity criterion in the search for an optimal forecasting model. The regularity and minimum-bias criteria, in addition to the noise immunity, also has the "consistency property" that permits the applicability of the inductive algorithms for complex problem-solving by using small as well as large samples of data observations. The reader can also refer to works on analogous results of Dyshin [13] [14] and Aksenova [2] for further study.

4.3 Calculation of locus of the minima

This section describes a procedure for calculating the locus of the minima (LM) for ideal criteria [121]. This is important because extrapolation of LM allows one to detect a true signal from the noisy data.

In the course of a numerical study of simulating properties of noise immunity, the following computational experiment is considered. An actual model of an object is given by $\overset{\circ}{y} = a_0^T \overset{\circ}{x}$, where input vector $\overset{\circ}{x} = (\overset{\circ}{x}_1, \overset{\circ}{x}_2, \cdots, \overset{\circ}{x}_{s^0})$. Based on this the noisy observations of the true output at N points are calculated; $y_i = \overset{\circ}{y}_i + \sigma\xi_i$, where σ is an arbitrarily selected variance of the noise and ξ_i are known realizations of uncorrelated noise with $E[\xi_i] = 0$, $E[\xi_i^2] = 1$. In the experiment, the number of points N, the realization of ξ, and the variance σ may vary. Moreover, it is assumed that there is an extended vector of input variables $x = (\overset{\circ}{x}, \tilde{x}^T)$ with the dimensionality of $m \geq s^0$. Models of different combinations of m input variables are compared; $\hat{y} = \hat{a}^T x$. This corresponds to an application of the combinatorial algorithm for modeling the actual signal $\overset{\circ}{y} = (\overset{\circ}{y}_1, \overset{\circ}{y}_2, \cdots, \overset{\circ}{y}_N)^T$ by comparing all the models of the above form or in the matrix notation

$$\hat{y}_s = X_s \hat{a}_s, \quad s = \overline{1, 2^m - 1}, \tag{3.110}$$

where parameters $\hat{a}$ are the least-squares estimates

$$\hat{a}_s = (X_s^T X_s)^{-1} X_s^T y, \tag{3.111}$$

calculated using the noisy output vector.

The aim of this computational experiment is to compare the efficiency of various criteria for selecting models in relation to the ideal criterion

$$\Delta_N(s) = \| \overset{\circ}{y} - \hat{y}_s \|^2 = \| \overset{\circ}{y} - X_s \hat{a}_s \|^2, \tag{3.112}$$

that gives a measure of precision in recovering the actual signal $\overset{\circ}{y}$ by means of the model $\hat{y}_s$ obtained using the noisy data for each s. By varying s one can obtain optimal value s^0 (optimal structure) and the corresponding minimum value of the criterion $\Delta_N(s^0)$ for different ξ and σ. It is convenient to pose the above problem according to the complexity of the models as per their number of input variables. Evidently, there are C_m^1 models of complexity for $s = 1$, C_m^2 models of complexity for $s = 2$, and so on to $C_m^m = 1$ models of complexity m. The minimum value of $\Delta_N(s)$ is determined for each s. It will then constitute the characteristic $\Delta_N(s)$ and $\Delta_N(s) \equiv \Delta_N(s^0)$, so that the optimal value s^0 corresponds to a model with the minimal variance.

Let us assume that the values of $\Delta_N(s)$ are obtained for $s = 1, 2, \cdots, m$ by successive inclusion of regressors x_s and that the properties of the functional J as the mathematical expectation of the $\Delta_N(s)$ is

$$J(s) = E \| \overset{\circ}{y} - X_s \hat{a}_s \|^2. \tag{3.113}$$

It is shown before that

$$J(s) = J^0(s) + J^*(s) = E \| \overset{\circ}{y} - X_s a_s \|^2 + \sigma^2 s, \tag{3.114}$$

where $a_s = (X_s^T X_s)^{-1} X_s \overset{\circ}{y} \equiv E(\hat{a}_s)$.

$J^0(s)$ is a monotonically decreasing function with complexity s, where $J(s|s \geq s^0) = 0$; $J(s)$ has a unique minimum for certain optimal complexity $s^* \leq s^0$. This minimum shifts to the left as σ increases (refer to Figure 5.3). $\Delta_N(s)$ possesses the same properties, as shown before (for regularity criterion).

$$\Delta_N(s) = \Delta_N^0(s) + \Delta_N^*(s) = J^0(s) + \sigma^2 \hat{a}_{\xi s} X_s^T X_s \hat{a}_{\xi s}, \tag{3.115}$$

where $\hat{a}_{\xi s} = (X_s^T X_s)^{-1} X_s^T \xi$, and $\Delta_N^*(s)$ is a monotonically increasing function.

Optimal intervals of each model can be calculated by using systematic algorithms.

The notion of the "locus of the minima" of a criterion is defined as a function $J_{min}(s)$ or $\Delta_{min}(s)$ whose value corresponds to the critical value $\sigma_{cr}(s)$ for which the model with complexity of $s - 1$ becomes optimal instead of the model with s.

Algorithm for calculating the LM of $J_{min}(s)$

Let us assume that the regressors are included in the model in order of the correlation coefficients between the regressors and the actual output.

$$r_{x_s \overset{\circ}{y}} = \frac{\sum_{j=1}^{N}(x_{js} - \bar{x}_s)(\overset{\circ}{y}_j - \bar{\overset{\circ}{y}})}{\left(\sum_{j=1}^{N}(x_{js} - \bar{x}_s)^2\right)^{\frac{1}{2}} \left(\sum_{j=1}^{N}(\overset{\circ}{y}_j - \bar{\overset{\circ}{y}})^2\right)^{\frac{1}{2}}}, \quad s = \overline{1, m}, \tag{3.116}$$

and the regressors are ranked in decreasing order of the correlation coefficients.
The algorithm consists of the following steps:

1. calculating the matrices $X^T X$, $X^T \overset{\circ}{y}$ for the Gaussian normal equations for the full model;
2. computing the least-squares estimates of the parameters a_s using the equation $X_s^T X_s a_s = X_s^T \overset{\circ}{y}$;
3. determining the quadratic error of the estimator $\overset{\circ}{y}$ using the least-squares method as

$$J^0(s) = \overset{\circ}{y}^T \overset{\circ}{y} - a_s^T X_s^T \overset{\circ}{y}; \tag{3.117}$$

4. calculating the estimate of a_{s+1};
5. determining the $J^0(s + 1)$;
6. calculating the decrease in error due to the inclusion of regressors by one:

$$\delta_{s+1}^2 = J^0(s) - J^0(s + 1), \ J^0(s^0) = 0; \tag{3.118}$$

7. determining the ordinate of LM of the ideal criterion at point s: $J_{min}(s) = J^0(s) + s\delta_{s+1}^2$;
8. increasing the complexity by one unit. Return to step 4.

Note that these calculations can also be conducted by using the recursive algorithm.

Algorithm for calculating the LM of $\Delta_{min}(s)$ for an individual realization of the noise vector

As in the above algorithm, the regressors are assumed to be ranked in decreasing order according to their correlation coefficients.
The algorithm consists of the following steps:

1. calculating the matrices $X^T X$, $X^T y$, $X^T \xi$;
2. determining the estimator a_s and the errors of this estimator $\hat{a}_{\xi s}$ due to the presence of noise. Here a_s is the solution of the equation $X_s^T X_s a_s = X_s^T \overset{\circ}{y}$, and $\hat{a}_{\xi s}$ is the solution of the equation $X_s^T X_s \hat{a}_{\xi s} = X_s^T \xi$;
3. calculating $J^0(s)$ using the formula $J^0(s) = \overset{\circ}{y}^T \overset{\circ}{y} - a_s^T X_s^T \overset{\circ}{y}$ as well as the quantity

$$\lambda_s^2 = \hat{\xi}_s^T \hat{\xi}_s = \hat{a}_{\xi s}^T X_s^T X_s \hat{a}_{\xi s}, \tag{3.119}$$

as it is given in the equation $\Delta_N(s) = \Delta_N^0(s) + \Delta_N^*(s) = J^0(s) + \sigma^2 \hat{a}_{\xi s}^T X_s^T X_s \hat{a}_{\xi s}$;
4. determining the estimators of a_{s+1}, and $\hat{a}_{\xi, s+1}^T$;

5. calculating the quantity $J^0(s+1)$ and $\lambda^2_{s+1} = \hat{\xi}^T_{s+1}\hat{\xi}_{s+1} = \hat{a}^T_{\xi,s+1}X^T_{s+1}X_{s+1}\hat{a}_{\xi,s+1}$;
6. calculating the increments of δ^2_{s+1} and the amount of increment of the random component λ of the criterion $\Delta_N(s)$ with complexity:

$$\delta^2_{\xi_{s+1}} = \lambda^2_{s+1} - \lambda^2_s; \tag{3.120}$$

7. obtaining the ordinate of the LM of the criterion $\Delta_N(s)$ as

$$\Delta_{min}(s) = J^0(s) + \delta^2_{s+1}\lambda^2_s/\delta^2_{\xi_{s+1}}; \tag{3.121}$$

8. increasing the complexity by one unit. Return to step 4.

Note: The above algorithms describe the calculating LM for two forms of an ideal criterion. Extrapolation of LM allows one to detect the true signal from the noisy data [45]. To develop an algorithm for calculating LM of the minimum-bias criterion, certain conditions are imposed on the subsets A and B. The criterion is represented in the form of a difference of LM of two ideal non-quadratic criteria as

$$\eta_{bs} = |\, \|\hat{y}^A - \mathring{y}\| - \|\hat{y}^B - \mathring{y}\| \,| \rightarrow \min. \tag{3.122}$$

In the same way, one can also eliminate $\mathring{y}$ for special data samples. If all these are possible, then the inductive learning algorithms can be replaced by analytical calculation of LM for number of criteria. This leads to additional investigation.

5 BALANCE CRITERION OF PREDICTIONS

The criterion of balance-of-variables is the first of several kinds developed as a balance criterion. It is the simplest criterion to use to find a definite relationship (a physical law) of several variables of the process being simultaneously predicted. This has opened the basis for long-range predictions using the ring of 'direct' and 'inverse' functions and is similar to the balance-of-variables criterion [117].

The balance criterion is designed to choose models of optimal complexity with respect to several interrelated variables being modeled. This occupies an important place among the external criteria because of its nature as a system criterion and because it is used in two-level algorithms. It is still in its basic form in the multilevel modeling of different practical problems. Let us give a general form of the criterion. Later, we should delve into the nature of change in position of the minimum with increase in noise intensity.

First, we give the balance criterion in a set of interrelated variables to be modeled. Let us assume that some connections are known or established between the variables at every instant of modeling; for example,

$$\phi_k = f(y_1, y_2, \cdots, y_L); \quad k \in W \tag{3.123}$$

is a known connection, where $y_1, y_2, \cdots, y_L$ are the interrelated variables which are independently identified. The balance criterion is written as

$$B^2_H = \sum_{k \in C} \left[\hat{\phi}_k - f(\hat{y}_{jk})\right]^2, \tag{3.124}$$

where $\hat{\phi}$ and $\hat{y}$ are the predicted values of ϕ and y. The established connection is a constraint that all the functions $y_{jk}, j = 1, 2, \cdots, L$ are assumed to satisfy both in the interpolation region $k = 1, 2, \cdots, N_W$, and in the prediction region $k = 1, 2, \cdots, N_C$.

The balance criterion B_H is intended to reflect either nonlinear or linear connection between the variables. The nonlinear balance connection is known in the form of the ring of differences of the "direct" and "inverse" functions. The linear relationship among the variables being modeled can be established as

$$\phi_k = \sum_{j=1}^{L} \beta_{jk} y_{jk}; \quad k = 1, 2, \cdots, N, \tag{3.125}$$

where β are the balance coefficients which are determined from the experimental data. This enables us to generate a linear balance criterion of the form

$$B^2_{(lin)} = \sum_{k \in G} [\hat{\phi}_k - \sum_{j=1}^{L} \beta_{jk} \hat{y}_{jk}]^2, \tag{3.126}$$

where G is the set that belongs to an arbitrary part or prediction part of initial measurements.

The linear type of the balance criterion is widely used in inductive learning algorithms. They are often based on a precisely known relationship; for example, the change in the population of a city is always to the population increment minus its decrement during a certain period; total biomass of a plant is always equal to the sum of the biomasses of the parts above and below the surface. In these examples, the balance coefficients are unity.

Second, given here is the balance criterion using the relationship of moving or sliding average as a variable and its elements. This can be used successfully in algorithms for separately predicting the chosen time functions defined from the series data $y_k, k = 1, 2, \cdots, N$. The balance connection is

$$\bar{y}_k = \frac{1}{L} \sum_{j=-\frac{1}{2}(L-1)}^{+\frac{1}{2}(L-1)} y_{k+j}; \quad k = 1, 2, \cdots, N_W. \tag{3.127}$$

The relationship holds between the measured and averaged values of length L. The balance criterion is written as

$$B^2 = \sum_{k \in C} [\hat{\bar{y}}_k - \frac{1}{L} \sum_{j=-\frac{1}{2}(L-1)}^{+\frac{1}{2}(L-1)} \hat{y}_{k+j}]^2, \tag{3.128}$$

which is based on the predictions of the $\bar{y}$ and $y_1, y_2, \cdots, y_L$ of the process.

The moving averages $\bar{y}_k, k = 1, 2, \cdots, N - L$ from the initial data $y_j, j = 1, 2, \cdots, N$ can be obtained by using the matrix $\sigma\, [N - L \times N]$ form [130] as

$$\bar{y} = \frac{1}{L} \sigma_{N-L,N} y, \tag{3.129}$$

where $y^T = (y_1 y_2 \cdots y_N)$; $\bar{y}^T = (\bar{y}_1 \bar{y}_2 \cdots \bar{y}_{N-L})$; and

$$\sigma_{N-L,N} = \begin{bmatrix} 11 \ldots 10 \ldots 00 \ldots 00 \\ 01 \ldots 11 \ldots 00 \ldots 00 \\ \ldots \quad \ldots \quad \ldots \\ 00 \ldots 00 \ldots 11 \ldots 10 \\ 00 \ldots 00 \ldots 01 \ldots 11 \end{bmatrix}.$$

The matrix used here has consecutive 1's of length L in each row. In adjacent rows the set of 1's is shifted one place to the right. The averaged vector $\bar{y}$ is of length $N - L$. The above criterion can be formally put in the form as below, though y_{ik} are not individual variables to be modeled.

$$B_{av}^2 = \sum_{k \in G} (\hat{Y}_k - \sum_{i=1}^{L} \beta_i \hat{y}_{ik})^2, \tag{3.130}$$

where $Y_k = \bar{y}_k$; $\beta_i = 1/L$; and $y_{ik} = y_{i-(k-(L-1)/2)}$.

The linear concept of the balance criterion is extended further as a balance-of-predictions criterion in modeling of time series data which is cyclic in nature. This is used in algorithms of two-level predictions, in which the connection between the predictions of artificial variables q_j, $j = 1, 2, \cdots, L$ is obtained from the time series data q_k, $k = 1, 2, \cdots$, by averaging on different time intervals; for example, season and year, month and year, and hour and day.

In general, we assume that a year contains L arbitrary intervals, the months. At the lower level of the algorithm one predicts the mean monthly values of the process and at the upper level, the mean yearly values. This means that we use two-dimensional time readout in months t and years T instead of actual one-dimensional time readout, in the usual manner of continuous mean monthly data. There is a unique pair of values (t, T); $t = 1, 2, \cdots, L$; $T = 1, 2, \cdots, N$, where L is number of months (twelve months), and N is number of years, for each observation corresponding to the original measurement. The average annual values Q_T and the monthly values $q_{t,T}$ are connected by the relationship called calender averaging.

$$Q_T = \frac{1}{L} \sum_{t=1}^{L} q_{t,T}. \tag{3.131}$$

The balance-of-predictions criterion has the form

$$B_{year}^2 = \sum_{T \in G} (\hat{Q}_T - \frac{1}{L} \sum_{t=1}^{L} \hat{q}_{t,T})^2. \tag{3.132}$$

This type of criterion is used in various applications; for example, predictions of river flows, air temperature [65], and the elements of the ecosystem of a lake [48].

The operation of calculating the mean annual values Q_T can be represented in the matrix form as

$$Q_T = \frac{1}{L} \sigma_{N,K} q, \tag{3.133}$$

where q is the vector of K elements, Q_T is the vector of N elements, and σ is the matrix of $[N \times K]$ as given below:

$$\sigma_{N,K} = \begin{bmatrix} 111 \ldots 100 \ldots 000 \ldots 000 \ldots 000 \ldots 000 \\ 000 \ldots 011 \ldots 100 \ldots 000 \ldots 000 \ldots 000 \\ 000 \ldots 000 \ldots 011 \ldots 100 \ldots 000 \ldots 000 \\ \ldots \quad \ldots \quad \ldots \quad \ldots \quad \ldots \\ 000 \ldots 000 \ldots 000 \ldots 000 \ldots 001 \ldots 111 \end{bmatrix}.$$

Each row of the matrix $\sigma_{N,K}$ contains L 1's and each column contains a single one; a fact that differentiates calender from the moving averages.

In this modeling, different monthly models are obtained with the consideration of both the other monthly values and the annual values; i.e., the delayed arguments in months as well as in years are considered in obtaining the monthly models. Therefore, the expression B_{year} has formal equivalence as in $B_{(lin)}$. Any balance criterion can be called a criterion of balance of predictions as long as the predictions of different variables are compared as in $B_{(lin)}$.

One can consider the general form of external balance criterion of linear type with the given balance coefficients β_j, $j = 1, 2, \cdots, L$ in vector notation as

$$\phi = (y_1|y_2|\cdots|y_L)\beta = Y\beta, \tag{3.134}$$

where ϕ is the N-dimensional vector; Y is the $[N \times L]$ matrix; and $\beta^T = (\beta_1, \beta_2, \cdots, \beta_L)$ is the vector of balance coefficients. The balance criterion is written as

$$B^2_{(lin)} = (\hat{\phi}_G - \hat{Y}_G\beta)^T(\hat{\phi}_G - \hat{Y}_G\beta) = \|\hat{\phi}_G - \hat{Y}_G\beta\|^2, \tag{3.135}$$

which is used on set G.

The important thing one has to note is that the balance criterion which is established among the variables ϕ and y_k, $k = 1, 2, \cdots, L$ indicates the linear dependence and has to be taken into account when using the balance criterion while process modeling.

5.1 Noise immunity of the balance criterion

Here we assume that the variable ϕ appearing in the balance criterion is physically measurable and that the balance coefficients are given (as a special case, $\beta_i = 1/L$).

Let us assume that the measurements of all the jointly modeled variables $\phi, y_1, y_2, \cdots, y_L$ are noisy.

$$\phi = \overset{\circ}{y}_0 + \xi_0; \quad y_i = \overset{\circ}{y}_i + \xi_i; \quad i = 1, 2, \cdots, L, \tag{3.136}$$

where $\overset{\circ}{y}_0, \overset{\circ}{y}_1, \cdots, \overset{\circ}{y}_L$ are the vectors of nonnoisy measurements, all the noise vectors ξ are independent of each other, and they normally have distributed independent components with mean zero and given variances.

$$\begin{aligned} &E[\xi_j] = 0, \; E[\xi_j\xi_j^T] = \sigma_j^2 I_N; \; j = 0, 1, \cdots, L, \\ &E[\xi_j^T\xi_i] = 0, \quad E[\xi_j\xi_i^T] = 0; \quad j \neq i; \; j, i = 0, 1, \cdots, L, \end{aligned} \tag{3.137}$$

where E is the mathematical expectation and I is the identity matrix. The exact models of the variables $\overset{\circ}{y}$ have the forms

$$\overset{\circ}{y}_j = \overset{\circ}{X}_j b_j^0; \quad j = 0, 1, 2, \cdots, L, \tag{3.138}$$

where $\overset{\circ}{X}_j$ are the $[N \times s_j^0]$ matrices of true independent arguments, b_j^0 are the $[s_j^0 \times 1]$ exact vectors of coefficients, and s_j^0 denote the complexities of the true models.

In self-organization modeling, one seeks for the optimal approximations to the true models from the noisy observations. The partial models are generated by sorting among the basis sets of $N \times m_j$ arguments X_j in which there are also, by assumption, true arguments $\overset{\circ}{X}_j$; that is, $m_j \geq s_j^0$, $j = 0, 1, \cdots, L$. Thus, in the sorting, one determines the coefficients in the partial models of differing complexity for each of the $L + 1$ variables in the conditional equations of the form

$$X_{j(s_j)} a_{j(s_j)} = y_j; \quad X_{j(s_j)} \; [N \times s_j], \; a_{j(s_j)} \; [s_j \times 1], \tag{3.139}$$

where s_j denotes the complexity of the partial model for the jth variable considering y_0 as ϕ for uniformity. All s_j vary independently, and the vectors of coefficients are determined by the least squares method using the noisy data.

$$\hat{a}_{j(s_j)} = [X^T_{j(s_j)} X_{j(s_j)}]^{-1} X^T_{j(s_j)} y_j. \tag{3.140}$$

For the existence of the inverse matrices, we assume that $N \geq \max(m_j, j = 0, 1, 2, \cdots, L)$ and that all the X_j have full rank.

The data sample (of length N) is not partitioned into number of sets because the balance criterion can be used on the interpolation region of the data. This means that all N points of the data are taken into consideration in all operations of the modeling.

The estimates for each variable being modeled can be written as

$$\begin{aligned}\hat{y}_{j(s_j)} &= X_{j(s_j)} \hat{a}_{j(s_j)} \\ &= X_{j(s_j)} [X^T_{j(s_j)} X_{j(s_j)}] X^T_{j(s_j)} y_j \\ &= P_{j(s_j)} y_j = P_{j(s_j)} (\overset{\circ}{y}_j + \xi_j); \; j = 0, 1, 2, \cdots, L,\end{aligned} \tag{3.141}$$

where $\hat{y}_{0(s_0)}$ and y_0 are considered as $\hat{\phi}_{(s_0)}$ and ϕ, respectively, for uniformity, and $P_{j(s_j)}$, $j = 0, 1, 2, \cdots, L$ are the projection matrices. The balance criterion can be obtained as

$$\begin{aligned}B^2_{(lin)} &= \|\hat{y}_{0(s_0)} - (\hat{y}_{1(s_1)} | \cdots | \hat{y}_{L(s_L)})\beta\|^2 \\ &= \|P_{0(s_0)} y_0 - (P_{1(s_1)} y_1 | \cdots | P_{L(s_L)} y_L)\beta\|^2 \\ &= \|[P_{0(s_0)} \overset{\circ}{y}_0 - (P_{1(s_1)} \overset{\circ}{y}_1 | \cdots | P_{L(s_L)} \overset{\circ}{y}_L)\beta] \\ &\quad + [P_{0(s_0)} \xi_0 - (P_{1(s_1)} \xi_1 | \cdots | P_{L(s_L)} \xi_L)\beta]\|^2.\end{aligned} \tag{3.142}$$

The objective of the balance criterion is to obtain consistent optimal predictive models for each of the $L + 1$ variables connected by the balance law. The criterion $B_{(lin)}$ is to be calculated for all variants of the partial models of varying complexity. The total amount of partial models is calculated as

$$p_B = \prod_{j=0}^{L} p_{m_j} = \prod_{j=0}^{L} (2^{m_j} - 1). \tag{3.143}$$

If $m_j = m$, then the complete sorting is proportional to 2^{mL}; obviously, in complex problems, for large m and L, complete sorting becomes impossible. We can tentatively assume that, as in the case of the combinatorial inductive approach, the complete sorting is efficient when $mL < 20$. For four seasons ($L = 4$), we can allow $m = 5$ arguments in each model of the seasons. For more complex problems, it is essential to apply proper way of sorting. Here let us assume that complete sorting is made. We shall seek the minimum of the balance criterion.

$$B^2_{(min)} = \min_{i=\overline{1,p_{m_j}};\; j=\overline{0,L}} B^2_{(lin)}[\hat{a}_{ij}(X_j, y_j)]. \tag{3.144}$$

The value of $B^2_{(min)}$ determines the set of models of optimal complexity s^*_j, $j = 0, 1, 2, \cdots, L$ for each of $L+1$ variables. Now let us see how the choice of models of optimal complexity changes with increase in the noise variances σ^2_j, $j = 0, 1, 2, \cdots, L$; i.e., let us investigate the noise immunity of the choice with respect to the balance criterion. This is analyzed in the mean value sense; i.e., with respect to its mathematical expectation.

Keeping in mind the noise properties, the fact that $P_j^T P_j = P_j^2 \equiv P_j$; $j = 0, 1, \cdots, L$, and applying the mathematical expectation to the above derived balance criterion

$$\begin{aligned}\bar{B}^2_{(lin)} = E[B^2_{(lin)}] &= \|P_{0_{(s_0)}} \overset{\circ}{y}_0 - (P_{1_{(s_1)}} \overset{\circ}{y}_1 \mid \cdots \mid P_{L_{(s_L)}} \overset{\circ}{y}_L)\beta\|^2 \\ &+ \sigma_0^2 s_0 + \sum_{i=1}^{L} \sigma_i^2 \beta_i^2 s_i = B^2_{\overset{\circ}{y}} + B^2_{\xi}. \end{aligned} \tag{3.145}$$

Thus, the expected value of the balance criterion $B^2_{(lin)}$ has two components: $B^2_{\overset{\circ}{y}}$, imbalance in the modeling of exact data, and B^2_{ξ} reflects the action of the noise.

First, let us look in greater detail at the component $B^2_{\overset{\circ}{y}}$ and then at the $B^2_{(lin)}$ as a whole.

$$\begin{aligned} B^2_{\overset{\circ}{y}} &= \|P_{0_{(s_0)}} \overset{\circ}{y}_0 - (P_{1_{(s_1)}} \overset{\circ}{y}_1 \mid \cdots \mid P_{L_{(s_L)}} \overset{\circ}{y}_L)\beta\|^2 \\ &= \|X_{0_{(s_0)}} \hat{b}_{0_{(s_0)}} - (X_{1_{(s_1)}} \hat{b}_{1_{(s_1)}} \mid \cdots \mid X_{L_{(s_L)}} \hat{b}_{L_{(s_L)}})\beta\|^2 \\ &= \|\hat{y}^{\circ}_{0_{(s_0)}} - (\hat{y}^{\circ}_{1_{(s_1)}} \mid \cdots \mid \hat{y}^{\circ}_{L_{(s_L)}})\beta\|^2 \\ &= \|\hat{y}^{\circ}_{0_{(s_0)}} - \hat{Y}^{\circ}\beta\|^2. \end{aligned} \tag{3.146}$$

Considered with the exact data, it is necessary to determine $B^2_{\overset{\circ}{y}} = 0$ and to check whether or not this corresponds to obtaining true models for which the balance relationship

$$\overset{\circ}{y}_0 = (\overset{\circ}{y}_1 \mid \overset{\circ}{y}_2 \mid \cdots \mid \overset{\circ}{y}_L)\beta = \overset{\circ}{Y}\beta; \quad \overset{\circ}{Y}\ [N \times L] \tag{3.147}$$

holds, and which is actually reflected in the exact initial data $\overset{\circ}{y}_0, \overset{\circ}{y}_1, \overset{\circ}{y}_2, \cdots, \overset{\circ}{y}_L$.

With increasing complexity of the models on the attainment of the true complexity s_j^0 (for each one of $L+1$ models), the coefficients are restored exactly to $\hat{b}_{j_{(s_j^0)}} \equiv \hat{b}_j^0$. Even with further increase in s_j, the coefficients $\hat{b}_{j_{(s_j|s_j \geq s_j^0)}} \equiv b_j^0$ do not change because the coefficients of the extra arguments are equal to zero. The models of all the variables are attained true and the value of the criterion $B_{\overset{\circ}{y}} = 0$. It is also possible, as it might turn out, that the criterion assumes the value $B_{\overset{\circ}{y}} = 0$ in the cases when the sorting among the different combinations of models for all the variables discloses partial models with coincidence structures such as $X_{0_{(s')}} = X_{1_{(s')}} = \cdots = X_{L_{(s')}} = X^*$. As $\overset{\circ}{y}_0 = \overset{\circ}{Y}\beta$, $P_{0_{(s')}} = P_{1_{(s')}} = \cdots = P_{L_{(s')}} = P^*$, $B_{\overset{\circ}{y}}$ becomes

$$B^2_{\overset{\circ}{y}} = \|P^*(\overset{\circ}{y}_0 - \overset{\circ}{Y}\beta)\|^2 \equiv 0 \tag{3.148}$$

for arbitrary values of the coefficients.

This property of the criterion is mentioned by Ihara [27] in his correspondence with the editor of the journal "avtomatika" (Soviet Journal of Automatic Control). Later, the idea arose as to nonuniqueness of the choice of models according to the balance criterion [36].

Theorem 1. Estimation of the coefficients of the prediction models with regard to the prediction balance criterion is an incorrect problem because it has an enormous set of solutions.

Proof: Considering β as a unit vector, the prediction balance criterion can be written as

$$B^2 = \|X_0 a_0 - (X_1 a_1| \cdots |X_L a_L)\|^2, \tag{3.149}$$

where $a_j, j = 0, 1, 2, \cdots, L$ are the vectors of coefficients in the models of averaged data and the detailed prediction models, correspondingly.

This is to be minimized with respect to the coefficients a_0 and a_j to obtain the optimal set of prediction equations. The system of normal equations in Gaussian form can be obtained as

$$\frac{\partial B^2}{\partial a_j} = 0; \ j = 0, 1, 2, \cdots, L, \tag{3.150}$$

Assuming that the structures of the models are already known,

$$\begin{aligned} 2X_0^T X_0 a_0 - 2X_0^T (X_1 a_1| \cdots |X_L a_L) &= 0, \\ -2X_j^T X_0 a_0 + 2X_j^T (X_1 a_1| \cdots |X_L a_L) &= 0. \end{aligned} \tag{3.151}$$

These matrix equations are linearly dependent. Each of the equations has an infinite set of solutions and the system of equations yields the trivial solution $\hat{a}_j = 0; \ j = 0, 1, 2, \cdots, L$.

Corollary 1. In modeling the exact data, it is necessary to obtain the value of $B^2_{\overset{\circ}{y}} = 0$, which is sufficient for structural identification of the true models ($\overset{\circ}{y}_j = \overset{\circ}{X}_j \ b_j^0; \ j = 0, 1, 2, \cdots, L$), if (i) the exact data of the variables $\overset{\circ}{y}_j; \ j = 0, 1, 2, \cdots, L$ satisfy the balance relationship ($\overset{\circ}{y}_0 = \overset{\circ}{Y} \beta$); (ii) the arguments of the matrices X_j contain all true arguments of the $\overset{\circ}{X}_j$, $j = 0, 1, 2, \cdots, L$; and (iii) the common basis of the arguments is nondegenerate; i.e., sorting does not reveal a complexity s' such that $X_{1_{(s')}} = \cdots = X_{L_{(s')}}$. These three conditions are neither excessively stringent nor idealized. The first two make the problem of modeling several variables connected by the linear balance relationship well posed and the third establishes the conditions for correct application of the balance criterion (uniqueness of choice of models), which can be ensured algorithmically.

Attainment of the value of the criterion ($B^2_{\overset{\circ}{y}} = 0$) is achieved with increase in the complexity, which is always monotonic. This can be explained by the calculation of $B^2_{\overset{\circ}{y}}$ on the same data as it is used to estimate the coefficients. This can be represented graphically as the dependence of the criterion on the complexity of the partial models of the different variables in a multidimensional space.

Let us look at the second part B^2_{ξ} of the balance criterion. The component of the criterion $\bar{B}^2_{(lin)}$ which reflects the influence of noise is

$$B^2_{\xi} = \sigma_0^2 s_0 + \sum_{j=1}^{L} \beta_j^2 \sigma_j^2 s_j. \tag{3.152}$$

This increases linearly with an increase in the complexity of the partial models $\hat{y}_{j(s_j)}; \ j = 0, 1, 2, \cdots, L$; i.e., it is the plane whose inclination in multidimensional space is determined by the noise variances $\sigma_j^2; \ j = 0, 1, 2, \cdots, L$. This inclination increases according to the increase in the noise variances.

Keeping in mind the properties of the component $B^2_{\overset{\circ}{y}}$, we conclude that (i) the criterion $B^2_{(lin)}$ being an $(L+1)$-dimensional function of the variables (in numbers) $s_j; \ j = 0, 1, 2, \cdots, L$

always has a unique minimum, (ii) this minimum is always in the hypercube $1 \leq s_j \leq s_j^0$; $j = 0, 1, 2, \cdots, L$ (i.e., the overly complex superfluous models are always weeded out), and (iii) with increase in the variances of the noises (at least in one of them), the minimum is displaced on the side of decrease in the complexity of the models (with respect to at least one of the variables).

These properties can be represented graphically. One can observe the decrease in the optimal complexity of the models, which is typical of external criteria, using a multi-dimensional surface whose sections (isolines) for different noise variances are σ_j^2; $j = 0, 1, 2, \cdots, L$.

Corollary 2. If the three conditions which are asserted in corollary 1 for $B^2_{\underset{y}{\circ}}$ also hold in the modeling of noisy variables y_j; $j = 0, 1, 2, \cdots, L$, then the following properties of the selected models in optimal complexity as per the balance criterion $\bar{B}^2_{(lin)}$ hold: (i) for arbitrary nonzero noise variances σ_j^2; $j = 0, 1, 2, \cdots, L$, the minimum of the criterion as a function of different complexities s_j; $j = 0, 1, 2, \cdots, L$ exists and is unique, (ii) the achieved minimum always lies in the bounded region $1 \leq s_j \leq s_j^0$, $j = 0, 1, 2, \cdots, L$, where s_j^0 is the complexity of the true models, and (iii) with increase in the variances of the noises σ_j^2; $j = 0, 1, 2, \cdots, L$ the minimum is displaced in the direction of decrease in the complexity of the optimal models.

Theorem 2. The problem of estimating the coefficients for the prediction models as per the balance criterion becomes correct (having a unique solution) if the quadratic stability criterion S is considered along with the balance criterion; i.e., by forming the combined criterion as

$$c5^2 = B^2 + S^2. \tag{3.153}$$

Proof: Let us consider the stability criterion as a stabilizing functional, the sum of the quadratic criteria giving the quality of the output vector error on each of the prediction levels.

$$S^2 = \sum_{j=0}^{L} \|y_j - X_j a_j\|^2. \tag{3.154}$$

The combined criterion $c5$ is the combination of "prediction balance plus stability criterion."

$$c5^2 = \|X_0 a_0 - (X_1 a_1 | \cdots | X_L a_L\|^2 + \sum_{j=0}^{L} \|y_j - X_j a_j\|^2. \tag{3.155}$$

Let us determine the estimates of the vectors a_j, $j = 0, 1, 2, \cdots, L$ by minimizing the combined criterion. We obtain the system of normal equations as

$$\frac{\partial c5^2}{\partial a_j} = 0; \; j = 0, 1, 2, \cdots, L. \tag{3.156}$$

Then, for the prediction model of the first level $\hat{y}_0$ or $\hat{\phi}$, we have:

$$\frac{\partial c5^2}{\partial a_0} = 2X_0^T X_0 a_0 - X_0^T (X_1 a_1 | \cdots | X_L a_L) - X_0^T y_0 = 0, \tag{3.157}$$

and from this,

$$\hat{a}_{0(c5)} = \frac{1}{2}(X_0^T X_0)^{-1} X_0^T [y_0 + (X_1 a_1 | \cdots | X_L a_L)]. \tag{3.158}$$

For the models of the second level $\hat{y}_j,\ j = 1, 2, \cdots, L$:

$$\frac{\partial c5^2}{\partial a_j} = 2(X_j^T X_1 a_1 | \cdots | X_j^T X_L a_L) - X_j^T X_0 a_0 - X_j^T (y_1 | \cdots | y_L) = 0; \quad j = 1, 2, \cdots, L \tag{3.159}$$

From this,

$$\hat{a}_{j(c5)} = \frac{1}{2}(X_j^T X_j)^{-1} X_j^T [(y_1 | \cdots | y_L) + X_0 a_0]; \ j = 1, 2, \cdots, L. \tag{3.160}$$

The matrices $X_j^T X_j,\ j = 0, 1, 2, \cdots, L$ are assumed to be nonsingular.

Solving the system of two matrix equations of $c5^2$ and $\hat{a}_0$, we obtain the estimates of the coefficients of the prediction models of both levels as

$$\begin{aligned} \hat{a}_{0(c5)} &= \frac{1}{3}(X_0^T X_0)^{-1} X_0^T [2y_0 + (y_1 | \cdots | y_L)] = \frac{2}{3}\hat{a}_0 + \frac{1}{3}(X_0^T X_0)^{-1} X_0^T (y_1 | \cdots | y_L) \\ \hat{a}_{j(c5)} &= \frac{1}{3}(X_j^T X_j)^{-1} X_j^T [2(y_1 | \cdots | y_L) + y_0] = \frac{2}{3}\hat{a}_j + \frac{1}{3}(X_j^T X_j)^{-1} X_j^T y_0; \\ & j = 1, 2, \cdots, L \end{aligned} \tag{3.161}$$

where $a_j,\ j = 0, 1, 2, \cdots, L$ are the estimates of the coefficients of respective models as per the least squares method. Thus, the goal of the regularization is achieved.

Corollary 3. On regularization of the problem of selection of structure for prediction models by the balance-of-prediction criterion with the help of stability criterion.

The problem of structure choice of prediction models by the balance criterion becomes correct (i.e., achieving a unique solution) if the quadratic stability criterion S^2 in the combined criterion $c5^2$ is used as regularizing the operator as

$$c5^2(s_0^*, s_j^*) = \min_{s_j \in \mathbf{m}_j;\, j=\overline{0,L}} [B^2(s_0^*, s_j^*) + S^2(s_0^*, s_j^*)], \tag{3.162}$$

where $\mathbf{m}$ denotes the set of arguments taking part in the predictive models of both levels, s_0^* and s_j^* are the notations for optimal structures.

The stability criterion in the above formulation makes it possible to reduce the region of solution of the problem of selecting structures by using the prediction balance criterion to a compact subset which leads to a unique solution of the problem.

Interpretation of the results in the case of B^2_{year}. It follows from the comparison of $B^2_{(lin)}$ and B^2_{year} that the number of variables is equal to, for example, the number of seasons ($L = 4$); the vector of connected variable is the vector of second-level variable $\phi = Q = (Q_1, Q_2, \cdots, Q_N)^T$. The remaining variables are the seasonal variables associated with the first level $y_j = q_t = (q_{t,1}, q_{t,2}, \cdots, q_{t,N})^T,\ t = 1, 2, \cdots, L$. All the balance coefficients are equal $\beta_i = 1/L,\ \beta = (1/L, \cdots, 1/L)^T$. However, for the results of the investigation of noise immunity of the criterion $B^2_{(lin)}$ to be applicable to the criterion B^2_{year} (or what amounts to the same thing,) to

$$B^2_{year} = \|\hat{Q} - (\hat{q}_1 | \cdots | \hat{q}_L)\beta\|^2, \tag{3.163}$$

it is necessary to show the validity of noise conditions specified before. The vectors Q and $q_t;\ t = 1, 2, \cdots, L$ are obtained from the measured time series data $q_k;\ k = 1, 2, \cdots, LN$. Let us suppose that a noise with the usual properties is imposed on these measurements,

$$q_k = \overset{\circ}{q}_k + \zeta_k; \ E[\zeta_k] = 0,\ E[\zeta_k^2] = \sigma^2,\ E[\zeta_i \zeta_j] = 0,\ i \neq j. \tag{3.164}$$

The components of each vector q_t are taken from q_k at a period of length L.

$$q_t = (q_{k=t}, q_{k=t+L}, \cdots, q_{k=t+(N-1)L})^T; \quad t = 1, 2, \cdots, L. \tag{3.165}$$

Therefore, for each q_t the noise vector ξ_t satisfies the original conditions.

$$q_t = \overset{\circ}{q}_t + \xi_t; \; E[\xi_t] = 0, \; E[\xi_t \xi_t^T] = \sigma^2 I_N, \; E[\xi_j^T \xi_i] = 0, \; j \neq i, \tag{3.166}$$

where, for all t vectors of the seasonal values, the noise variances are equal to σ^2 and are independent of t. Further, with reference to the mean annual values, we obtain

$$Q = \overset{\circ}{Q} + \xi_0, \; \overset{\circ}{Q} = (\overset{\circ}{q}_1 \mid \cdots \mid \overset{\circ}{q}_L)\beta,$$

$$\xi_0 = (\xi_1 | \cdots | \xi_L)\beta, \; E[\xi_0] = 0, \; E[\xi_0 \xi_0^T] = \frac{\sigma^2}{L} I_N; \tag{3.167}$$

i.e., the noise variance for the second-level mean annual variable is $1/L$ times the variance of the original noise $\sigma_0^2 = \sigma^2/L$ which increases the noise immunity of modeling at the second level. The components of the noise vector ξ_0 are also independent. This means that one of the conditions fails to be satisfied; i.e., the condition of independence of ξ_0 and all ξ_t; $t = 1, 2, \cdots, L$.

$$\begin{aligned} E[\xi_0^T \xi_t] &= \beta^T E[(\xi_1 | \cdots | \xi_L)^T \xi_t] \\ &= \frac{1}{L} E[\xi_t^T \xi_t] = \frac{\sigma^2}{L}. \end{aligned} \tag{3.168}$$

This reflects on the calculation of the mathematical expectation of the criterion B^2_{year}, which is obtained in somewhat different form

$$\begin{aligned} \bar{B}^2_{year} &= E[B^2_{year}] = \|\hat{Q}^\circ - (\hat{q}_1^\circ | \cdots | \hat{q}_L^\circ)\beta\|^2 \\ &+ \frac{\sigma^2}{L} s_0 + \frac{\sigma^2}{L} \sum_{i=1}^{L} s_i - 2\frac{\sigma^2}{L} \sum_{i=1}^{L} \mathrm{tr}\left(P^T_{0(s_0)} P_{i(s_i)}\right). \end{aligned} \tag{3.169}$$

The trace of the $(P_0^T P_i)$ can be written as

$$\mathrm{tr}(P_0^T P_i) = \mathrm{tr}[X_0 (X_0^T X_0)^{-1} X_0^T X_i (X_i^T X_i)^{-1} X_i^T],$$

which cannot be calculated in the general case.

It is difficult to determine how $\bar{B}^2_{year}$ behaves with complication of the models of both levels. It may fail to be unimodal, or its global minimum may not be displaced in the direction of simplification of models with increase in the noise variances.

The criterion $\bar{B}^2_{year}$ will have the same properties as $\bar{B}^2_{(lin)}$ when the matrices X_0 and X_i are orthogonal; i.e., $X_0^T X_i = 0$; $i = 1, 2, \cdots, L$. Then

$$\bar{B}^2_{year} = \|\hat{Q}^\circ - (\hat{q}_1^\circ | \cdots | \hat{q}_L^\circ)\beta\|^2 + \frac{\sigma^2}{L}\left(s_0 + \sum_{i=1}^{L} s_i\right). \tag{3.170}$$

Although this condition usually contradicts the condition of linear connection of variables of the two levels ($Q_T = \frac{1}{L}\sum_{t=1}^{L} q_{t,T}$), one can interpret it as the specific nature of the two-level modeling problem of a single variable given by its seasonal and annual values. The orthogonality condition $X_0^T X_i = 0$ will in fact hold when the seasonal models of the first level are not used at the second level. Here, the mean annual models are constructed independently of the seasonal models, and these are used in the selection of the best two-level models according to the balance criterion.

Corollary 4. The problem of selecting structures for predictive models by the prediction balance criterion becomes correct if a different (in nature and composition) set of arguments is used for constructing models of the two levels.

Indeed, in reconciliation of seasonal and annual predictions using the same source of measured data, the balance criterion is inefficient and leads to trivial results. On the other hand, if one uses a different set of arguments for constructing two-level models, then the balance criterion in the choice of structure becomes efficient.

In practice, such a case is ensured, for example, by constructing the seasonal models in the form of a system of L difference equations and the annual models in the form of a harmonical functions [48]. This corresponds to the condition of independence of the informational bases of the models of different levels [36], which is necessary for efficiency of algorithms for multilevel modeling.

6 CONVERGENCE OF ALGORITHMS

First we give the definition for canonical formulation of the external criteria [135] and then proofs for internal convergence of two multilayer algorithms; one is the original multilayer algorithm; the other is the algorithm with propagating errors [42], [79] [134], [136].

6.1 Canonical formulation

The canonical form of the external criteria is an analytical tool to investigate various properties of the criteria. This is not convenient from a practical standpoint for calculating the value of a criterion in cases involving large numbers of observations, but it can be used directly for model selection of a small number of observations.

Definition

The canonical form of the criterion is defined as the expression $y^T Dy$, where D is a symmetric strictly positive-semidefinite matrix—strictly in the sense that (a) $\forall y \neq \Theta,\ y^T Dy \geq 0$ and (b) $\exists y \neq \Theta,\ y^T Dy = 0$.

The matrix D is determined by the corresponding criterion and the partitioning of data.

Residual sum of squares

We give here the canonical form of residual sum of squares (RSS) used in the least-squares method. Suppose we have a system of conditional equations of the form $y = Xa$. The parameters a are estimated as

$$a = (X^T X)^{-1} X^T y. \tag{3.171}$$

The RSS is calculated as

$$\varepsilon^2 = (y - Xa)^T (y - Xa). \tag{3.172}$$

This can be written in the canonical form using the notation $P_{NN} = X(X^TX)^{-1}X^T$ as

$$\varepsilon^2 = y^T(I - P_{NN})y = y^T D_{LS} y, \tag{3.173}$$

where $D_{LS} \hat{=} (I - P_{NN})$ is a symmetric positive semidefinite matrix for $N > m$, I is the unit matrix, and N indicates the total number of data points.

Regularity criterion

This is given as

$$\Delta^2(B) = (y_B - \hat{y}_B)^T(y_B - \hat{y}_B), \tag{3.174}$$

where $\hat{y}_B = X_B(X_A^TX_A)^{-1}X_A^Ty_A$. Using the notation $P_{BA} \hat{=} X_B(X_A^TX_A)^{-1}X_A^T \hat{=} (p_{ij})$, the criterion can be written as

$$\Delta^2(B) = \sum_{i \in B}(y_{B_i} - \sum_{j \in A} p_{ij} y_{A_j})^2, \tag{3.175}$$

where A and B are the training and the testing sets correspondingly. By expanding this algebraically, we get

$$\Delta^2(B) = \sum_{i \in B} y_{B_i}^2 - 2\sum_{i \in B}\sum_{j \in A} y_{B_i} p_{ij} y_{A_j} + \sum_{i \in B}\sum_{j \in A}\sum_{k \in A} p_{ij} p_{ik} y_{A_j} y_{A_k}, \tag{3.176}$$

or the matrix form

$$\Delta^2(B) = \left(y_A | y_B\right) \left(\begin{array}{c|c} \sum_{i \in B} p_{ij} p_{ik} & (-p_{ij}) \\ --- & --- \\ (-p_{ij}) & I \end{array}\right) \left(\begin{array}{c} y_A \\ --- \\ y_B \end{array}\right)$$

$$= y^T \left(\begin{array}{c|c} P_{BA}^T P_{BA} & -P_{BA}^T \\ --- & --- \\ -P_{BA} & I \end{array}\right) y = y^T D_{reg} y. \tag{3.177}$$

This is the canonical form for the regularity criterion. The matrix D_{reg} depends on the sequencing of the training and testing sets—so does vector y.

Minimum bias criterion

This is given as

$$\eta_{bs}^2 = (\hat{y}^A - \hat{y}^B)^T(\hat{y}^A - \hat{y}^B)_W, \tag{3.178}$$

where W indicates that the criterion is computed on the set W; $\hat{y}_W^G = X_W(X_G^TX_G)^{-1}X_G^Ty_G$; G corresponds to either A or B and $W = A \cup B$.

Let us define the notations as

$$X_W(X_G^TX_G)^{-1}X_G^Ty_G \hat{=} P_{WG}y_G, \quad G = A \text{ or } B. \tag{3.179}$$

The criterion can be rewritten as

$$\eta_{bs}^2 = (P_{WA}y_A - P_{WB}y_B)^T(P_{WA}y_A - P_{WB}y_B). \tag{3.180}$$

The canonical form can be obtained as

$$\eta_{bs}^2 = y^T \left(\begin{array}{c|c} P_{WA}^T P_{WA} & -P_{WA}^T P_{WB} \\ --- & --- \\ -P_{WB}^T P_{WA} & P_{WB}^T P_{WB} \end{array}\right) y = y^T D_{bs} y. \tag{3.181}$$

This is the canonical form for the minimum bias criterion.

Analogously, one can obtain canonical forms for other criteria.

6.2 Internal convergence

Defining multilayer algorithm with propagating outputs

Let us assume that there are m input variables of x $(x_1, x_2, \cdots, x_m)$, y is the output variable, G_r is the set of q input variables at the rth layer $(z_1, z_2, \cdots, z_q)$, $(q \geq m)$, N is the number of initial data points. Mapping $\mathcal{R}$ takes place from layer r to the layer r+1; i.e., $\mathcal{R} : G_r \rightarrow G_{r+1}$.

First, the elements $z_k^{(r)}, k = 1, 2, \cdots, F$ are the column vectors of the matrix $z^{(r)}$ of the transformed experimental data. They are determined from the condition

$$z_k^{(r)} = P_k y, \tag{3.182}$$

where $P_k = (z_i^{(r-1)}|z_j^{(r-1)})[(z_i^{(r-1)}|z_j^{(r-1)})^T(z_i^{(r-1)}|z_j^{(r-1)})]^{-1}(z_i^{(r-1)}|z_j^{(r-1)})^T$, is the projection operator of the least-squares method; and y is the observation vector of output variable.

Second, the N-dimensional vector z is a partial description of the rth layer as its kth component is expressed by

$$z_{(k)}^{(r)} = g\left(z_{i(k)}^{(r-1)}, z_{j(k)}^{(r-1)}\right), \tag{3.183}$$

where i,j vary as per their representation from the $(r-1)$st layer. The partial polynomial in its simplest form is

$$g(z_{i(k)}, z_{j(k)}) = a_1 z_{i(k)} + a_2 z_{j(k)}; \quad i = 1, 2, \cdots, q-1; \; j = i+1, i+2, \cdots, q \tag{3.184}$$

where a_1 and a_2, as the arbitrary coefficients, assumes an iterative process.

Finally, from the set of elements $z_{(k)}^{(r)}$ of the following layer that is obtained, a subset $z^{(r)}$ is singled out according to an external criterion. The external criterion gives to these solutions qualitatively new properties that the modeler finds desirable.

Suppose the regularity criterion $\Delta^2(B)$ is considered as the external criterion that has the sum of squares of the deviations on the testing set B

$$y = \begin{pmatrix} y_A \\ --- \\ y_B \end{pmatrix}, \quad Z^{(r-1)} = \begin{pmatrix} Z_A^{(r-1)} \\ --- \\ Z_B^{(r-1)} \end{pmatrix}. \tag{3.185}$$

The algorithm stops when the criterion achieves the minimum in the layer r compared with the layer $r+1$ for a particular component; it is then said that it is converged; i.e.

$$y^T D_{reg}^{(r)} y \leq y^T D_{reg}^{(r+1)} y, \tag{3.186}$$

where $D_{reg}^{(r)}$ and $D_{reg}^{(r+1)}$ are the positive-semidefinite canonical matrices formed based on the components at r and $r+1$ layers, correspondingly.

Internal convergence is an especially important property of multilayer algorithms. If the external criterion becomes the internal criterion (i.e., the regularity criterion $\Delta^2(B)$ becomes the residual sum of squares (RSS) ε^2), the result of the algorithm must be equivalent to the result of multiple regression analysis, at least when the function of y is linear in variables and coefficients.

Here the internal convergence is considered (i) towards a solution and (ii) with respect to the structures.

Convergence to a solution. Suppose that stopping is not envisioned and the class of functions formed by superposition of the function g includes a function $h(x_{(k)}) = y_{(k)}$, $k = 1, 2, \cdots, N$ where $x_{(k)} = (x_{1(k)}, x_{2(k)}, \cdots, x_{m(k)})$, then the algorithm converges to a solution if the sequence of vectors $z_i^{(r)}$ has a limit as $r \rightarrow \infty$ and if this limit is y.

Convergence with respect to structures. Suppose that stopping is not envisioned and the class of functions formed by superposition of the function g includes a unique function $h(x)$ such that $h(x_{(k)}) = y_{(k)}$, $k = 1, 2, \cdots, N$, then the algorithm converges with respect to the structure if the sequence of functions $z_i^{(r)}(x)$ has a limit as $r \to \infty$ and if it is equal to $h(x)$. Unlike the above case, here it takes the measure of distance between the functions. In the class of linear polynomials, a natural measure for distance between two functions is the sum of squares of the distances between similar terms involved in them. The distance between two arbitrary functions is measuerd as the sum of the squares of distances between their values from the initial data. Based on this, the definitions of convergence to a solution and with respect to structure are equivalent.

Definition 1. An algorithm converges in a finite number of steps if, beginning with some layer, $z_l^{(r)}$ are equal to their limiting value.

Definition 2. There is effective convergence if the algorithm converges in a finite number of steps; i.e., the layer with which $z_l^{(r)}$ is the first one and equal to its limiting value; the next layer has the divergent characteristics.

Definition 3. It is referred to the convergence under the condition that $\Delta^2(B) = \text{RSS}$, where RSS is calculated on the initial data, as internal convergence.

The internal convergence to the solution and in structure is ensured by the following theorem.

Theorem 3. Suppose that y^* is the projection of the vector y on to the linear space $L(X)$, formed by the columns of the matrix X. Suppose the criterion is calculated on the set W. Then, F number of partial descriptions with the sequence of vectors $z_k^{(r)}$ converges to y^* as $r \to \infty$. If the X^TX is nonsingular, the model corresponding to the limiting vector coincides with the regression equation for y as a function of X.

Assume that the best model in optimal complexity is being sought, that means it is the case of $F = 1$.

Let us look at the numerical sequence of $\|y - z^{(r)}\|$, which can be shown as nonincreasing. In the multilayer algorithm with the propagating outputs, the vector $z^{(r+1)}$ is formed by

$$z^{(r+1)} = a_1 z_i^{(r)} + a_2 z_j^{(r)}, \tag{3.187}$$

where a_1 and a_2 are found by minimizing the quantity $\|y - a_1 z_i^{(r)} - a_2 z_j^{(r)}\|$. It follows that the vector $z^{(r+1)}$ is the projection y onto $L(z_i^{(r)}|z_j^{(r)})$; i.e., the linear hull of the vectors $z_i^{(r)}$ and $z_j^{(r)}$.

$$\|y - z^{(r+1)}\| \leq \|y - z^{(r)}\|, \quad r = 0, 1, \cdots \tag{3.188}$$

Thus, the sequence $\|y - z^{(r)}\|$ is nonincreasing and as a sequence of norms it is lower bounded. Therefore it has a limit that is denoted by ϱ.

Let us look at the sequence $\|z^{(r)}\|$. By the definition, $z_k^{(r)} \in L(X)$ for all r. Consequently, $z \in L(X)$. Further more, $(y - z)$ is orthogonal to $L(X)$; i.e., $(y - z)^T X = 0$. It follows from the above that $\|z^{(r+1)}\| \geq \|z^{(r)}\|$ and one can easily see that $\|z^{(r)}\| \leq \|y\|$. Thus, the sequence $\|z^{(r)}\|$ is nondecreasing and higher bounded. It has a limit, which is denoted by τ.

Let us look at the sequence $z^{(r)}$. The existence of the limits of the sequences $\|y - z^{(r)}\|$ and $\|z^{(r)}\|$ implies that with increasing r, the vectors $z^{(r)}$ become arbitrarily closer to the manifold defined by the system of equations

$$\begin{aligned} \|y - z\| &= \varrho \\ \|z\| &= \tau. \end{aligned} \tag{3.189}$$

It is shown that there exists a unique vector z^* belonging to this manifold, which is the limiting vector of the sequence $z^{(r)}$. It follows that

$$\begin{aligned} z^{(r+1)} &= (z_i^{(r)}|z_j^{(r)})\,[(z_i^{(r)}|z_j^{(r)})^T(z_i^{(r)}|z_j^{(r)})]^{-1}(z_i^{(r)}|z_j^{(r)})^T y \\ &= \frac{z_j^{(r)^T} z_j^{(r)} z_i^{(r)} z_i^{(r)^T} - z_i^{(r)^T} z_j^{(r)} z_i^{(r)} z_j^{(r)^T} - z_i^{(r)^T} z_j^{(r)} z_j^{(r)} z_i^{(r)^T} + z_i^{(r)^T} z_i^{(r)} z_j^{(r)} z_j^{(r)^T}}{z_i^{(r)^T} z_i^{(r)} z_j^{(r)^T} z_j^{(r)} - (z_i^{(r)^T} z_j^{(r)})^2} y \\ &= P_{ir} y, \end{aligned} \tag{3.190}$$

where P_{ir} denotes the corresponding projection matrix.

It follows from the convergence of the sequence $\|y - z^{(r)}\|$ that by choosing r suitably, the equation $\|y - P_{ir}y\| = \varrho$ can be satisfied to any desired closeness for all $i = 1, 2, \cdots, m$.

From the above, the following equation

$$\begin{aligned} (y^T y - \varrho^2)[z_i^{(r)^T} z_i^{(r)} z_j^{(r)^T} z_j^{(r)} - (z_i^{(r)^T} z_j^{(r)})^2] = y^T(& z_j^{(r)^T} z_j^{(r)} z_i^{(r)} z_i^{(r)^T} - z_i^{(r)^T} z_j^{(r)} z_i^{(r)} z_j^{(r)^T} \\ & - z_i^{(r)^T} z_j^{(r)} z_j^{(r)} z_i^{(r)^T} + z_i^{(r)^T} z_i^{(r)} z_j^{(r)} z_j^{(r)^T})y \end{aligned} \tag{3.191}$$

will be satisfied for any desired accuracy by noting that $z_i^{(r)^T} z_i^{(r)} z_j^{(r)^T} z_j^{(r)} - (z_i^{(r)^T} z_j^{(r)})^2 \neq 0$.

The unknowns $z_i^{(r)^T} z_j^{(r)}$ can be determined to an arbitrary degree of accuracy from the above equation because of its dependence on coefficients in terms of the unknowns. The solution can be found as

$$z_i^{(r)^T} z_j^{(r)} = z_i^{(r)^T} y, \quad i = 1, 2, \cdots, m \tag{3.192}$$

using the relationships $\varrho^2 + \tau^2 = y^T y$, and $y^T z^{(r)} = z^{(r)^T} z^{(r)}$. This is satisfied with an arbitrary accuracy as the quantities $\|y - z^{(r)}\|$ and $\|z^{(r)}\|$ tending to their limits ϱ and τ, respectively. This determines uniquely the limiting vector $z^* \in L(X)$. It can be written as

$$X^T(z^* - y) = 0. \tag{3.193}$$

Thus, z^* is the orthogonal projection of y on to $z(X)$ or, what amounts to the same thing, $z^* = y^*$.

Let us look at the case $F > 1$. It shows that the distances between the partial descriptions belonging to the same layer get arbitrarily smaller as $r \to \infty$; i.e., $\|z_k^{(r)} - z_l^{(r)}\|$, $k = 1, \cdots, F-1$, $l = k+1, k+2, \cdots, F$ gets arbitrarily small. Let us define $\|y - z_1^{(r)}\| = \varrho + \delta_r$, and it leads to

$$\|y - z_F^{(r+1)}\| \leq \|y - z_1^{(r)}\|. \tag{3.194}$$

We consider F partial descriptions of the form $a_1 z_i^{(r)} + a_2 z_j^{(r)}$ at the $(r+1)$st layer, for which the above inequality holds. Therefore,

$$\varrho + \delta_r \leq \|y - z_k^{(r)}\| \leq \varrho + \delta_{r-1}, \quad k = 1, 2, \cdots, F. \tag{3.195}$$

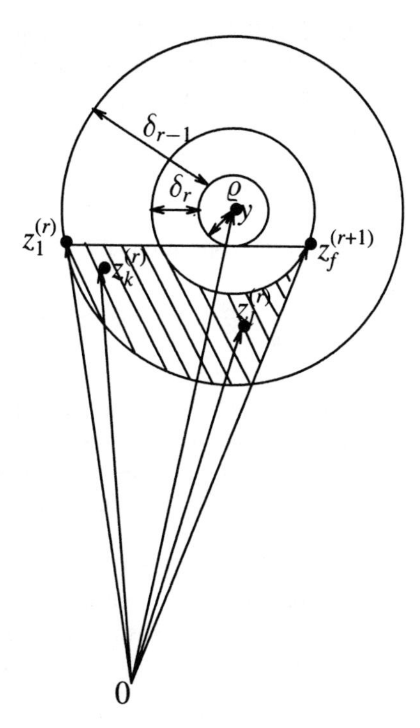

Figure 3.7. Geometrical interpretation of the sequences used in the internal convergence

Also, for arbitrary values of a_1 and a_2 we can have the relation $\|y - a_1 z_k^{(r)} - a_2 z_l^{(r)}\| \geq \varrho$. From the above inequalities, one can obtain the estimate of (Figure 3.7)

$$\|z_k^{(r)} - z_l^{(r)}\| \leq \sqrt{(\varrho + \delta_{r-1})^2 - \varrho^2} + \sqrt{(\varrho + \delta_r)^2 - \varrho^2}. \tag{3.196}$$

The right side quantity of the inequality can become arbitrarily small for a suitable r. This completes the proof for the internal convergence of the algorithm to the solution and in structure.

Defining multilayer algorithm with propagating errors

The function g has the form

$$g(x_i, x_j) = x_i + a x_j, \quad i = 1, 2, \cdots, F, \quad j = F+1, F+2, \cdots, Q, \tag{3.197}$$

where $Q = F + m$ and a is determined by the least-squares method using the set A or W.

Here the algorithm is described in its simplest way. In the first step the partial descriptions of the form

$$z_i = a x_i, \quad i = 1, 2, \cdots, m \tag{3.198}$$

of which the residual errors are computed as $\Delta^{(1)} z_{ij} = z_i - y$ and F best of the descriptions are chosen.

In the rth step ($r > 1$), the partial descriptions of the form

$$z_{ij}^{(r+1)} = z_i^{(r)} + a_j x_j, \quad i = 1, 2, \cdots, F, \qquad j = 1, 2, \cdots, m \tag{3.199}$$

of which the residuals are computed as $\Delta^{(r+1)}z_{ij} = z_{ij}^{(r+1)} - z_i^{(r)}$, and F best models are chosen.

The process continues until the value of the criterion decreases significantly. Suppose we are required to reproduce a dependence of the form $y = h(x) + \xi$. The approximation is achieved as

$$h(x) = g_1(x) + g_2(x) + \cdots, \tag{3.200}$$

where $g_i(x),\ i = 1, 2, \cdots$ correspond to the chosen equations at each step.

The internal convergence of the algorithm to the solution and in the structure is ensured by the following theorem.

Theorem 4. Suppose y^* is the projection of y onto $L(X)$ and the criterion is computed on the set W. For any F number of partial descriptions, the sequence of vectors $z^{(r)}$ converges to y^* as $r \to \infty$. If the matrix X^TX is nonsingular, the model corresponding to the limiting vector coincides with the regression equation for y as function of x.

The proof of this theorem differs from the theorem 1 because the vectors $(y - z^{(r)})$ and $z^{(r)}$ are not orthogonal in this case. We shall follow the preceding scheme.

When $F = 1$, the sequence $\|y - z^{(r)}\|$ is nonincreasing and lower bounded; it is denoted by the limit ϱ. As per the step-by-step iterations in the algorithm, we have

$$\|y - z^{(r)}\| - \|y - z^{(r+1)}\| \leq \delta, \tag{3.201}$$

and we note that

$$z^{(r+1)} = z^{(r)} + ax_i = z^{(r)} + \frac{x_i x_i^T}{x_i^T x_i}(y - z^{(r)}). \tag{3.202}$$

From the above inequality, $\|x_i^T y - x_i^T z^{(r)}\| \leq \delta(2\|y - z^{(r)}\| - \delta),\ i = 1, 2, \cdots, m$ can be obtained. Thus, the sequence $z^{(r)}$ has a limit τ.

The rest of the proof is analogous to the preceding one.

Chapter 4
Physical Fields and Modeling

Cybernetical systems are natural systems with complex phenomena in a multi-dimensional environment. The concept of a physical field is given here as a three-dimensional field of (x, y, z), where x, y are considered a surface coordinates and z, a space coordinate. Our main task is to identify a system in a physical field using our knowledge of certain variables and considering their interactions in the environment and with physical laws. Researchers are experimenting to predict the behavior of various complex systems by analyzing emperical data using advanced techniques. Resulting mathematical models must be able to extrapolate the behavior of complex systems in (x, y) coordinates, as well as predict in time t another dimension in the coordinate system. The possibility of better modeling is related through the use of heuristic methods based on sorting of models, pretendents in the form of finite difference equations, empirical data, and selection criteria developed for that purpose.

Examples of physical fields may be fields of air pollution, water pollution, meteorological systems and so on. Observations of various variables—such as data about distributed space, intensity, and period of variable movement—are used for identifying such fields. It corresponds to the observations from control stations corresponding to input and output arguments. The problem goal may be interpolation, extrapolation or prediction, where the area of interpolation lies within the multi-bounded area, and the area of extrapolation or prediction lies outside the area of interpolation process. Models must correspond to the future course of processes in the area. Problems can be further extended to short-range, long-range or combined forecasting problems depending on principles and selection of arguments. A model must correspond to the function (or solution of differential equation) that has the best agreement with future process development. A physical model can be point-wise or spatial (one-, two-, three-, or multi-dimensional). It can be algebraic, harmonic, or a finite-difference equation. A model with one argument is called single-dimensional and multi-dimensional when it has more than one argument. If the model is constructed from the observed data in which the location of the sensors is not known, then it is point-wise. If the data contain the information concerning the sensor locations, then the model is spatial or distributive parametric. Spatial models require the presence of at least three spatial locations on each axis.

In the theory of mathematical physics, physical field is represented with differential or integro-differential equations; linear differential equations have nonlinear solutions. For solving such equations numerically, discrete analogues in the form of finite difference equations are built up. This is done by considering two subsequent cubicles for analogue of first derivative, three cubicals for analogue of second derivative, and so on. As the higher analogues are taken into consideration, the number of arguments in the model structure are correspondingly increased. In other words, the physical field is discretized in terms of the discrete analogues or patterns. To widen the sorting, it is worthwhile to adopt different patterns (consisting of arguments) starting from simple two-cubical patterns to patterns with the possibility of all polynomials. Higher-ordered arguments and paired sorting of patterns and nonlinear polynomials give the possibility of fully reexamining the majority of partial polynomials for representing the physical field. By sorting, it is easier to "guess" the linear character of a finite-difference equation rather than the nonlinearity of its solution. This reduces the sorting of basis functions. The collection of data with regard to the pattern structures, presentation to the algorithm, and evaluation of the patterns are considered as important aspects of the inductive modeling.

1 FINITE-DIFFERENCE PATTERN SCHEMES

Discrete mathematics is based on replacing differentials by finite differences measured at the mesh points of a rectangular spatial mesh or grid. For example, the axes of the three dimensional coordinates x, y, z are discretized into equal sections (steps), usually taken as the unit measurement of $\Delta x = 1$, $\Delta y = 1$, and $\Delta z = 1$. The building up of finite difference equations are based on the construction of patterns or elementary finite difference schemes.

A geometric pattern that indicates the points of the field used to form the equation structure is called elementary pattern. A pattern is a finite difference scheme that connects the value of a given function at the kth point with the value of several other arguments at the neighboring points of the spatial mesh. The pattern for the solution of a specific problem can be determined in two ways: (i) by knowing the physics of the plant (the deductive approach), (ii) by sorting different possible patterns to select the best suited one by an external criterion (the inductive approach). The former is out of the scope of this book and emphasis is given to the latter through the use of inductive learning algorithms.

In a system where y is an output variable and x is an independent variable, a pattern with mesh points within a step apart is shown in Figure 4.1. The general form of the equation representing the complete pattern is

$$\begin{aligned} y_{i,j,k} = f(&y_{i+1,j,k}, y_{i-1,j,k}, y_{i,j+1,k}, y_{i,j-1,k}, y_{i,j,k+1}, y_{i,j,k-1},\\ &x_{i+1,j,k}, x_{i-1,j,k}, x_{i,j+1,k}, x_{i,j-1,k}, x_{i,j,k+1}, x_{i,j,k-1}). \end{aligned} \tag{4.1}$$

This will be more complicated if the delayed arguments are considered by introducing the time axis t as a fourth dimension.

$$\begin{aligned} y^t_{i,j,k} = f(&y^t_{i+1,j,k}, y^t_{i-1,j,k}, y^t_{i,j+1,k}, y^t_{i,j-1,k}, y^t_{i,j,k+1}, y^t_{i,j,k-1}, y^{t-1}_{i,j,k}\\ &x^t_{i+1,j,k}, x^t_{i-1,j,k}, x^t_{i,j+1,k}, x^t_{i,j-1,k}, x^t_{i,j,k+1}, x^t_{i,j,k-1}, x^{t-1}_{i,j,k}). \end{aligned} \tag{4.2}$$

In actual physical problems, most of these arguments are absent because they do not influence the dependent variable. This is the difference between the actual pattern and the complete pattern.

For example, in the linear problem of two-dimensional (x and t) turbulant diffusion we have

$$\frac{\partial q}{\partial t} + u\frac{\partial q}{\partial x} - K\frac{\partial^2 q}{\partial x^2} = 0, \tag{4.3}$$

where u is the flow velocity and K is the diffusion coefficient. The discrete analogue of this equation can be written as

$$(q_i^{t+1} - q_i^t) + \gamma_1(q_{i+1}^t - q_{i-1}^t) - \gamma_2(q_{i+1}^t - 2q_i^t + q_{i-1}^t) = 0, \tag{4.4}$$

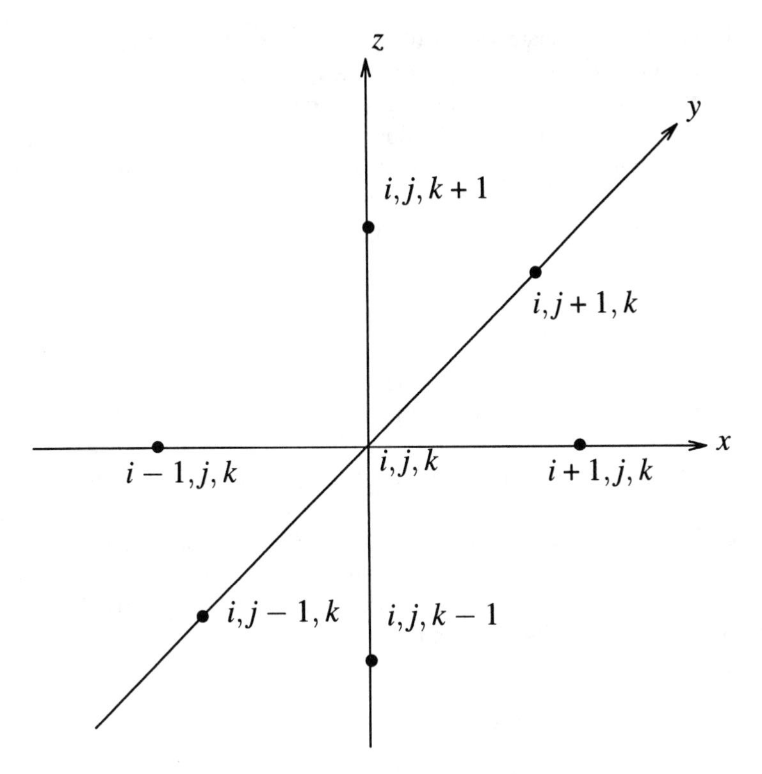

Figure 4.1. "Complete" pattern in (x, y, z) coordinates

where $\gamma_1 = \frac{\tau u}{2h}$, and $\gamma_2 = \frac{\tau K}{h^2}$. In other words, we use a pattern with three arguments in the functional form of

$$q_i^{t+1} = f(q_i^t, q_{i+1}^t, q_{i-1}^t). \tag{4.5}$$

If we consider the Fokker-Planck equation, which takes the above diffusion equation with variable diffusion coefficient K as

$$\frac{\partial q}{\partial t} + u\frac{\partial q}{\partial x} - \frac{\partial^2 Kq}{\partial x^2} = 0, \tag{4.6}$$

then the discrete counterpart is taken as

$$(q_i^{t+1} - q_i^t) + \gamma_1(q_{i+1}^t - q_{i-1}^t) - \gamma_2(K_i^{t+1} q_{i+1}^t - 2K_i^t q_i^t + K_{i-1}^t q_{i-1}^t) = 0, \tag{4.7}$$

where $\gamma_1 = \frac{\tau u}{2h}$, and $\gamma_2 = \frac{\tau}{h^2}$. The pattern consists of the functional form of

$$q_i^{t+1} = f(q_i^t, q_{i+1}^t, q_{i-1}^t, K_i^t, K_{i+1}^t, K_{i-1}^t). \tag{4.8}$$

Usually the dynamic equation is expressed in the form of a sum of two parts: the left side "operator" and the right side "source function" or "remainder." In the problem of turbulant diffusion we can write the equation as

$$\frac{\partial q}{\partial t} + u\frac{\partial q}{\partial x} - K\frac{\partial^2 q}{\partial x^2} = f(x, t), \tag{4.9}$$

where the left side is the "operator" and $f(x, t)$ is the "remainder." The discrete analogue of this equation takes the form of

$$q_i^{t+1} = f(q_i^t, q_{i+1}^t, q_{i-1}^t) + f(x, t), \tag{4.10}$$

where x and t are the coordinate values of q_i^{t+1} on the grid. $f(x,t)$ can be considered as a linear trend in x and t; for example, $f(x,t) = a_0 + a_1x + a_2t$.

For solving very complex problems using the inductive approach, complete polynomials with a considerable number of terms should be used. Usually if the reference function or "complete" polynomial has less than 20 arguments, the combinatorial algorithm is used to select the best model. If it has more than or equal to 20 arguments, the multilayer algorithm is used, depending on the capacity of the computer.

1.1 Ecosystem modeling

The following examples illustrate the identification of one-dimensional and multi-dimensional physical fields related to the processes in the ecosystem.

Example 1. Usually model optimization refers to the choice of the number of time delays considering a one-dimensional problem in time t. For the synthesis of the optimal model, the number of time delays must be gradually increased until the selection criterion decreases. The optimal model corresponds to the global minimum of the external criterion.

Let us consider identification of concentration of dissolved oxygen (DO) and biochemical oxygen demand (BOD). The discrete form of the Streeter-Phelps law [9] is taken along with the experimental data as

$$\begin{aligned} q^{t+1} &= k_1 q_{max} + (1 - k_1)q^t - k_2 u^t \\ u^{t+1} &= u^t - k_2 u^t, \end{aligned} \tag{4.11}$$

where q^t is the DO concentration in mg/liter at time t; q_{max} is the maximum DO concentraion; u^t is the BOD in mg/liter at time t; k_1 is the rate of reaeration per day; and k_2 is the rate of BOD decrease per day.

Complete polynomials are considered as

$$\begin{aligned} q^{t+1} &= f(q^t, q^{t-1}, q^{t-2}, \cdots, q^{t-\tau_1}, u^t, u^{t-1}, u^{t-2}, \cdots, u^{t-\tau_2}) \\ u^{t+1} &= f(q^t, q^{t-1}, q^{t-2}, \cdots, q^{t-\tau_1}, u^t, u^{t-1}, u^{t-2}, \cdots, u^{t-\tau_2}), \end{aligned} \tag{4.12}$$

where τ_1 and τ_2 are time delays taken as three. The combinatorial algorithm is used to generate all possible combinations of partial models. The data is collected in daily intervals—65 data points are used in training and 15 points are kept for examining the predictions. The combined criteria of "minimum bias (η_{bs}) plus prediction (i)" is used for selecting the best model in optimal complexity. The optimal models obtained are

$$\begin{aligned} q^{t+1} &= 1.3350 + 0.8142q^t - 0.00001u^t \\ u^{t+1} &= u^t - 0.2545u^t + 0.1471q^{t-3}. \end{aligned} \tag{4.13}$$

The prediction errors for the model of DO concentrations is 7% and for the model of BOD, 14%. This shows how a physical law can be discovered using the inductive learning approach.

The interpolation region is the space inside the three-dimensional grid with points located at the measuring stations and which lay inside the time interval of the experimental data. The extrapolation region in general lies outside the grid, and the prediction region lies in the future time outside the interpolation region. Usually, the interpolation region is involved in the training of the object. According to the Weiesstrass theorem, the characteristic feature of the region is that any sufficiently complicated curve fits the experimental data with any desired accuracy. In the extrapolation and prediction regions, the curves quickly diverge, forming so-called "fan" of predictions. The function with optimal complexity must have the best agreement with the future process development.

The following example illustrates modeling of a two-dimensional (x, t) physical field of an ecosystem for identification, prediction, and extrapolation. This shows that the optimal pattern and optimal remainder can be found by sifting all possible patterns, with the possible terms of "source function" using the multilayered inductive approach and the sequential application of minimum bias and prediction criteria.

Example 2. The variables (i) dissolved oxygen q^t, (ii) biochemical oxygen demand u^t, and (iii) temperature T^t are measured at three stations of a water reservoir at a depth of 0.5 m. The measurements are taken eight times at 4-week intervals. As a first step, with the measured data, a uniform two-dimensional grid (16×16) of data is prepared by using quadratic interpolation and algebraic models [46].

Here two types of problems are considered: prediction and extrapolation problems. The model formulations are considered as combination of source and operator functions with the following arguments.

(i) Prediction problem:

$$\begin{aligned} q_i^{t+1} &= f_1(x,t) + f_2(q_i^t, q_i^{t-1}, q_i^{t-2}, q_{i-1}^t, q_{i+1}^t, u_i^t, u_i^{t-1}, u_i^{t-2}, u_{i-1}^t, u_{i+1}^t, T_i^t) \\ u_i^{t+1} &= f_3(x,t) + f_4(q_i^t, q_i^{t-1}, q_i^{t-2}, q_{i-1}^t, q_{i+1}^t, u_i^t, u_i^{t-1}, u_i^{t-2}, u_{i-1}^t, u_{i+1}^t, T_i^t). \end{aligned} \tag{4.14}$$

(ii) Extrapolation problem:

$$\begin{aligned} q_{i+1}^t &= f_5(x,t) + f_6(q_i^t, q_{i-1}^t, q_{i-2}^t, q_i^{t-1}, q_i^{t+1}, u_i^t, u_{i-1}^t, u_{i-2}^t, u_i^{t-1}, u_i^{t+1}, T_i^t) \\ u_{i+1}^t &= f_7(x,t) + f_8(q_i^t, q_{i-1}^t, q_{i-2}^t, q_i^{t-1}, q_i^{t+1}, u_i^t, u_{i-1}^t, u_{i-2}^t, u_i^{t-1}, u_i^{t+1}, T_i^t). \end{aligned} \tag{4.15}$$

The data tables are prepared in the order of the output and input variables in the function. Each position of the pattern gives one data measurement of the initial table.

The complete polynomial in each case is considered second-degree polynomial. For example, the complete polynomial for prediction of DO concentration is

$$\begin{aligned} q_i^{t+1} = {} & (a_0 + a_1 x + a_2 t) \\ & + (a_3 q_i^t + a_4 q_i^{t-1} + \cdots + a_7 q_{i+1}^t + a_8 u_i^t + a_9 u_i^{t-1} + \cdots + a_{12} u_{i+1}^t + a_{13} T_i^t \\ & + a_{14} q_i^{t^2} + \cdots + a_{24} T_i^{t^2} + a_{25} q_i^t q_i^{t-1} + \cdots + a_{79} u_{i+1}^t T_i^t). \end{aligned} \tag{4.16}$$

This has 80 terms: 14 linear terms, 11 square terms, and 55 covariant terms. A multilayer algorithm is used. In the first layer, C_{80}^2 (=3160) partial models are formed and the best 80 of them are selected using the minimum bias criterion. It is repeated layer by layer until the criterion decreases. At the last layer 20 best unbiased models are selected for considering long-term predictions. Finally, the optimal models for each problem are chosen with regard to the combined criterion "minimum bias (η_{bs}) plus prediction ($c3$)."

For prediction:

$$\begin{aligned} q_i^{t+1} &= 12.4306 - 4.6477 u_i^t + 0.1615 q_i^{t-2} u_i^t + 0.8896 u_i^{t^2} - 0.0035 u_i^t u_i^{t+1} + 0.0004 t; \\ c3 &= 0.00012 \\ u_i^{t+1} &= 1.1149 + 0.0200 u_i^{t-1} T_i^t + 0.0042 u_{i+1}^t T_i^t; \\ c3 &= 0.0003. \end{aligned} \tag{4.17}$$

For extrapolation:

$$
\begin{aligned}
q_{i+1}^{t} &= 9.5237 + 0.0937u_i^t u_{i-2}^t + 0.0031u_i^t T_i^t \\
&\quad +0.2500u_{i-1}^t u_{i-2}^t - 0.1403u_{i-1}^t u_i^{t+1} - 0.0271u_{i-2}^{t^2} \\
&\quad +0.0392u_{i-2}^t q_i^{t-1} - 0.1782u_{i-2}^t u_i^{t-1} - 0.0288u_i^{t-1^2} \\
&\quad -0.1046u_i^{t-1} u_i^{t+1}; \\
c3 &= 0.00015 \\
u_{i+1}^t &= -0.1798 + 1.1573u_i^t - 0.1124u_i^{t^2} + 0.000966u_i^{t-1} T_i^t; \\
c3 &= 0.000075.
\end{aligned} \tag{4.18}
$$

The accuracy of these models is considerably higher for long-term predictions or extrapolations of up to 10 to 20 steps ahead (the error is not over 20%).

In the literature, "Cassandra predictions" (prediction of predictions) are suggested under specific variations in the data [1], [30]. As we all know, the fall of Troy came true as predicted by Cassandra, the daughter of King Priam of Troy, while the city was winning over the Hellenes. It is important that the chosen model must predict a drop/rise in the very near future on the basis of monotonically increasing/ decreasing data, correspondingly. If the model represents the actual governing law of the system, it will find the inflection point and predict it exactly. Usually, the law connecting the variables is trained in the interpolation region to represent the predicting variable. This does not remain constant in the extrapolation region. "Cassandra predictions" explains that it is possible to identify a governing law within the reasonable noise levels on the basis of past data using inductive learning algorithms. For example, let us consider the model formulation as

$$
\begin{aligned}
q &= f(u, t) \\
u_j &= f_j(u), \quad j = 1, 2, \cdots, m,
\end{aligned} \tag{4.19}
$$

where q is the output variable, u is the vector of input variables, and t is the current time. The secret of obtaining the "Cassandra predictions" is to build up the function that has the characteristics of variable coefficients.

To identify a gradual drop/rise in the data at a later time by predicting q, one has to obtain the predicted values of u_j using the second function and use these predicted values in predicting q. In other words, it works as prediction of predictions. However, the "Cassandra predictions" demand more unbiased models ($0 \leq \eta_{bs} \leq 0.05$). For an unbiased equation $q = f(u, t)$ to have an extremum at an prediction point (u'^n, t_n), where t_n is the time the prediction is made, it is expected that either a decrease or increase occurs in the value of q. If the data is too noisy, it restricts the interval length of the prediction time.

Here another example is given to show that the choice of a pattern and a remainder uniquely determines the "operator" and the "source function" of a multidimensional object.

Example 3. Identification of the mineralization field of an artesian aquifer in the steppe regions of the Northern Crimea is considered [56], [57].

We give a brief description of the system; a schematic diagram of the object with observation net of wells is shown in Figure 4.2. The coordinate origin is located at an injection well. The problem of liquid filtration from a well operating with a constant flow rate is briefed as below: an infinite horizontal seam of constant power is explored by a vertical well of negligibly small radius. Initially the liquid in the seam is constant and the liquid begins to flow upwards at a constant volumetric rate. From a hydrological point of view, the object of investigation is a seam of water-soaked Neocene lime of 170 m capacity,

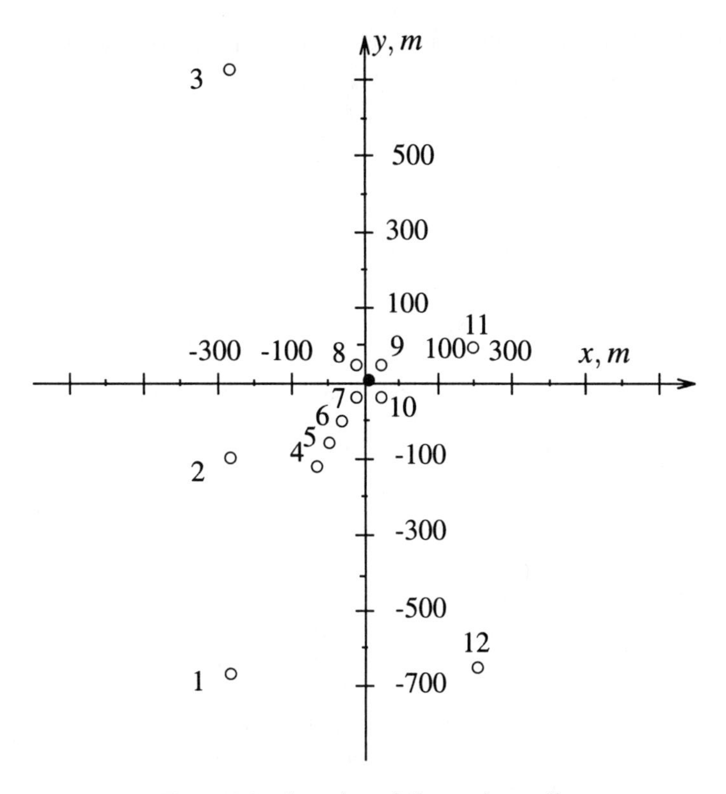

Figure 4.2. Location of observation wells

bounded from above and below by layers of clay that are assumed to be impermeable to water—in the sense that it does not permitting significant passage of liquid. The average depth of the seam is 60 m. The piezometric levels used for exploring the seam are fixed at a depth ranging from 0 to 7 m below the earth's surface. Their absolute markings relative to sea level vary between 0.8 and 4.0 m. In the experimental region, the water flow has a minor deviation in the northernly direction—this agreeing with the regional declination of the seam in the direction of the flow of subterranean waters of this area. According to the prevailing hypothesis, the Black sea is regarded as a run-off region—this is confirmed by the intrusion of salty waters into the aquifer, accompanied by a lowering of the water head in the boundary region as a result of high water extraction for consumption. The aquifer has an inhomogeneous structure, that consists of porous lime with cracks whose permeability varies along the vertical from 8 to 200 m per 24-hour period. The mineralization of the water varies along the vertical from 2 to 3 g/l (as the surface of the seam) to 6 g/l at a depth of 100 m from the surface.

The physical law that is considered as a dynamic model representing the mineralization is the conservation of mass. In hydrodynamics this principle is called continuity law or "principle of close action." The equation is expressed as

$$\frac{\partial q}{\partial t} + u\frac{\partial q}{\partial x} + v\frac{\partial q}{\partial y} + w\frac{\partial q}{\partial z} - [\frac{\partial}{\partial x}(K_x\frac{\partial q}{\partial x}) + \frac{\partial}{\partial y}(K_y\frac{\partial q}{\partial y}) + \frac{\partial}{\partial z}(K_z\frac{\partial q}{\partial z})]$$

$$= Q_q(x, y, z, t) + P(x, y, z), \tag{4.20}$$

where $u, v,$ and w are the velocity components, $K_x, K_y,$ and K_z are the diffusion coefficients,

Q_q is a source function for the ith element, and P is a function representing the interaction of the terms (it is called "remainder").

This can be expressed in the discrete analogue as follows:

$$(q_{i,j,k}^{t+1} - q_{i,j,k}^{t}) + \gamma_1[u(q_{i+1,j,k}^{t} - q_{i,j,k}^{t}) + v(q_{i,j+1,k}^{t} - q_{i,j,k}^{t}) + w(q_{i,j,k+1}^{t} - q_{i,j,k}^{t})]$$

$$-\gamma_2[K_x(q_{i+1,j,k}^{t} - 2q_{i,j,k}^{t} + q_{i-1,j,k}^{t}) + K_y(q_{i,j+1,k}^{t} - 2q_{i,j,k}^{t} + q_{i,j-1,k}^{t}) +$$

$$K_z(q_{i,j,k+1}^{t} - 2q_{i,j,k}^{t} + q_{i,j,k-1}^{t})] = f(x, y, z, t), \tag{4.21}$$

where $\gamma_1 = \tau/h$, $\gamma_2 = \tau/h^2$; $f(x, y, z, t)$ is taken in the general form as

$$f(x, y, z, t) = P(x, y, z) + Q_q(x, y, z, t)$$

$$= a_0 + a_1 x + a_2 y + a_3 z + a_4 q_{i,j-1,k}^{t-1} + a_5 q_{i,j,k}^{t} + a_6 \frac{Qe^{-\frac{0.35R_i}{t}}}{\sqrt{t4\pi R_i}} \tag{4.22}$$

in which Q is the water flow rate in cubic meters during time Δt, R is the distance of a point with coordinates x, y, z from the injection well, $R = \sqrt{(x^2 + y^2 + z^2)}$, $z = z - z_0$, t is the running time from the beginning of the operation (in 24-hour periods), and 0.35 is the optimal value determined for porosity of the medium.

There exists a unique correspondence between the adopted pattern and dynamic equation of the physical field. The choice of pattern determines the structure of the dynamic equation, but only of its left side operator and not of the right-side part of the equation. The optimal pattern is determined by the inductive approach using an external criteria. The pattern must yield the deepest minimum of the criteria. In other words, the optimization problem is reduced to a selection of a pattern. The inductive approach is of interest because it leads to discovery of new properties of the system. Simulation of complex systems by this approach is very convenient for examining a large number of percolation hypotheses and selecting the best one. The selection of the arguments in the algorithm is directly related to the percolation hypothesis to be adopted and must have a sufficiently wide scope. In this example, the optimal selection of arguments is based on sorting of a large number of patterns.

The above finite difference equation is considered a reference function representing the "complete" pattern. All the partial models corresponding to the partial patterns can be obtained by zeroing in the terms of the reference function as is done in the "structure of functions." This means that a specified pattern determines the operator of the left side equation, and not the remainder. For example, for pattern no. 1 the partial function is given as

$$(q_{i,j,k}^{t+1} - q_{i,j,k}^{t}) + \frac{\tau w}{h}(q_{i,j,k+1}^{t} - q_{i,j,k}^{t}) + \frac{\tau u}{h}(q_{i+1,j,k}^{t} - q_{i,j,k}^{t}) -$$

$$K_x(q_{i+1,j,k}^{t} - 2q_{i,j,k}^{t} + q_{i-1,j,k}^{t}) = f(x, y, z, t). \tag{4.23}$$

Overall, there are 13 coefficients for the complete pattern. $2^{13} - 1$ partial models are generated if the combinatorial algorithm is used. It is equivalent to the optimal selection of arguments based on sorting a sufficiently large number of patterns. The difference data is measured from the given region with interpolation of the q value at the intermediate points of the mesh. The problem is reduced to the selection of an optimal pattern among a set of

Table 4.1. Values of the minimum bias criterion

No.	Value	No.	Value	No.	Value
1	0.08239	8	0.04669	15	0.04669
2	0.08239	9	0.04669	16	0.04669
3	0.08239	10	0.06546	17	0.04669
4	0.11075	11	0.06545	18	0.04669
5	0.04669	12	0.06545	19	0.04669
6	0.04669	13	0.04669	20	0.04669
7	0.09652	14	0.04669	21	0.09674

patterns; i.e., a unique model that yields the deepest minimum of the combined criterion is selected.

$$c1^2 = \bar{\eta}_{bs}^2 + \bar{\Delta}^2(W/A), \tag{4.24}$$

where $\bar{\eta}_{bs}$ is the normalized minimum bias criterion and $\bar{\Delta}(W/A)$ is the normalized regularity criterion.

The total number of feasible patterns is $2^6 - 1 = 63$. Some of the patterns are shown in the Figure 4.3. Table 4.1 exhibits the values of the minimum bias criterion for these patterns. The optimal pattern with regard to the combined criterion $c1$ is found to be pattern 9. The optimal equation is

$$q_{i,j,k}^{t+1} = 1.325q_{i,j,k}^{t} + 14.13\frac{q_{i,j+1,k}^{t} - 2q_{i,j,k}^{t} + q_{i,j-1,k}^{t}}{h^2} - 2.15\frac{q_{i,j+1,k}^{t} - q_{i,j-1,k}^{t}}{2h}$$
$$- 0.08747q_{i,j-1,k}^{t-1} - 0.002716\frac{Qe^{-\frac{0.35R_i}{t}}}{\sqrt{t4\pi R_i}}. \tag{4.25}$$

The last two terms in the equation correspond to the remainder function.

Stability analysis

The stability analysis of equations of the form above was carried out. It was proved that stability with regard to the initial data can be realized under the conditions

$$(i)\ \frac{\tau K_y}{h^2} < \frac{1}{2},$$
$$(ii)\ v^2 < \frac{2K_y}{\tau}\left(1 - \frac{2\tau K_y}{h^2}\right), \tag{4.26}$$

where the former is the well-known stability condition and the latter is the condition for interconnection of the coefficients of the finite-difference equation.

2 COMPARATIVE STUDIES

As a continuation of our study on elementary pattern structures, some examples of correspondence between linear differential equations and their finite-difference analogues are given in Tables 4.2 and 4.3. Here time t is shown as one of the axes (see Figure 4.4).

For physical fields some deterministic models are usually known, they are given by differential or integro-differential equations. Such equations from the deterministic theories

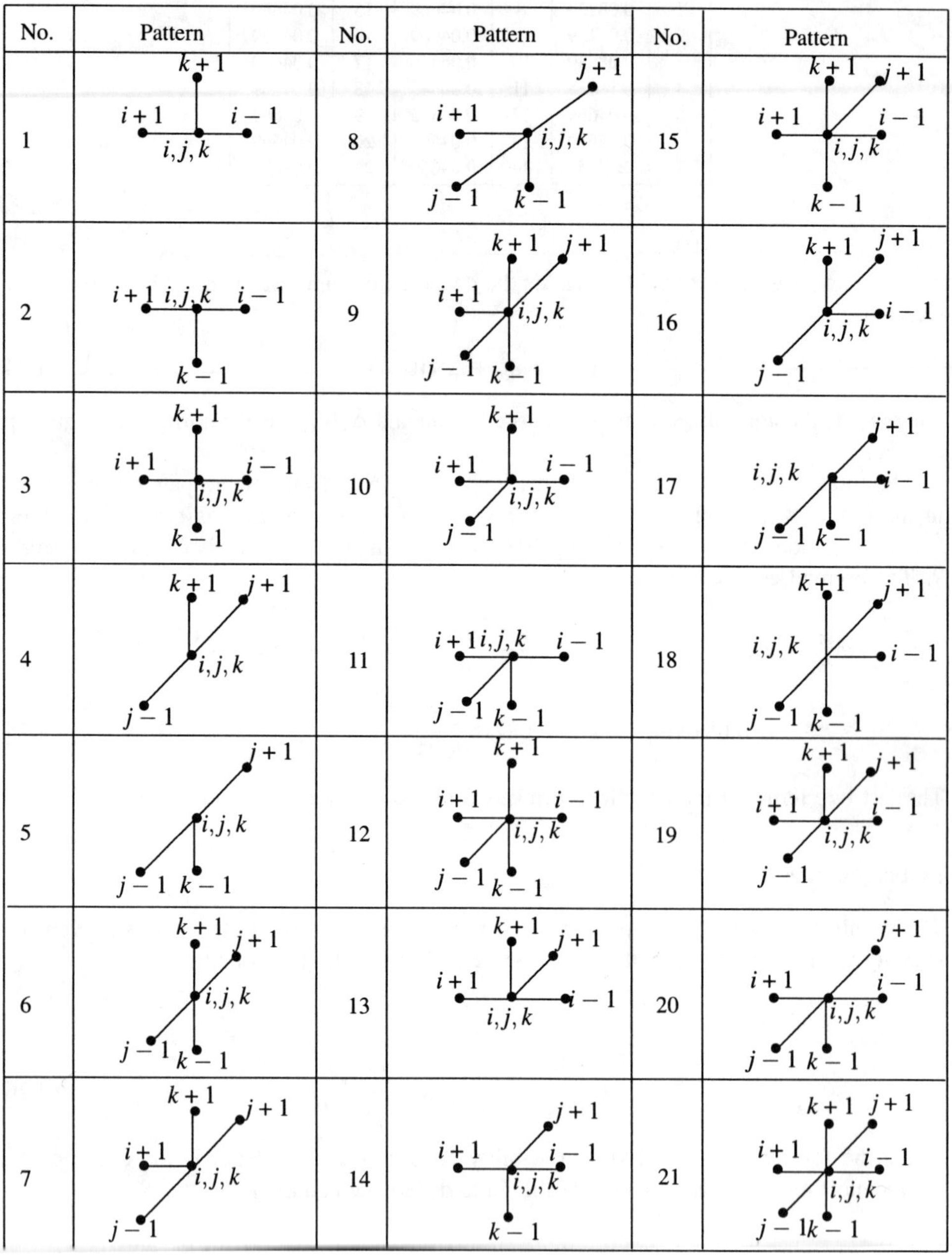

No.	Pattern	No.	Pattern	No.	Pattern
1	$k+1$, $i+1$, $i-1$, i,j,k	8	$j+1$, $i+1$, i,j,k, $j-1$, $k-1$	15	$k+1$, $j+1$, $i+1$, $i-1$, i,j,k, $k-1$
2	$i+1$, i,j,k, $i-1$, $k-1$	9	$k+1$, $j+1$, $i+1$, i,j,k, $j-1$, $k-1$	16	$k+1$, $j+1$, $i-1$, i,j,k, $j-1$
3	$k+1$, $i+1$, $i-1$, i,j,k, $k-1$	10	$k+1$, $i+1$, $i-1$, i,j,k, $j-1$	17	$j+1$, i,j,k, $i-1$, $j-1$, $k-1$
4	$k+1$, $j+1$, i,j,k, $j-1$	11	$i+1$, i,j,k, $i-1$, $j-1$, $k-1$	18	$k+1$, $j+1$, i,j,k, $i-1$, $j-1$, $k-1$
5	$j+1$, i,j,k, $j-1$, $k-1$	12	$k+1$, $i+1$, $i-1$, i,j,k, $j-1$, $k-1$	19	$k+1$, $j+1$, $i+1$, $i-1$, i,j,k, $j-1$
6	$k+1$, $j+1$, i,j,k, $j-1$, $k-1$	13	$k+1$, $j+1$, $i+1$, $i-1$, i,j,k	20	$j+1$, $i+1$, $i-1$, i,j,k, $j-1$, $k-1$
7	$k+1$, $j+1$, $i+1$, i,j,k, $j-1$	14	$j+1$, $i+1$, $i-1$, i,j,k, $k-1$	21	$k+1$, $j+1$, $i+1$, $i-1$, i,j,k, $j-1$, $k-1$

Figure 4.3. Certain patterns among 63 feasible patterns

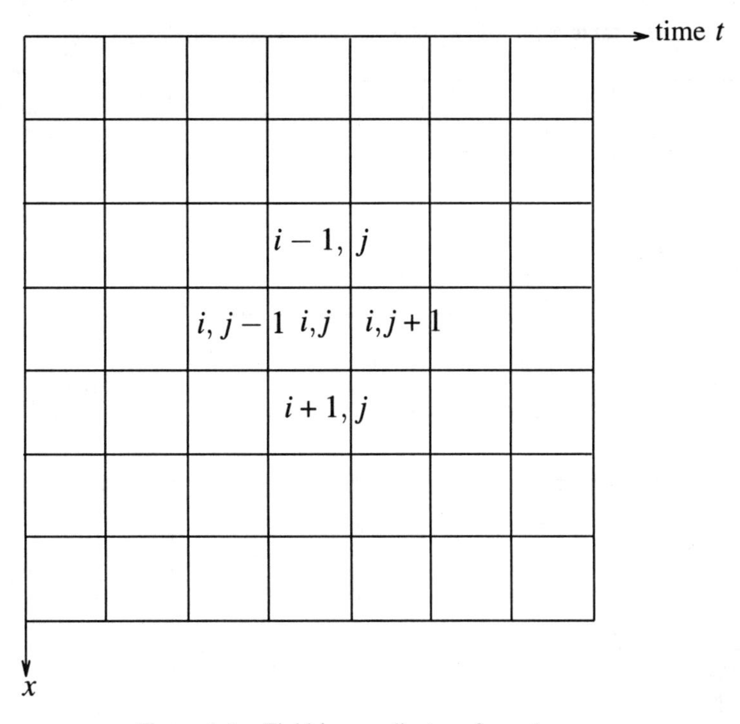

Figure 4.4. Field in coordinates of x and t

may be used for choosing the arguments and functions for a "complete" reference function. A complete pattern is made from the deterministic equation pattern by increasing its size by one or two cells along all axes; i.e., the equation order is increased by one or two to let the algorithm choose a more general law.

2.1 Double sorting

There are two ways of enlarging the sorting of arguments. One way is as shown in Tables 4.2 and 4.3 and starts from the simplest to the more complex pattern. Another way is by considering higher-order arguments for each pattern and sorting them. The polynomials with higher-order terms provide a more complete view of the set of possible polynomials. The complexity of the polynomials increases as the delayed and other input variables are added to them. For example, shown are the pointwise models of a variable q using simple patterns. Without delayed arguments, it is

$$q^{t+1} = f(q^t) = a_0 + a_1 q^t + a_2 q^{t^2}; \tag{4.27}$$

with one delayed argument it is

$$q^{t+1} = f(q^t, q^{t-1}) = a_0 + a_1 q^t + a_2 q^{t-1} + a_3 q^{t^2} + a_4 q^{t-1^2} + a_5 q^t q^{t-1}, \tag{4.28}$$

and with two delayed arguments,

$$\begin{aligned} q^{t+1} &= f(q^t, q^{t-1}, q^{t-2}) \\ &= a_0 + a_1 q^t + a_2 q^{t-1} + a_3 q^{t-2} + a_4 q^{t^2} + a_5 q^{t-1^2} + a_6 q^{t-2^2} \\ &\quad + a_7 q^t q^{t-1} + a_8 q^t q^{t-2} + a_9 q^{t-1} q^{t-2}. \end{aligned} \tag{4.29}$$

Table 4.2. Sorting of elementary patterns and data Tables

Pattern	Model and its Discrete Analogue	Data Table Representation
i,j — i,j+1	$\frac{\partial q}{\partial t} + a_1 q = f(x,t),$ $q_i^{t+1} = f_1(x,t) + f_2(q_i^t)$	q_i^{t+1} q_i^t
i-1,j i,j i,j+1	$\frac{\partial q}{\partial t} + a_1 \frac{\partial q}{\partial x} + a_2 q = f(x,t),$ $q_i^{t+1} = f_1(x,t) + f_2(q_i^t, q_{i-1}^t)$	q_i^{t+1} q_i^t q_{i+1}^t
i-1,j i,j i,j+1 i+1,j	$\frac{\partial^2 q}{\partial t^2} + a_1 \frac{\partial^2 q}{\partial x^2} + a_2 \frac{\partial q}{\partial x} + a_3 q = f(x,t),$ $q_i^{t+1} = f_1(x,t) + f_2(q_i^t, q_{i+1}^t, q_{i-1}^t)$	q_i^{t+1} q_i^t q_{i+1}^t q_{i-1}^t
i-1,j i,j i,j-1 i,j+1 i+1,j	$\frac{\partial^2 q}{\partial t^2} + a_1 \frac{\partial q}{\partial t} + a_2 \frac{\partial^2 q}{\partial x^2} + a_3 \frac{\partial q}{\partial x}$ $+a_4 q = f(x,t),$ $q_i^{t+1} = f_1(x,t) + f_2(q_i^t, q_{i+1}^t, q_{i-1}^t, q_i^{t-1})$	q_i^{t+1} q_i^t q_{i+1}^t q_{i-1}^t q_i^{t-1}
i-1,j i,j i,j-2 i,j-1 i,j+1 i+1,j	$\frac{\partial^3 q}{\partial t3} + a_1 \frac{\partial^2 q}{\partial t^2} + a_2 \frac{\partial q}{\partial t} + a_3 \frac{\partial^2 q}{\partial x^2} + a_4 \frac{\partial q}{\partial x}$ $+a_5 q = f(x,t),$ $q_i^{t+1} = f_1(x,t)$ $+f_2(q_i^t, q_{i+1}^t, q_{i-1}^t, q_i^{t-1}, q_i^{t-2})$	q_i^{t+1} q_i^t q_{i+1}^t q_{i-1}^t q_i^{t-1} q_i^{t-2}

Similarly, in case of two variables q and x, the formulations

$$
\begin{aligned}
q^{t+1} &= f(q^t, x^t), \\
q^{t+1} &= f(q^t, x^t, q^{t-1}, x^{t-1}), \\
q^{t+1} &= f(q^t, x^t, q^{t-1}, x^{t-1}, q^{t-2}, x^{t-2}),
\end{aligned} \tag{4.30}
$$

and so on, gradually increase their complexity. In the same way, spatial models can be developed by considering the delayed and higher-order terms. Sorting of all partial polynomials means generation of all combinations of input arguments for "structure of functions" using the combinatorial algorithm. One can see that the sorting is done in two aspects: one is pattern-wise sorting and the other is orderwise sorting. This is called "double sorting." These are used below for modeling of simulated air pollution fields in the example given.

One should distinguish between Tables of measuring stations and interpolated initial data. Different patterns result in different settings of numerical field of the Table. The measurement points are ordered as shown in the "data representation" (Tables 4.2 and 4.3). Each position of a pattern on the field corresponds to one measurement point in the data table. Each pattern results in its own data table; there are as many tables as there are patterns compared. Tables resulting from the displacement of patterns with respect to the numerical

Table 4.3. Sorting of "diagonal" type patterns and data tables

'Diagonal' Type Pattern	Model and its Discrete Analogue	Data Table Representation
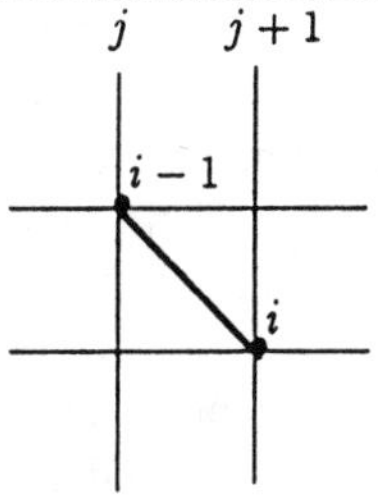 j $j+1$ $i-1$ i	$\frac{\partial q}{\partial t} + a_1\frac{\partial q}{\partial x} + a_2 q = f(x,t)$, $q_i^{t+1} = f_1(x,t) + f_2(q_{i-1}^t)$	q_i^{t+1} | q_{i-1}^t
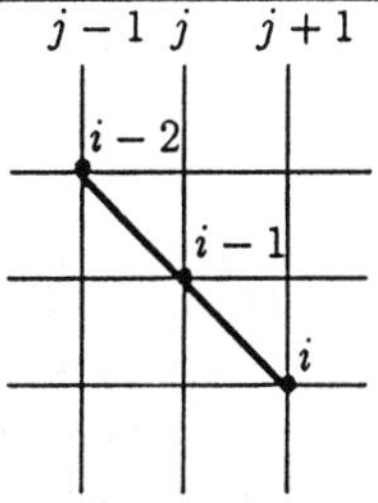 $j-1$ j $j+1$ $i-2$ $i-1$ i	$\frac{\partial^2 q}{\partial t^2} + a_1\frac{\partial^2 q}{\partial x^2} + a_2\frac{\partial q}{\partial t} + a_3\frac{\partial q}{\partial x}$ $+a_4 q = f(x,t)$, $q_i^{t+1} = f_1(x,t) + f_2(q_{i-1}^t, q_{i-2}^{t-1})$	q_i^{t+1} | q_{i-1}^t | q_{i-2}^{t-1}
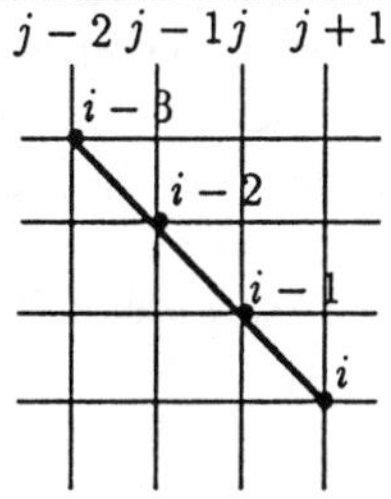 $j-2$ $j-1$ j $j+1$ $i-3$ $i-2$ $i-1$ i	$\frac{\partial^3 q}{\partial t^3} + a_1\frac{\partial^3 q}{\partial x^3} + a_2\frac{\partial^2 q}{\partial t^2} + a_3\frac{\partial^2 q}{\partial x^2} + a_4\frac{\partial q}{\partial t}$ $+a_5\frac{\partial q}{\partial x} + a_6 q = f(x,t)$, $q_i^{t+1} = f_1(x,t) + f_2(q_{i-1}^t, q_{i-2}^{t-1}, q_{i-3}^{t-2})$	q_i^{t+1} | q_{i-1}^t | q_{i-2}^{t-1} | q_{i-3}^{t-2}

field of data are divided for external criteria. The best pattern provides the deepest minimum of the criteria.

2.2 Example—pollution studies

Example 4. Modeling of air pollution field. Three types of problems are formulated [47] for modeling of the pollution field using: (i) the data of a single station, (ii) the data about other pollution components, and (iii) combining both.

In the first problem, the finite-difference form of the model is found by using experimental data through sorting the patterns and using the higher-ordered arguments. The number of terms of the "complete" equation is usually much greater than the total number of data points. In the second problem, the arguments in the finite difference equations are chosen as they correspond to the "input-output matrix" [122] of pollution components; whereas in the third problem, it corresponds to the "input-output matrix" of pollution components and sources. Three problems can be distinguished based on the choice of arguments. The first problem is based on the "principle of continuity or close action;" the second, which is opposite to the first, is based on the "principle of remote action." The third is based on both principles "close and remote actions."

The number of stations that register pollution data increases each year, but sufficient data are still not available. The inductive approach requires a relatively small number of data points and facilitates significant noise stability according to the choice of an external criterion. The mathematical formulations of a physical field described in connection with the above problems compare the different approaches. Additional measurements are used for refinement of each specific problem. In representing the pollution field, station data, data about location, intensity and time of pollutions are used. The choice of output quantity and input variables determines the formulations. This depends on the problem objective (interpolation, extrapolation, or prediction) and on availability of the experimental data.

Before explaining the problem formulations, a brief description about the formation of "input-output matrix" is given here.

Input-output matrix

The "input-output matrix" is estimated based on the linear relationships between the pollution sources u and pollution concentrations q using the observation data at the stations. The matrix is used as a rough model of the first approximation and the differences between the actual outputs q and estimated outputs using the inductive algorithm. The pollution model in vector form is given as $q = f.u$, where q is the pollution concentration at a station, u is the intensity of the pollution source, and f is a coefficient that accounts for various factors relating to the source and diffusion fields—f is regarded as a function of the relative coordinates between pollution source and the observation station. Other factors, such as terrain and atmospheric count, are implicitly taken into consideration in determining f on the basis of observation data.

For a set of sources u_j, $j = 1, 2, \cdots, m$, the pollution concentration for each observation station q_i, $i = 1, 2, \cdots, n$ is represented by

$$q_i = \sum_{j=1}^{m} f_i(x_{ij}, y_{ij})u_j; \quad i = 1, 2, \cdots, n, \tag{4.31}$$

where

$$\begin{aligned} x_{ij} &= x_i^r - x_j^s; \quad i = 1, 2, \cdots, n; \; j = 1, 2, \cdots, m \\ y_{ij} &= y_i^r - y_j^s; \quad i = 1, 2, \cdots, n; \; j = 1, 2, \cdots, m. \end{aligned}$$

q_i is the pollution concentration at the ith station; u_j is the intensity of the jth source; x_i^r, y_i^r are the coordinates of the ith station; x_j^s, y_j^s are the coordinates of the jth source; n is the number of stations and m is the number of sources.

This can be written in matrix form as

$$q = F.u, \tag{4.32}$$

where

$$q^T = (q_1, q_2, \cdots, q_n), \quad u^T = (u_1, u_2, \cdots, u_m) \text{ and}$$

$$F = \begin{pmatrix} f_{11} & f_{12} & \cdots & f_{1m} \\ f_{21} & f_{22} & \cdots & f_{2m} \\ & \cdots & & \\ f_{n1} & f_{n2} & \cdots & f_{nm} \end{pmatrix}.$$

Here F is called the "input-output matrix." Each element f_{ij} can be described by

$$f_{ij} = a_{0j} + a_{1j}x_{ij} + a_{2j}y_{ij} + a_{3j}x_{ij}^2 + a_{4j}y_{ij}^2 + a_{5j}x_{ij}y_{ij} + a_{6j}e^{-\alpha_j\sqrt{x_{ij}^2+y_{ij}^2}}, \tag{4.33}$$

which is estimated by using spatially distributed data. The equation obtained for one source can be used for all other pollution sources. The matrix F is determined by applying f_{ij} to each source. This is treated as a "rough" model because of its dependence on the coordinate distances in the field. This is used to estimate the linear trend part of the system, the remainder part, which is the unknown nonlinear part of the system, is described by

$$\Delta q_i = \frac{1}{m}\sum_{j=1}^{m} g_j(x_{ij}, y_{ij}); \quad i = 1, 2, \cdots, N, \tag{4.34}$$

where

$$x_{ij} = x_i^r - x_j^s; \quad i = 1, 2, \cdots, N; \; j = 1, 2, \cdots, m$$
$$y_{ij} = y_i^r - y_j^s; \quad i = 1, 2, \cdots, N; \; j = 1, 2, \cdots, m.$$

$\Delta q_i\,(= q_i - \hat{q}_i)$ is the remainder at the ith point; N is the total number of points on the (x, y) grid; the function $g_j(x_{ij}, y_{ij})$ is described by a polynomial of a certain degree in x_{ij} and y_{ij}. The remainder equation is estimated as an average of m source models that is identified by using an inductive algorithm. The predictions obtained from the linear trend or rough model are corrected with the help of a remainder model.

Problem formulations

The first problem is formulated to model the pollution field by using only the data of a few stations; this is denoted as I-1. Here the emphasis of modeling is to construct the pollution field not only in the interpolation region, but also to extrapolate and predict the field in time. Pollutants are assumed to change slowly in time so that complete information about them is not used. Only the arguments from the stations data are included in the formulation.

I-1.

(i) for prediction

$$\begin{aligned} q_{i,j}^{t+1} = f_j(&q_{i,j}^t, q_{i-1,j}^t, q_{i+1,j}^t, q_{i,j}^{t-1}, q_{i,j}^{t-2}, \cdots, q_{i,j}^{t-\tau}, \\ &q_{i,k}^t, q_{i-1,k}^t, q_{i+1,k}^t, q_{i,k}^{t-1}, q_{i,k}^{t-2}, \cdots, q_{i,k}^{t-\tau}); \end{aligned} \tag{4.35}$$

(ii) for extrapolation

$$\begin{aligned} q_{i+1,j}^t = f_j(&q_{i,j}^t, q_{i,j}^{t-1}, q_{i,j}^{t+1}, q_{i-1,j}^t, q_{i-2,j}^t, \cdots, q_{i-\tau,j}^t, \\ &q_{i,k}^t, q_{i,k}^{t-1}, q_{i-1,k}^t, q_{i,k}^{t+1}, q_{i-2,k}^t, \cdots, q_{i-\tau,k}^t), \end{aligned} \tag{4.36}$$

where $q_{i,j}^t$ is the pollution parameter j measured at the station i at the time t; f_j indicates the vector of polynomial functions corresponding to j parameters. The input variables may include delayed and higher-ordered arguments; for example, q_j^{t-1}, q_k^{t-1}, $\cdots$, $q_j^{t^2}$, $q_k^{t^2}$, $\cdots$, $(q_j^t q_k^t)$, $(q_j^t q_j^{t-1})$, $\cdots$ at station i.

One can encounter the influence of the phenomena considering the settling of polluting particles. External influences with the above diffusion process and source function are introduced. The source function includes perturbations such as the wind force vector P and its projection on x−axis V'. In general, the formulation for prediction looks like

$$q_{i,j}^{t+1} = f_j(\cdots) + f(x,t) + Q(P, V'), \quad (4.37)$$

where $f(x,t)$ is the trend function with the coordinates of x and t is the pollution component; similarly one can write for the extrapolation.

The second problem is formulated to model the physical field by using the "input-output matrix" along with the above turbulant diffusion equations. This is usually recommended when forecasting of the pollution changes in time. This has three formulations; these are denoted by II-1, II-2, and II-3 as given below.

II-1. In the first formulation, the "input-output matrix" uses only information from the stations. The prediction equation at the station i is

$$q_{i,j}^{t+1} = \sum_{s=1;s\neq i}^{n} f_{s,j}(q_s), \quad (4.38)$$

where q_s denotes the vector of $[q_{s,1}^t, q_{s,1}^{t-1}, \cdots, q_{s,1}^{t-\tau}, \cdots, q_{s,j}^t, q_{s,j}^{t-1}, \cdots, q_{s,j}^{t-\tau}]$; $j = 1, \cdots, m$ are pollution parameters; n is number of stations; f is a polynomial function operator; m is the number of components. The pollution at the ith station (or field point) depends on the values measured at the neighboring points. For example, $n = 3$, $m = 2$, and $\tau = 2$

$$q_{1,1}^{t+1} = f_{2,1}(q_{2,1}^t, q_{2,1}^{t-1}, q_{2,1}^{t-2}, q_{2,2}^t, q_{2,2}^{t-1}, q_{2,2}^{t-2}) + f_{3,1}(q_{3,1}^t, q_{3,1}^{t-1}, q_{3,1}^{t-2}, q_{3,2}^t, q_{3,2}^{t-1}, q_{3,2}^{t-2}). \quad (4.39)$$

II-2. In the second formulation, it uses the "input-output matrix" containing only information about the pollutants. The prediction equation for station i is

$$q_{i,j}^{t+1} = \sum_{s=1}^{p} f_{s,j}(u_s), \quad (4.40)$$

where u_s denotes the vector of pollutants $[u_{s,1}^t, u_{s,1}^{t-1}, \cdots, u_{s,1}^{t-\tau}, \cdots, u_{s,m}^t, u_{s,m}^{t-1}, \cdots, u_{s,m}^{t-\tau}]$; p is the number of sources. For example, $m = 2$, $\tau = 2$, and $p = 2$

$$q_{1,1}^{t+1} = f_{1,1}(u_{1,1}^t, u_{1,1}^{t-1}, u_{1,1}^{t-2}, u_{1,2}^t, u_{1,2}^{t-1}, u_{1,2}^{t-2}) + f_{2,1}(u_{2,1}^t, u_{2,1}^{t-1}, u_{2,1}^{t-2}, u_{2,2}^t, u_{2,2}^{t-1}, u_{2,2}^{t-2}). \quad (4.41)$$

II-3. In the third formulation, it uses the "input-output matrix" containing the information of neighboring stations and the pollution sources—both q and u appear in the matrix. The prediction for station i is

$$q_{i,j}^{t+1} = \sum_{s=1;s\neq i}^{n} f_{s,j}(q_s) + \sum_{s=1}^{p} f_{s,j}(u_s). \quad (4.42)$$

It is good practice to add a source function Q to the above formulations in order to consider external influences like wind force, temperature, and humidity. The complete descriptions are obtained as sums of polynomials as was the case in the first problem. The formulations with the source function may also be considered for multiplicative case; for example,

$$q_{t+1} = Q_1(P, V') + Q_2(P, V')[\sum_{s=1;s\neq i}^{n} f_{s,j}(q_s) + \sum_{s=1}^{p} f_{s,j}(u_s)]. \quad (4.43)$$

This is done if it provides a deeper minimum of the external criterion.

The third problem is formulated to model the pollution field by using the principles of "close action" and "remote action." This has three formulations.

III-1. This uses the "close action" principle as well as information of stations forming the "input-output matrix;" this means that a combination of I-1 and II-1 is used in its formulation.

III-2. This uses the "principle of close action" and information of pollutants forming the "input-output matrix;" thus, a combination of I-1 and II-2 is used in its formulation.

III-3. This uses the "principle of close action" and information of stations and sources of pollutions from the extended "input-output matrix;" this means that a combination of I-1 and II-3 is used in its formulation.

The above seven types of formulations are synthesized and compared for their extrapolations and predictions by using a simulated physical field. The field is constructed using a known deterministic formula that allows changes of pollution without wind and that assumes that particles diffusion in space.

$$q = \frac{2\pi R}{kx} \int_{x=\frac{R}{kt}}^{\infty} \frac{e^{-x}}{x} dx, \tag{4.44}$$

where k is the turbulant diffusion coefficient, R is the distance between station and source, and t is time from the start of pollution to the time of measuring. The number of sources is assumed to be one. The change of pollution source and concentration of polluting substances are shown in Figure 4.5; the above formula is used to obtain the data. Integral values serve as the arguments.

All polynomials are evaluated by the combined criterion $c3$, "bias plus prediction error."

$$c3^2 = \bar{\eta}_{bs}^2 + \bar{i}^2(W), \tag{4.45}$$

where $\bar{\eta}_{bs}^2$ and $\bar{i}^2(W)$ are the normalized minimum bias and prediction criteria, respectively. For extrapolation error $\bar{\Delta}^2(C)$ is used instead of step-by-step prediction errors.

$$\eta_{bs}^2 = \frac{\sum_{p=1}^{\delta N_W} (\hat{q}_p^A - q_p)(\hat{q}_p^B - q_p)}{\sum_{p=1}^{\delta N_W} q_p^2}, \tag{4.46}$$

where δ is the noise immune coefficient that varies from 1.5 to 3.0, and

$$i^2(W) = \frac{\sum_{p \in N_W} (\hat{q}_p - q_p)^2}{\sum_{p \in N_W} q_p^2}. \tag{4.47}$$

The solutions of the first and second problems allow one to construct the field, extrapolate, and predict along the spatial coordinates. The solution of the second problem also allows one to interpolate, extrapolate, and predict pollution parameters at the stations. The results show that the model, based on the "principle of close action," cannot survive alone for better predictions compared with the model that are based on the "principle of remote action" (II-3) and on the "combined principle" (III-2).

Model II-3.

$$\begin{aligned} q_1^{t+1} &= 2.0361 - 2.1815 q_2^{t-1} - 0.2102 q_2^{t-3} + 0.00754 u^{t^2} + 0.1099 q_2^t q_2^{t-2} + 0.3924 q_2^{t-1} \\ &\quad + 0.00002 q_2^{t^2} - 0.000002 q_2^t q_3^{t-1} - 0.000001 q_3^{t^2}. \end{aligned} \tag{4.48}$$

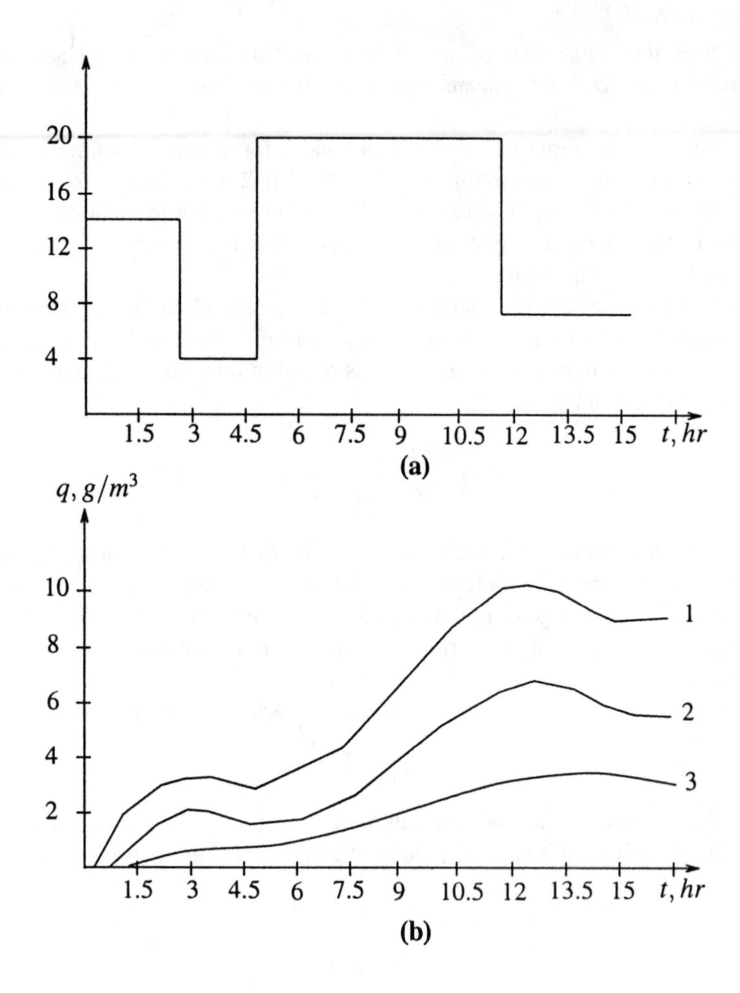

Figure 4.5. (a) Change of pollution discharge in time (from the experiment) and (b) change in concentrations of polluting substances at stations 1, 2, 3.

Model III-2.

$$
\begin{aligned}
q_j^{t+1} &= 0.4228 - 0.7792q_j^t - 0.5797q_j^{t-1} - 0.4908q_j^{t-3} + 0.94q_{j-1}^t + 1.1502q_{j+1}^t \\
&\quad + 0.0442u_j^{t-2} + 0.0047q_j^{t^2} + 0.0017q_j^t q_j^{t-1} + 0.0018q_j^t q_j^{t-2} + 0.0074q_j^{t-1} q_j^{t-2} \\
&\quad - 0.0211q_j^{t-1} q_j^{t-3} + 0.0045q_j^{t-1} q_{j+1}^t - 0.0066q_j^{t-2} q_{j-1}^t - 0.0021q_j^{t-2} q_{j+1}^t \\
&\quad + 0.0187q_j^{t-3} q_{j-1}^t - 0.0032q_{j-1}^{t^2} - 0.002q_{j-1}^t q_{j+1}^t,
\end{aligned}
\tag{4.49}
$$

where j indicates the pollution component pertaining to the station 1.

Figures 4.6 to 4.8 illustrate the step-by-step predictions of all formulations. Table 4.4 gives the performance of these formulations on the given external criteria.

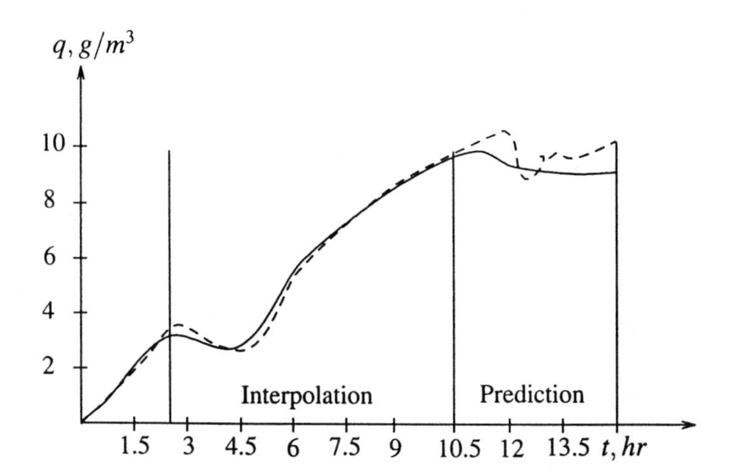

Figure 4.6. Performance of model I-1 ("close" action principle)

Table 4.4. Performance of the formulations

Formulation	$c3$	η_{bs}	$\Delta(C)$	i
I-1	0.032	0.017	0.027	0.182
II-1	0.061	0.046	0.040	0.188
II-2	0.089	0.082	0.036	0.169
II-3	0.080	0.054	0.059	0.151
III-1	0.064	0.063	0.026	0.176
III-2	0.033	0.009	0.031	0.149
III-3	0.115	0.050	0.040	0.246

3 CYCLIC PROCESSES

We have studied the formulations based on the "principle of continuity or close action," the "principle of distant or remote action," and, to some extent the "principle of combined action" using a combination of formulations. The "close action principle" is realized by considering nearby cells and delayed arguments in the finite-difference analogues. The "remote action principle" is arrived at by constructing the "input-output matrix," which is one way of realizing this principle. The elements in the "input-output matrix" can be the values of perturbations or values of variables in distant cells. The "combined action" gives the way to consider the influence of both principles on the output variable.

Many processes in nature that have characteristic cyclic or seasonal trend are oscillatory. For example, the mean monthly air temperature has characteristic maxima during the summer months and minima during the winter months. These values of maxima and minima do not coincide with one another from year to another. Therefore, processes with seasonal fluctuations of this kind are called cyclic in contrast with the strictly periodic processes. They include all natural processes with constant duration—a cycle (year or day). The variations in these processes are determined by the influence of supplementary factors. Certain agricultural productions, economical processes (sale of seasonal goods, etc.), and technological processes might be classified as cyclic. These are described by integro-differential

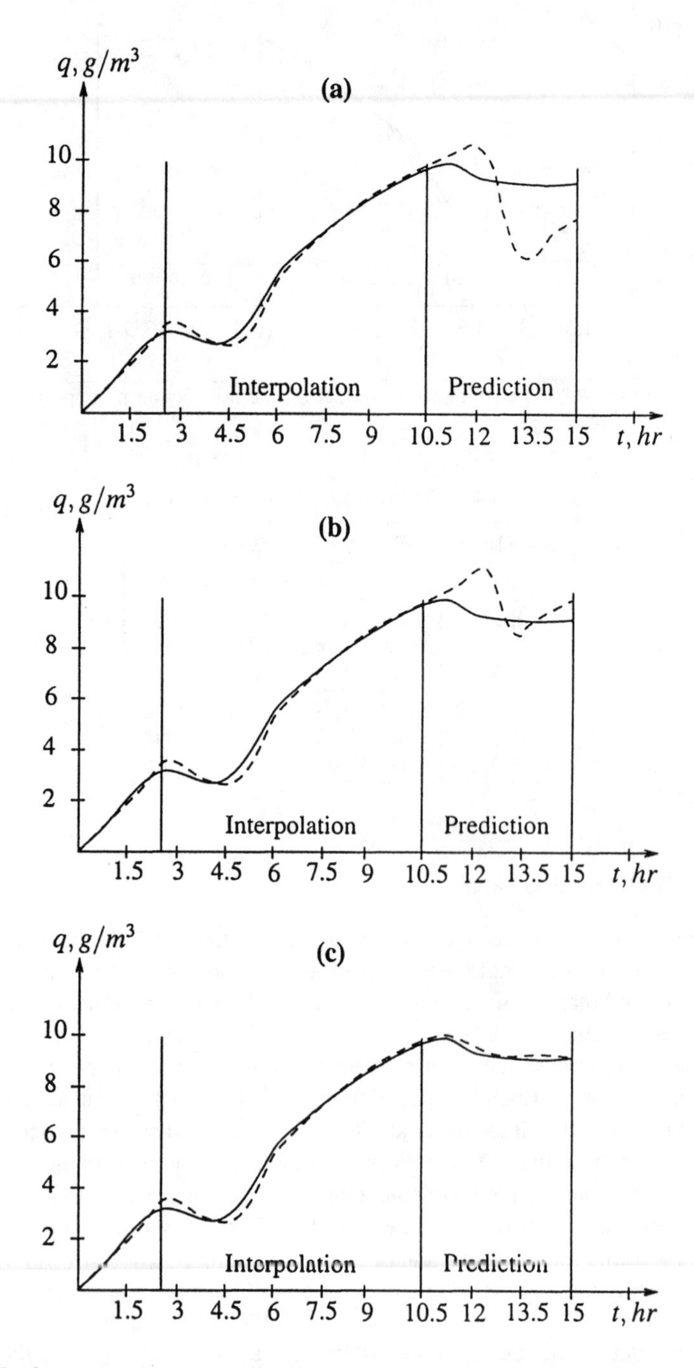

Figure 4.7. Performance of (a) model II-1, (b) model II-2, and (c) model II-3 for "remote" action principle

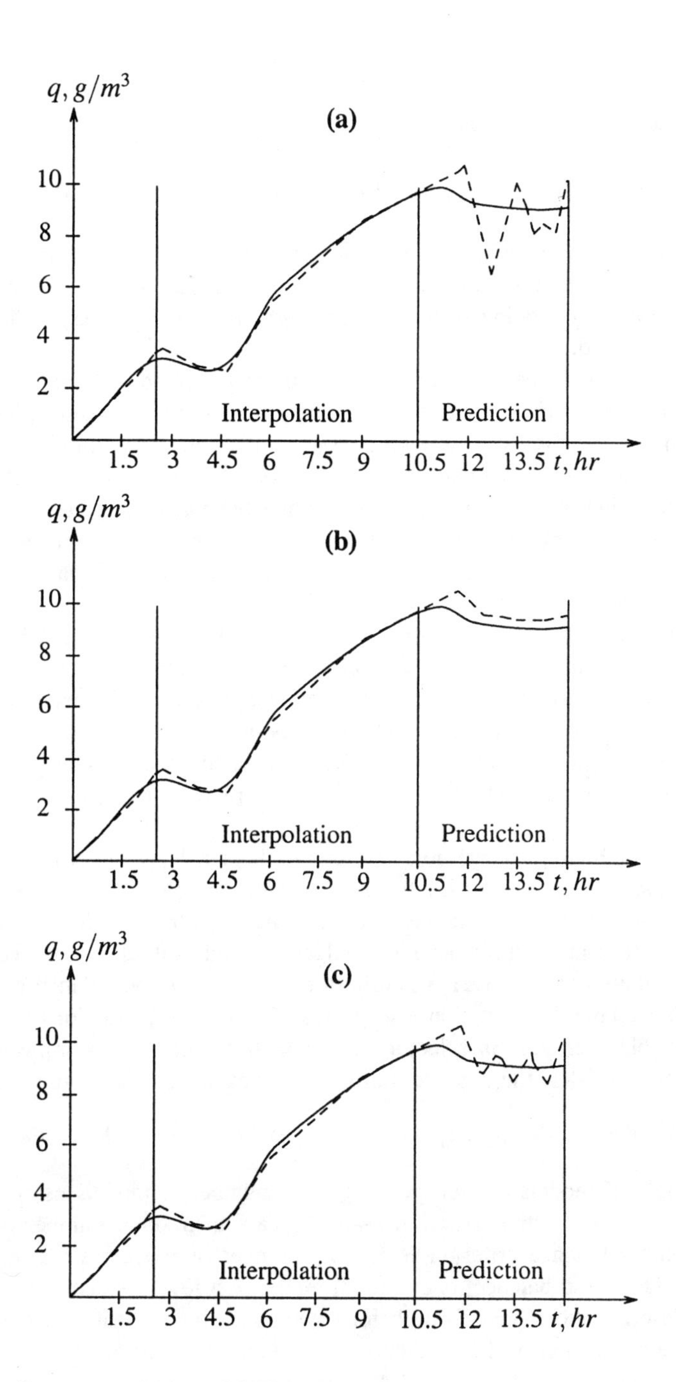

Figure 4.8. Performances of (a) model III-1, (b) model III-2, and (c) model III-3 for "close" and "remote" action principles

equations—among such processes are the non-Markov processes. Such equations contain terms such as moving averages (sometimes referred to as "summation patterns"). For example, an equation of the form

$$\int_{t_k}^{0} q dt + \frac{dq}{dt} + q = 0 \tag{4.50}$$

has a finite-difference analogue as

$$a_1 \frac{1}{k+1}(q^{t-k} + \cdots + q^{t-1} + q^t) + a_2(q^t - q^{t-1}) + a_3 q^t = 0. \tag{4.51}$$

The "summation pattern" represents the moving average of k cells in the interval of integration. In training the system, the moving averages take place along with the other arguments of the model. For each position of the pattern on the time-axis, corresponding summation patterns are considered.

The use of summation patterns for obtaining predictive models implies a change from the principle of close- or short-range action to the principle of combined action because the general pattern of the finite-difference scheme is doubly connected. In other words, during self-organization modeling, two patterns are used: one for predicting the output value and the other for the value of the sum. Predictive models have a single pattern that is based on the "principle of close action" are suitable only for short-range predictions. For example, weather forecasting for more than 15 days in advance using hydrodynamic equations (the principle of close action) is impossible.

Long-range predictions require a transfer to equations based on the principle of long-range action and combined models. In a specific sense, such models are a result of using the interior of balance of variables based on the combined principle. The external criterion that is based on a balance law allows specification of a point in the distant future, through which the integral curve of stepwise prediction passes, and selects the optimal prediction model. It enables overcoming the limit of prediction characteristic of the principle of short-range action.

The criterion of balance-of-variables (refer to Chapter 1) is the simplest way to find a definite relationship (a physical law) among several variables being simultaneously predicted. This is the basis of long-range prediction using the ring of "direct" and "inverse" functions. The ring can be applied both for algebraic and finite-difference equations. The second form of the balance-of-variables criterion is the prediction balance criterion, which fulfills the balance law. This simultaneously uses two or more predictions that differ in the interval of variable averaging in selecting the optimal model. For example, in choosing a system of monthly models the algorithm utilizes the sequence of applying the criteria

$$F_0 \to F_1(\eta_{bs}) \to F_2(B_{month}) \to F_3(B_{year}); \quad F_3 \ll F_2 \ll F_1 \ll F_0, \tag{4.52}$$

where F_1 number of models are selected out of F_0 number of models using the minimum bias criterion η_{bs} or prediction criterion i—in the case of a small number of data points. Using the monthly balance criterion B_{month}, F_2 number of models are selected from F_1. Finally, using the annual balance, one optimal model or a few models (F_3) are chosen.

Here we describe the model formulations with one-dimensional and two-dimensional readout and the realization of the prediction balance criterion for cyclic processes.

3.1 Model formulations

One-dimensional and two-dimensional models are given for comparison.

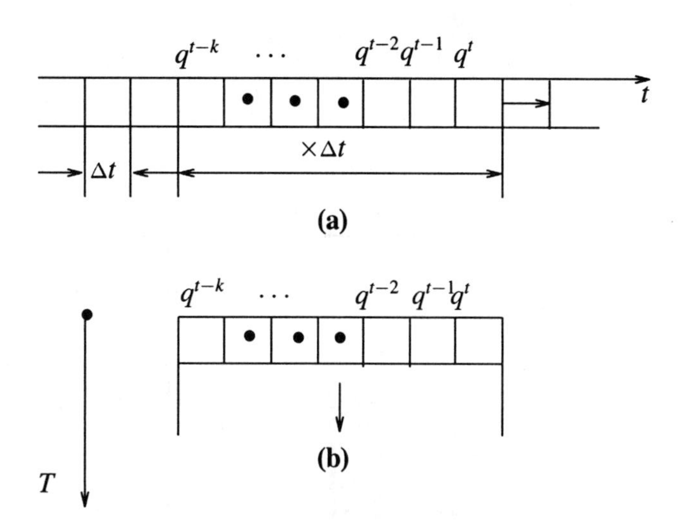

Figure 4.9. Pattern movement. The arrow indicates movement during training along (a) a $t-$axis, and (b) a $T-$axis

One-dimensional time readout

Let us assume that given a sampled data, q^t is the output value at time t depending on its delayed values $q^{t-1}, q^{t-2}, \cdots$. We have

$$q^t = f_1(t) + f_2(q^{t-1}, q^{t-2}, q^{t-3}, \cdots), \tag{4.53}$$

where f_1 is the source function, which is a trend equation as $q^t = f_1(t)$. The data, given in discrete form, is designated at equal intervals of time (Figure 4.9).

Two-dimensional time readout

If the process has an apparent repetitive (seasonal, monthly) cycle, one can also apply a two-dimensional readout. For example, let t be time measured in months and T the time measured in years. The experimental data takes the shape of a rectangular grid (Figure 4.10). The model includes the delayed arguments from both the monthly and yearly dimensions in the two-dimensional fields,

$$q_{t,T} = f_1(t,T) + f_2(q_{t-1,T}, q_{t-2,T}, \cdots, q_{t,T-1}, q_{t,T-2}, \cdots), \tag{4.54}$$

where $f_1(t,T)$ is the two-dimensional "source function"—considered two-dimensional time trend equation.

The trend functions are obtained through self-organization modeling by using the minimum bias criterion. With the one-dimensional time readout, the training of the data is carried out using its transposition along the horizontal axis t. With the two-dimensional time readout, training is done by transposing the pattern along the vertical axis T (Figure 4.11) for individual columnwise models or along the both axes (t, T) for a single model. Connecting the participating delayed arguments of the output variable provides the shape of the pattern used in the formulation.

One advantage with the data of two-dimensional time readout is that it can be used to build up a system of equations (the seasonal fluctuations in the data are taken care of by

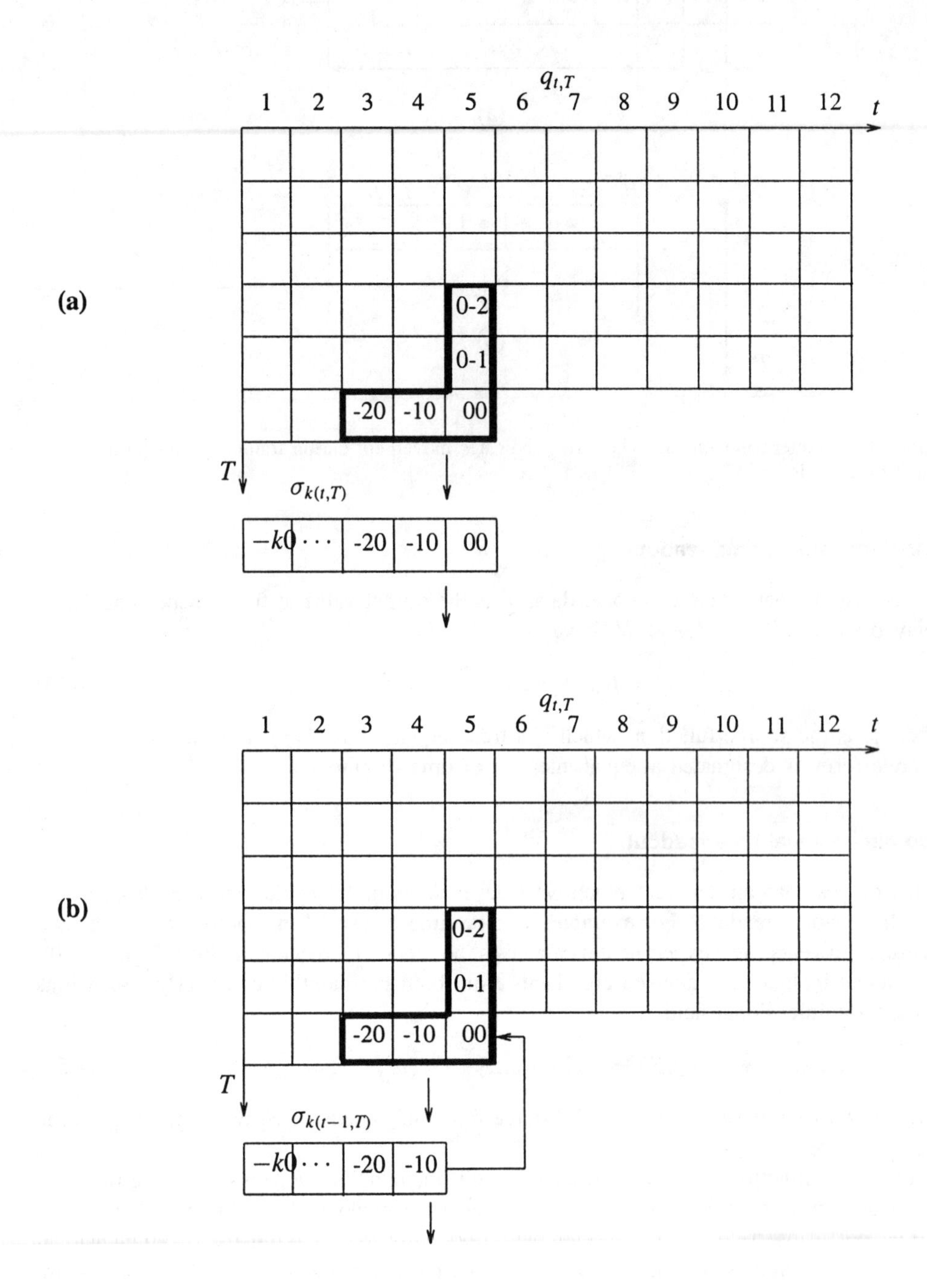

Figure 4.10. Scheme for two-dimensional time readout: (a) a model using predictions of moving averages $\sigma_{k(t,T)}$ and (b) a model using the averages $\sigma_{k(t-1,T)}$ as arguments.

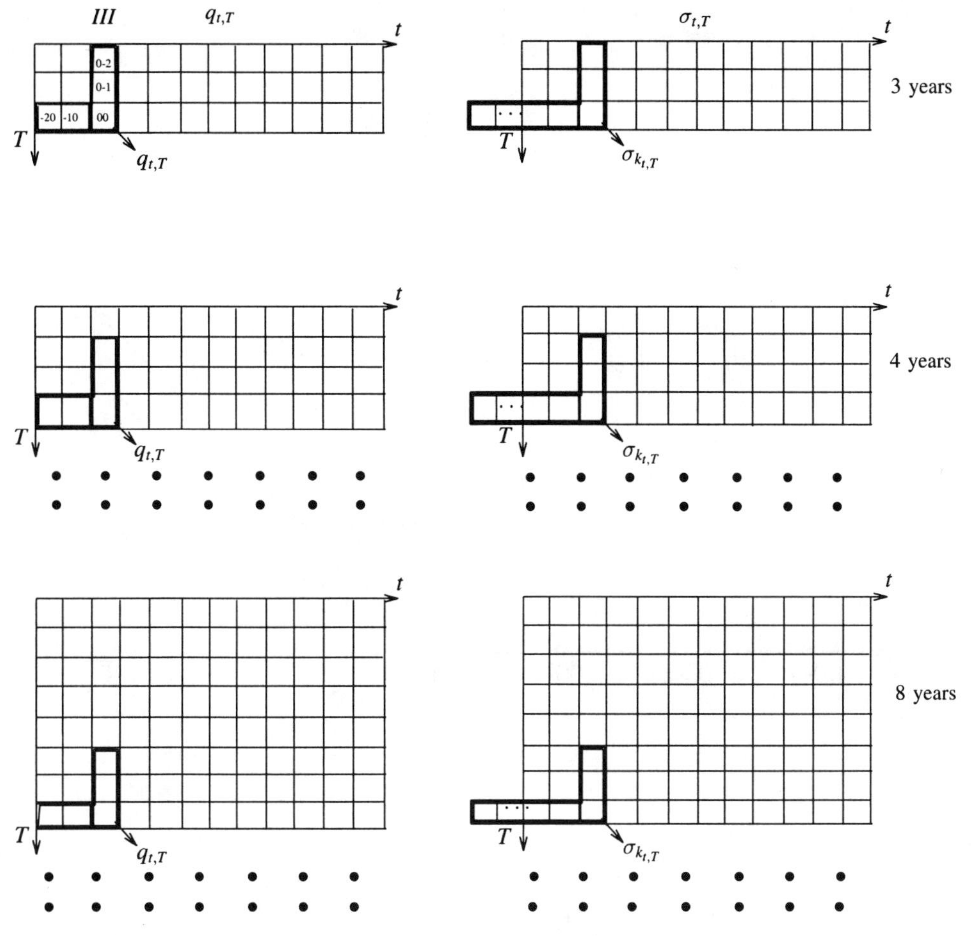

Figure 4.11. Schematic diagram for training of the model for the month of March by transposing patterns $q_{t,T}$ and $\sigma_{k(t,T)}$ along the $T-$ axis.

the system of equations). Each model in the system of equations is valid only for the given month and the system of equations (twelve monthly models) for the whole process. For a long-range prediction with stepwise integration, a transition is realized from one month"s model to the next month's model. Similarly, the idea of three-dimensional time readout can be realized in modeling cyclic processes (for example, period of solar activity; see Figure 4.12).

Moving averages In modeling of cyclic processes, one or more of the following moving averages are considered arguments of the model [65].

$$\sigma_{k_{t},T} = \frac{1}{k}(q_{t,T} + \cdots + q_{t-(k-1),T}); \quad k = 2, 3, \cdots, 12 \tag{4.55}$$

When one moving average is used, it is reasonable to select precisely that moving average which ensures the deepest minimum to the model. If all possible moving averages are used, there remain only the most significant ones—usually two averages σ_3 and σ_{12} corresponding to season and year remain more frequently than others. Moving averages can also be considered by giving weights to the individual elements.

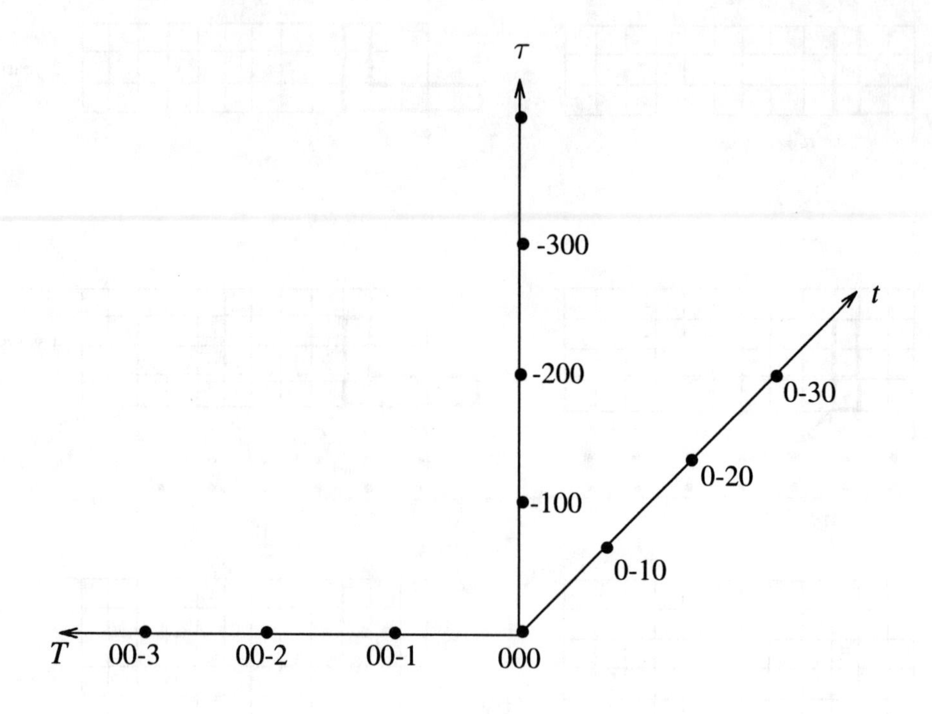

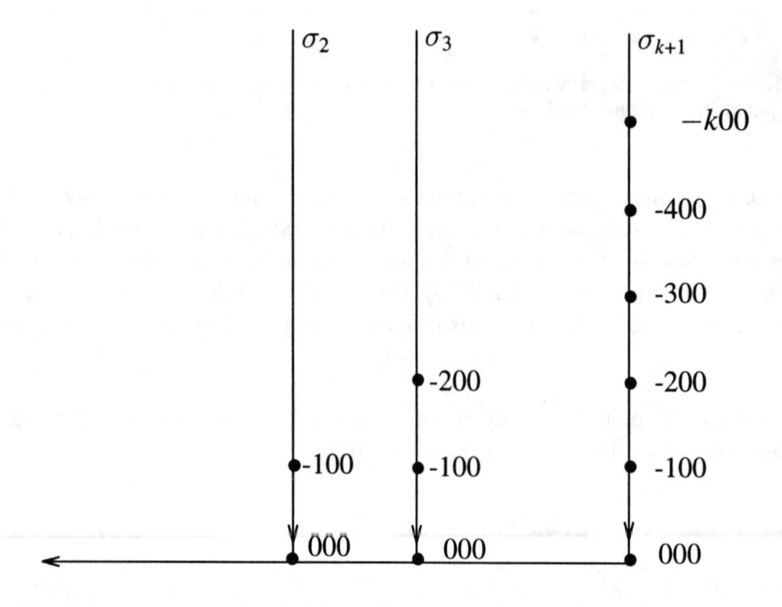

Figure 4.12. Pattern representation for three-dimensional time readout, where t represents months, T years, and τ units of 11.2 years (in case of solar activity).

Monthly models

In the two-dimensional time readout, each cell of the numeric grid (t, T) is represented through the output value q and the estimated value of a moving average σ_k. For example, a monthly prediction model has the form

$$q_{t,T} = f_1(t, T) + f_2(q_{t-1,T}, q_{t-2,T}, \cdots, q_{t,T-1}, q_{t,T-2}, \cdots,$$
$$\hat{\sigma}_{k_{t,T}}, \sigma_{k_{t-1,T}}, \sigma_{k_{t-2,T}}, \cdots, \sigma_{k_{t,T-1}}, \sigma_{k_{t,T-2}}, \cdots). \quad (4.56)$$

The estimated values $q_{t,T}$ and $\sigma_{k_{t,T}}$ are not known in the process of prediction, but the others can be determined from the initial data or by predictions. The monthly prediction model (full description) for $\sigma_{k_{t,T}}$ is

$$\hat{\sigma}_{k_{t,T}} = f_3(t, T) + f_4(\hat{q}_{t,T}, q_{t-1,T}, q_{t-2,T}, \cdots, q_{t,T-1}, q_{t,T-2}, \cdots), \quad (4.57)$$

where f_1,f_2,f_3, and f_4 are the polynomials. There are auxiliary variables that can be used in the complete descriptions.

3.2 Realization of prediction balance

The balance relation b for the prediction of sth year is expressed by

$$b_s = [\hat{\sigma}_{k_{t,T}}]_s - \frac{1}{k+1}(\hat{q}_{t,T} + q_{t-1,T} + q_{t-2,T} + \cdots + q_{t-k,T})_s, \quad (4.58)$$

where $s = 1, 2, \cdots, N$ and N is the number of years of observing process. The criterion of monthly prediction balance for each month is written as

$$B_{month} = \sum_{s=1}^{N} b_s^2. \quad (4.59)$$

It is difficult to see the feasibility of the criterion in this form because we need to know $\hat{\sigma}_{k_{t,T}}$ to predict $\hat{q}_{t,T}$. We need to know $\hat{q}_{t,T}$ to predict $\hat{\sigma}_{k_{t,T}}$. This requires a recursive procedure. Assuming the initial value $\hat{q}_{t,T} = 0$, we find $\hat{\sigma}_{k_{t,T}}$, the second value $\hat{q}_{t,T}$, and so on until the value of the criterion B_{month} decreases. It is necessary to eliminate either $\hat{\sigma}_{k_{t,T}}$ or $\hat{q}_{t,T}$ from the composition of the arguments. (Possible simplification follows below.)

The monthly prediction model for q is

$$\hat{q}_{t,T} = \hat{\sigma}_{1_{t,T}} = f_1(t, T) + f_2(\hat{\sigma}_{k_{t,T}}, \sigma_{k_{t-1,T}}, \sigma_{k_{t-2,T}}, \cdots, \sigma_{k_{t,T-1}}, \sigma_{k_{t,T-2}}, \cdots,$$
$$q_{t-1,T}, q_{t-2,T}, \cdots, q_{t,T-1}, q_{t,T-2}, \cdots). \quad (4.60)$$

The monthly prediction model for σ_k is

$$\hat{\sigma}_{k_{t,T}} = f_3(t, T) + f_4(q_{t-1,T}, q_{t-2,T}, \cdots, q_{t,T-1}, q_{t,T-2}, \cdots,$$
$$\sigma_{k_{t-1,T}}, \sigma_{k_{t-2,T}}, \cdots, \sigma_{k_{t,T-1}}, \sigma_{k_{t,T-2}}, \cdots). \quad (4.61)$$

The criterion of monthly balance remains unchanged and is in usable form with the simplification in the formulations.

$$B_{month} = \sum_{i=3}^{N} b_i^2. \quad (4.62)$$

The sequential application of criteria is according to the scheme $F_0 \rightarrow F_1(\eta_{bs}) \rightarrow F_2(B_{month}) \rightarrow F_3(B_{year})$.

The patterns of the above models are doubly connected (Figure 4.10). One can use the expanded set of arguments and can also eliminate the predicted value of $\sigma_{k_{t,T}}$. One can use the combined criterion of "minimum bias plus prediction" in place of minimum bias criterion. When a small number of data points are used, minimum bias criterion can be replaced by the prediction criterion for step-by-step predictions of λ months ahead.

$$
\begin{aligned}
i^2(\lambda) &= \frac{\sum_{k=1}^{\lambda}(q_k - \hat{q}_k)^2}{\sum_{k=1}^{\lambda} q_k^2} \\
I^2(\lambda) &= \frac{1}{N}\sum_{s=1}^{N}[i^2(\lambda)]_s.
\end{aligned}
\tag{4.63}
$$

For example, let us assume that $\lambda = 3$. To select models for the month of March, one must obtain all possible models for March, April, and May. The predictions of these models are used sequentially in computing the prediction criterion error. To obtain the data, the patterns are used along the t, T-coordinate field as indicated in Figure 4.11.

$$
\begin{aligned}
[i^2(3)]_s &= [\frac{\sum_{i=III}^{V}(q_i - \hat{q}_i)^2}{\sum_{i=III}^{V} q_i^2}]_s \\
I^2(3) &= \frac{1}{N}\sum_{s=1}^{N}[i^2(3)]_s.
\end{aligned}
\tag{4.64}
$$

The criterion $I(3)$ demands that the average error in predictions that consider a three-month model should be minimal. This determines the optimal March model; F_1 number of March models are selected. Usually F_1 is not greater than two to three models.

The criterion of yearly balance is used in selecting all 12 models; one model for each month is selected such that the system of 12 models would give the maximum assurance of the most precise prediction for the year.

$$
B_{year}^2 = \frac{1}{N}\sum_{s=1}^{N}[\bar{q}_{year} - \frac{1}{12}(\hat{q}_I + \hat{q}_{II} + \cdots + \hat{q}_{XII})]_s^2, \tag{4.65}
$$

where $\bar{q}_{year}$ is the average yearly value computed directly and used in training. The predictions ($\hat{\bar{q}}_{year}$) can be obtained by using a separate algorithm, such as a harmonic algorithm, while the B_{year} is calculated.

Various sequences of applying criteria can be written as

$$
\begin{aligned}
&F_0 \rightarrow F_1(\eta_{bs}) \rightarrow F_2(B_{month}) \rightarrow 1(B_{year}), \\
&F_0 \rightarrow F_1(c3) \rightarrow F_2(B_{month}) \rightarrow 1(B_{year}), \\
&F_0 \rightarrow F_1[I(\lambda)] \rightarrow F_3(B_{year}), \\
&F_0 \rightarrow F_1(c3) \rightarrow F_3(B_{year}),
\end{aligned}
\tag{4.66}
$$

and so on. The selection of sequence differs in a number of ways depending on the mathematical formulation, availability of data, and user's choice.

3.3 Example—Modeling of tea crop productions

Example 5. Modeling of a cyclic process such as tea crop production is considered here [59], [87].

First we give a brief description of the system. The cultivation of tea on a large scale is only about 100 years old. North Indian tea crop production accounts for 5/6 percent of the country's tea output. Tea is cultivated in nearly all the subtropics and mountainous regions of the tropics. When dormant, the tea shrub withstands temperatures considerably below freezing point, but the northern and southern limits for profitable tea culture are set by the freezing point. A well-distributed annual rain fall of 150 to 250 cms. is good for satisfactory growth. Well drained, deep friable loam or forest land rich in organic matter is ideal for growing the tea crop. Indian tea soils are low in lime content and therefore somewhat acidic. The subsoil should not be hard or stiff. The fertilizer mixtures of 27 kg. of N, 14 kg. of P_2O_5 and 14 kg. of K_2O per acre are applied in one or two doses.

In North India tea leaves are plucked at intervals of seven to ten days from April to December; whereas in the South plucking is done throughout the year at weekly intervals during March to May (the peak season) and at intervals of 10 to 14 days during other months. The average yield per acre is about 230 to 280 kg. of processed tea. Vegetatively propagated clones often give as much as 910 kg. of tea per acre. The quality of tea depends not only on the soil and the elevation at which the plant is grown, but also on the care taken during its cultivation and processing.

Here two cases are considered: one for modeling of North Indian tea crop productions and another for South Indian tea crop productions. The weather variables, such as mean monthly sunshine hours, mean monthly rain fall, and mean monthly water evaporation (data collected from the meteorological stations during the same period), can be used in the modeling.

The following sets of variables are considered for the model formulations.

$$\begin{aligned}
(t, T) &\in \mathcal{T} \\
(q_{t-1,T}, q_{t-2,T}, \cdots, q_{t,T-1}, q_{t,T-2}, \cdots) &\in \mathcal{P} \\
(\sigma 2_{t-1,T}, \sigma 3_{t-2,T}, \cdots, \sigma 12_{t-1,T}) &\in \sigma \\
(S_{t,T}, S_{t-1,T}, S_{t-2,T}, \cdots, S_{t,T-1}, S_{t,T-2}, \cdots,& \\
R_{t,T}, R_{t-1,T}, R_{t-2,T}, \cdots, R_{t,T-1}, R_{t,T-2}, \cdots,& \\
E_{t,T}, E_{t-1,T}, E_{t-2,T}, \cdots, E_{t,T-1}, E_{t,T-2}, \cdots) &\in \mathcal{Z},
\end{aligned} \tag{4.67}$$

where t and T are the time coordinates measured in months and years, respectively; $q_{t,T}$ is considered the output variable measured at the coordinates of (t, T); $q_{t-i,T}$ and $q_{t,T-j}$ are the delayed arguments at i units in months and j units in years, correspondingly; $\sigma k_{t-1,T} = \frac{1}{k}(q_{t-1,T} + q_{t-2,T} + \cdots + q_{t-k,T})$ are the moving averages of length k. The weather variables $S, R,$ and E represent sunshine hours, rainfall, and water evaporation, correspondingly.

In modeling North Indian tea crop productions the following model formulation is adopted for each month.

I-1.

$$q_{t,T} = f(\mathcal{P}, \sigma). \tag{4.68}$$

Because of a small number of data points, the complete polynomial below is used as reference function for each month.

$$q_{t,T} = a_0 + a_1 q_{t-1,T} + a_2 q_{t,T-1} + a_3 \sigma 3_{t-1,T} + a_4 \sigma 6_{t-1,T}. \tag{4.69}$$

The sequence of criteria, which has shown better performance than other sequences, is shown here.

$$F_0 \rightarrow F_1(\eta_{bs}) \rightarrow F_2[I(3)] \rightarrow 1(B_{year}). \tag{4.70}$$

The total number of data points correspond to eleven years; $N_A = 5, N_B = 5$, and $N_C = 1$. The coefficient values of the best system of monthly models are given as

Month i	a_0	a_1	a_2	a_3	a_4
1	0.318	0.026	−0.366		
2	−0.010	0.022	−0.384	0.013	
3	−6.730				0.452
4	−0.620		−0.084		1.309
5	0.276	−1.040			4.335
6	35.350	−0.820	−0.174	2.289	
7	18.010		0.321		1.227
8	67.730	−1.124		−1.459	4.571
9	18.110	0.313			0.576
10	−10.340	1.110			
11	−23.850	−1.017	−0.293		2.485
12	−10.530			3.498	−2.933

The blank spaces indicate that the corresponding variable does not participated in the model. The prediction error on the final-year data is computed as 0.0616. The system of monthly models is checked for stability in a long-range perspective.

In modeling South Indian tea crop productions, five types of model formulations are considered as complete polynomials that are studied independently.

Different formulations

II-1.

$$q_{t,T} = f(\tau, \mathcal{P}, \sigma, \mathcal{Z}), \tag{4.71}$$

where f is a single function that considers all variables. It is considered a one-dimensional model that represents the system.

II-2.

$$q_{t,T} = f_1(\tau) + f_2(\mathcal{P}, \sigma, \mathcal{Z}), \tag{4.72}$$

where f_1 is the trend function in two time coordinates; f_2 is the function of delayed arguments, moving averages, and other input variables. Use of the two-dimensional time trend function is preferred when the initial data is noiseless and when individual components of the cyclic processes that have a character of time variation have no effect. The behavior of f_2 is supposed to be effected by these variables.

This formulation is evaluated in two levels. First, the trend function is estimated based on whole data, residuals are computed, and the function f_2 is estimated using the residuals. The final prediction formulation will be the summation of both.

II-3.

$$q_{t,T} = f_i(\tau, \mathcal{P}, \sigma); \;\; i = 1, 2, \cdots, 12. \tag{4.73}$$

This is similar to the formulation II-1, but represents the system of 12 monthly models; 12 separate prediction formulas f_i for each month.

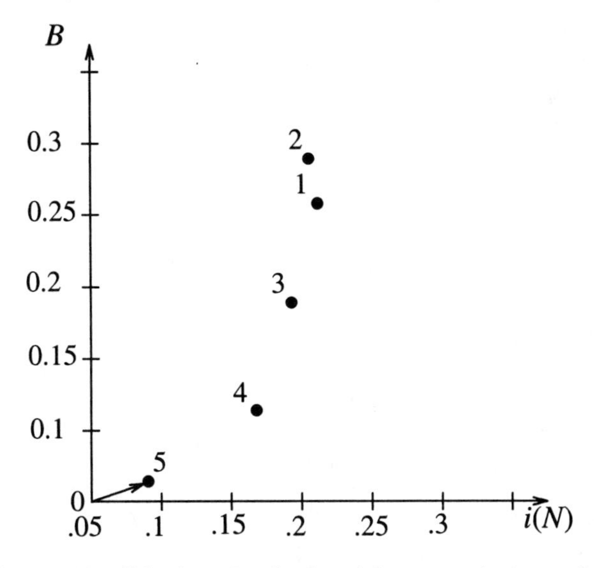

Figure 4.13. Selection of optimal model on two criterion analysis

II-4.

$$q_{t,T} = f_1(\tau) + f_{2_i}(\mathcal{P}, \sigma); \quad i = 1, 2, \cdots, 12. \tag{4.74}$$

This is similar to the formulation II-2, but has a system of 12 monthly models at the second level. The trend function f_1 is a single formula, as in the formulation II-2. The residuals are computed on all data; this data is used for identifying the system of 12 monthly models f_{2_i}.

II-5.

$$q_{t,T} = f_{1_i}(\tau) + f_{2_j}(\mathcal{P}, \sigma); \quad i, j = 1, 2, \cdots, 12. \tag{4.75}$$

Time-trend equations for each month are separately identified; in other words, the function $f_{1_i}(\tau)$ is considered a function of T for each month. The residuals are computed and the second set f_{2_j} of the system of monthly models are obtained. This makes a set of combined models for the system.

Each formulation is formed for its complete polynomial; combinatorial algorithm is used in each case for sorting all possible combinations of partial polynomials as "structure of functions." The optimal models obtained from each case are compared further for their performance in predictions. The scheme of the selection criteria is

$$F_0 \rightarrow F_1(c3) \rightarrow F_2[i(N)] \rightarrow 1(B_{year}), \tag{4.76}$$

where $c3$ is the combined criterion with "minimum bias (η_{bs}) plus prediction ($i(W)$)," $i(W)$ being the prediction criterion used for step-by-step predictions on the set W, and $i(N)$—the whole data set N.

The data used in this case belong to ten years; $N_A = 4$, $N_B = 4$, and two years data is preserved for checking the models in the prediction region. The simplest possible pattern is considered for the formulations II-3, II-4, and II-5, because of the availability of a few collected data. In the monthly models the weather variables are not considered for simplicity. One can see the influence of such external variables in the analysis of cyclic processes. All

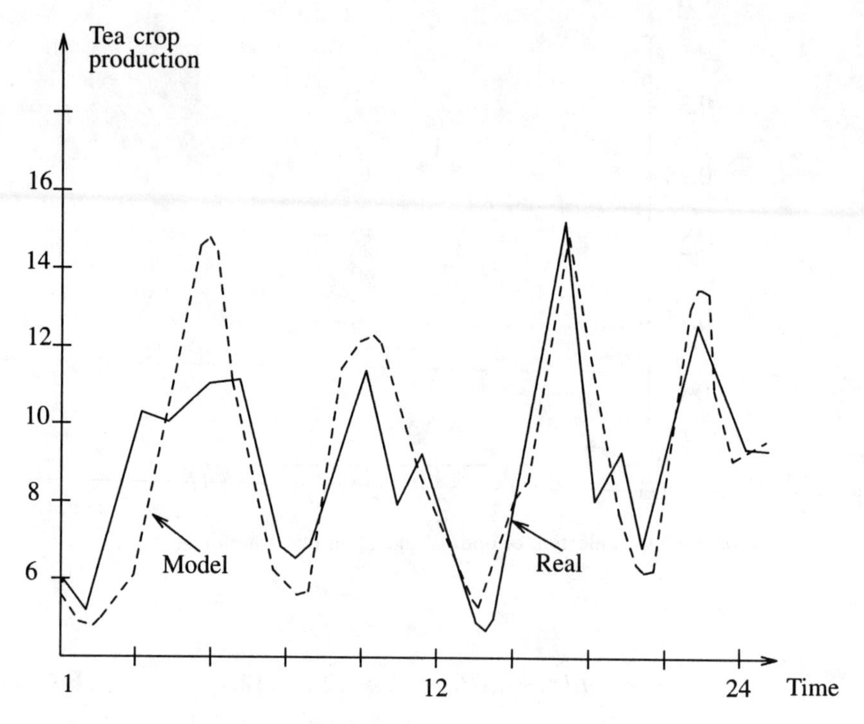

Figure 4.14. Performance of the best model

optimal models are compared for their step-by-step predictions of up to ten years and tested for their stability in long-range actions. The results indicate that the formulation II-5 has optimal ability in characterizing the stable prdictions (shown in Figures 4.13 and 4.14). The system of monthly models in an optimum case is given below; first, the set of time trend models is

$$f_{1_i}(T) = a_0 + a_1 T + a_2 T^2 + a_3 T^3 + a_4 T^4, \tag{4.77}$$

where

Month i	a_0	a_1	a_2	a_3	a_4
1	5.801	0.048			
2	12.380	−1.587		0.070	−0.006
3	6.289				−.000012
4	9.454	−0.082			
5	10.364	0.368			
6	8.141	0.937	−.070		
7	6.009	0.057			
8	5.665	0.076			
9	7.678	0.316	−.013		
10	6.588	1.917	−.258		
11	6.288				−.000012
12	5.659	0.076			

and the set of remainder models is

$$f_{2_i}(\mathcal{P}, \sigma) = b_1 q_{t-1,T} + b_2 q_{t,T-1} + b_3 \sigma_{3_{t-1,T}} + b_4 \sigma_{12_{t-1,T}}, \tag{4.78}$$

where

Month i	b_1	b_2	b_3	b_4
1	0.626		−0.317	−0.260
2	0.168		−0.923	0.292
3	1.489		−1.501	−2.611
4	−0.774		−1.976	3.481
5	1.189		−2.196	1.215
6	−0.514		3.176	−1.588
7	−0.162		−0.647	1.188
8	0.355		0.109	−0.587
9	0.334	−0.034		−0.190
10	−0.298	−0.039		0.132
11	1.630		0.931	−3.238
12	−0.069			0.205

The blank space indicates that the corresponding variable does not participate in the monthly model. These two sets of monthly model systems form the optimal model for an overall system.

I-2 & II-6. Here is another idea for forming a model formulation which is not discussed above. This considers a harmonical trend at the first level instead of time trend.

$$q_{t,T} = f_1(\sin wt, \cos wt) + f_{2_i}(\mathcal{P}, \sigma, \mathcal{Z}), \tag{4.79}$$

where f_1 represents a single harmonic function for the whole process with the arbitrary frequencies and f_{2_i} is the system of monthly models. At the first level the harmonic trend is obtained as $q_t = f_1(\sin wt, \cos wt)$ using the harmonical inductive algorithm. Residuals ($\Delta q_t = q_t - \hat{q}_t$) are computed using the harmonic trend, then the system of monthly models are estimated as in the above cases.

The first level of operation for obtaining the harmonical trend of tea crop productions is shown. The data q_t is considered a time series data of mean monthly tea crop productions. The function f_1 is the sum of m harmonic components with pairwise distinct frequencies w_k, $k = 1, 2, \cdots, m$.

$$f_1 = \sum_{k=1}^{m} (A_k \sin w_k t + B_k \cos w_k t), \tag{4.80}$$

where $w_i \neq w_j$, $i \neq j$; $0 < w_k < \pi$, $k = 1, 2, \cdots, m$. The function is defined by its values in the interval of data length $N (1 \leq t \leq N)$.

The initial data is divided into training N_A, testing N_B, and examining N_C points. The maximum number of harmonics is $m_{max} lN/3$ $(1 \leq m \leq m_{max})$. The sorting of the partial trends that are formed based on the combination of harmonics is done by the multilayer selection of trends. In the first layer, the freedom of choice F best harmonics are obtained by the selection criterion on the basis of the testing sequence, the remainders are then calculated. In the second layer, the procedure is continued using the data of remainders and is repeated in all subsequent layers. Finally F best harmonics are selected. The complexity of the trends increases as long as the value of the "inbalance" decreases (refer to Chapter 2 for details on the harmonic algorithm). In the last layer, the unique solution corresponding to the minimum of the criterion is selected. As this algorithm is based on the data of remainders, the sifting of harmonics can be stopped usually at the second or third layer.

The data is separated into $N_A = 90\%$, $N_B = 6\%$, and $N_C = 4\%$, and m_{max} is considered as eight in these cases.

In North Indian tea crop productions model, the structure of the optimal harmonic trend is obtained as

$$\hat{q}_t = \sum_{i=1}^{l}\sum_{j=1}^{m_i}(A_{ij}\sin w_{ij}t + B_{ij}\cos w_{ij}t), \tag{4.81}$$

where $\hat{q}_t$ is the estimated output, l is the number of layers, m_i are the number of harmonic components at each layer, and the parameters for $l = 3$ are given as

Layer	Components	Frequency	Coefficients	
i	m_i	w_{ij}	A_{ij}	B_{ij}
1	7	0.523	−24.64	−13.09
		0.693	−0.64	0.23
		1.052	−1.24	−2.13
		1.570	−0.60	−2.50
		1.988	0.28	−0.08
		2.285	−0.63	0.16
		2.775	0.53	−0.13
2	1	4.598	0.20	0.18
3	6	0.458	−0.68	0.69
		0.917	−0.37	0.07
		1.278	−0.32	0.16
		1.847	−0.15	0.82
		2.203	−0.12	0.22
		2.699	−0.21	−0.23

The root mean square (RMS) error on overall data is achieved as 0.0943.

In South Indian tea crop productions modeling, the data is initially smoothed to reduce the effect of noise by taking moving averages as

$$\bar{q}_t = \frac{1}{L}\sum_{k=1}^{L} q_{t+k-1}. \tag{4.82}$$

This transformation acts as a filter that does not change the spectral composition of the process, but changes only the amplitude relation of the harmonic components [130]. The harmonic trend for $\bar{q}_t$ can be written as

$$\bar{q}_t = \frac{1}{L}\sum_{k=1}^{L}\sum_{j=1}^{m}[A_j \sin w_j(t+k-1) + B_j \cos w_j(t+k-1)]. \tag{4.83}$$

After simple transformations, this can be reduced to the form:

$$\bar{q}_t = \sum_{j=1}^{m}(\bar{A}_j \sin w_j t + \bar{B}_j \cos w_j t). \tag{4.84}$$

The filtered data is used for obtaining the harmonic trend. For fixing the optimal smoothing interval, the length of the summation interval L was varied from one to ten. For $L < 3$, the algorithm was not effective. L_{opt} is achieved at 4 because it is not expedient to greatly increase the value of L (Table 4.5). The optimal harmonic components for $l = 3$ and $L = 4$ are listed as

$$\hat{\bar{q}}_t = \sum_{i=1}^{l}\sum_{j=1}^{m_i}(A_{ij}\sin w_{ij}t + B_{ij}\cos w_{ij}t), \tag{4.85}$$

Table 4.5. Effect of smoothing interval on the noisy data

L	RMS error
3	0.06779
4	0.05579
5	0.07321
6	0.06375
7	0.06648
8	0.05730
9	0.05647
10	0.04957

where $\hat{q}_t$ is the estimated filtered output;

Layer	Components	Frequency	Coefficients	
i	m_i	w_{ij}	A_{ij}	B_{ij}
1	4	0.486	−0.25	0.70
		0.846	−0.01	0.08
		1.073	0.08	−0.38
		2.371	−0.002	0.001
2	5	0.282	0.002	0.33
		0.508	0.309	0.03
		0.721	−0.26	0.04
		1.016	0.11	0.32
		1.193	−0.005	−0.03
3	3	0.452	0.41	−0.21
		0.853	0.03	−0.01
		1.236	−0.03	0.05

The RMS error on the filtered data is achieved as 0.05579. Part of the prediction results are shown in Figure 4.15.

3.4 Example—Modeling of maximum applicable frequency (MAF)

Example 6. Modeling of maximum applicable frequency (MAF) of the reflecting ionospheric layer [43].

This example shows the applicability of self-organization method using the two-level prediction balance criterion for constructing short-range hourly forecasting models for the process of MAF variations at a preassigned point of the reflecting ionospheric layer. The general formulation of the models for the process of MAF variations can be set down as follows:

$$q^{t+1} = f(q^{t-\tau}, t) \tag{4.86}$$

$$q^{t+1} = f(q^{t-\tau}, t, u), \tag{4.87}$$

where q^{t+1} is the MAF value at the time $t+1$ in MHz; $q^{t-\tau}$ is the delayed argument of q at the time $t-\tau$; t is the time of the day and u is the vector of the external perturbations. The size of the MAF is influenced by a large number of external perturbations, such as solar activity, agitation of the geomagnetic field, interplanetary magnetic field, cosmic rays, and

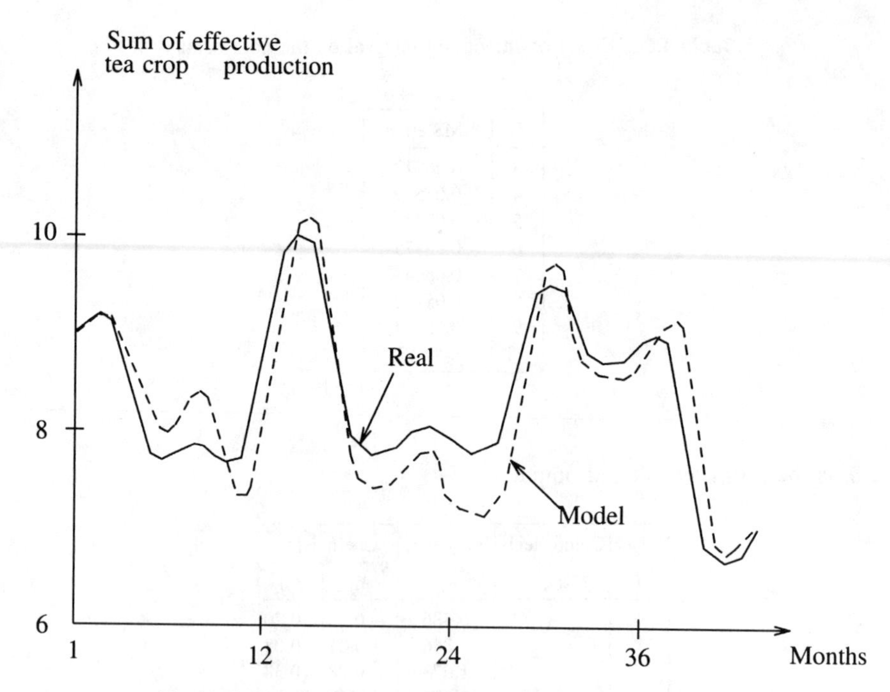

Figure 4.15. Performance of the harmonic model with $L = 4$

so on. These perturbations are estimated by several indices, such as the K- and A-indices and the geomagnetic field components H, F, D, etc.

Here the scope of the example is limited to the use of first formulation to compare the performances of individual models and system of equations. The combinatorial inductive algorithm is used in synthesizing the models.

Experiment 1. Because MAF variations depend on the time of the day, time of day is considered one of the arguments. The following complete polynomial is considered in the first experiment.

$$\begin{aligned} q^t &= a_0 + a_1 t_t + a_2 q^{t-1} + a_3 q_{t-2} + \cdots + a_k q^{t-\tau} \\ &\quad + a_{k+1} q^{t-1} t^{t-1} + a_{k+2} q^{t-2} t^{t-2} + \cdots + a_{2k} q_{t-\tau} t^{t-\tau}, \end{aligned} \tag{4.88}$$

where $t^t, t^{t-1}, \cdots, t^{t-\tau}$ are the time values corresponding to the output variable and its delayed arguments.

Observations are made for five days and 65 data points were tabulated. Two series of data are made up: one for interval of small variations (from 8AM to 8PM), another for interval of sharp variations (from 8PM to 8AM). For these two types intervals of data, individual models are constructed considering $\tau = 5$. The prediction criterion i is used to select these models; for an interval of small MAF variation

$$q^t = 18.13 + 0.0229 q^{t-3} t^{t-3} - 0.0183 q^{t-5} t^{t-5}. \tag{4.89}$$

For an interval of sharp MAF variation

$$\begin{aligned} q^t &= 10.26 + 0.6554 t^t + 0.3299 q^{t-1} + 0.1109 q^{t-2} - 0.1802 q^{t-5} \\ &\quad + 0.0013 q^{t-1} t^{t-1} - 0.0078 q^{t-2} t^{t-2} + 0.0106 q^{t-3} t^{t-3} \\ &\quad - 0.0138 q^{t-4} t^{t-4} + 0.0152 q^{t-5} t^{t-5}. \end{aligned} \tag{4.90}$$

In addition to the above, another model is constructed without having to divide the data into separate segments.

$$q^t = 8.55 + 0.0413t^t + 0.359q^{t-1} + 0.0289q^{t-2} + 0.2283q^{t-3} - 0.0351q^{t-4}$$
$$0.0032q^{t-1}t^{t-1} + 0.0066q^{t-5}t^{t-5}. \qquad (4.91)$$

Figure 4.16a demonstrates the performance of predictions of these models. The thin line indicates the actual MAF variations for 12 hours ahead, the thick line is for predictions using two individual models, and the broken line is the predictions using the single model. Two individual models are considerably more accurate in comparison to the single model.

Experiment 2. Here two-dimensional readout (t, T) is used—t indicates the time in hours and T indicates the time in days. The value of the process output variable q is taken as the average for each hour. The complete polynomial is considered as:

$$q_{t,T} = f_t(q_{t-1,T}, q_{t-2,T}, \cdots, q_{t-\tau_t,T}, q_{t,T-1}, q_{t,T-2}, \cdots,$$
$$q_{t,T-\tau_T}, \sigma_{2_{t-1,T}}, \sigma_{3_{t-1,T}}, \cdots, \sigma_{L_{t-1,T}}), \qquad (4.92)$$

where $t = 1, \cdots, 24$; τ_t and τ_T are the limits of the delayed arguments on both directions t and T, correspondingly. $\sigma_{k_{t-1,T}}, k = 2, 3, \cdots, L$ are the moving averages, maximum length of L considered.

Combinatorial algorithm is used to select the F variants of 24 models in relation to the combined criterion of "minimum bias plus regularity." From these F variants of 24 hourly models, one model—the best set of 24 models—is chosen according to the prediction balance criterion,

$$B^2_{day} = \sum_{j=1}^{N}(\bar{q}_j - \frac{1}{24}\sum_{i=1}^{24}\hat{q}_{i,j})^2, \qquad (4.93)$$

where $\bar{q}_j, j = 1, 2, \cdots, N$ are the daily averages of MAF variations for N days; $\hat{q}_{i,j}, i = 1, 2, \cdots, 24,\ j = 1, 2, \cdots, N$ are the estimated values of the hourly values using the hourly models by step-by-step predictions given the initial values. The hourly data was collected for 25 days and arranged in two-dimensional readout. The system of equations obtained are

$$q_{1,T} = -0.298 + 0.45q_{24,T} + 0.459\sigma_{6_{24,T}},$$
$$q_{2,T} = 0.497 + 0.892q_{1,T},$$
$$q_{3,T} = 1.929 + 0.8q_{2,T},$$
$$q_{4,T} = -0.208 + 0.289q_{3,T} - 0.399q_{2,T} + \sigma_{3_{3,T}},$$
$$q_{5,T} = 2.006 + 1.715q_{4,T} - 0.979q_{3,T} - 2.052\sigma_{3_{4,T}} + 0.199\sigma_{6_{4,T}},$$
$$q_{6,T} = 3.13 + 0.814q_{5,T} - 0.057q_{6,T-1},$$
$$q_{7,T} = 2.58 + 0.619q_{6,T} - 0.263q_{5,T},$$
$$q_{8,T} = 1.51 + 0.072q_{8,T-1} + 1.11\sigma_{3_{7,T}},$$
$$q_{9,T} = -0.607 - 0.308q_{8,T} + 1.644q_{7,T} + 0.636\sigma_{3_{8,T}} - 0.773\sigma_{6_{8,T}},$$
$$q_{10,T} = 3.596 + 0.788q_{9,T},$$
$$q_{11,T} = 0.816 + 1.084q_{10,T} - 0.156q_{11,T-1},$$
$$q_{12,T} = 3.54 - 0.309q_{12,T-1} + 1.107\sigma_{3_{11,T}},$$
$$q_{13,T} = -0.857 + 1.089\sigma_{3_{12,T}},$$
$$q_{14,T} = -0.824 + 0.069q_{14,T-1} + 2.584\sigma_{3_{13,T}} - 1.576\sigma_{6_{13,T}},$$

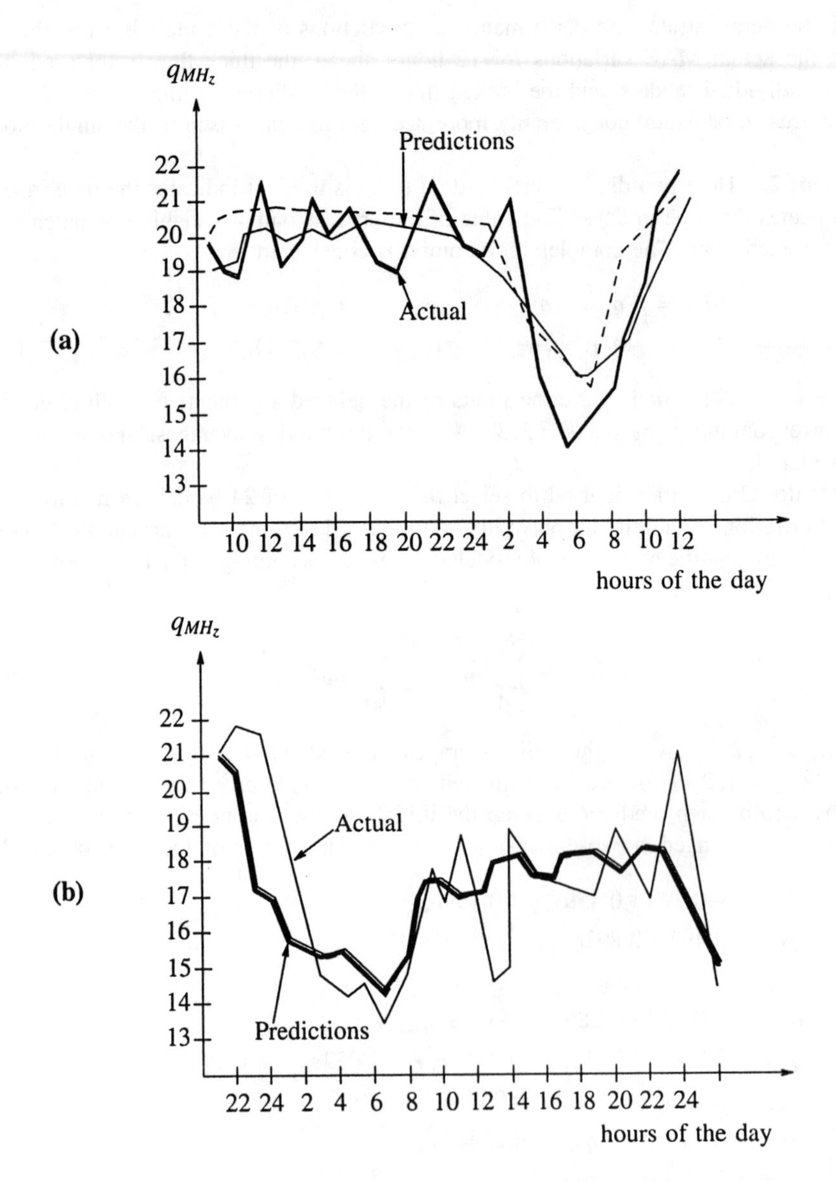

Figure 4.16. (a) Predictions using individual models and (b) predictions using system of equations

$$\begin{aligned}
q_{15,T} &= 5.12 + 0.8q_{14,T} - 0.107q_{15,T-1}, \\
q_{16,T} &= 5.104 + 0.806q_{15,T} - 0.988\sigma_{3_{15,T}} + 0.887\sigma_{6_{15,T}}, \\
q_{17,T} &= 0.822 + 1.151q_{16,T} - 0.058q_{17,T-1} + 0.493\sigma_{3_{16,T}} - 0.598\sigma_{6_{16,T}}, \\
q_{18,T} &= 2.976 + 1.172\sigma_{3_{17,T}} - 0.311\sigma_{6_{17,T}}, \\
q_{19,T} &= -9.478 + 3.079q_{18,T} + 0.302q_{19,T-1} - 1.908\sigma_{3_{18,T}}, \\
q_{20,T} &= -2.436 + 1.119q_{19,T}, \\
q_{21,T} &= 3.056 + 2.592q_{20,T} + 2.774q_{19,T} - 5.734\sigma_{3_{20,T}} + 1.225\sigma_{6_{20,T}}, \\
q_{22,T} &= -4.317 - 0.727q_{20,T} + 0.023q_{22,T-1} + 1.595\sigma_{3_{21,T}}, \\
q_{23,T} &= 10.57 + 2.316q_{22,T} - 0.85\sigma_{3_{22,T}} - 1.087\sigma_{6_{22,T}}, \\
q_{24,T} &= 1.683 + 0.797\sigma_{3_{23,T}}
\end{aligned} \tag{4.94}$$

Figure 4.16b exhibits the actual and forecast values of the MAF variations on 24-hour duration of the interval considered. It shows that the models of this class select a basically regular cyclical component in the process.

Inductive algorithms make it possible to synthesize more universal models to forecast both regular and abrupt irregular MAF variations by providing the information on external perturbations. This also makes it possible to raise forecast accuracy and anticipation time by using prediction balance criterion with two-dimensional time readout.

Chapter 5
Clusterization and Recognition

1 SELF-ORGANIZATION MODELING AND CLUSTERING

The inductive approach shows that the most accurate predictive models can be obtained in the domain of nonphysical models that do not possess full complexity. This corresponds to Shannon's second limit theorem of the general communication theory. The principle of self-organization is built up based on the Gödel's incompleteness theorem. The term "self-organization modeling" is understood as a sorting of many candidates or partial models by the set of external criteria with the aim of finding a model with an optimal structure.

A "fuzzy" object is an object with parameters that change slowly with time. Let us denote N as a number of data points and m as a number of variables. For $N < m$, the sample is called short and the object "fuzzy" (under-determined). The greater the ratio m/N, the "fuzzier" the object.

By describing the relationships, clustering is considered a model of an object in a "fuzzy" language. Sorting of clusters with the aim of finding an optimal cluster is called "self-organization clustering." Although self-organization clustering has not yet been developed in detail, it has adapted the main principles and practical procedures from the theory of "self-organization modeling." This chapter presents the recent developments of self-organization clustering and nonparametric forecasting and explains how the principles of self-organization theory are applicable for identifying the structure of the most accurate and unbiased clusterizations.

Analogy with Shannon's approach

Structural identification by self-organization modeling is directed not only toward obtaining a physical model, but also toward obtaining a better, and not overly complicated, prediction model. The theoretical basis of this statement is taken from the communication theory by Shannon's second-limit theorem for transmission channels with noise. The optimal complexity of clusterizations is required as the optimal frequency passband in a communication system. Complexity must decrease as the variance of noise increases. The complexity of the models to be evaluated is often measured by the number of parameters and the order of the equation. The complexity of clusterization is usually measured by the number of clusters and attributes. The complexity of a model or clusterization is determined by the magnitude of the minimum-bias of the criterion as minimum of the Shannon-bias. The greater the bias, the simpler the object of investigation. The measurement of bias represents the difference of the abscissa of the characteristic point of the physical model. Bias is measured for different models of varying complexities. However, without Shannon's approach, it would be incomprehensible why one cannot find a physical model for noisy data and why a physical model is not suitable for predictions. This is analogous to the noise immunity of the criteria for template sorting in cluster analysis.

Gödel and non-Gödel types of systems

The inductive approach is fundamentally a different approach. It has a completely opposite assertion to the deductive opinion of "the more complex the model, the more accurate it is" with regard to the existence of a unique model with a structure of optimal complexity. It is possible to find an optimal model for identification and prediction only by using the external criteria.

The concept of "external criteria" is connected with the Gödel's incompleteness theorem. This means that the Gödel type systems use a criterion realizing the support of the system on an external medium, which is like an external controller in a feedback control system. There is no such controller in the non-Gödel type systems. Usually, the controller is replaced by a differential element for comparison of two quantities without any explicit reference to the external medium.

Let us recall some of the basic propositions of these theories of modeling. In case of ideal data (without noise), both approaches produce the same choice of optimal models or clustering with the same optimal set of features. In case of noisy data, the advantage with Gödel's approach is that although the method is robust compared to the non-Gödel type, it captures the optimal robust model or clustering with its basic features. It conveys to the modeler that it is simpler to follow traditional approaches without taking any complicated paths with inductive approaches. However, an obvious affirmative solution to this question, in which the training data sample does not participate, must be sought among external criteria.

One important feasibility of such a criterion that possesses the properties of an external controller is the partitioning of data sample into two subsets A and B by the subsequent comparison of the modeling or clustering results obtained for each of them. Various examples of constructing the criteria differ according to the initial requirement and in the degree of fuzziness of the mathematical language.

Division of data as per dipoles

In self-organization modeling, usually the data points with a larger variance of the output quantity are taken into the training set A and the points with a smaller variance are taken into the testing set B. Such a division is not applicable in self-organization clustering because "local clusters" of points for the subsamples are destroyed. The "dipoles" of the data sample as point separations allow us to find $(N/2-1)$ pairs of points nearest to one another, where N denotes the total number of points in the sample. Figure 5.1 depicts six "dipoles" whose vertices are used to form the sets A and B, as well as C and D. The points located closer to the observation point I are taken into the set A, while those closer to the observation point II are taken into the set B. The other vertices of the dipoles respectively form the sets C and D. This is also demonstrated in one of the examples given in this chapter.

Clusterization using internal and external criteria

Cluster analysis is usually viewed as a theory of pattern recognition "without teacher"; i.e., without indication of a target function. The result of the process is called clusterization. We know that the theory of clustering is not a new one. One can find a number of clustering algorithms existing in pattern-recognition literature that allow clusterization to be obtained;

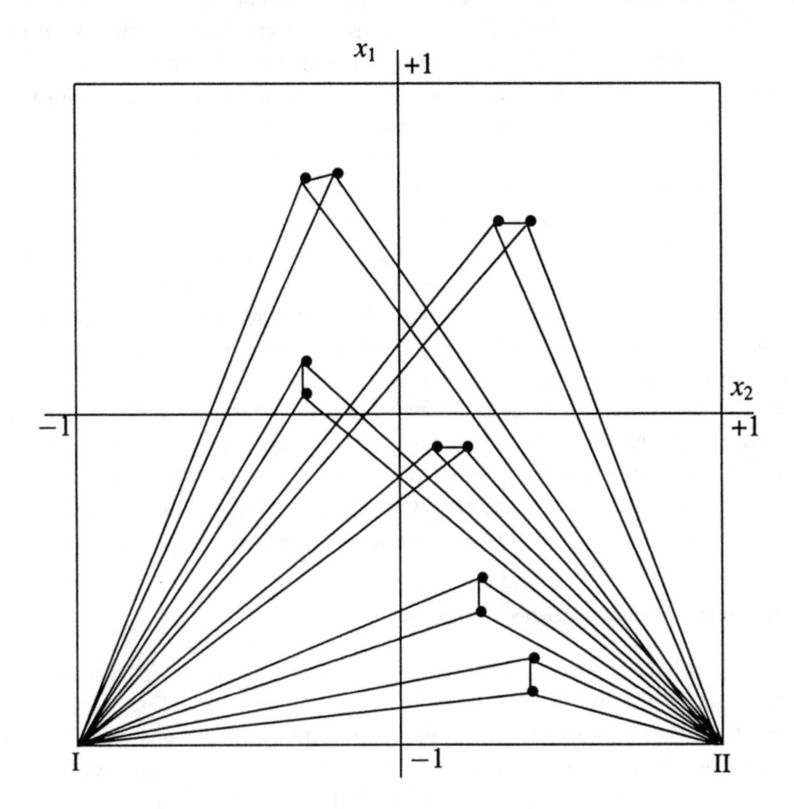

Figure 5.1. Partitioning of data sets A, B from observation I and C, D from observation II

namely, to divide a given set of objects represented by data points in a multi-dimensional space of attributes into a given number of compact groups or clusters. Most of the traditional algorithms are used in the formation of clusters and in the determination of their optimal number by using a single internal criterion having a meaning related to its accuracy or information. With a single criterion, we obtain "the more clusters—the more accurate the clusterization." It is needed for specifyng either a threshold or some constraints when the choice of the number of clusters is made.

Here it describes algorithms for objective computer clusterization (OCC) in which clusters are formed according to an internal, minimum-distance criterion. Their optimal number and the composition of attributes are determined by an external, minimum-bias criterion called a consistency or non-contradictory criterion. Any criterion is said to be external when it does not require specification of subjective thresholds or constraints. The criteria regularity (called precision or accuracy here), consistency, balance-of-variables, and so on, serve as examples of external criteria. Internal criteria are those that do not form the minimum, and therefore exclude the possibility of determining a unique model or clusterization in optimal complexity corresponding to global minimum.

Explicit and implicit templates

The main difference between self-organization modeling and self-organization clustering is the degree of detail of the mathematical language. In clustering analysis, one uses the language of cluster relationships for representing the symptoms and the distance measurements as objective functions instead of equations. The synthesis of models in the implicit form $f(x) = 0$ corresponds to the procedure of unsupervised learning (without teacher, in the literature it is also notified as competitive learning) and in the explicit form $y = f(x)$ it corresponds to the procedure of supervised learning (with teacher).

The objective system analysis (OSA) algorithm usually chooses a system that contains three to five functions which are clearly insufficient for describing large scale systems. Such "modesty" of the OSA algorithm is only superficial. Indeed, a small system of equations is basic, but the algorithm identifies many other systems which embrace all the necessary variables using the minimum-bias criterion. The final best system of equations is chosen by experts or by further sorting of the best ones. What one really has to sort in the inductive approach is not models, plans, or clusterings, but their explicit or implicit templates (Figure 5.2). This helps in the attainment of unimodality of the "criterion-template complexity." If the unimodality is ensured, then the characteristics look as they do in Figure 5.3 for different noise levels. The figures demonstrate the results of sorting of explicit and implicit templates; i.e., in single and system models, correspondingly. These are obtained by computational experiments that use inductive algorithms with regularity and consistent criteria. "Locus of the minima" represents the path across the minimum values achieved at each noise level.

Self-organization of clusterization systems

The types of problems we discuss here—one is the sorting of partial models and other is sorting of clusters—can be dealt with with some care and modeling experience. Figure 5.3b shows the curves that are characteristic for objective systems analysis. Here the model is represented not by a single equation, but by a system of equations, and one can see a gradual widening of the boundaries of the modeling region. There is a region which is optimal with respect to the criterion. The problem of convolution of the partial criteria of individual equations are encountered into a single system criterion.

The theory behind obtaining the system of equations also applies to clusterization in the form of partial clusterization systems that differ from one another in the set of attributes and output target functions. For example, in certain properties of the object, two independent autonomous clusterizations of the form

$$< y_1 > \leftrightarrow < x_{11}x_{12}x_{13}\cdots x_{1m} >, \quad < y_2 > \leftrightarrow < x_{21}x_{22}x_{23}\cdots x_{2m} >$$

have to be replaced by a system of two clusterizations being jointly considered

$$< y_1 > \leftrightarrow < x_{11}x_{12}x_{13}\cdots x_{1m}y_2 >, \quad < y_2 > \leftrightarrow < x_{21}x_{22}x_{23}\cdots x_{2m}y_1 >,$$

where y_1 and y_2 are the output components corresponding to certain properties of the object and x_{ij}, $(i = 1, 2$ and $j = 1, 2, \cdots, m)$ first denote two data points corresponding to the m input attributes.

This is analogous to the operation of going from explicit to implicit templates. The optimal number of partial clusterizations forming the system is determined objectively according to the attainable depth of the minimum of the criteria as achieved in the OSA algorithm.

Figure 5.4 illustrates the results of self-organization in sorting of clusterings by showing a special shape of curve using two criteria: consistency and regularity. The objective based self-organization algorithms are oriented toward the search for those clusterizations that are unique and optimal for each noise level, although the overall consistency criterion leads to zero as the noise variance is reduced. It is helpful to have some noise within the limits in the data; however, the greater the inaccuracy of the data, the simpler the optimal clusterization.

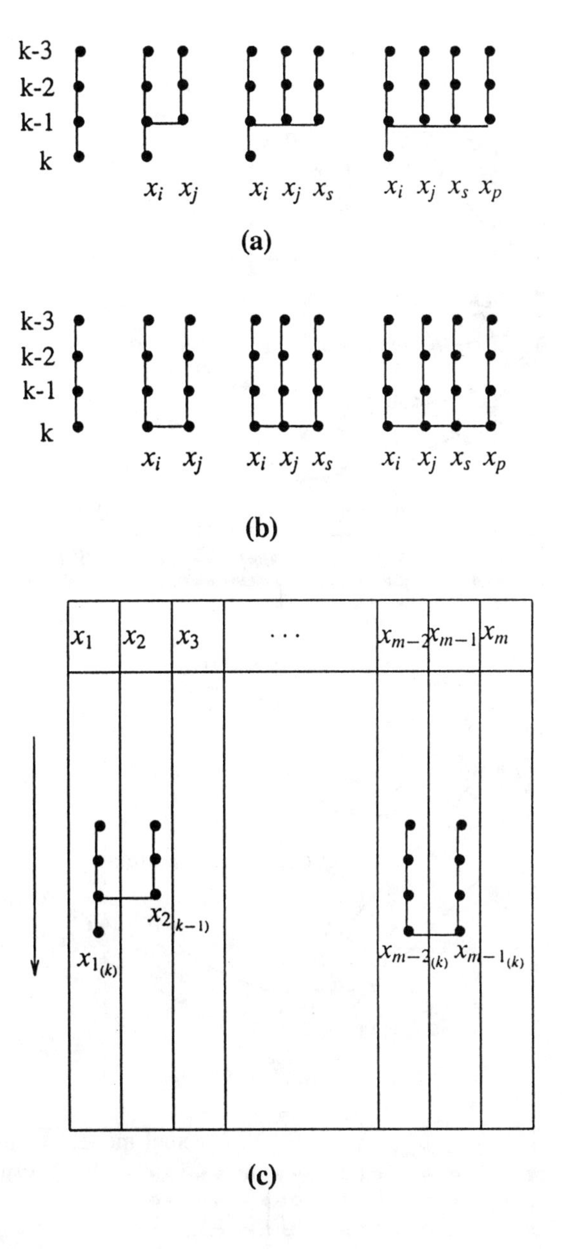

Figure 5.2. Representation of increase in complexity of (a) explicit, (b) implicit templates, and (c) their movement in the data table (k indicates delayed index)

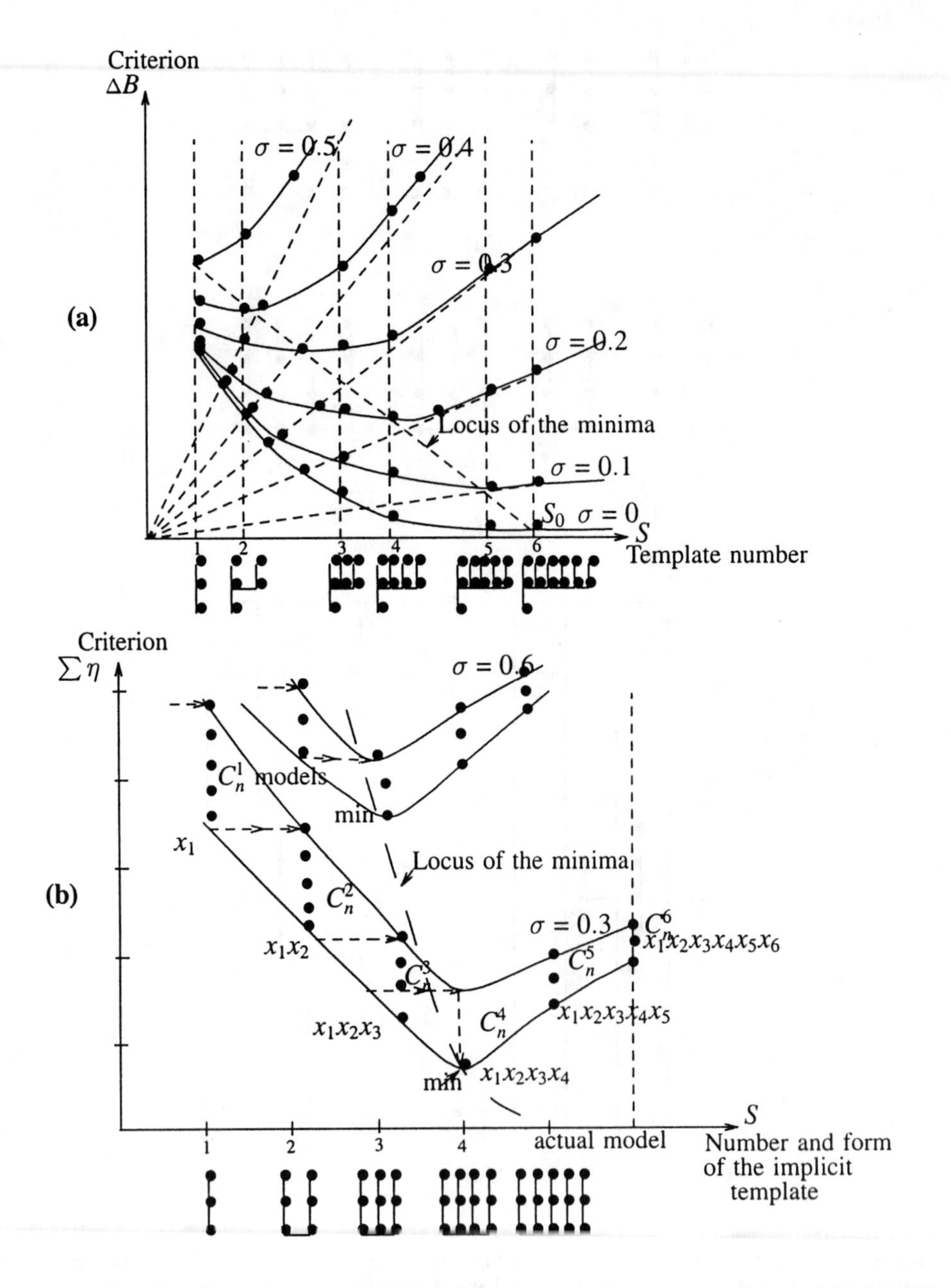

Figure 5.3. Results of experiments with (a) explicit patterns using vector models and (b) implicit patterns using objective systems analysis algorithm

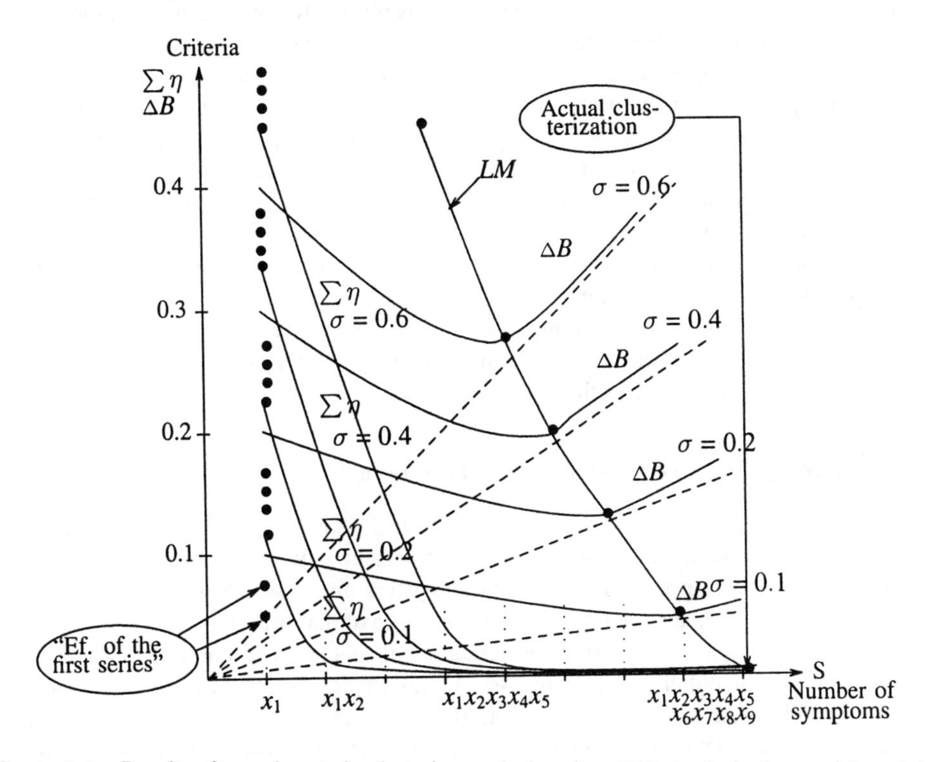

Figure 5.4. Results of experiments in clustering analysis, where LM stands for locus of the minima

Clusterization as investigation of a model in a "fuzzy" language

Clusterization algorithms differ according to their learning techniques that are categorized as learning "without teacher" and learning "with teacher." This means that in the latter case, the problem consists not only of the spontaneous division of the attribute space into clusters, but also of establishing the correspondence of each cluster with some point or region in the target function space. These algorithms are described for both the techniques as different stages "with teacher" and "without teacher." In other words, it leads to clusterization not only with the space of attributes X but also of the target function space Y, or of the united space XY where the target function is one of the attributes. As a result, clusterization $< X > \leftrightarrow < Y >$ or $< XY > \leftrightarrow < Y >$ is obtained—considered a certain "fuzzy" analogue of the model $\hat{y} = f(\hat{x})$ of the object under investigation. The obtained model is optimal with respect to the criteria used and is unique for each object. In ideal data (without noise), it corresponds to the true target of the physical model. In noisy data, it corresponds to the nonphysical model—unique for that level of noise variance. Stability is considered according to the Darwin's classification of species and Mendelev's table of elements which confirm the uniqueness of classifications.

Artificial analogue of the target function

When the target function is not specified, it is sometimes necessary to visualize the output or target function through certain analysis. Visualization here means to make visible that which objectively exists but is concealed from a measurement process. This can refer to a person making a choice of initial data, not intentionally making it nonrepresentative, arranging it along certain axes—"weak-strong," "many-few," "good-bad," etc.—even when the target function is not completely known. A sample of conventionally obtained measurements thus contains information about the target function. Therefore all clusters must be represented in a sample for it to be representative. This is verified in various examples: in water quality problems, samples without any direct indication of the quality spanned the entire range from "purest" to "dirtiest" water. In tests of a person's intelligent quotient (IQ), it represents a broad range of values ($IQ = 70 - 170$). Since it is also determined by experts, it is always possible to check the idea of visualization of the target function. As results indicate, the experimentally measured target function correlates with its artificial analogue of correlation function (value ranges from 0.75 to 0.80), which is considered as adequate. Even for some experiments these are of higher values. The component analysis or Karhunen-Loeve transformation which is used to determine the analogue of the target function can be scalar, two-dimensional or three-dimensional (not more than three) corresponding to visualization of a scalar or a vector target function.

True, undercomplex, and overcomplex clusterizations

The view of clusterization as a model allows us to transfer the basic concepts and procedures of self-organization modeling theory into the self-organization theory of clusterization. A true clusterization corresponds to the so-called physical model which is unique and can be found in ideal and complete data using the first-level external criteria.

The consistent criterion expresses the requirement of clusterization structures as unbiased. Clusterization obtained using the set A must differ as little as possible from the clusterization obtained using the set B ($A \cup B = W$). The simplest among the unbiased (overcomplex) clusterizations is called true clusterization—the point with the optimum set of features denoted as "actual model" in Figure 5.3b. The overcomplex ones are located to the right of that point. Optimal clusterization corresponding to the minimum of the criterion is also unique, but only for a certain level of noise variance (the trivial consistent clusterization where the number of clusters is equal to the number of given points is not considered here). It is determined according to the objectives of the clusterization, and it cannot be specified. This explains the word "objective" in "objective computer clusterization." Optimal clusterizations are found by searching the set of candidate clusterizations differing from one another in the number of clusters and attribute ensembles. The first-level external criteria are explained previously in self-organization modeling. The basic criteria for clusterizations are defined analogously.

The consistency criterion of clusterizations is given as

$$\eta_c = (p - \Delta k)/p, \tag{5.1}$$

where p is the number of clusters or the number of individual points subject to clusterization in the subsets A and B; Δk is the number of identical clusters in A and B [70]. The regularity criterion of clusterizations is measured by the difference between the number of clusters (k_B) of the attribute space in the subset B and their actual number (k) indicated by the teacher. This is represented as $\Delta B = (k_B - k)$.

It has been established that in the problem of sorting models the values of the minimum-bias criterion depend on the design of the experiment and on the method of its partitioning into two equal parts. For an ideal data (without noise), the criterion is equal to zero both for the physical model and for all the overcomplicated models. The greater the difference between the separated sets A and B, the greater the value of the criterion. It is recommended that one can range the data points according to the variance of the output variable, then partition the series into equal parts of A and B. In clustering (delayed arguments are not considered), it is recommended that one choose a sufficiently small difference between the sets to preserve the characteristics of different clusters. If the clusters on the sets A and B are not similar, it is not worth using the consistent criterion. We cannot expect a complete coincidence of subsets A and B, which is inadmissible. Consequently, the problem of sorting clusters becomes a delicate one.

The consistent criterion is almost equal to zero for all the ensembles when the data are exact. It is recommended that the data be partitioned in such a manner that the criterion does not operate on the exact data. However, one can use various procedures to find the unique consistent cluster: (i) according to regularity criterion, (ii) according to system criterion of consistency $\sum \eta_c = \frac{1}{s}(\eta_{c(1)} + \eta_{c(2)} + \cdots + \eta_{c(s)})$ by forming more supplementary consistent criteria computed on other s partitions, (iii) by adding noise to the data and from there finding the most noise-immune clustering, or (iv) by involving experts.

Necessity for regularization

Mathematical theory so far has not been able to suggest an expression for a consistency criterion indicating the closeness of all properties of models and clusterizations for the subsets A and B. The most widely used form of the criterion (minimum-bias criterion) stipulates the idea that the number of clusters ($k_A = k_B$) be equal and that there be no clusters containing different points ($\Delta k = 0$). The patterns of point divisions into A and B must coincide completely in the case of consistent clusterization. The consistent criterion is a criterion that is necessary but not sufficient to eliminate "false" clustrizations. This means that a circumstance might occur that leads to nonuniqueness of the selection. Several "false" clusterizations will be chosen along with the required consistent clusterizations. In these situations, regularization is necessary to filter out false clusterizations.

When the consistency criterion is used in sorting, a small number of clusterizations is found from which the most consistent one is selected—unique for each level of noise variance. For regularization, it is suggested that one use the consistent criterion once more, but employ a different method of forming it. To obtain a unique sample while sorting and using the consistency criterion, only a small number of clusterizations should be taken—chosen by an auxiliary unimodal criterion. Such an auxiliary, regularizing criterion is provided by a consistency criterion calculated on the other data sets C and D. For consistency of clusterizations, the patterns of point divisions into A and B, as well as C and D must completely coincide. In addition to this, the optimal consistent clusterization must be unique. If more than one clusterization are obtained, then the regularization must be continued by introducing another two-subselections until a single answer is obtained. If the computer declares that there are no consistent clusterizations, then the sorting domain is extended by introducing new attributes and their covariances (higher order of the terms), introducing their values with delayed values in order to find a unique consistent clusterization.

High effectiveness of inductive algorithms

As in self-organization modeling, the model with optimal complexity does not coincide with the expert's opinions. The best cluster, being consistent and optimal according to the regularity precision, does not coincide with *a priori* specified expert decisions. Expert decisions are related to complete and exact data. The self-organization clustering that considers the effect of noise in the data, reduces the number of symptoms in the ensemble and the number of clusters. The greater the noise variance, the greater will be the reduction in the number. The computer takes the role of arbiter and judge in specific decisions concerning the results of modeling, predictions and clustering analysis of incomplete and noisy data. This explains the presence of the word "computer" in the name "objective computer clusterization."

It is simply amazing how much world-wide effort has been spent on building the most complex theories oriented toward, surely, the hopeless business of finding a physical model and its equivalent exact clusterizations by investigating only the domain of overcomplex structures. The revolution associated with the emergence of the inductive learning approach consists of the problem of identification of a physical model and clusterization. The problems of prediction are solved in the other direction—of proceeding from undercomplex biased estimates and structures. Optimal biased models and clusterizations are directly recommended for prediction. Advancements in this direction propose a procedure for plotting the "locus of the minima" (LM) of external criteria for identification of the physical model and true clusterization.

Calculation and extrapolation of locus of the minima

The analogy between the theory of self-organization modeling and the theory of self-organization clustering can be continued to find optimal undercomplex clusterizations. One can use either search for variants according to external criteria or calculation of the locus of the minima of these criteria.

The calculation and extrapolation of the locus of the minima of external criteria is an effective method of establishing true clusterization from noisy or incomplete data. A special procedure for extrapolating the locus of the minima or the use of the canonical form of the criterion is recommended in various works [138] and [45] for finding a physical model or an exact clusterization. (Refer to Chapter 3 for the procedures in case of ideal criteria.) One can only imagine the effect of the analytical calculation of the locus of the minima on various criteria. This is calculated for a number of values of the variance and for various distributions of perturbation probabilities.

Usage of canonical form of the criterion for extrapolating LM. All the quadratic criteria can be transformed into a normalized canonical form by dividing the trace of the matrix of the criterion. The criterion is expressed as follows.

$$CR = Y^T S_{0-m} Y, \tag{5.2}$$

where CR indicates an external criterion in the canonical form. Y and Y^T are the output vector and its transpose, correspondingly. S_{0-m} is the canonical matrix of the criterion for different structural complexities.

The mathematical expectation of the criterion for all the models is

$$\overline{CR} = Y^T S_{0-m} Y + \sigma^2 \, \mathrm{tr} S_{0-m}, \quad \text{where } S_{0-m} = \overline{S_0, S_m}. \tag{5.3}$$

For example, S_0 corresponds to a physical model, then

$$\frac{\overline{CR}}{\mathrm{tr} S_0} = \sigma^2, \tag{5.4}$$

and S_s corresponding to a nonphysical model, then

$$\frac{\overline{CR}}{\mathrm{tr} S_j} = \sigma^2 + \frac{Y^T S_j Y}{\mathrm{tr} S_j} = \sigma^2 + A, \quad A \geq 0. \tag{5.5}$$

Hence, $\frac{\overline{CR}}{\mathrm{tr} s_j} \geq \frac{\overline{CR}}{\mathrm{tr} s_0}$.

Theorem. The minimum of the mathematical expectation of the criterion in canonical form for nonphysical models is greater than it is for a physical model [138].

It is shown that all the criteria in canonical form create LM which coincides with the ordinate of the physical model (Figure 5.5). From a geometric point of view, transformation of the criterion to canonical form means rotation of the coordinate axes around the point S_0 and some small nonlinear transformation of the coordinate scale. Figure 5.5 exhibits the locus of the minima: (a) for an external criterion with the usual form and (b) for its canonical form taking the values of $\overline{CR}/\ \mathrm{tr}S$. This shows that with the use of the canonical form of the criteria, one can find a model in optimal complexity without adding any auxiliary noise to the data.

The choice of a rule for restoring the actual or physical model depends on the number of candidate models subject to descrimination, the perturbation level, and the type of criterion.

First rule. If the number of candidate models and the perturbation level are so small that the noise level σ^2 is not exceeded; there is no need for special procedures. The actual clustering is found by using the consistency criterion.

Second rule. If the number of models or candidate clusterings and the perturbation level are comparatively large, a "jump" to the left by the locus of the minima is observed (Figure 5.5a). By imposing supplementary noise on the data sample, one can find several points of the envelope of locus of the minima and use its extrapolation to determine the physical model or actual clustering [45].

Third rule. Addition of auxiliary noise is not needed if the criterion is transformed into canonical form. The ordinate of the minimum of the canonical criterion will indicate the optimal structure (or template) of the physical model or of the clustering if the perturbation variance is within considerable limits (Figure 5.5b).

Asymptotic theory of criteria and templates

In Chapter 3, we discussed the asymptotic properties of certain external criteria. For the mathematical expectation of the external criterion with an infinitely long data sample, the characteristic of the criterion-template sorting is unimodal which is required according to the principle of self-organization. One should not conclude from this result that every time-averaging of the criteria is well only in asymptotic behavior. But unimodality is attained considerably within the limits for a sample length of five to ten correlation intervals; however, a more accurate estimate of the required time-averaging of the criteria is to be found analytically—a subject of theoretical interest.

Asymptotic theory of templates is also not yet developed, although it has been established experimentally. The gradual increase in the number of models according to a specific template leads to an increase in the probability (number of occurrences) of attaining unimodality. Figure 5.6 demonstrates the proposed dependence using the consistency criterion in the plane of "perturbation variants-template complexity."

The future asymptotic theory of templates requires the investigation of the behavior not of the average line of criterion variation, as one selects out of each cluster of feature variants that comes for sorting only one model—the best. This is done by distinguishing among the patterns of variation using a partial, solitary, and overall consistency criteria. For features with noiseless data in clusterizations, the partial nonoverall consistency criterion is identically equal to zero for the entire duration of sorting if the subsamples A and B are close to each other, but nonetheless distinct. The interval of the zero values of the consistency criterion shrinks with sufficiently high probability as the perturbation variance

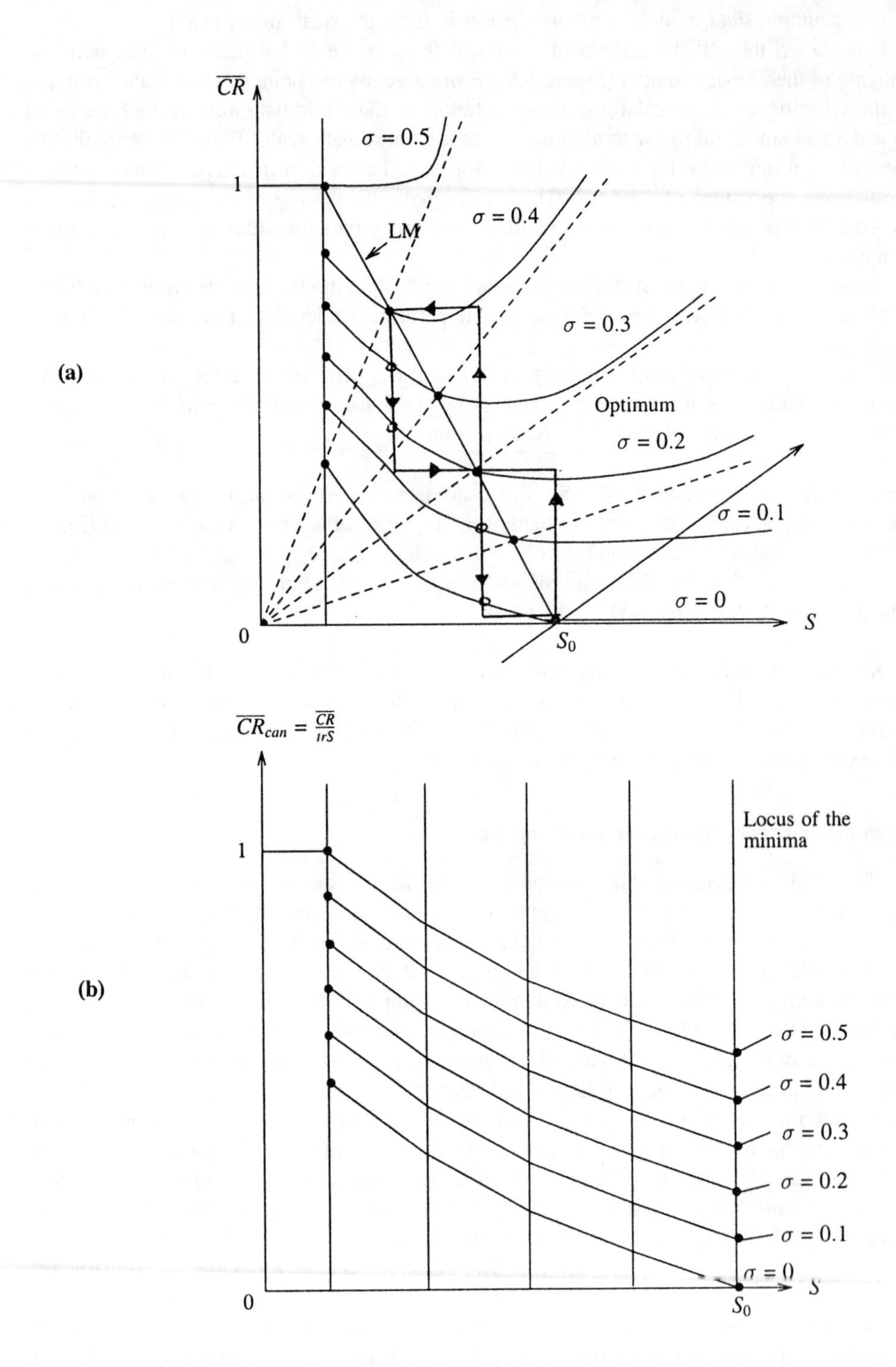

Figure 5.5. Locus of minima (LM) in transition (a) to the ordinary, and (b) to the canonical forms of the criteria, depending on model complexity S and noise dispersion σ.

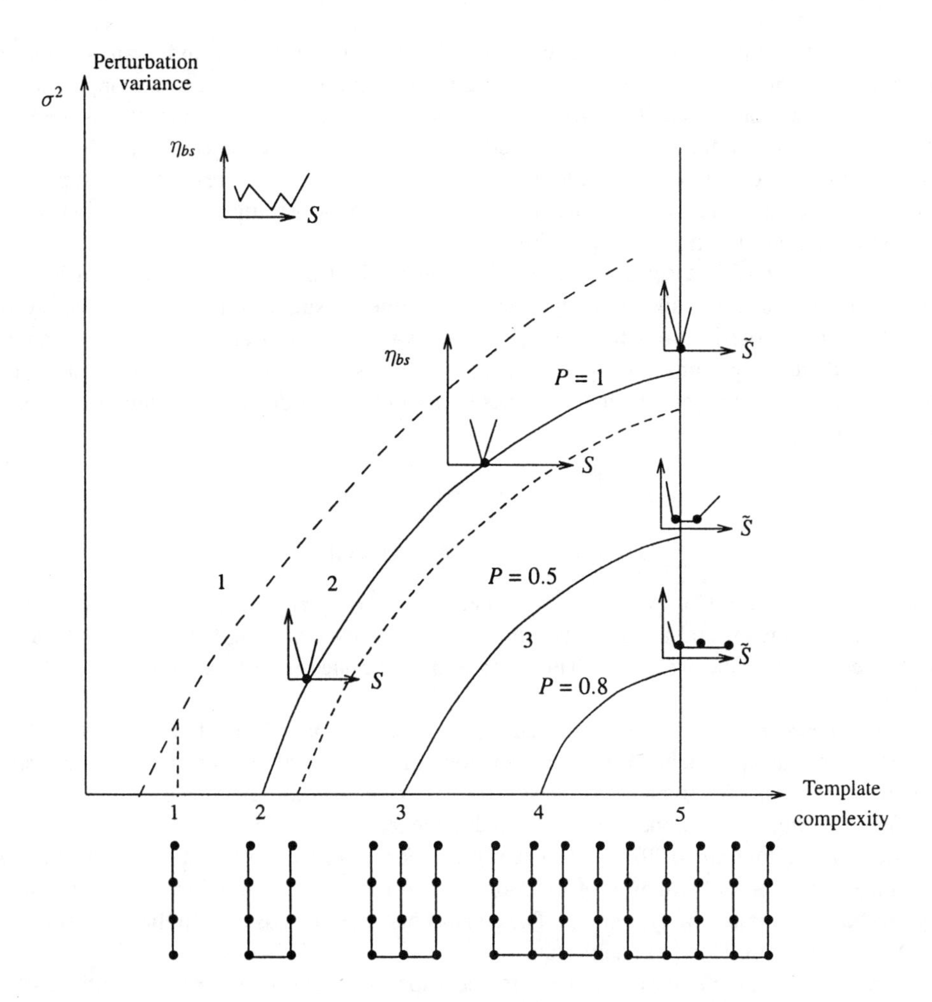

Figure 5.6. Proposed change in probability P of attaining unimodality of the consistency criterion: (1) region of loss of unimodality, (2) region of unimodality without extension of determination, (3) region in which extension of determination required

increases. When it becomes sufficiently small to distinguish between the templates, it becomes expedient to extend the sorting by using an accuracy criterion or a series of consistency criteria calculated for various partitions of data sample. For a larger perturbation variance, it will be in the region of unimodality of a solitary criterion, where a larger perturbation variance is required for more complex templates. Strictly speaking, this serves as the basis for the asymptotic theory of templates. For excessively large perturbations, it becomes impossible to find an optimal consistent model or clusterization, since the regular nature of the curve disappears (Figure 5.6).

2 METHODS OF SELF-ORGANIZATION CLUSTERING

Unlike the sorting of partial models, which is almost always obtained, the sorting of clusters can be implemented only for a sufficiently large number of points that are located favorably in the symptoms (variables). The importance of special experimental designs are enhanced in this section.

If there are m symptoms, one can construct 2^m different ensembles and evaluate them by a suitable external criterion; for example, regularity criterion for an accurate approach and the system criterion of consistency for a robust approach. This corresponds to unsupervised learning because of the absence of specific objectives. If the objective is specified as the ensembles are grouped to a known target function, then it corresponds to supervised learning. The self-organization clustering methods vary according to the techniques used for the reduction of computational volume.

The *first method* is a selection-type of sorting method based on unsupervised learning [39]. At the first step, all the symptoms at the time of succession are evaluated by the specified basic criterion and the best of F (freedom-of-choice) are chosen (for example, $F = 3$ and the symptoms are x_1, x_7 and x_9). At the second step, all the ensembles that contain two symptoms are evaluated. These ensembles include all the symptoms selected at the first step.

$$\begin{array}{c|c|c|c|c|c|l} - & x_1x_2 & x_1x_3 & x_1x_4 & x_1x_5 & x_1x_6 & \\ x_7x_1 & x_7x_2 & x_7x_3 & x_7x_4 & x_7x_5 & x_7x_6 & \cdots \\ x_9x_1 & x_9x_2 & x_9x_3 & x_9x_4 & x_9x_5 & x_9x_6 & \end{array} \tag{5.6}$$

The F best ensembles (for example, $F = 3$, and they are $x_1x_7, x_3x_7,$ and x_1x_4) are selected. At the third step, the ensembles that have three symptoms by including the ensembles selected at the second step are evaluated. This evaluation continues until the $3 \times m$ ensembles are selected.

The *second method*, which is based on correlation analysis [70], is suitable for the precision in the approach. Here, one can obtain a series of m symptoms which range according to their effectiveness; only m different ensembles are evaluated by the criterion.

The *third method* uses one of the basic inductive learning algorithms, either combinatorial or multi-layer, to find m effective ensembles. For example, one can use a device like combinatorial type of "structure of functions" for generating all combinations of ensembles by limiting the number of symptoms. The consistent criterion is used with the data sequences of A and B that are close to each other.

The latter two methods correspond to the supervised learning (learning with teacher) because they use information about the output vector Y based on the comparison among the actual and the estimated data. One way of doing this is by specifying the output data from the experiment and another way is by using the orthogonal Karhunen-Loeve projection method for obtaining the artificial data.

The above methods does not limit the scope of all possibilities. They are feasible only when the unimodality characteristic of the "criterion-clustering complexity" is ensured. These we see in detail below.

2.1 Objective clustering—case of unsupervised learning

There are various computer algorithms that have been proposed for separating a set of ensembles or clusters given in a multidimensional space of variables or symptoms. This includes the classical algorithm of ISODATA (Iterative Self-Organizing Data Analysis Techniques Algorithm) [124] that is based on comparing all possible clusters using the minimum distance criterion. In this program, the number of clusters are specified in advance by the expert.

Objective clustering is envisaged by the inductive approach in which a gradual increase in the number of clusters is specified to the computer and are compared according to the consistent criterion. In separating a multidimensional data space into clusters, the consistent criterion may, for example, stipulate that the partitioned clusters differ from one another as little as possible as they are partitioned according to the odd and even-indexed points of initial data. As is well known, typographical images of some pattern consist of dots. Even when the even or odd dots are excluded, it preserves the image with large numbers of initial data points. If the original image is chaotic; i.e., even if it contains no information conforming to some law, the criterion allows discovery of a physical law.

The object or image is given in a multidimensional space represented in the form of observation data with symptoms $x_1, x_2, \cdots, x_m$. The first part of the problem consists of dividing the space into a specified number of regions or clusters using the measurements of distance between the points [124]. The number of clusters is specified in advance by the experts. Self-organization involves iteration of such clusterings for various numbers of clusters from $k = 2$ to $k = N/2$, where N is the number of data points. It also invloves comparison of results by the consistent criterion—non-contradictory clusters are selected. A single-valued choice is achieved by regularization. Here regularization is selecting the single most appropriate cluster from several non-contradictory clusterings indicated by the computer. The role of regularization criterion is to use the minimizing function which takes into account the number of k and number of variables or symptoms m according to the computer's and expert's clusterings.

$$\rho = [(k_{exp} - k_{comp})^2 + (m_{exp} - m_{comp})^2], \tag{5.7}$$

where k_{exp} is the number of clusters specified by the expert and k_{comp} is the number of clusters in the process of computer clustering.

If k_{exp} is known, then the computer completes the determination of clusters—for example, by using the function $L = k/m$. This is also determined by other relations, in case it is required by agreeing results on three equal parts of the selection.

Even if the k_{exp} is not known, one can use the consistent criterion calculated in other parts of the data sample. It evaluates the degree of non-contradiction on various clusters and helps to choose the best one.

Example 1. Clustering of water quality indices (one-dimensional problem).

The initial data contain the following variables: x_1—suspended matter in mg/liter, x_2—chemical consumption of oxygen (CCO), x_3—mineralization in mg/liter, x_4—carbohydrates in mg/liter, and x_5—sulphates in mg/liter. The data is normalized according to the formula $x_{i_{norm}} = \frac{x_i - x_{i_{min}}}{x_{i_{max}} - x_{i_{min}}}$. The measurements are averaged on seven years of data for each station. The data sets A and B include all stations with even and odd numbers, respectively.

The algorithm is confronted with the problem of isolating all non-contradictory clusterings using the given set of variables and all subsets which could be obtained from them. Thus, the water quality expert could choose the most valid clustering and find the number of clusters and the set of variables that are optimal under given conditions. It computes the value of the criterion for all possible combinations of the set of given variables. In this case the validity of clustering is not verified because of the absence of expert clustering. The sorting process showed that it is not possible to obtain a non-contradictory cluster using all five variables. For each identified cluster, the centers and boundaries are found and the water quality at thc given station using the corresponding variables from the cluster is computed.

Example 2. Clustering of water quality along the series of water stations along a river system.

In this case, expert clustering is known. It is established based on the information available on ecologic-sanitary classification of the quality of surface waters of dry land. It differs from certain variables which are absent from the data (out of total of 21 variables, only 14 participated in the example). The data of 14 variables is normalized and separated into two sets A and B.

The number of clusters specified by experts is $k = 9$ with the variables $m = 14$. There is no single set of variables chosen from the given 14 variables which would yield a non-contradictory partition of the stations into nine clusters as required by the experts. This means that the expert cluster is contradictory.

Non-contradictory partitions into eight clusters are given by a comparatively small number of variables which include x_{14}, x_6x_{14}, $x_2x_7x_{10}x_{14}$ and $x_2x_4x_6x_{10}x_{14}$. Many sets of variables give non-contradictory partitions into seven clusters; eight such sets are x_1x_2, x_1x_4, x_1x_5, x_1x_6, x_1x_7, x_1x_{12}, x_1x_{14}, x_4x_{13}, and 22 sets—each having three variables (from $x_1x_2x_4$ to $x_1x_{12}x_{14}$). The following three sets each with 10 variables give a partition which is closest to one of the expert's clusterings:

$$x_1x_2x_3x_4x_5x_6x_7x_9x_{10}x_{12},$$

$$x_1x_2x_3x_4x_5x_6x_7x_9x_{10}x_{14},$$

$$x_1x_2x_3x_4x_5x_6x_7x_9x_{12}x_{14}.$$

The sets with higher number of variables (11, 12, 13 and 14) do not increase the number of clusters. The set of variables $m = 9$ is denoted as optimal in this example which gives a non-contradictory partition into seven clusters. The boundaries, the stations making up their composition, and the cluster centers are indicated for all non-contradictory clusters for further analysis of water quality.

2.2 Objective clustering—case of supervised learning

Classification, recognition, and clusterization of classes are similar names given for processing a measured input data. The space of measured data for input attributes $X\,(x_1, x_2, \cdots, x_m)$ with a given space of output $Y\,(y_1, y_2, \cdots, y_l)$ representing a target or goal function (where $l \leq m$) is common in these algorithms. The problem task is to divide both spaces into certain subspaces or clusters to establish a correspondence between the clusters of the attribute space and goal function space $X \leftrightarrow Y$.

Unlike in traditional subjective algorithms, the number of clusters are not specified in advance in objective clustering, but the number of clusters is chosen by the computer so that clusterization is consistent. This means that it remains the same in different parts of the initial input data. This number is reduced to preserve the consistency in case of noisy and incomplete data.

As it is mentioned earlier, the objective computer clustering is based on the search for the variants of ensemble of attributes and the number of clusters using the consistency criterion on the given measured data assuming certain errors. The algorithm gives the consistent clusterizations while all existing measurements are distributed over the clusters. The new measurements that do not participate in the clustering also belong to certain cluster, according to the nearest neighbor rule, or according to the minimum-distance rule from the center of the cluster.

The search for the attribute ensembles and for the number of clusters leads to multiple solutions: several variants of ensembles giving consistent clusterizations are found on the plane "ensemble of attributes-number of clusters." This is solved by further determination of consistent clusterings using some second-level criterion or by inquiring from experts.

Table 5.1. Initial Data

No.	x_1	x_2	x_3	x_4	x_5	y
1	2.131	10.41	69.22	73.52	4.43	12.23
2	2.031	9.797	69.26	74.10	4.84	11.86
3	2.076	9.892	69.06	73.42	4.36	11.72
4	2.084	10.09	69.02	73.36	4.34	11.83
5	2.057	9.816	68.97	73.32	4.45	11.47
.	.	.	.	.	.	.
.	.	.	.	.	.	.
19	2.109	10.05	68.81	73.16	4.31	12.05
20	2.143	10.52	68.76	73.01	4.25	12.48
21	2.115	10.24	68.77	73.07	4.30	12.22
22	2.150	10.45	68.71	73.10	4.39	12.38
23	1.919	9.295	68.66	73.06	4.40	10.96
24	2.046	9.840	68.63	73.06	4.43	11.64
.	.	.	.	.	.	.
.	.	.	.	.	.	.
37	2.005	9.631	68.01	72.33	4.32	11.50
38	2.047	9.937	68.06	72.43	4.37	11.67
39	2.013	9.864	68.06	72.42	4.36	11.60
40	2.123	10.37	68.03	72.42	4.39	12.30

Example 3. Objective clustering of the process of rolling of tubes [71].

Here the problem of objective partitioning of an m-dimensional space of features x_1, x_2, $\cdots$, x_m into clusters corresponding to compact groups of images is considered; each image is defined by a data sample of observations.

Objective clustering of images (data points) is done based on sorting a set of candidate clusterings using the consistency criterion to choose the optimally consistent clusterings. The data is divided into four subsets: $A \cap B$ and $C \cap D$. Here the concept of dipoles (pairs of points close to each other) is used; one vertex of a dipole goes into one subsample and the other into another. Thus, the greatest possible closeness of points forming the subsamples is achieved. This example demonstrates the various stages of self-organization clustering algorithm which does not require computations of the mean square distances between the points.

The table of initial data is given (Table 5.1), where x_1 is the length of the blank, x_2 is the length of the tube after the first pass, x_3 and x_4 are the distances between the rollers in front of the two passes, x_5 $(= x_4 - x_3)$ is the change in distance between the rollers, and $y = f(x_1, x_2, \cdots, x_5)$ is the length of the tube.

The objective clustering is conducted in the five-dimensional space of the features $x_1, x_2, \cdots, x_5$. The clustering for which we obtain the deepest minimum of the consistency criterion is the optimal one. The stage-wise analysis of the algorithm is shown below.

Stage 1. To compute the table of interpoint distances. The first $N = 34$ data points from the 40 points of the original sample are used to form the subsets $A \cap B$ and $C \cap D$. The remaining six points are kept as testing sample to check the final results of clustering and for establishing the connection between the output variable y and the cluster numbers. The initial data table is represented as a matrix $X = [x_{ij}]$; $i = 1, 2, \cdots, N$ and $j = 1, 2, \cdots, m$ (here $N = 34$ and $m = 5$).

Table 5.2. Interpoint distances between dipoles

No.	1	2	3	4	5	6	...	32	33	34
1	0	1.015	0.310	0.171	0.484	0.344	...	3.779	1.952	3.484
2		0	0.743	0.943	0.845	1.434	...	5.211	2.339	5.376
3			0	0.449	0.0325	0.399	...	2.547	1.195	2.561
4				0	0.092	0.636	...	2.503	1.150	2.391
5					0	0.318	...	2.115	0.895	2.169
6						0	...	1.990	1.465	2.361
.							.	.	.	.
.							.	.	.	.
.							.	.	.	.
32								0	0.966	0.111
33									0	0.954
34										0

The interpoint distances are calculated as

$$d_{ik} = \sum_{j=1}^{m}(x_{ij} - x_{kj})^2, \quad i = \overline{1,N}; \; k = \overline{i+1,N}. \tag{5.8}$$

The results are shown in the Table 5.2.

Stage 2. To determine the pairs of closest points and partition into subsets. The clusterings are to be identified in the two subsets of $A \cap B$ and $C \cap D$. Thus, the coincidence of clusters is required, indicating that they are consistent. This leads to the attainment of a unique choice of consistent clustering.

The subsets $A \cap B$ and $C \cap D$ are formed using the values of the dipoles. The dipoles are arranged in increasing length: for $N = 34$, there are $N(N-1)/2 = 561$ dipoles. The shortest dipoles are exhibited as

1) 11 0.0020 14, 2) 12 0.0038 13, 3) 23 0.0850 25, ...

To form the subsets A and B, the first $(\frac{N}{2} - 1) = 16$ shortest dipoles are chosen in such a way that the data points are not repeated. In this specific example, it turns out that these 16 dipoles are obtained from the first 389 dipoles; the 17th dipole which satisfies the condition is obtained at the end of the series; i.e., the 561st dipole connects the points 2 and 34 at a length of $d_{2,34} = 5.376$ units.

The following 16 shortest dipoles belong to the subsamples A and B.

1) 11 – 14	2) 12 – 13	3) 23 – 25	4) 26 – 27
5) 16 – 19	6) 10 – 15	7) 5 – 8	8) 17 – 24
9) 20 – 22	10) 3 – 7	11) 31 – 34	12) 29 – 33
13) 6 – 18	14) 9 – 21	15) 1 – 4	16) 28 – 30

From the remaining dipoles, the 16 shortest dipoles are chosen in an analogous manner to form the subsamples C and D.

1) 18 – 23	2) 13 – 21	3) 16 – 17	4) 8 – 10
5) 14 – 15	6) 12 – 19	7) 3 – 5	8) 9 – 22
9) 30 – 31	10) 11 – 24	11) 4 – 7	12) 32 – 34
13) 20 – 27	14) 6 – 25	15) 26 – 33	16) 1 – 2

The dipoles obtained in this way enable the formation of the set of points into the subsets $A, B, C,$ and D.

$$
\begin{aligned}
A:&\quad 11, 12, 25, 26, 16, 15, 8, 24, 20, 7, 34, 29, 18, 21, 4, 30;\\
B:&\quad 14, 13, 23, 27, 19, 10, 5, 17, 22, 3, 31, 33, 6, 9, 1, 28;\\
C:&\quad 23, 21, 16, 10, 14, 19, 5, 22, 31, 24, 7, 34, 27, 25, 33, 1;\\
D:&\quad 18, 13, 17, 8, 15, 12, 3, 9, 30, 11, 4, 32, 20, 6, 26, 2.
\end{aligned}
$$

Stage 3 To sort the clusterings according to the consistency criterion.
The following steps are followed:

1. *Grouping the subsets into 16 clusters* ($k = 16$). The points in subsets A and B are indexed from 1 to 16 as vertex numbers, indicating a group of 16 clusters shown below:

$$
\begin{array}{c} A \\ k=16 \\ B \end{array}
\left\{
\begin{array}{cccccccccccccccc}
11 & 12 & 25 & 26 & 16 & 15 & 8 & 24 & 20 & 7 & 34 & 29 & 18 & 21 & 4 & 30\\
\dot{1} & \dot{2} & \dot{3} & \dot{4} & \dot{5} & \dot{6} & \dot{7} & \dot{8} & \dot{9} & \dot{1}0 & \dot{1}1 & \dot{1}2 & \dot{1}3 & \dot{1}4 & \dot{1}5 & \dot{1}6\\
14 & 13 & 23 & 27 & 19 & 10 & 5 & 17 & 22 & 3 & 31 & 33 & 6 & 9 & 1 & 28\\
\dot{1} & \dot{2} & \dot{3} & \dot{4} & \dot{5} & \dot{6} & \dot{7} & \dot{8} & \dot{9} & \dot{1}0 & \dot{1}1 & \dot{1}2 & \dot{1}3 & \dot{1}4 & \dot{1}5 & \dot{1}6.
\end{array}
\right.
$$

Number of corresponding vertices or clusters:

$$\Delta k = 1 + 1 + 1 + 1 + 1 + 1 + 1 + 1 + 1 + 1 + 1 + 1 + 1 + 1 + 1 + 1 = 16.$$

In each subset A or B, the upper row denotes the actual data point and the lower row denotes the number of the vertex of the dipole. If the number of the vertices coincide, then those vertices are called "corresponding" vertices. Here, all vertices of subset A correspond to the vertices of the subset B. The consistency criterion is computed as $\eta_c = (p - \Delta k)/p = (16 - 16)/16 = 0$, where p is considered the total number of vertices and Δk is the corresponding vertices which coincide.

2. *Grouping the subsets into 15 clusters* ($k = 15$). Tables of interpoint distances are to be compiled for the points of each subset A and B (Tables 5.3 and 5.4, correspondingly). Points 2-14 in subset A and points 1-8 in subset B are the closest to each other.

   For the evaluation of the consistency criterion, it is grouped into 15 clusters in the following form.

$$
\begin{array}{c} A \\ k=15 \\ B \end{array}
\left\{
\begin{array}{l}
\dot{1}\ \boxed{\begin{matrix}\dot{2}\\14\end{matrix}}\ \dot{3}\ \dot{4}\ \dot{5}\ \dot{6}\ \dot{7}\ \dot{8}\ \dot{9}\ \dot{1}0\ \dot{1}1\ \dot{1}2\ \dot{1}3\ \boxed{\begin{matrix}\dot{2}\\14\end{matrix}}\ \dot{1}5\ \dot{1}6\\
\boxed{\begin{matrix}\dot{1}\\8\end{matrix}}\ \dot{2}\ \dot{3}\ \dot{4}\ \dot{5}\ \dot{6}\ \dot{7}\ \boxed{\begin{matrix}\dot{1}\\8\end{matrix}}\ \dot{9}\ \dot{1}0\ \dot{1}1\ \dot{1}2\ \dot{1}3\ \dot{1}4\ \dot{1}5\ \dot{1}6.
\end{array}
\right.
$$

   Number of corresponding vertices:

$$\Delta k = 0 + 0 + 1 + 1 + 1 + 1 + 1 + 0 + 1 + 1 + 1 + 1 + 1 + 0 + 1 + 1 = 12.$$

   The double number of the vertices indicate the formation of a cluster consisting of two points. Having the corresponding vertices as $\Delta k = 12$, the consistency criterion is $\eta_c = (16 - 12)/16 = 0.25$.

3. *Grouping the subsets into 14 clusters* ($k = 14$). Again the tables of interpoint distances are compiled, considering the formed clusters from the previous step. According to the nearest neighbor method, the distance from a cluster to a point is taken to be the

Table 5.3. Interpoint distances for subset A

No.	11	12	25	26	16	15	8	24	20	7	34	29	18	21	4	30
11	0	0.184	0.363	0.219	0.098	0.032	0.044	0.067	0.572	0.107	1.510	0.929	0.290	0.221	0.193	0.952
12		0	1.021	0.141	0.047	0.189	0.216	0.216	0.142	0.238	1.766	1.213	0.915	0.022	0.077	1.028
25			0	0.782	0.694	0.343	0.429	0.358	1.697	0.614	1.424	0.711	0.026	1.047	1.005	1.222
26				0	0.052	0.205	0.342	0.093	0.223	0.465	0.956	0.600	0.791	0.075	0.359	0.425
16					0	0.081	0.151	0.065	0.248	0.225	1.385	0.803	0.637	0.046	0.145	0.740
15						0	0.039	0.047	0.597	0.135	1.586	0.782	0.301	0.234	0.179	0.965
8							0	0.139	0.683	0.042	1.934	1.119	0.307	0.295	0.130	1.283
24								0	0.524	0.264	1.109	0.521	0.348	0.201	0.314	0.598
20									0	0.716	1.649	1.467	1.631	0.085	0.386	0.939
7										0	2.298	1.417	0.451	0.346	0.108	1.568
34											0	0.671	1.726	1.461	2.391	0.136
29												0	0.897	1.078	1.519	1.376
18													0	0.979	0.827	1.337
21														0	0.171	0.799
4															0	1.544
30																0

Table 5.4. Interpoint distances for subset B

No.	14	13	23	27	19	10	5	17	22	3	31	33	6	9	1	28
14	0	0.173	0.315	0.173	0.065	0.048	0.078	0.027	0.446	0.152	1.156	0.739	0.296	0.623	0.685	1.021
13		0	0.938	0.108	0.029	0.305	0.214	0.102	0.092	0.208	1.404	0.918	0.849	0.144	0.292	0.946
23			0	0.734	0.647	0.243	0.455	0.438	1.392	0.672	1.156	0.802	0.185	1.764	1.823	1.546
27				0	0.081	0.338	0.334	0.099	0.134	0.452	0.738	0.533	0.900	0.291	0.746	0.522
19					0	0.167	0.128	0.028	0.182	0.158	1.220	0.765	0.608	0.288	0.427	0.924
10						0	0.034	0.077	0.640	0.125	1.412	0.723	0.218	0.841	0.761	1.362
5							0	0.084	0.531	0.0325	1.718	0.895	0.318	0.663	0.484	1.503
17								0	0.288	0.160	1.099	0.580	0.475	0.451	0.590	0.958
22									0	0.542	1.296	0.862	1.442	0.036	0.446	0.801
3										0	2.086	1.195	0.399	0.612	0.310	1.745
31											0	0.606	1.966	1.729	2.968	0.410
33												0	1.465	1.231	1.952	1.108
6													0	1.691	0.344	2.014
9														0	0.323	1.040
1															0	2.117
28																0

smaller of the two distances. For example, the distance from point 1 to cluster 2,14 is the smaller of the two quantities $d_{1-2} = 0.184$ and $d_{1-14} = 0.221$; ie., $d_{1-2,14} = 0.184$. Thus, the closest points to each other are 3-13 (subset A) and 5-1,8 (subset B).

The third candidate is grouped into 14 clusters of the form

$$
\begin{array}{c} A \\ k=14 \\ B \end{array}
\left\{
\begin{array}{l}
\dot{1}\ \boxed{\begin{matrix}\dot{2}\\14\end{matrix}}\ \boxed{\begin{matrix}\dot{3}\\13\end{matrix}}\ \dot{4}\ \dot{5}\ \dot{6}\ \dot{7}\ \dot{8}\ \dot{9}\ \dot{1}\dot{0}\ \dot{1}\dot{1}\ \dot{1}\dot{2}\ \boxed{\begin{matrix}\dot{3}\\13\end{matrix}}\ \boxed{\begin{matrix}\dot{2}\\14\end{matrix}}\ \dot{1}\dot{5}\ \dot{1}\dot{6} \\
\boxed{\begin{matrix}\dot{1}\\5\\8\end{matrix}}\ \dot{2}\ \dot{3}\ \dot{4}\ \boxed{\begin{matrix}\dot{1}\\5\\8\end{matrix}}\ \dot{6}\ \dot{7}\ \boxed{\begin{matrix}\dot{1}\\5\\8\end{matrix}}\ \dot{9}\ \dot{1}\dot{0}\ \dot{1}\dot{1}\ \dot{1}\dot{2}\ \dot{1}\dot{3}\ \dot{1}\dot{4}\ \dot{1}\dot{5}\ \dot{1}\dot{6}.
\end{array}
\right.
$$

Number of corresponding vertices:

$$\Delta k = 0+0+0+1+0+1+1+0+1+1+1+1+0+0+1+1 = 9$$

and $\eta_c = (16-9)/16 = 0.437$.

4. *Fourth and subsequent steps.* Continuation of the partitioning of the subsets into clusters and evaluation by consistency criterion is followed from $k = 13$ to $k = 2$.
   For the last two clusterings; i.e., in case of $k = 2$, $\eta_c = (16-16)/16 = 0$, and in case of $k = 3$, $\eta_c = (16-16)/16 = 0$.

All groupings of the clusterings is complete. From the above evaluation, the consistent clusterings for $k = 2$, 3, and 16 can be chosen because $\eta_c = 0$ in these groupings.

One can note that if the table of interpoint distances consists of two equal numbers, then the number of clusters changes by two units. To avoid this, one must either raise the accuracy of the measurement distances in such a way that there will not be equal numbers in the table, or skip the given step of sorting of clusterings in one of the subsets. The consistency criterion is used only when the number of clusters is the same on two subsets $A \cap B$ and $C \cap D$; otherwise, the amount of sorting increases and it ends up with bad results.

To reduce the computational time of the algorithm, the comparison of the variants of the clusterings can be started with eight clusters instead of 16 clusters. This means that at the first step the points are not combined by two, but by eight points.

Stage 4. Repetition of clustering analysis on subsets A and B for all possible sets of variable attributes (scales) and compilation of the resulting charts (Figure 5.7a).

The cluster analysis described above should be repeated for all possible compositions of the variable attributes. As there are $m = 5$ attributes, there are altogether $2^5 - 1 = 31$ variants. The dots in the figure indicate the most consistent clusterings which are obtained on the subsets A and B.

Stage 5. To single out the unique consistent clustering with the aid of experts or by using the subsets C and D (regularization).

It is desirable to choose a single most consistent one from the clusterings obtained on the subsets A and B. This can be done in two ways: One way of singling out is with the help of experts for whom examination of a small number of variants of clusterings does not constitute any great difficulty. The unique clustering suggested by the expert might not be the most consistent clustering, but merely one of the sufficiently consistent clustering. Another way is by repeating the clustering analysis on subsets C and D to obtain a clustering

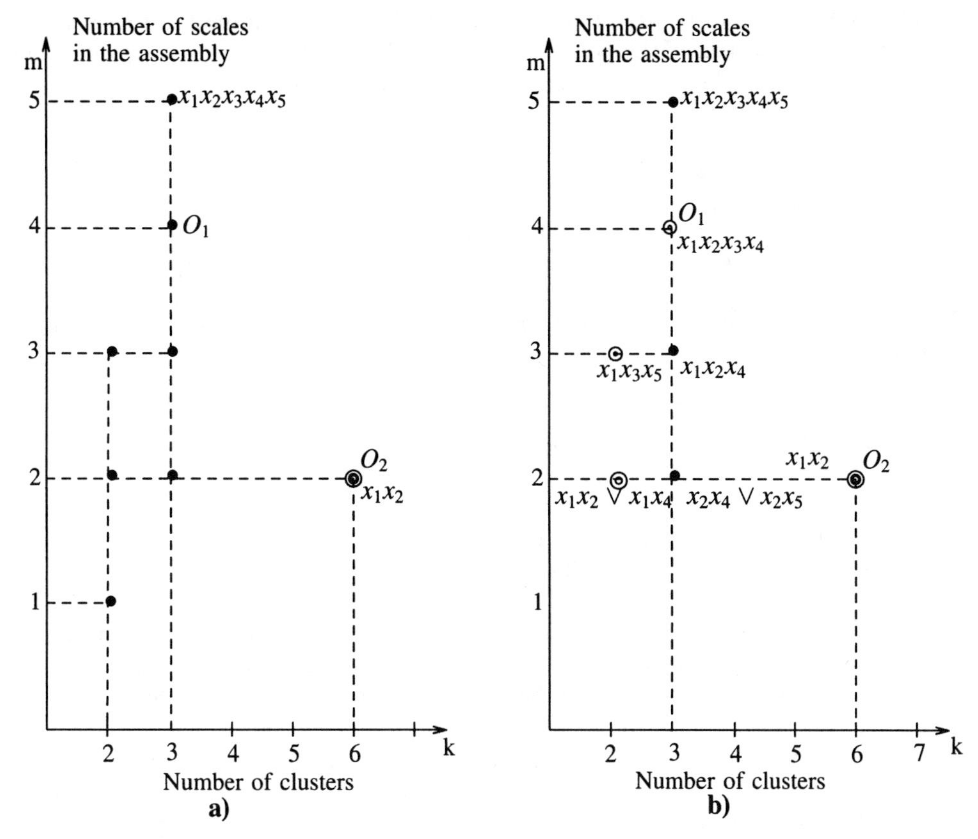

Figure 5.7. Results of search for the most consistent clusterings on (a) subsamples A and B and (b) subsamples C and D

that will prove to be sufficiently consistent both for the subsets $A \cap B$ and $C \cap D$. Figure 5.7b shows the results of choice of consistent clusterings on subsets C and D. The value of the consistency criterion for the clustering corresponding to the point O_2 is zero both on the subsets $A \cap B$ and $C \cap D$. For the clustering O_1, it is zero only for $C \cap D$. Here clustering O_2 is considered to be the true most consistent ones.

If unique clustering is not obtained, the points are further divided into three equal subsets, thus forming another consistency criterion and so on until the goal of the regularization—a single consistent clustering—is achieved.

Figure 5.7 shows less than eight clusters (out of the 16 possible ones) along the abscissa, since further increase in their number yields an inadmissibly small mean number of points in each of them (total 34 points are subjected to grouping in clusters).

For *reducing the sorting of attributes*, it is recommended that

1. the attribute sets for which half or more of the dipoles on $A \cap B$ (or $C \cap D$) do not coincide are not considered, and
2. for analysis on subsets C and D, one considers only those attribute sets for which small values of the criterion during the analysis on the subsets A and B are obtained.

Stage 6. Results of the two clusterings corresponding to O_1 and O_2.

Corresponding to the point O_1, three clusters are obtained with respect to four scales of attributes x_1, x_2, x_3, and x_4. The points of the original data sample are distributed among the clusters as below (the point numbers and the mean values of the output variable y are given):

$$\begin{aligned}
&\text{1st cluster}: \ 6, 18, 23, 25; \ \bar{y} = 10.99m; \\
&\text{2nd cluster}: \ 30, 31, 32, 34; \ \bar{y} = 11.58m; \\
&\text{3rd cluster}: \ 1, 2, 3, 4, 5, 7, 8, 9, 10, 11, 12, 13, 14, 15, \\
&\qquad 16, 17, 19, 20, 21, 22, 24, 26, 27, 28, 29, 33; \ \bar{y} = 11.799m.
\end{aligned}$$

Corresponding to the point O_2, six clusters are obtained with respect to the two scales of attributes x_1 and x_2.

$$\begin{aligned}
&\text{1st cluster}: \ 6, 18, 23, 25; \ \bar{y} = 10.99m; \\
&\text{2nd cluster}: \ 29, 32; \ \bar{y} = 11.20m; \\
&\text{3rd cluster}: \ 1, 2, 3, 5, 7, 8, 9, 10, 11, 14, 15, 17, 24, 31, 33, 34; \ \bar{y} = 11.47m; \\
&\text{4th cluster}: \ 4, 16, 19, 26, 27, 30; \ \bar{y} = 11.83m; \\
&\text{5th cluster}: \ 12, 13, 21, 28; \ \bar{y} = 11.93m; \\
&\text{6th cluster}: \ 20, 22; \ \bar{y} = 12.43m.
\end{aligned}$$

Stage 7. To check the optimal clustering using the checking sample of data points (35 to 40) according to the prediction accuracy of required quality of the tube length.

The single consistent clustering can be used to predict the output variable y from the cluster number. For example, let us consider the three clusters corresponding to the point O_1 with the attributes x_1, x_2, x_3, and x_4 (the three clusters with the point numbers and mean values of the variable y are given above). The mean values of y are arranged in an increasing order and the regression line for y according to the groupings of clusters N is given in Figure 5.8. A new point belongs to the cluster for which the distance from it to the closest point of the cluster is least; knowing the cluster, the estimated value of y can be obtained from the figure. This type of prediction is checked for the testing sample points 35 to 40. Out of six points, five are correctly predicted.

2.3 Unimodality—"criterion-clustering complexity"

We understand that the experimental design is feasible only when the unimodality of the "criterion-clustering complexity" characteristic is ensured. This can be done in three ways to determine the optimal consistent clustering: (i) extend the cluster analysis using a regularity criterion for further precision, (ii) design the cluster analysis for using a overall or system criterion of consistency by increasing the number of summed partial consistency criteria, and (iii) design the experiment by applying a supplementary noise to the data.

The applicability of the first method is demonstrated in the preceding example.

The second method of attaining unimodality is when an increase in the number of partial criteria which constitute the overall consistency criterion reduces the number of consistent clusterings from which an optimal one is to be selected. Specially designing the experiment can make this method very efficient in yielding a single consistent clusterization. The following example demonstrates the usefulness of this method.

Example 4. Investigation of the consistent criterion by computational experiments [69].

Here is a test example to clarify whether (i) it is possible to select a data sample such

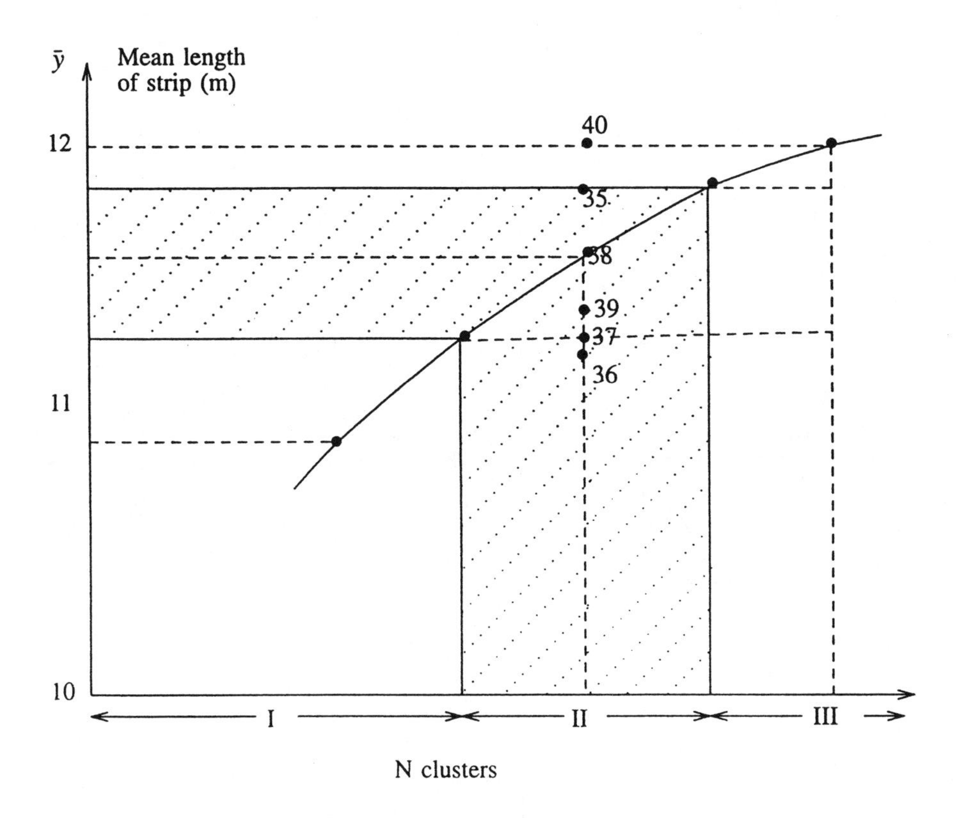

Figure 5.8. Regression line for prediction of mean strip length for the cluster number N for the set of x_1, x_2, x_3, x_4

that sorting of clusterings by the consistency criterion yields a unique solution and (ii) the overall consistency criterion leads to a unique solution.

The consistency criterion is expressed as $\eta_c = (k - \Delta k)/k$, where k is the number of clusters and Δk is the number of identical clusters in the subsets A and B.

According to the procedure involved in the experimental design of cluster analysis, the original data sample is divided into two equal parts by ranking their distances from the coordinate origin. Then the consistent clusterings are found by complete sorting of hypotheses about the number of clusters, proceeding from $k = N/2$ to a single cluster, where N is the total number of points in the data sample. The initial data sample along with their ranked distances are given in Table 5.5 and in Figure 5.9, where, for simplicity, two variants of ten points ($N = 10$) on the plane of two attributes x_1 and x_2 are shown.

Figure 5.10 shows the procedure for sorting of clusters using the tables of interpoint distances for subsets A and B.

For each transition from one number of clusters to another, the tables of interpoint distances for each subset are rewritten such that the newly formed row in the table contains (when the poles of the dipoles are united) the shortest distance in the two cells of the preceding table. The poles of the dipoles are united in pairs for each hypothesis according to the minimum of the criterion of interpoint distance in this example.

The subsets A and B are taken into two equal parts. This is represented as an original

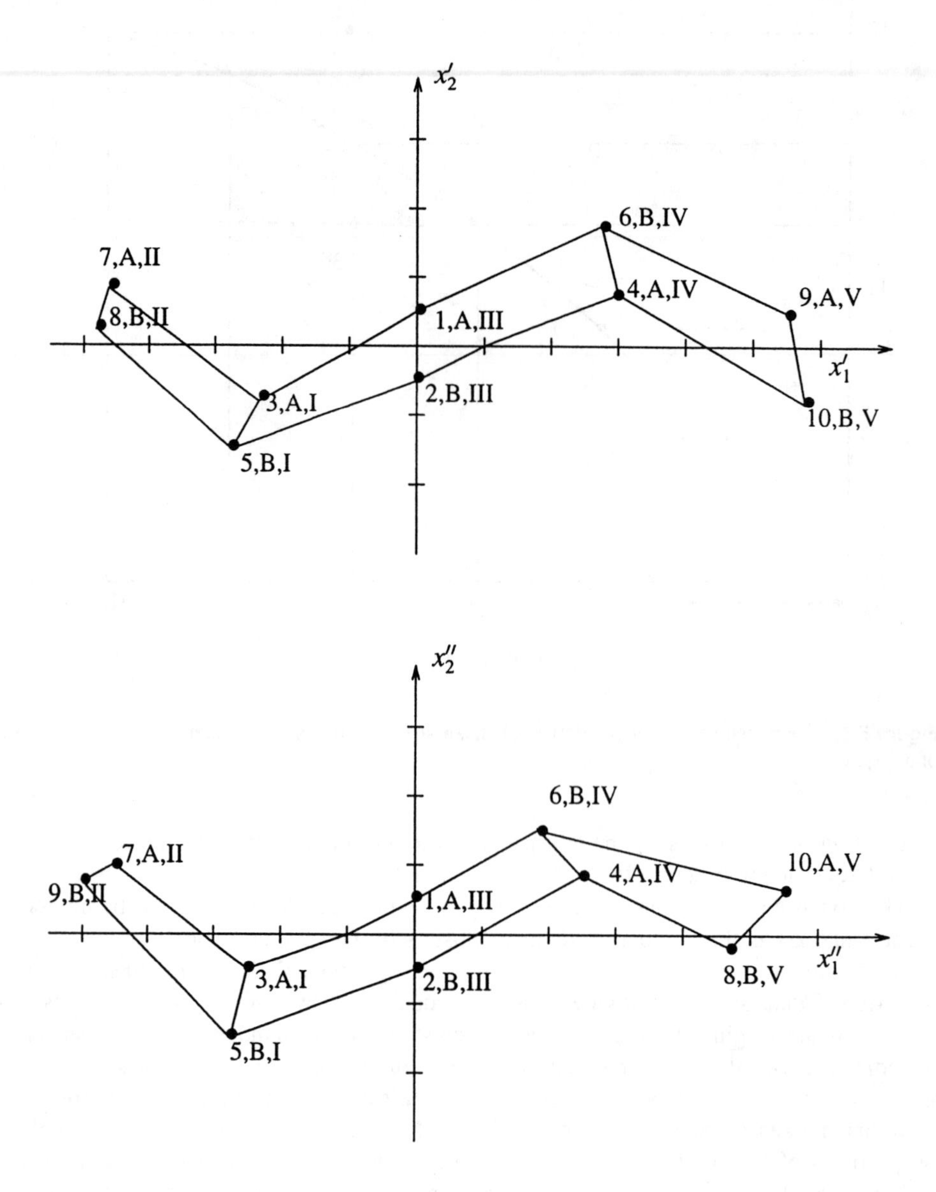

Figure 5.9. Location of the points of the two samples A, B in the plane; I, II, ..., V are the address of dipoles

Table 5.5. Two samples of initial data ranked by distances

No.	First sample of points			Second sample of points		
	x_1'	x_2'	$x_1'^2 + x_2'^2$	x_1''	x_2''	$x_1''^2 + x_2''^2$
1	0.00	0.40	0.16	0.00	0.40	0.16
2	0.00	−0.40	0.16	0.00	−0.40	0.16
3	−2.32	−0.69	5.86	−2.48	−0.69	6.62
4	2.80	0.68	8.30	2.54	0.785	7.07
5	−2.70	−1.25	8.85	−2.76	−1.32	9.36
6	2.60	1.60	9.32	2.52	1.78	9.52
7	−4.61	0.93	22.12	−4.40	0.90	20.17
8	−4.70	0.25	22.15	4.76	−0.10	22.67
9	5.50	0.60	30.61	−4.99	0.99	25.88
10	5.85	−0.75	34.78	5.44	0.75	30.16

code:

$$\begin{array}{lccccc}
\text{Code} & 0 & 0 & 0 & 0 & 0 \\
 & 3 & 7 & 1 & 4 & 9 \\
 & \vdots & \vdots & \vdots & \vdots & \vdots \\
 & | & | & | & | & | \\
 & \vdots & \vdots & \vdots & \vdots & \vdots \\
 & 5 & 8 & 2 & 6 & 10 \\
 & I & II & III & IV & V
\end{array}$$

(a) $k = 4$:

$$\begin{array}{llccccc}
 & & I & II & III & IV & V \\
 & & 3 & 7 & 1 & 4 & 9 \\
I & 3 & 0 & 7.9 & 0 & 21 & 63 \\
II & 7 & & 0 & 22 & 55 & 102 \\
III & 1 & & & 0 & 8 & 30 \\
IV & 4 & & & & 0 & 7 \\
V & 9 & & & & & 0
\end{array}
\qquad
\begin{array}{llccccc}
 & & I & II & III & IV & V \\
 & & 5 & 8 & 2 & 6 & 10 \\
I & 5 & 0 & 2 & 6 & 36 & 73 \\
II & 8 & & 0 & 22 & 51 & 112 \\
III & 2 & & & 0 & 11 & 34 \\
IV & 6 & & & & 0 & 16 \\
V & 10 & & & & & 0
\end{array}$$

(b) $k = 3$:

$$\begin{array}{lcccc}
 & I,III & II & IV & V \\
I,III & 0 & 7.9 & 8 & 30 \\
II & & 0 & 55 & 102 \\
IV & & & 0 & 7 \\
V & & & & 0
\end{array}
\qquad
\begin{array}{lcccc}
 & I,II & III & IV & V \\
I,II & 0 & 6 & 36 & 73 \\
III & & 0 & 11 & 34 \\
IV & & & 0 & 16 \\
V & & & & 0
\end{array}$$

(c) $k = 2$:

$$\begin{array}{lccc}
 & I,III & II & IV,V \\
I,III & 0 & 7.9 & 8 \\
II & & 0 & 55 \\
IV,V & & & 0
\end{array}
\qquad
\begin{array}{lccc}
 & I,III & IV & V \\
I,III & 0 & 11 & 34 \\
IV & & 0 & 16 \\
V & & & 0.
\end{array}$$

It is known that the consistency criterion indicates the false consistent clusterings with the actual consistent clusterings. The false consistent clusterings; i.e., false zeros of the

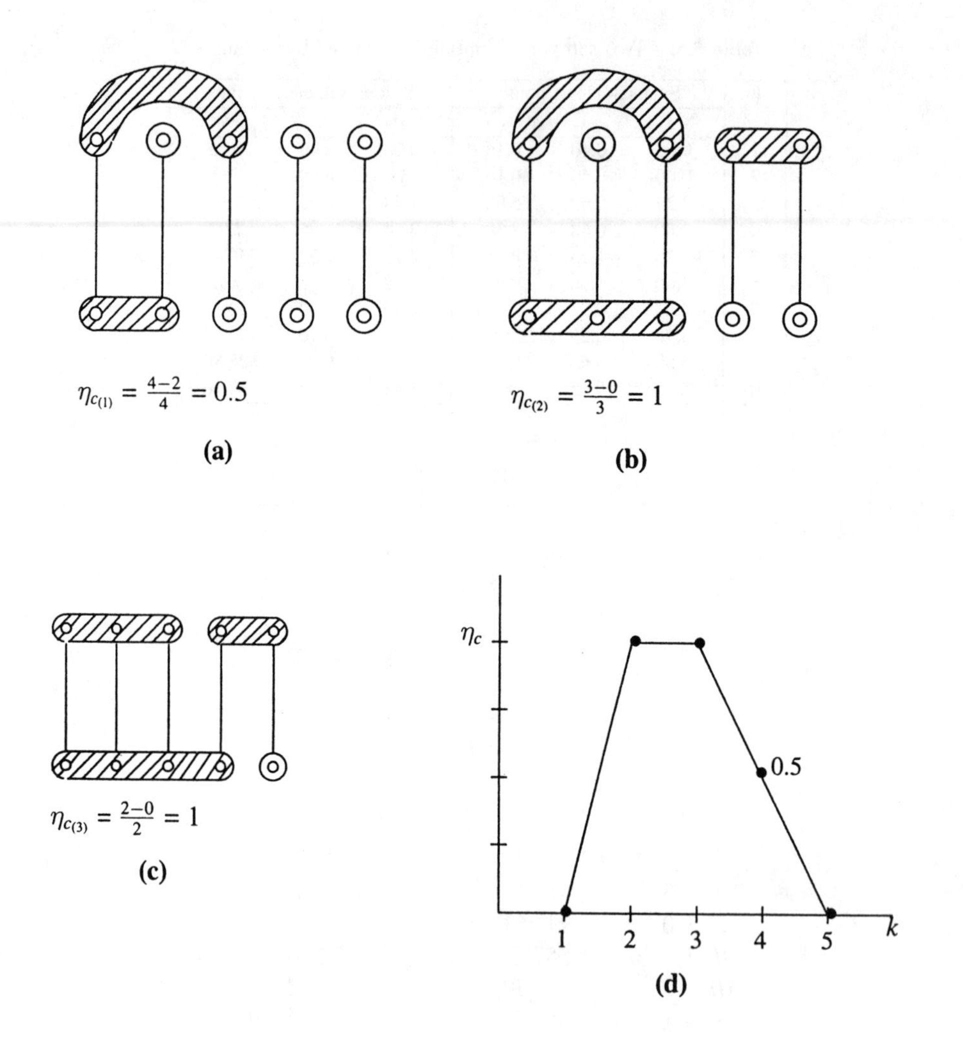

Figure 5.10. Calculation of consistency criterion on the two equal parts of the data sample

criterion can be removed by (i) a special experimental design, the purpose of which is to form a data sample for which the criterion does not indicate false zeros and (ii) using the overall consistent criterion, which is equal to the sum of partial criteria obtained for different compositions of subsets A and B.

To sort among the hypotheses, the notations are introduced for the original data sample and to the subsets (vertex numbers) as below:

Code	0	0	0	0	0	
	3	7	1	4	9	subset*A*
	.	.	.	.	.	
Dipoles	|	|	|	|	|	
	.	.	.	.	.	
	5	8	2	6	10	subset*B*
	I	*II*	*III*	*IV*	*V*	

where $I - V$ are the dipole addresses and 00000 is the initial code for the sample. A dipole is a two-point subsample. Selected dipoles have the shortest dimension of all the feasible points of the considered sample. The code changes if the corresponding dipole changes the pole addresses in the subsets. For example,

$$\begin{array}{lccccccl} \text{Code} & 0 & 1 & 1 & 0 & 0 & \\ & 3 & 8 & 2 & 4 & 9 & \text{subset}A \\ & \cdot & \cdot & \cdot & \cdot & \cdot & \\ \text{Dipoles} & | & | & | & | & | & \\ & \cdot & \cdot & \cdot & \cdot & \cdot & \\ & 5 & 7 & 1 & 6 & 10 & \text{subset}B \\ & I & II & III & IV & V & \end{array}$$

The partial consistency criteria are calculated for all the variants of subset composition, and their dependencies on the number of clusters are constructed. As shown in Figure 5.11, some partitioning variants for the first sample of data points do indeed yield false zeros. This gives rise to the problem of removing false zeros of the false clusterings. Repetition of the experiment with the second sample of the data points showed that none of the 16 characteristics yields false zeros.

In this example, the consistency criterion for the selected original data sample is unimodal. One can see from Figure 5.9 that a very small variation in the locations of the sample points disturbs the unimodality. So, the above experimental design aimed at attaining criterion unimodality may lead to the required result, although it is still very sensitive. This means that a small deviation in the data leads to the formation of false value of the criterion.

Overall consistency criterion

The overall consistency criterion is the sum of the values of the partial criteria obtained for all possible compositions of subsets A and B.

$$\sum \eta = \frac{1}{L}(\eta_{c(1)} + \eta_{c(2)} + \cdots + \eta_{c(L)}), \tag{5.9}$$

where $L = 2^{k-1}$.

Figure 5.11 demonstrates the performance of the overall consistency criterion, which does not lead to the formation of false zeros for various numbers of clusters. The experiment explains the physical meaning of the stability of the overall criterion and substantiates the basic conclusions of the coding theory as follows:

- if the overall criterion does not lead to the formation of complete zeros, then among the partial codes there is at least one that ensures the same result;
- if at least one of the codes does not form false zeros, then the overall code will also be effective; and
- for a complete sorting of the codes, one necessarily finds a partitioning into parts that leads to false zeros (the unsuccessful partitioning).

Apparently, one can apply the optimal coding theory, developed in the communication theory, for determining the optimal partitioning of a data sample into subsamples.

The goal of the experimental design is to attain the global minimum among the models. The high sensitivity to small variations in the input data and absence of unimodality

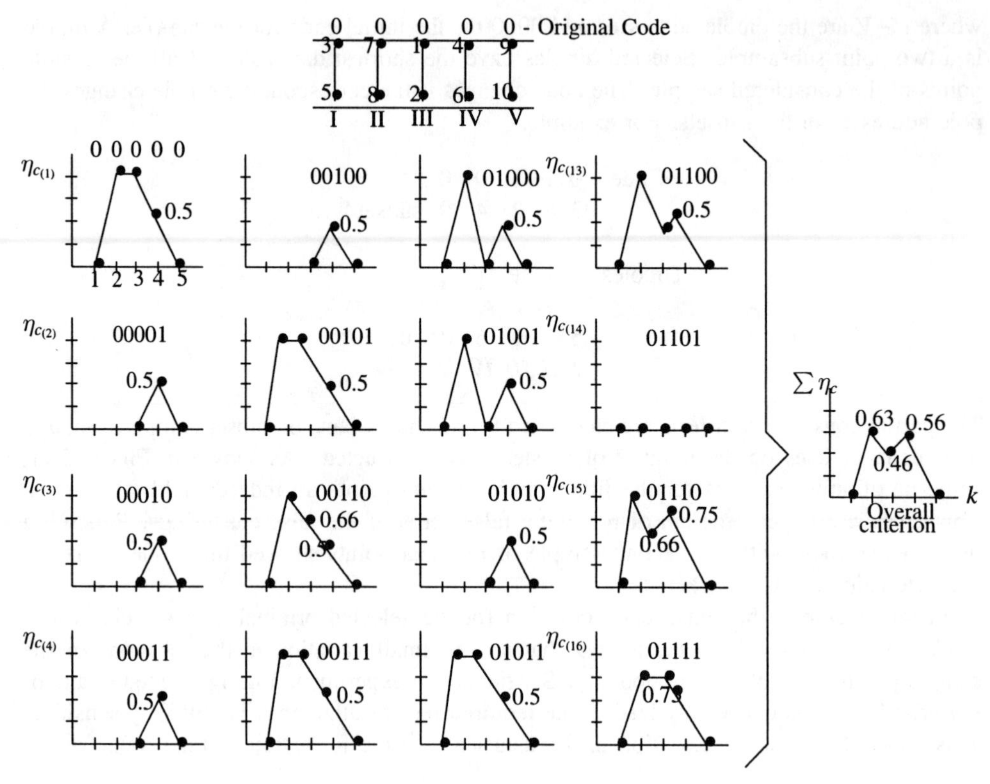

Figure 5.11. Dependence of the criterion on the number of clusters for various compositions of subsamples A and B

are characteristic symptoms of the noncorrectness of the problem of selecting a model or clusterization on the basis of a single consistency criterion. The transition to an overall consistency criterion can be viewed as one possible regularization method. With a robust approach as demonstrated above, the main goal must be the attainment of the unimodality of the consistency criterion. Sometimes, the use of the overall criterion might be insufficient in removing all the composite zeros, even for all possible partitions of the data sample into two subsets. This can be avoided by further splitting the data into subsets.

The third method of attaining unimodality consists of superimposing an auxiliary normal noise to the data sample. Its variance is increased until the most noise-immune consistent clusterization as the "locus of the minima" is achieved. One can obtain consistent clusterization without extending the experiment for regularization by the precision criterion or by experts.

Further development of this method is done by appling the canonical form of the external criterion. The locus of the minimum of the criterion coincides with the coordinates of the optimal design of the experiments and the optimal model structure. The Shannon-bias as displacement of the criterion becomes zero for all the designs and structures. This leads to a new dimension of research which will be discussed in detail in our future works.

3 OBJECTIVE COMPUTER CLUSTERING ALGORITHM

The objective computer clustering (OCC) algorithm in a generalized form is given here. The algorithm consists of the following blocks.

Block 1. Normalization of variables

Normalization is done here for the input variables $\tilde{x}_1, \tilde{x}_2, \cdots, \tilde{x}_m$, measured at N time instances as

$$x_{1i} = \frac{\tilde{x}_{1i} - \bar{x}_1}{\tilde{x}_{1max} - \tilde{x}_{1min}}, \; x_{2i} = \frac{\tilde{x}_{2i} - \bar{x}_2}{\tilde{x}_{2max} - \tilde{x}_{2min}}, \; \cdots, x_{mi} = \frac{\tilde{x}_{mi} - \bar{x}_m}{\tilde{x}_{mmax} - \tilde{x}_{mmin}}, \tag{5.10}$$

where $\bar{x}_j$, $j = 1, 2, \cdots, m$ are the mean values of corresponding variables; x_{ji}, $j = 1, 2, \cdots, m$; $i = 1, 2, \cdots, N$ are the normalized values. This can be done not only from the mean value but also from a trend of the variable. It is also useful to extend the table of attributes with the additional generalized attributes such as

$$\tilde{x}_{ij} = \frac{1}{2}(\tilde{x}_i + \tilde{x}_j), \text{ or } \tilde{x}_{ij} = \sqrt{(\tilde{x}_i \tilde{x}_j)}, \text{ or } \tilde{x}_{ij} = \sqrt{[\frac{1}{2}(\tilde{x}_i^2 + \tilde{x}_j^2)]}, \tag{5.11}$$

where $i = 1, 2, \cdots, m$; $j = i + 1, i + 2, \cdots, m$.

In addition to the input attributes, information about the goal function can be included into the original data in the form of columns with the deviated data of the output variables $y_1, y_2, \cdots, y_l$, where $l \leq m \leq m1$; and $m1$ is the total number of primary and generalized attributes. The information about the goal function is very useful for reducing the amount of cluster search. In many clustering problems the dimension of the space l of the goal function is known: l = constant. If it is not specified, it can be determined by the successive test of Karhunen-Loeve projection on to an axis, a plane, a cube, etc. or by means of the component analysis.

This is justified as follows: The modeler, while compiling the table of data, knows the goal function without fully realizing it. There necessarily exists certain axes like "good-bad," "strong-weak," "much-little," etc. These correspond to the axes serving as orthogonal projection. The space of the goal function in certain cases is two-dimensional or three-dimensional. For example, clustering of atmospheric circulation, is distinguished between two axes: the "form" and "type" of circulation; the Karhunen-Loeve orthogonal projection is applied on two variables $Y(y_1, y_2)$.

Sub-block 1a: Choose dimension of goal function

The clustering target function may be expressed by a particular vector of qualities, rather than by a scalar value. In most complex clustering problems, it is necessary to derive a complete quality vector $Y(y_1, y_2, \cdots, y_l)$.

There is a sample of observations $X(x_1, x_2, \cdots, x_m)$. Experts maintain that the target function (at any rate, one of its components—the target index) may be determined from the variance formula:

$$y = \sqrt{[\frac{1}{m}\sum_{i=1}^{m}(x_i - \bar{x}_i)^2]}, \tag{5.12}$$

where $\bar{x}_i$ is the mean value of the ith attribute.

The above formula represents the Karhunen-Loeve discrete transformation in the case where m-dimensional space of factors is mapped into one average point ("center of gravity" point, if each of the constituents has an identical mass), and the target formula is represented as a single scalar value [137]. This way, more information is retained in projecting points of an m-dimensional space onto a single axis y, although it remains a scalar quantity. The y-axis is chosen in such a way that (i) it passes through the "center of gravity" of points that is the origin of the attributes x_i, and (ii) the axis direction in the m-dimensional space is such that the points have minimum moment of inertia around the y-axis.

In the same way, even more information is retained in projecting the m-dimensional measurement space onto a two-, three-, or more dimensional spaces, to the state of projecting it on itself and not loose information. To reduce the number of computations involved in these operations, one can limit the comparisons of Karhunen-Loeve transformations to the final stage at the point on the axis or on the two-dimensional plane. The target function will be two-dimensional $Y(y_1, y_2)$, which is enough for many problems. The joint space attributes correspond to the vector of $XY(x_1, x_2, \cdots, x_m, y_1, y_2)$. This might be excessive for the optimal number of dimensions of the goal space in specific practical purposes. An optimal number of measurements for the target function space is determined by comparing the versions of the best number of coordinates that leads to consistent and accurate clusters, and by positioning these closer to the number of clusters E specified by an expert.

A way of estimating the target index. An estimation method for a single dimensional axis is developed as given below. The equation for the y-axis takes the form

$$\frac{x_1}{l_1} = \frac{x_2}{l_2} = \cdots = \frac{x_m}{l_m}, \tag{5.13}$$

where $l_1, l_2, \cdots, l_m$ are the components of the unique target vector. The moment of inertia is computed using the following criterion as

$$J_{mi} = \sum_{i=1}^{N} \sum_{j=1}^{m} x_{ij}^2 - (\sum_{i=1}^{N} (l_j x_{ij})^2 / \sum_{j=1}^{m} l_j^2) \rightarrow \min, \tag{5.14}$$

which amounts to the selection of $l_1, l_2, \cdots, l_m$. The second term in the criterion J_{mi} is maximal as $\sum_{i=1}^{N} \sum_{j=1}^{m} l_j x_{ji} \rightarrow \max$, with the constraints $\sum_{j=1}^{m} l_j^2 = 1$. The parameters $l_1, l_2, \cdots, l_m$ are found iteratively using the initial approximation of $l_1 = l_2 = \cdots = l_m = 1/\sqrt{m}$. This gives an equation for the y-axis. The projection of data points on the y-axis are then found. The hyperplane passing through the ith point perpendicular to the y-axis takes the orthogonal form

$$\sum_{j=1}^{m} l_j (x_j - x_{ij}) = 0, \quad i = 1, 2, \cdots, N. \tag{5.15}$$

The coordinates for the projection x_{ij_y} are determined while solving the above equation along with the equation for the y-axis. The function for allocating the projections along the i-axis is found as

$$y_i = \sqrt{[} \sum_{j=1}^{m} (x_{ij_y} - x_i)^2], \quad i = 1, 2, \cdots, N. \tag{5.16}$$

This is considered a target function and recorded in the input data.

For example, the input data corresponding to the nodes of a three-dimensional cube are shown in the Figure 5.12. The minimum value of the criterion J_{mi} corresponds to the maximum value of the function

$$(l_1 x_{11} + l_2 x_{21} + l_3 x_{31}) + (l_1 x_{12} + l_2 x_{22} + l_3 x_{32}) + \cdots + (l_1 x_{18} + l_2 x_{28} + l_3 x_{38}) \rightarrow \max. \tag{5.17}$$

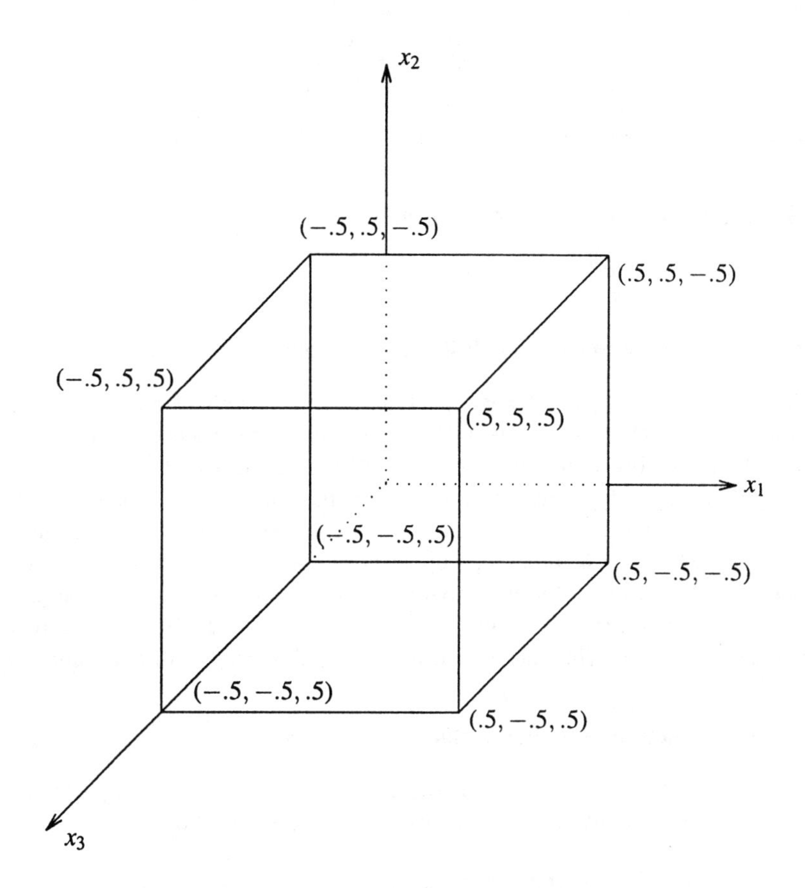

Figure 5.12. data for the given example

By iteration, $l_1 = 1$, $l_2 = 1$, and $l_3 = 1$ are found. The equation of the y-axis is $x_1 = x_2 = x_3$. Projections are allocated along the y-axis; at point 1, $y = +\sqrt{3/2}$, at point 8, $y = -\sqrt{3/2}$; at points 2, 3, and 4, $y = +\sqrt{3/6}$. At points 5, 6, and 7, $y = -\sqrt{3/6}$. Here, it is better not to use the Karhunen-Loeve transformation on the axis of the plane because of overlappings of many point projections. Only two projections coincide on the plane. This is solved in a different way in [124].

There is much in common between the successive application of Karhunen-Loeve projection and the method of principal components of factor analysis. The variance decreases continuously as components are isolated. Specifing a threshold is required for choosing the number of components. According to Shannon's second-limit theorem, there exists an optimal number of factors which are to be isolated. In self-organization clustering, the consistent criterion is recommended to select the optimal number of principal components; consequently, the dimension of the goal function $Y(y_1, y_2, \cdots, y_l)$ is determined.

Block 2. Calculation of variances and covariances

The data sample is given in matrix form as $X = [x_{ij}]$, $Y = [y_i]$; $i = 1, 2, \cdots, N$, $j = 1, 2, \cdots, m$. The matrix of variances and covariances $G = \frac{1}{N}X^TX$ has the elements

$$g_{ij} = \text{cov}(x_i, x_j) = \frac{1}{N}\sum_{k=1}^{N} x_{ki}x_{kj}, \tag{5.18}$$

where x_i and x_j are the columns i and j of the matrix X.

Block 3. Isolation of effective ensembles

This is done in one of the following three ways:

Sub-block 3a. Full search over all attribute ensembles

This refers to clustering without goal function. A full search of all possible clusterings differing by the contents of the set is to be carried out in the absence of the numerical data on the goal function. For each value of the number of clusters k, 2^{m1} clusterings are to be tested using the consistency criterion, where $m1$ is the number of attributes—including the paired or generalized attributes. This type of cluster analysis is feasible for a small number of attributes of up to $m1 = 6$. In a larger dimension of the attribute space, effective attribute ensembles are selected using the inductive learning algorithms or correlation analysis. At the same time, the goal function (scalar or vector form) must be determined experimentally by orthogonal projection. This means that it leads to clustering with goal function.

Sub-block 3b. Selection by inductive learning algorithms

This is done by using the inductive learning algorithms. The consistency criterion is used in selecting the effective attribute ensembles. The models are of the form:

$$\begin{aligned}
y_{11} &= f_{11}(x_1x_2\cdots x_{m1}), y_{21} = f_{21}(x_1x_2\cdots x_{m1}), \cdots, y_{l1} = f_{l1}(x_1x_2\cdots x_{m1}),\\
y_{12} &= f_{12}(x_1x_2\cdots x_{m1}), y_{22} = f_{22}(x_1x_2\cdots x_{m1}), \cdots, y_{l2} = f_{l2}(x_1x_2\cdots x_{m1}),\\
y_{1F} &= f_{1F}(x_1x_2\cdots x_{m1}), y_{2F} = f_{2F}(x_1x_2\cdots x_{m1}), \cdots, y_{lF} = f_{lF}(x_1x_2\cdots x_{m1}),
\end{aligned} \tag{5.19}$$

where F denotes the quantity of "freedom-of-choice." It is the number of models selected on the last layer. This indicates an ensemble of attributes for which we have to seek the most consistent clustering.

Sub-block 3c. Selection by correlation algorithm

If there are many attributes (m is large) and the number of measurements are small ($N \leq 2m$), then it is better to use the correlation algorithm (also called "Wroslaw taxonomy") instead of inductive learning algorithms. Initially, a table of correlation coefficients of paired attributes (G) is set up. Using this matrix, the graphs of interrelated attributes for different limit values of the correlation coefficient are set up. One attribute that is correlated least with the output quantity is chosen from each graph. Ultimately, an ensemble of attributes which are correlated as little as possible with the output are determined. The limit of the correlation coefficient is gradually reduced commencing from $r_{xx} = 1$ until all attributes fall into a single path; i.e., until an ensemble containing a single attribute $y = f(x_i)$ is obtained. This way, discriminant functions which indicate effective ensembles of attributes are found:

$$y_1 = f_1(x_1 x_2 \cdots x_{m1}), y_2 = f_2(x_1 x_2 \cdots x_{m1}), \cdots, y_l = f_l(x_1 x_2 \cdots x_{m1}).$$

Block 4. Division of data points

The ensembles obtained for different values of the correlation coefficient are subjected to a search for consistent clusterings. All ensembles are processed using the same search algorithm [124]. A square table of distances between points (with a zero diagonal) corresponding to the attributes is set up. Segments connecting any two points in the attribute space is called dipoles. These are arranged according to their length to form a full series of dipoles.

The next step is to select dipoles whose nodes form the subsets $A - B$, and $C - D$. The two nodes of the shortest dipole go into A and B; the next in magnitude go into C and D, and so on, until all nodes are investigated. Alternatively, first dipoles are chosen for A and B, and the remaining dipoles are chosen for C and D.

Half of the nodes of the dipoles go into A, while the other half go into B; subsets C and D are simply different division of the same full set of points. Conventionally, the nodes of dipoles located nearer to the coordinate origin are introduced into A and C, while those more remote are into B and D.

Block 5. Search for clusterings by consistency criterion

The next step is to carry out a search for all clusterings on the subsets A and B. Nodes belonging to the same dipole are considered equivalent. Commencing from the division of subsets into $N/2$ clusters, the number of clusters decreases to unity. The subsequent clusterings are formed by uniting into a single cluster of two points located closest to one another. The consistency criterion is determined for all clusterings by $\eta_c = (p - \Delta k)/p$, where p is the number of clusters or the number of individual points subject to clusterization, and Δk is the number of identical clusters in the subsets A and B. As a result, all clusterings for which $\eta_c = 0$ are identified. The search is repeated for all possible attribute ensembles and a map is obtained, in which consistent clusterings are denoted by dots (for example, Figure 5.8).

Additional analysis and exclusion of clusters with single dipoles. The clusters containing more than two points and the clusters containing two points belonging to the same dipole are obtained from the search of consistency criterion. The latter ones are better assigned to other clusters, or excluded from the analysis because they can represent long dipoles. Such clusters containing a single dipole are located at the end of the series of the dipoles ordered according to their length.

If the initial data table is sufficiently large (for example, $N \geq 100$, in order to avoid formation of two-points clusters), it is sufficient to use $N/3$ points instead of $N/2$ points and leave the rest of them for examining the clustering results.

Block 6. Regularization

The search is repeated on subsets C and D for further confirmation. Only those clusterings that are consistent both on A and B and on C and D are in fact considered. If we again find not one but several of the consistent clusterings, then the clustering closest to the clustering recommended by the experts is chosen. Usually, the clustering recommended by the experts turns out to be contradictory.

Block 7. Formation of output data table

The output data table that contains the division of the points of the original table into an optimal number of clusters is formed.

Block 8. Recognition

At this step, assignment of new points (images) to some cluster with the indication of the value of the goal function is carried out according to the "nearest neighbor" rule. This means that this is based on the minimum distance from the image to a point belonging to a set indicated in the initial data table.

Here we can say that the two-stage algorithm in image recognition is established in the OCC algorithms. At the first stage (teaching) of $y_i = f(x_1 x_2 \cdots x_{m1})$, the data about the space of measurements (attributes) and about the space of the goal function is used to obtain the discriminant functions with the objective of dividing the space into clusters. At the second stage (recognition), new points are assigned to some class or cluster. The number of clusters and the attribute ensemble are identified objectively using a variant search according to the consistent criterion. All the blocks given above form a schematic flow of the OCC algorithm.

Calculation of membership function of a new image to some cluster. A membership function (taken from the theory of fuzzy sets of Zadey) is given as

$$z = \frac{d_{x,i}^{-1}}{d_{x,1}^{-1} + d_{x,2}^{-1} + \cdots + d_{x,k}^{-1}} .100\%, \tag{5.20}$$

where $d_{x,i}$ is the distance from the image to the center of the cluster x; $d_{x,j}, j = 1, 2, \cdots, k$ are the distances to the centers of all clusters measured; k is the number of clusters.

The greater the membership function of an image to a cluster, the smaller is the distance from the image to the center of the cluster. The measurement of distances is carried out in the space of an effective attribute ensemble.

Example 5. Application of OCC algorithm.

The objective clustering of the rolling conditions of steel strip is considered. The original variables (x_1, x_2, x_3, x_4, and x_5) and the goal function (strip length, y) are given. It is expanded to other sets of generalized paired variables ($x_6 - x_{15}$).

Block 1. Table 5.6 has been obtained as a result of normalization of the variables as deviations from their mean values.

Block 2. The matrix of variances and covariances is given in Table 5.7.

Block 3c Isolation of the effective attribute ensembles by the correlation algorithm of "Wroslaw taxonomy" yielded the 15 effective ensembles shown in Table 5.8.

Block 4 Division of the data according to the dipole search for the ensemble $x_5 x_{11} x_{12} x_{13}$ is as follows:

subset A: 12, 23, 38, 37, 14, 27, 15, 24, 39, 19, 28, 11, 16, 29, 20, 34, 3, 22, 25, 40;
subset B: 13, 18, 31, 32, 8, 26, 10, 17, 35, 4, 30, 7, 21, 33, 9, 36, 5, 1, 6, 2;
subset C: 32, 14, 23, 21, 38, 24, 16, 31, 22, 13, 17, 8, 34, 28, 26, 20, 18, 10, 33, 7;
subset D: 36, 11, 25, 12, 39, 15, 27, 35, 9, 4, 19, 3, 37, 40, 30, 1, 6, 5, 2, 29.

Block 5. The cluster search is carried out using the consistency criterion by dividing the subsets into eight.

Table 5.6. Normalized initial data

No.	x_1	x_2	x_3	x_4	x_5	$x_6 - x_{13}$	x_{14}	x_{15}	y
1	0.286	0.374	0.302	0.303	-.137	...	-.077	-.081	0.338
2	-.055	-.091	0.322	0.619	0.706	...	0.766	0.771	0.111
3	0.098	-.019	0.222	0.249	-.043	...	0.003	0.001	0.026
4	0.125	0.131	0.202	0.216	-.075	...	-.033	-.035	0.093
5	0.034	-.077	0.675	0.195	0.097	...	0.236	0.127	-.127
.	.	.	.	.	.	.	.	.	.
.	.	.	.	.	.	.	.	.	.
.	.	.	.	.	.	.	.	.	.
37	-.144	-.217	-.300	-.343	-.106	...	-.166	-.157	-.110
38	0.000	0.016	-.275	-.280	-.026	...	-.082	-.073	-.005
39	-.117	-.040	-.276	-.205	-.043	...	-.098	-.089	-.048
40	0.250	0.346	-.200	-.205	0.004	...	-.054	-.044	0.381

Table 5.7. Matrix of variances and paired variances

Attributes	x_1	x_2	x_3	x_4	x_5	...	x_{14}	x_{15}	y
x_1	0.0518	0.0525	0.0086	0.0054	-.0078	...	-.0059	-.0064	0.0515
x_2		0.0602	0.0051	0.0022	-.0077	...	-.0066	-.0070	0.0553
x_3			0.0492	0.0433	-.0028	...	0.0072	0.0045	0.0047
x_4				0.0519	0.0114	...	0.0200	0.0194	0.0044
x_5					0.0481	...	0.0474	0.0474	-.0058
.						.	.	.	.
.						.	.	.	.
.						.	.	.	.
x_{14}							0.0482	0.0479	-.0049
x_{15}								0.0480	-.0048
y									0.0569

Table 5.8. Effective attribute ensembles

No.	Ensembler
1	x_{11}
2	$x_5\ x_{11}$
3	$x_5\ x_{11}\ x_{13}$
4	$x_5\ x_{11}\ x_{12}\ x_{13}$
5	$x_5\ x_9\ x_{11}\ x_{12}\ x_{13}$
6	$x_5\ x_9\ x_{11}\ x_{12}\ x_{13}\ x_{14}$
7	$x_5\ x_9\ x_{11}\ x_{12}\ x_{13}\ x_{14}\ x_{15}$
8	$x_1\ x_5\ x_9\ x_{11}\ x_{12}\ x_{13}\ x_{14}\ x_{15}$
9	$x_1\ x_5\ x_8\ x_9\ x_{11}\ x_{12}\ x_{13}\ x_{14}\ x_{15}$
10	$x_1\ x_5\ x_7\ x_8\ x_9\ x_{11}\ x_{12}\ x_{13}\ x_{14}\ x_{15}$
11	$x_1\ x_5\ x_6\ x_7\ x_8\ x_9\ x_{11}\ x_{12}\ x_{13}\ x_{14}\ x_{15}$
12	$x_1\ x_3\ x_5\ x_6\ x_7\ x_8\ x_9\ x_{11}\ x_{12}\ x_{13}\ x_{14}\ x_{15}$
13	$x_1\ x_3\ x_4\ x_5\ x_6\ x_7\ x_8\ x_9\ x_{11}\ x_{12}\ x_{13}\ x_{14}\ x_{15}$
14	$x_1\ x_2\ x_3\ x_4\ x_5\ x_6\ x_7\ x_8\ x_9\ x_{11}\ x_{12}\ x_{13}\ x_{14}\ x_{15}$
15	$x_1\ x_2\ x_3\ x_4\ x_5\ x_6\ x_7\ x_8\ x_9\ x_{10}\ x_{11}\ x_{12}\ x_{13}\ x_{14}\ x_{15}$

Block 6. The consistent clusters are further determined by the condition of their presence on the maps obtained for the subsets A, B and C, D and summarized on the summary map as shown in Figure 5.13. The clustering marked C in the figure is the most effective one.

Block 7. The following data points are grouped into clusters according to the mean strip length by using the above result of objective clustering.

Cluster 1: Points 6, 18, 23, and 25 for $\bar{y} = 10.99$;
Cluster 2: Points 2, 29, and 33 for $\bar{y} = 11.63$; and
Cluster 3: Points 1, 3, 4, 5, 7, 8, 9, 10, 11, 12, 13, 14, 15, 16, 17, 19, 20, 21, 22, 24, 26, 27, 28, 30, 31, 32, 34, 35, 36, 37, 38, 39, and 40 for $\bar{y} = 11.77$.

Block 8. In the recognition stage, let us assume that a new image is obtained with the attribute values of $x_5 = 4.373$, $x_{11} = 26.986$, $x_{12} = 6.631$, and $x_{13} = 70.202$. Then the distances from the point obtained to all 40 initial points are calculated. The nearest point is located as the point 30 with the attribute values of $x_5 = 4.410$, $x_{11} = 26.96$, $x_{12} = 6.65$, and $x_{13} = 70.28$. This point belongs to the third cluster; consequently, the new point image belongs to the third cluster. The values of the membership function reveal that the first cluster $z = 0.203$, the second cluster $z = 0.240$, and the third cluster $z = 0.553$; i.e., the input image affiliates more to the third cluster.

4 LEVELS OF DISCRETIZATION AND BALANCE CRITERION

The criteria of differential type are quite varied, but they, nonetheless, ensure the basic requirement of Gödel's approach. They are a clustering found by sorting according to a criterion using a new data set which is not used with the internal criterion. In the algorithms described above, the basic criterion used is consistency. Here is another form of differential criterion: the criterion of balance of discretization is proposed for selecting optimal clusterings in self-organization clustering algorithms for a varying degree of fuzziness of the mathematical description language [34]. The principle behind this criterion is that the overall picture of the arrangement of the clusters in the multidimensional space of features must not differ greatly from the type of discretization of the variable attributes. The optimal clustering (the number of clusters and the set of features) must be the same—independent of the number of levels of discretization of the variables indicated in the data sample.

Initial data sample is discretized into various levels on the coordinate axes to find the optimal clustering. Hierarchical trees for sorting the number of clusters are set up from the tables of interpoint distances. The optimal number of clusters coincides at the higher levels of hierarchy of reading variables. The balance of discretization criterion is used like the criterion of consistency; i.e., according to the number of identical clusters.

In self-organization modeling the criterion of consistency, which is called the minimum-bias criterion to estimate the balance of structures, is computed according to the formula $\eta_{bs} = \sum_{i=1}^{N}(\hat{y}^A - \hat{y}^B)^2$. The criterion requires that the model obtained for the subset A ($\hat{y}^A$) differs as little as possible from the model obtained for the subset B ($\hat{y}^B$). If the criterion has several equal minima (balances), then we have to apply some method of regularization.

In self-organization clustering, the data sample is discretized into different numbers of levels according to the coordinates of the points for obtaining subsets A and B. It is then sorted among the hypotheses as to the number of clusters for each of the subsets and the results compared with one another. The optimal clustering corresponds to the minimum of the consistency criterion; usually its zero value resembles the balance of clusterings on both the subsets.

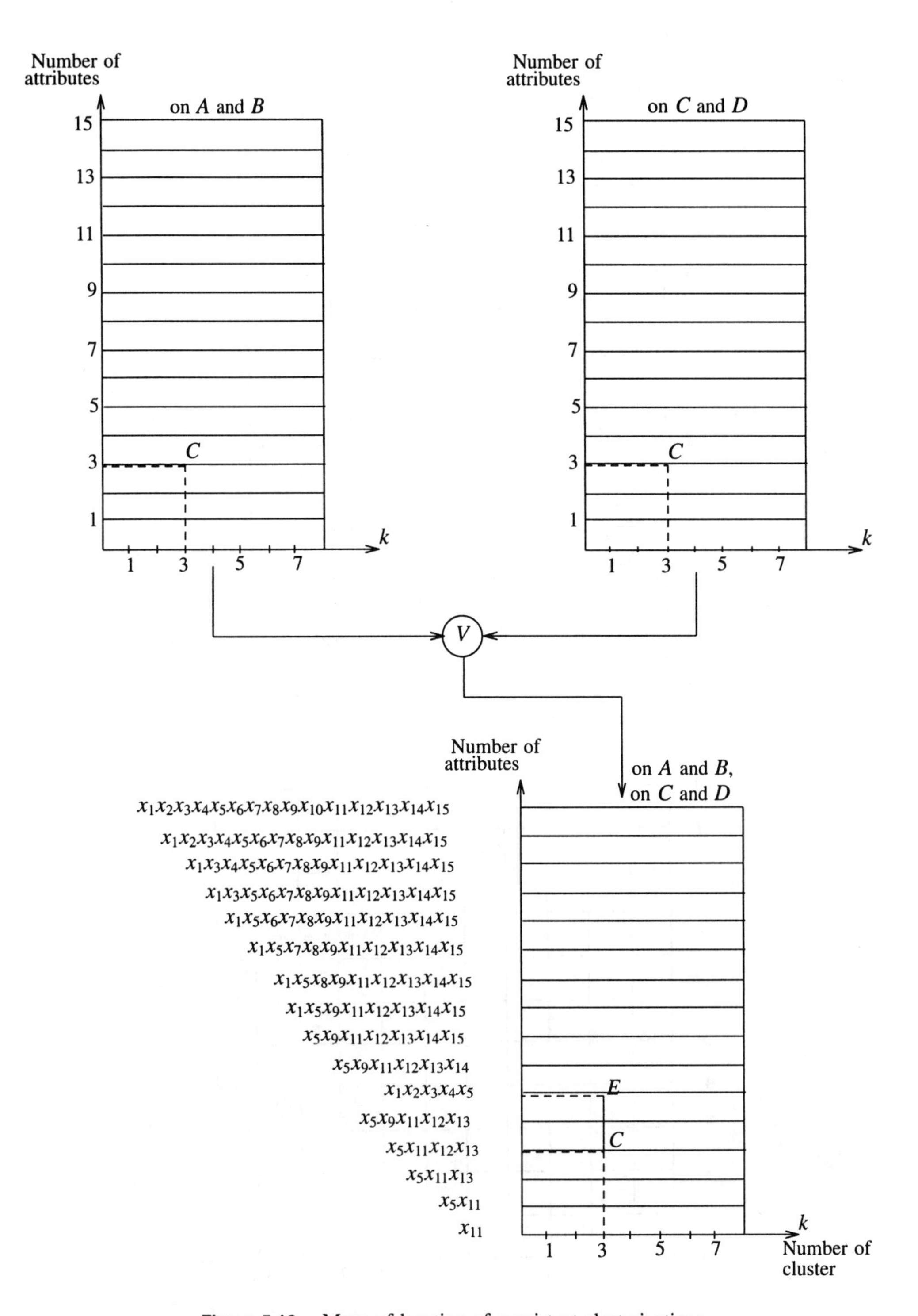

Figure 5.13. Maps of location of consistent clusterizations

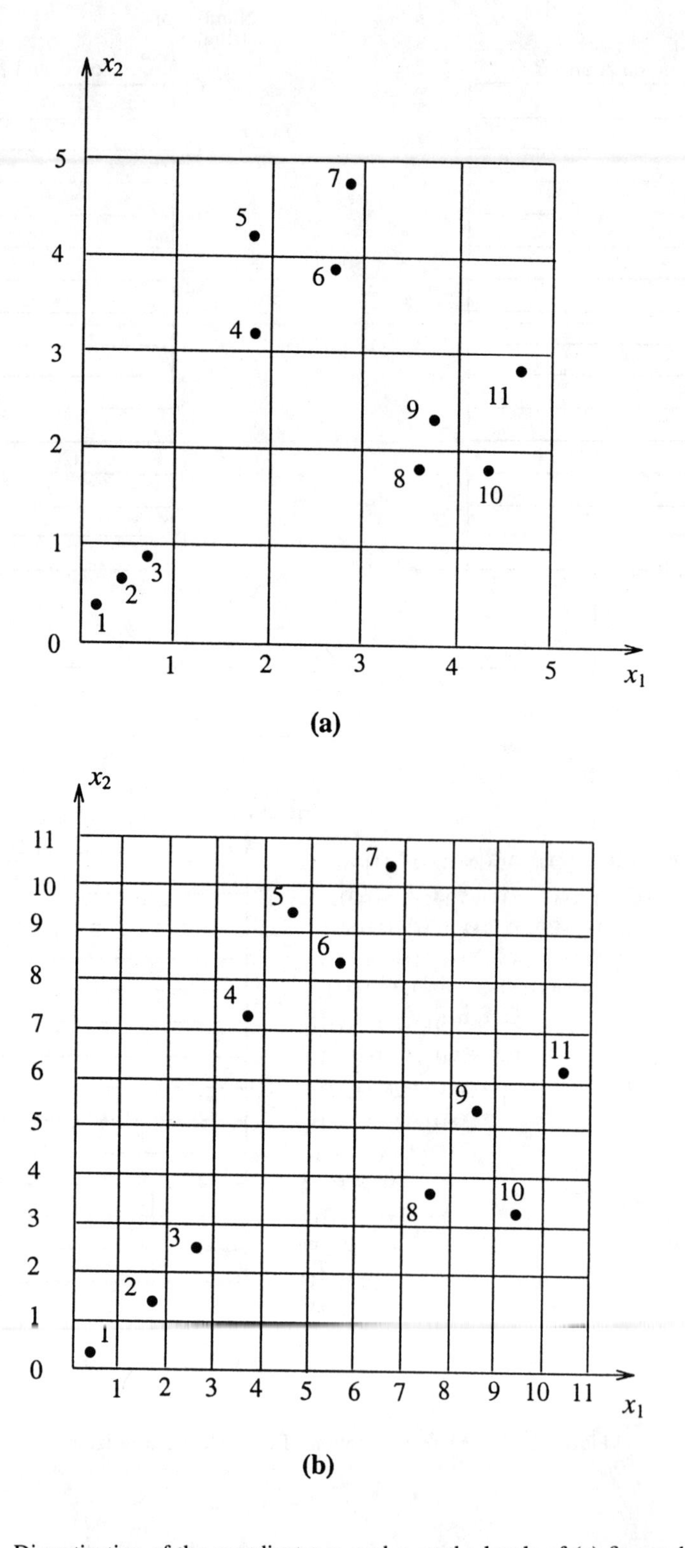

Figure 5.14. Discretization of the coordinates x_1 and x_2 at the levels of (a) five and (b) eleven

Levels of discretization

Figure 5.14 illustrates the different levels of discretization of the coordinates of the points x_1 and x_2 according to Widrow's recommendations. It is suggested that the number of discretization levels of the multiples correspond to obtaining the false zeros of the criterion; for example, here it is $N_1 = N = 11$ and $N_2 = N/2 - 1 = 5$ levels.

In computing the criterion of consistency or balance of discretizations, one has to carry out a special procedure of superimposing square matrices of interpoint distances. The following matrices are obtained according to the 11th and 5th levels of discretizations.

	1	2	3	4	5	6	7	8	9	10	11
1	0	2	4	10	13	13	16	10	13	13	16
2		0	2	8	11	11	14	8	11	11	14
3			0	6	9	9	12	6	9	9	12
4				0	3	3	6	8	7	9	8
5					0	2	3	9	8	10	9
6						0	3	7	6	8	7
7							0	8	7	9	8
8								0	3	3	6
9									0	2	3
10										0	3
11											0

	1	2	3	4	5	6	7	8	9	10	11
1	0	0	0	4	5	5	6	4	5	5	6
2		0	0	4	5	5	6	4	5	5	6
3			0	4	5	5	6	4	5	5	6
4				0	1	1	2	4	3	5	4
5					0	2	1	5	4	6	5
6						0	1	3	2	4	3
7							0	4	3	5	4
8								0	1	1	2
9									0	2	1
10										0	1
11											0

The following matrix shows the inter-cluster distances of clusters from both of the above tables. The table for five levels does not differ essentially from the table for eleven levels.

	1, 2, 3	4, 5, 6, 7	8, 9, 10, 11
1, 2, 3	0	6	6
4, 5, 6, 7		0	6
8, 9, 10, 11			0

Calculation of the criterion

The criterion of balance of discretization is calculated in a special way, which is very convenient for programming. This is done at each step of the construction of hierarchical trees for sorting hypotheses as to the number of clusters. The points that make a cluster are marked with indices (vertices) in a space of $N \times N$ matrices for subsets A and B. The criterion is computed as

$$B_L = \frac{(k - \Delta k)}{k}, \tag{5.21}$$

where $k = N^2$ and Δk is the number of coincidence points or indices on the marking spaces.

The final values are trivial and always hold good. It gives $B_L = 0$ for the optimal clusters, which corresponds to our human impressions when looking at the given arrangement of points.

Regularization

If in the interval from $k = 1$ to $k = N/2$ several zero values of the criterion B_L are formed (excluding ends of the interval), it is necessary to determine which of the "zeros" are false and which are true. This can be checked by repeating the construction of the sorting tree for the hypotheses from some intermediate number of levels (for example, seven or eight if it was checked for 11 before). The whole procedure does not cause any special difficulties for larger number of points and levels.

Example 6. Optimal clustering using the criterion of balance of discretization.

The data is given in Figure 5.14b for the attributes x_1 and x_2 at the discretization level of 11. The table of interpoint distances for the entire sample is measured as given in the matrix

	1	2	3	4	5	6	7	8	9	10	11
1	0	1	2	7	9	9	11	7	8	9	11
2		0	1	6	8	8	10	7	8	9	10
3			0	6	8	7	10	6	7	8	9
4				0	2	2	4	5	4	7	7
5					0	2	2	7	6	8	7
6						0	3	5	4	6	4
7							0	7	6	7	5
8								0	2	2	3
9									0	2	3
10										0	2
11											0

The dipoles are constructed so that they start with the shortest until all the points are in the subsets A and B without repeating them. The following dipoles are obtained and formed into subsets A and B.

subset A: 1 2 1 4 4 5 5 8 8 9 10

subset B: 2 3 3 5 6 6 7 9 10 10 11

They are addressed as $I, II, III, IV, V, VI, VII, VIII, IX, X, XI$. The matrices of interpoint distances are compiled for the subsets A and B separately as below:

		I	*II*	*III*	*IV*	*V*	*VI*	*VII*	*VIII*	*IX*	*X*	*XI*
	A	1	2	1	4	4	5	5	8	8	9	10
I	1	0	1	0	7	7	9	9	7	7	8	9
II	2		0	1	6	6	8	8	7	7	8	9
III	1			0	7	7	9	9	7	7	8	9
IV	4				0	0	2	2	5	5	4	7
V	4					0	2	2	5	5	4	7
VI	5						0	0	7	7	6	8
VII	5							0	7	7	6	8
VIII	8								0	0	2	2
IX	8									0	2	2
X	9										0	2
XI	10											0

		I	*II*	*III*	*IV*	*V*	*VI*	*VII*	*VIII*	*IX*	*X*	*XI*
	B	2	3	3	5	6	6	7	9	10	10	11
I	2	0	1	1	8	8	8	10	8	9	9	10
II	3		0	0	8	7	7	10	7	8	8	9
III	3			0	8	7	7	10	7	8	8	9
IV	5				0	2	2	2	6	8	8	7
V	6					0	0	3	4	6	6	4
VI	6						0	3	4	6	6	4
VII	7							0	4	7	7	3
VIII	9								0	2	2	2
IX	10									0	0	2
X	10										0	2
XI	11											0

Two hierarchical trees of sorting hypotheses as to the clusters (figure 5.15) are built up using the compiled interpoint distance matrices. The criterion of balance of discretization is calculated at each step of constructing the hierarchical trees. The vertices of the dipoles are combined in the tree into a cluster. The elements of the clusters are marked with indices or circles in the matrix form as mapped out in Figure 5.16. Superimposition of the matrix constructed for subset A on the matrix constructed for subset B makes it possible to compute the criterion $B_L = (k - \Delta k)/k$, where $k = N^2 = 121$ and Δk is the number of cells that are coinciding in the matrices.

The "zero" values for the criterion are found for $k_{(A)} = k_{(B)} = 1, 3, 5,$ and 11 by comparing both the trees.

If there are several "zero" values of the criterion, then one has to "invert" certain dipoles and calculate the overall criterion of consistency or one has to repeat the procedure with the different number of levels of discretization.

The examples described in this chapter show that sorting according to the differential criteria (having the properties of the external criteria), consistency, and balance of discretization can replace a human expert in arriving at subjective notions regarding the number and composition of points of the clusters.

5 FORECASTING METHODS OF ANALOGUES

In the traditional deductive methods of modeling, specifying the output and input variables is usually required. The number of variables is equal to or less than the number of data

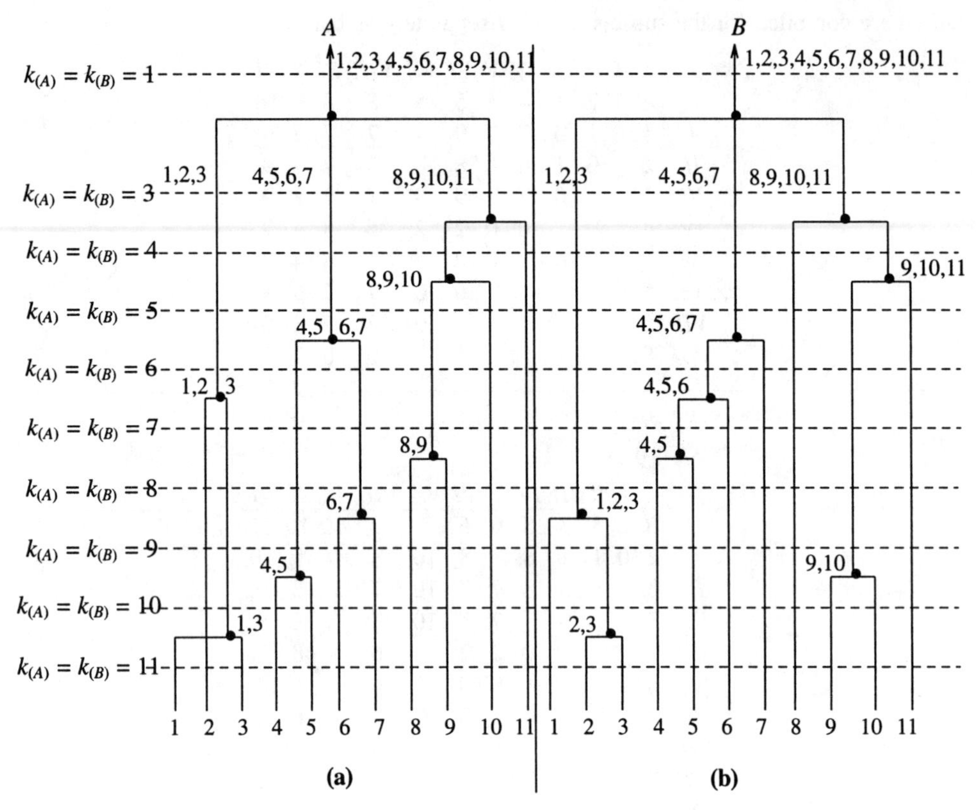

Figure 5.15. Hierarchical trees of sorting hypotheses as to the number of clusters using different discretization levels

measurements. In regression analysis, there are additional limitations, such as the noise factor affecting the output variable, the regressor set being complete, and the regressors not taking into account the equation operate as additional noise. The theories of principal component analysis and pattern analysis for predicting biological, ecological, economic, and social systems which have proven to be possible in a fuzzy language are not new. Again, this is based on the deductive principle that the more fuzzy the mathematical language of prediction, the longer its maximum achievable anticipation time.

Unlike deductive algorithms, the objective system analysis (OSA) algorithm has additional advantages. This does not require an output variable to be specified. In turn, all variables are considered as output variables and the best variant is chosen by the external criterion. The weak point of the inductive learning algorithms is that the estimate of parameters is done by means of the regression analysis. The limitations of the regression analysis cannot be overcome even by using the orthogonal polynomials. The resultant expectations of estimators are biased both by noise in the initial data and the incomplete number of input variables. A physical model is the simplest one among unbiased ones derived with the exact data or with the infinitely large data sample.

Nonparametric inductive learning algorithms offer another possibility and promise to be more effective than the deductive and parametric inductive ones. Its approach is to clarify that in the area of complex systems modeling and forecasting where objects and their mathematical models are ill-defined, the optimum results are achieved as the degree of "fuzzyness" of a model is adequate to the "fuzzyness" of an object. This means that the

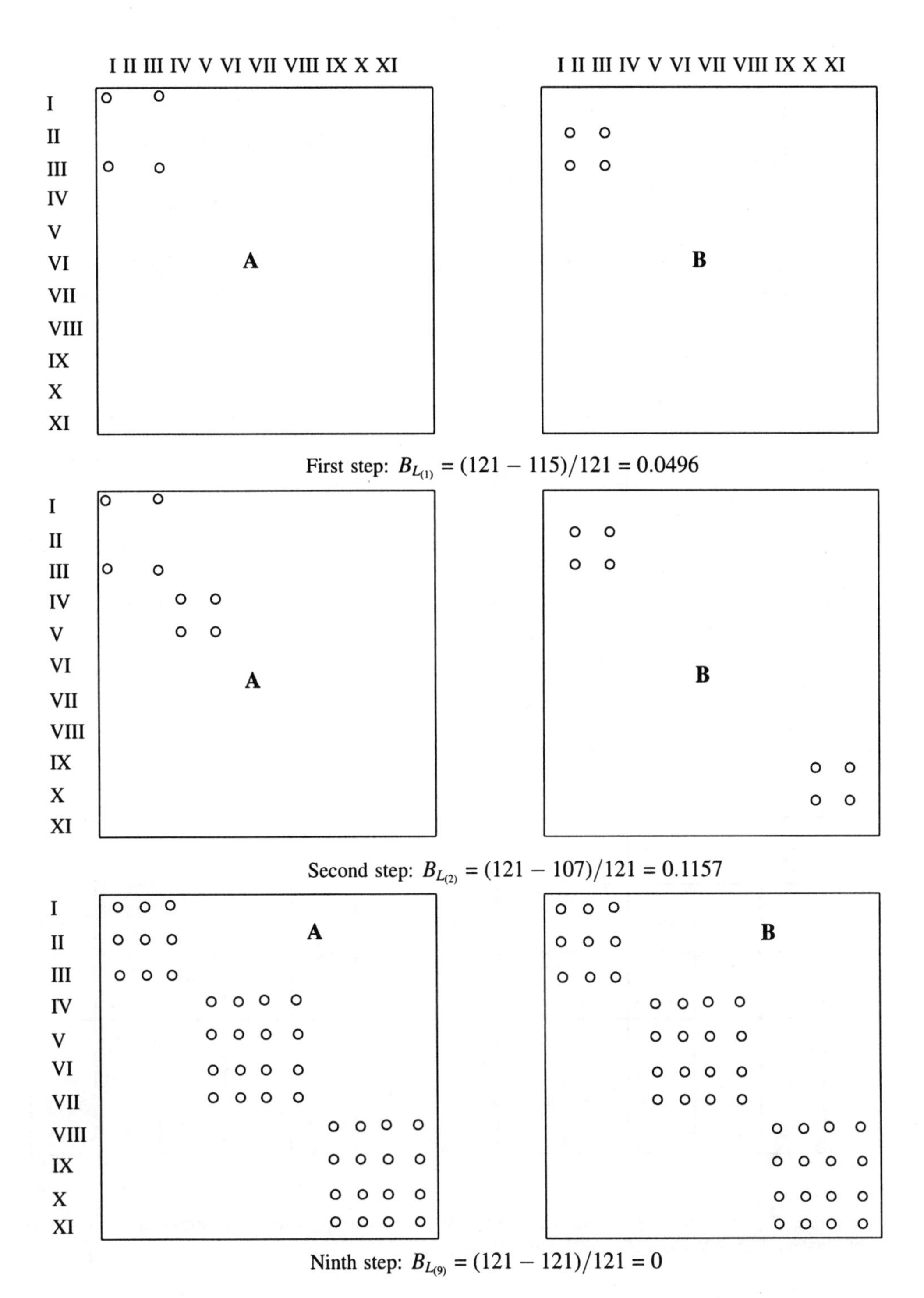

Figure 5.16. Calculation of the consistency criterion from the mappings

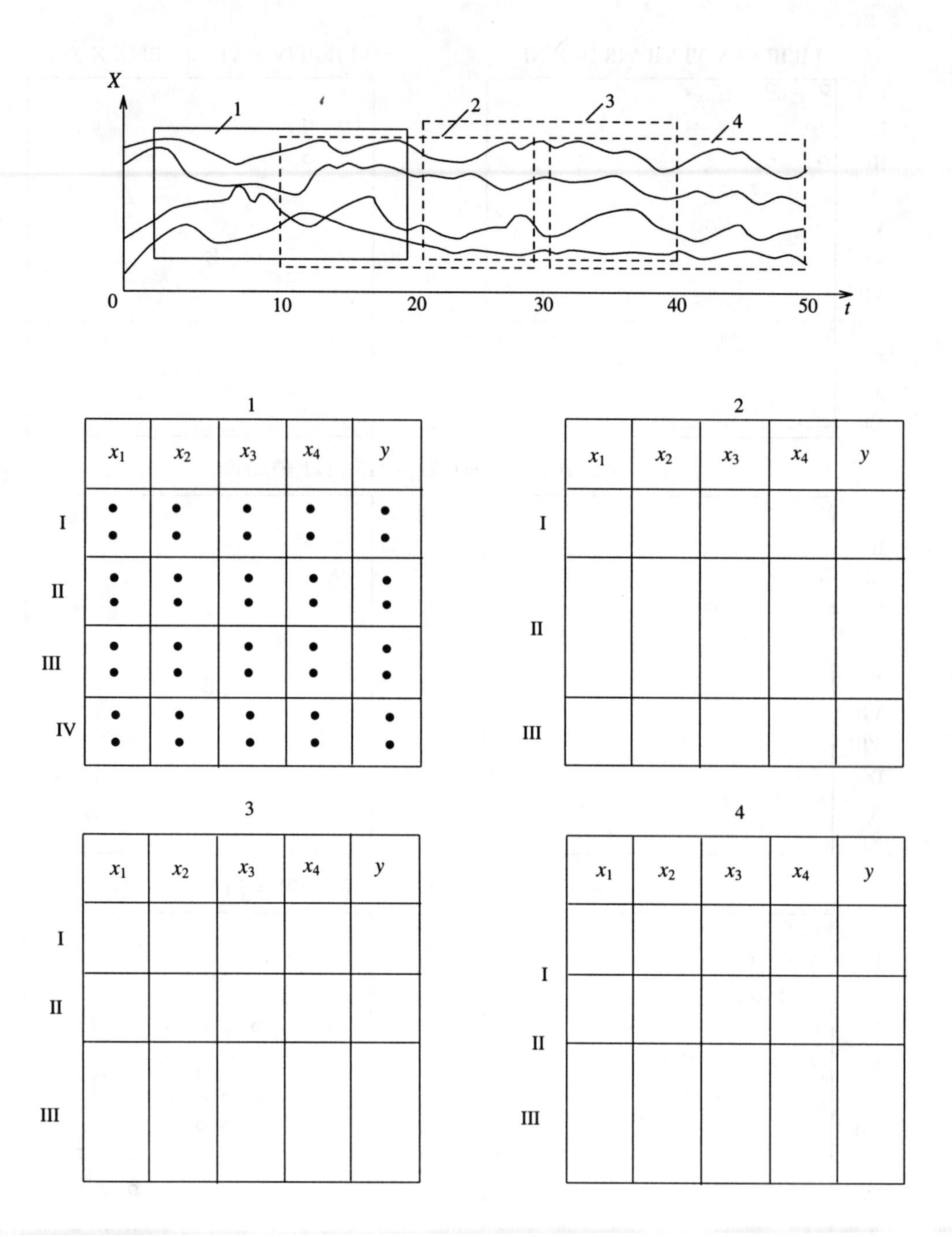

Figure 5.17. Four positions of the "sliding window" and coresponding four clusterizations (number of clusters decreases from four to three)

equal "fuzzyness" is reached automatically if the object itself is used for forecasting. This is done by searching analogues from the given data sample as the clusterizations are tracked using a "sliding window" that moves along the data sample in time axis. For example, the data sample for the ecosystem of Lake Baykal contains measurements over an interval of 50 years (Figure 5.17). One can obtain 40 clusterization forms used to track how the ecological system varies by moving a 10-year wide sliding window in order to predict its further developement. The longest anticipation time of a prediction is obtained without using any polynomial formulations. The objective clusterization of the given data sample is used to calculate the graph of the probability of transition from one class to another. This makes it possible to find an analogue of the current state of the object in prehistory and, consequently, to indicate the long-term prediction. It follows that the choice of the number of clusters is a convenient method of changing the degree of fuzziness in the mathematical language description of the object. By varying the width of the "sliding window," one can realize an analogous action in the choice of the patterns. This approach has an advantage over the clustering analysis given by the OCC algorithm and also the OSA algorithm for having a minimum number of points.

5.1 Group analogues for process forecasting

The method of group analogues leads to the solution of the forecasting problem of a multidimensional process by pattern and cluster search with a subsequent development of a weak into a detailed forecast by the forecasting method of analogues. A sample of observations (N) of a multidimensional process serves as the initial data, and the set of measured variables $(x_1, x_2, \cdots, x_m)$ is sufficiently representative; i.e., it characterizes the state of the observed object and what has occurred in the past is repeated in the present if the initial state has been analogous.

In the problems of ecology, economics, or sociology the available sample size is usually small. The number of forecast characteristic variables m is significantly larger than the number of sample points N ($N \ll m$). Nevertheless, the forecasts are necessary and are of the basic means of increasing their effectiveness through the use of the "method of group analogues."

Forecasts are not calculated, but selected from the table of observation data. This opens up the possibility of more successful forecasting of multidimensional processes.

Formula for forecast measure

The forecasting accuracy of each variable is characterized by the forecast variation of

$$\delta^2_{i_{(k)}} = \sum_{k=1}^{N_C} (x_{i_{(k)}} - \hat{x}_{i_{(k)}})^2 / (x_{i_{(k)}} - \bar{x}_{i_{(k)}})^2, \tag{5.22}$$

where $x_{i_{(k)}}$ is the actual value of the ith variable, $\hat{x}_{i_{(k)}}$ is the forecast obtained as explained below, and $\bar{x}_{i_{(k)}}$ is the mean value (for a quasi-stationary process) without taking the forecast point into account. If the process is nonstationary; i.e., if some of the variables have a clear expression of trend (they increase or decrease continuously), then $\bar{x}_{i_{(k)}}$ equals the value of the trend at each forecasting step. The above formula compares the average error of the forecast by the analogues method with respect to the average error of the forecast as the mean value or trend value.

The forecast of each variable is considered to be successful if the variation $\delta^2_{i_{(k)}} \leq 1.0$ (or in percentage, $\leq 100\%$). Usually, only some variables forecast well. In the best case for all m variables $\delta^2_{i_{(k)}} = 1.0$ (or = 100%). To successfully increase this percentage of forecast variables for a short sample of initial data, one has to go from a search for one analogue in prehistory to the problem of combining several analogues.

Forecast space of several analogues

Here x_i is the point in the multidimensional (Euclidean) space of variables and $\hat{x}_i$, in the space of forecasts, corresponds to each row of the table of initial data sample. The former space is used for computing the interpoint distances, while the latter is used to approximate the forecasts by splines or polynomial formulations.

The point B of the multidimensional spaces x_i and $\hat{x}_i$ is denoted as the output point for forecasting. This is either the last point of the sample in time or the last one that would be possible in estimating the variation of the obtained forecast by the last row. The distances between the point B and all other points measured in the space x_i determine the possibility of using them as analogues. The closest point A_1 is called the first analogue, the next one in distance A_2 is called second analogue, and so on until the last analogue A_F ($F \leq N$). A specific forecast $\hat{x}_i$ corresponds in the forecast space to each analogue. The number of analogues are combined—either specified by an expert or determined according to an inductive algorithm. Various methods can be proposed. Here the method based on extrapolating the forecast space by splines is considered. It is assumed that some forecast value, which is determined by using the forecasts at adjacent points of the space, exists at each point of the forecast space $\hat{x}_i$.

"Combining" forecasts by splines

Here "combining" means approximating the data by splines or polynomial equations with a subsequent calculation of the forecast at the point B. The forecast is defined with the help of weighted summing of forecast analogues using spline equations

$$\begin{aligned} \hat{x}_{i_{(B)}} &= f(\hat{x}_{i_{(A_1)}}, \hat{x}_{i_{(A_2)}}, \cdots, \hat{x}_{i_{(A_F)}}) \\ &= a_0 + a_1\hat{x}_{i_{(A_1)}} + a_2\hat{x}_{i_{(A_2)}} + \cdots + a_F\hat{x}_{i_{(A_F)}}. \end{aligned} \tag{5.23}$$

The splines are selected such that the point B approaches the optimal set of analogues A_s ($1 < s < F$); i.e., the difference between their forecasts decreases. The closer the points in the forecast space $\hat{x}_i$ are, the closer are the forecasts themselves at these points.

Distances between points for a short-range one-step forecast are measured in the space $\hat{x}_i$ as below:

$$d_j = \sqrt{[(x_1^{(A_j)} - x_1^{(B)})^2 + (x_2^{(A_j)} - x_2^{(B)})^2 + \cdots + (x_m^{(A_j)} - x_m^{(B)})^2]}; \quad \text{where } j = 1, 2, 3, \cdots, F, \tag{5.24}$$

where d_j are the Euclidean distances of the point B from the analogues A_j, $j = 1, 2, \cdots, F$; A_1 is the first analogue (closest), A_2 is the second more distant analogue, A_3 is the third even more distant analogue, and so on.

The Euclidean distance is a convenient measure of proximity of a point, but only for a one-step forecast. The repetitive procedure of stepwise forecast can be used to obtain a long-range forecast with a multi-step lead, in which a "correlative measure" is estimated for the proximity of groups of points. The canonical correlation coefficient [104] is also recommended as a proximity measure for forecasting more than four steps.

The interpoint distances d_j, $j = 1, 2, \cdots, F$ are used for calculating the coefficients of the following splines;

1. for one analogue ($F = 1$):

$$\hat{x}_{i_{(B)}} = \hat{x}_{i_{(A_1)}}; \tag{5.25}$$

2. when two analogues are taken into consideration ($F = 2$):

$$\hat{x}_{i_{(B)}} = (d_1^{-1}\hat{x}_{i_{(A_1)}} + d_2^{-1}\hat{x}_{i_{(A_2)}})/(d_1^{-1} + d_2^{-1}); \tag{5.26}$$

3. when the forecasts of three analogues are taken into account ($F = 3$):

$$\hat{x}_{i_{(B)}} = (d_1^{-1}\hat{x}_{i_{(A_1)}} + d_2^{-1}\hat{x}_{i_{(A_2)}} + d_3^{-1}\hat{x}_{i_{(A_3)}})/(d_1^{-1} + d_2^{-1} + d_3^{-1}); \tag{5.27}$$

4. when the forecasts of F analogues are taken into account:

$$\hat{x}_{i_{(B)}} = (\sum_{\alpha=1}^{F} d_\alpha^{-1}\hat{x}_{i_{(A_\alpha)}})/\sum_{\alpha=1}^{F} d_\alpha^{-1}. \tag{5.28}$$

   The largest number of analogues that are taken into account is $F \leq N$. Here F behaves like the "freedom-of-choice."

Alternatively, one can use a parametric inductive algorithm for combining the forecast analogues in which a complete polynomial of the form

$$\hat{x}_{i_{(B)}} = a_0 + a_1\hat{x}_{i_{(A_1)}} + a_2\hat{x}_{i_{(A_2)}} + \cdots + a_F\hat{x}_{i_{(A_F)}} \tag{5.29}$$

is used instead of the splines.

The following choices are to be considered to provide the most accurate forecasting process:

- choice of the optimal number of complexed analogues $F = F_{opt}$;
- choice of optimal set of features $m = m_{opt}$; and
- choice of the permissible variable measurement step width $h = h_{max}$.

Method of reducing variable set size

The two-stage method given below enables us to find the optimal set of effective features.

Stage 1. Variables are ordered according to their efficiency $F = 1, 2, 3, \cdots$ (not more than five) using the partial cross- validation criterion $CV_j \to \min$, defined with the help of moving a so-called "sliding window" (which is equal to one line) along the data sample (Figure 5.18). For each position of the "sliding line" its analogues are found in prehistory and the common analogue forecast is calculated using the splines. The discrepancy between the "sliding line" and the forecast analogue defines a forecast error for each variable. The error is found for all positions of the "sliding line" in the sample. The results are summed and averaged according to the following formulae:

$$\Delta\hat{x}_{ij} = |\hat{x}_{ij}(B) - \hat{x}_{ij}(A_1, A_2, \cdots, A_F)|$$

$$CV_j = \frac{1}{N}\sum_{i=1}^{N}|\Delta\hat{x}_{ij}|; \quad 1 \leq j \leq m.$$

$$CV_{\min} = \frac{1}{N}\sum_{i=1}^{N}|\Delta\hat{x}_{ij}|_{\min} \tag{5.30}$$

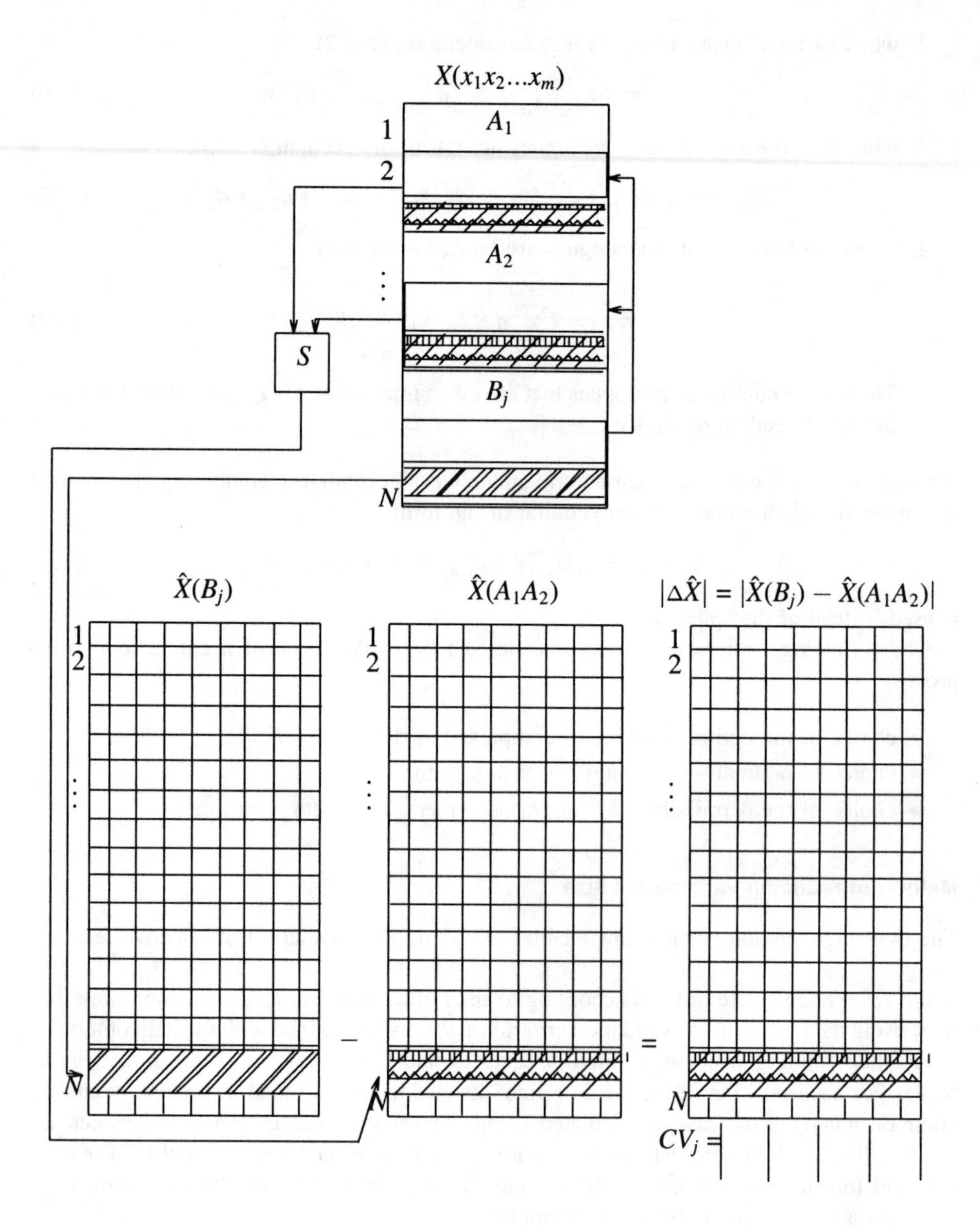

Figure 5.18. Schematic flow of the algorithm corresponding to process forecasts for calculating the cross-validation criterion CV_j when two analogues complexed, where B-current position of sliding window, S-spline, and $|\Delta\hat{x}|$—absolute errors.

where i,j are numbers of data rows and columns respectively ($1 \leq i \leq N$, $1 \leq j \leq m$), $CV_{\min}$ is the cross-validation criterion for choosing optimal set of input variables (features), $|\Delta\hat{x}_{ij}|$ are the absolute values of errors, and $|\Delta\hat{x}_{ij}|_{\min}$ are the minimal value of $|\Delta\hat{x}_{ij}|$ in the lines of sample. In general, a different series of features ordered according to the criterion CV_j are produced for different numbers F of complex analogues. This is analyzed on a plane of F versus m.

Stage 2. The feature series are arranged as per the values of the criterion CV_j. A small number of feature sets are selected from all possible sets for further sorting out using the complete cross-validation criterion,

$$CV = \frac{1}{m} \sum_{j=1}^{m} CV_j \rightarrow \min. \tag{5.31}$$

The ordered feature set shows which sets should remain and which should be excluded. The complete set of feature sets is divided into groups, containing an equal number of features. Only one set, in which less efficient features are absent, remains in each group.

For example, there exists an ordered feature series of $x_3x_2x_1x_4$ (the best feature is x_3, the worst one is x_4); then the following sets are to be sorted out:

one set containing all four features: $x_3x_2x_1x_4$ (all included);
one set containing three features: $x_3x_2x_1$ (x_4 excluded);
one set containing two features: x_3x_2 (x_1x_4 excluded); and
one set consisting of one variable: x_3 ($x_2x_1x_4$ excluded).

The whole number of sets tested is equal to four, being equal to the number of features.

Algorithm for optimal forecast analogue

The schematic flow of mode of operation of the algorithm for optimal forecast analogue is illustrated in Figure 5.19. The overall algorithm consists of two levels: the first one corresponds to obtaining the optimal parameter set by using the two- stage method and the second one corresponds to the process forecasting. Figure 5.19a illustrates the analogue search $\hat{x}$ and evaluation of the forecast error δ for each position of the "sliding window" and the process observation. Figure 5.19b illustrates the efficiency estimation and ordering of variables using the criterion $CV_j \rightarrow \min$. Figure 5.19c illustrates how to obtain F_{opt} and m_{opt} with the help of the criterion $CV \rightarrow \min$. The variable sets are obtained using the criterion $CV_j \rightarrow \min$, and the complete cross-validation criterion $CV \rightarrow \min$ is calculated for them as explained above. The results are plotted on the plane of $F - m$, where the minimum value of the criterion is found. Optimization of the criterion for set of variables is evaluated as

$$\sum CV = \left(\frac{CV}{CV_{\min}} + \frac{CV_{\min}}{(CV_{\min})_{\min}} \right) \longrightarrow \max. \tag{5.32}$$

The point of the plane which gives the criterion minimum, defines the optimal parameters $F = F_{opt}$ and $m = m_{opt}$ sought for.

Variable set optimization enables the so called "useful" and "harmful" features in an initial sample to be highlighted; i.e., it makes possible the exclusion of some data sample columns. The forecast sought for is then read out from the sample using only those optimal parameter values. Figure 5.19d illustrates the forecast at the output position of the "sliding window" B.

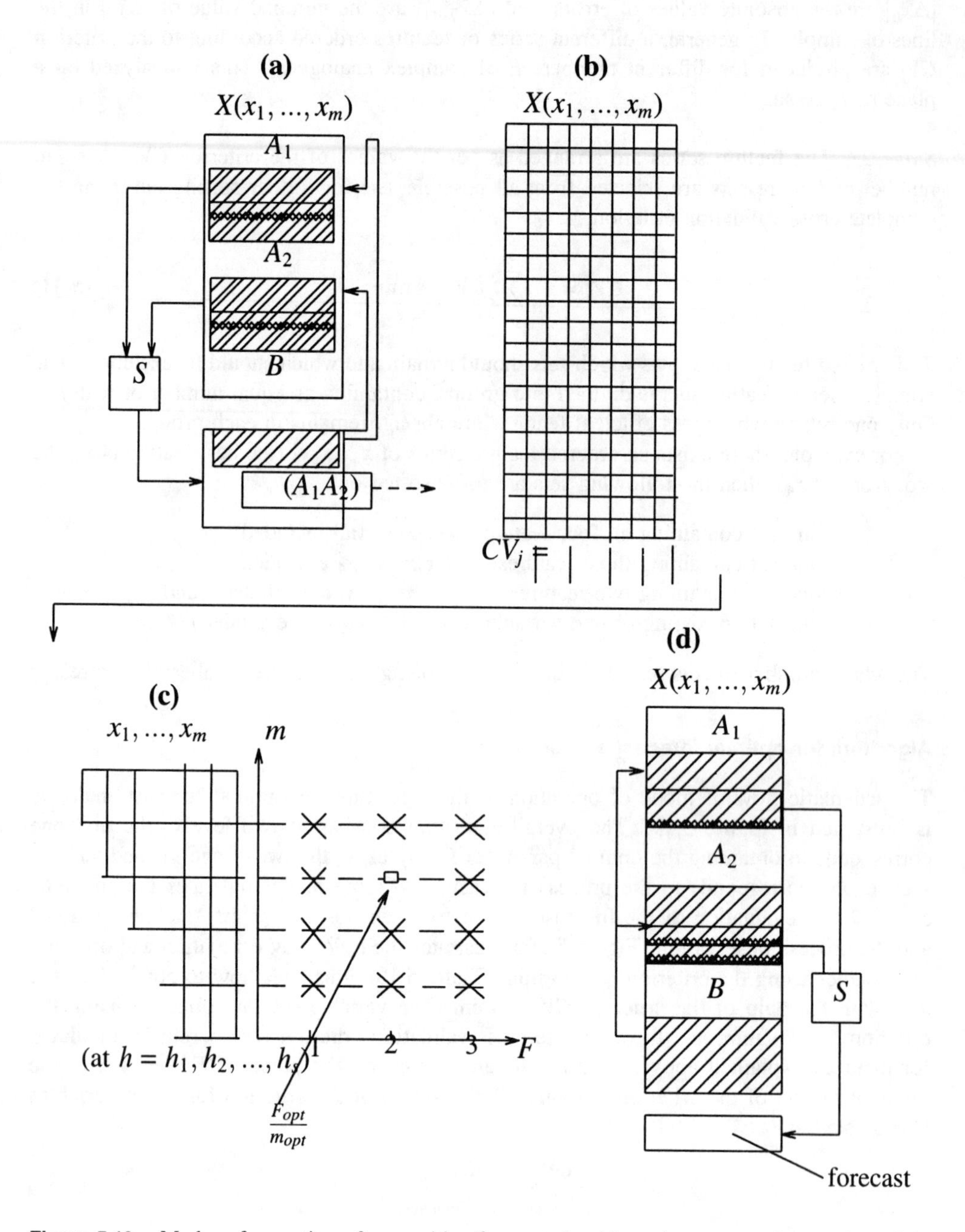

Figure 6.10. Modes of operation of recognition/forecast algorithm when two analogues A_1 and A_2 are complexed; (a) and (b) calculation of errors and criteria, (c) optimization of the criterion $\sum CV$, and (d) application mode; where $\sum CV = \left(\frac{CV}{CV_{\min}} + \frac{CV_{\min}}{(CV_{\min})_{\min}}\right) \longrightarrow \max$.

Pattern width optimization

This concerns the choice of permissible variable measurement width h_{max}. One observation point in the data table is called a pattern—in other words, it is a complete line of expansion. These lines of expansions can be transformed by summing up two, three, etc. adjoining lines and averaging the result. Due to overlapping of the number of lines in each junction, it is only reduced by unit; i.e., a sample containing twenty lines can be transformed into a sample containing nineteen doubled lines, or a sample containing eighteen tripled lines, and so forth. The sorting out of data sample makes it possible to select a permissible pattern width. Thus, the amount of sorting of the ensemble variants is reduced substantially if one succeeds in ranking the predictor-attributes (placing them in a row according to their effectiveness) in advance. The solution for the problem becomes simple. When the algorithm for optimal forecast analogue is used, one estimates each predictor separately according to the forecast measure ($\delta_i^2(x_i) \leq 1$). This simplifies substantially the problem of choosing an effective ensemble of predictor-attributes. This means that one should identify the pattern width which provides a forecast variance value $\delta_i^2(x_i)$ less than unity for all variables treated. To estimate the value of $\delta_i^2(x_i)$, the forecast is to be calculated for the penultimate pattern.

We conclude that, in general, the optimization of the process forecast analogue algorithm is done in a three-dimensional space of the choices (F, m, h) for $Y = 0$, where F is the number of complexed analogues, m is the number of features taken into account, h is the data sample pattern width, and Y is the target function which is not specified.

5.2 Group analogues for event forecasting

The above procedure of process forecasting is described without specifying the output vector Y (target function); i.e., it deals only with the data sample of the variable attributes of X.

We extend this problem to a forecasting event where the output vector Y is defined as an event. In solving this type of problem, it is important that there be a correlation between the columns of the samples X and Y. However, it is usually absent. For successful events forecasting, samples X and Y must be complete and representative. In other words, the data sample has to contain a complete set of events of all types. For instance, when a crop harvest is forecasted, examples of "bad," "mean" and "good" harvests should be represented in Y. The data is complete if it contains a complete set of typical classes of observed functions. In addition, the sample should be representative. This means that clusters of matrices X and Y must coincide in time.

One of the tests for completeness and representativeness is that the matrices X and Y be subjected to cluster analysis using one of the known criteria. If identical correspondence clusters are obtained on the matrices (for example, good harvest has to correspond to good weather conditions and proper cultivation), then the sample is representative.

The problem of event forecasting is formulated in a more specific cause and effect manner and it has wider field of applications. In the formulation, the sample of attribute variables X is given in $(N+1)$ time intervals, and the event factor Y is given in N intervals, if forecast of event Y in the $(N+1)$st step is required. Some of the examples are:

1. **sample** X —observations of cultivation modes and weather conditions for $(N+1)$ years.
   sample Y —harvest data for N years.
   It is necessary to predict the harvest for $(N+1)$st year.
2. **sample** X —design and production features of $(N+1)$ electronic devices.

sample Y —"life-time" and damage size data for N devices.
It is to predict the duration of uninterrupted operation of the $(N+1)$st device.

3. to forecast the result of a surgical cancer treatment;
sample Y —used as a loss vector containing three binary components; y_1 (recovery), y_2 (relapse), y_3 (metastases), and y_4 (the extent of disease, evaluated by the experts as a continuous quality).
matrix X —includes various features (about 20), describing the state and method of surgical treatment for 31 patients.
The results are known for 30 patients. These results are then used to predict the surgical treatment result for the recently operated 31st patient after the operation.

These are some typical examples of the event forecasting.

In order to predict the events, it is necessary to consider the following aspects to provide the accurate event forecasting;

- choice of the optimal number of complexed analogues $F = F_{opt}$;
- choice of the optimal set of features $m = m_{opt}$; and
- choice of the optimal target function vector $Y = Y_{opt}$.

The first two entities describe the process forecasting algorithm, whereas the latter is a specific aspect of the event forecasting problem.

Here, the pattern width (measurement step) $h = 1$ should not be changed. It is strictly equal to one line of an initial sample and the data sample cannot be transformed as explained before. Instead it is expedient to sort out the components of the vector Y (output value). For example, the harvest can be represented in the data sample not only by crops weight, but also by its sort and quality. The sorting out procedure allows only those components which give the minimal value for the criterion CV leading to a more accurate forecast to remain.

First, it is necessary to reduce the number of feature sets involved in the sorting. This is demonstrated in the Figure 5.20. The distinction from the method described in Figure 5.18 is that here two matrices X and Y are participating. Instead of getting the difference between sliding line and complexed analogue forecast, the differences of the vectors Y (not their forecasts) are calculated as

$$|\Delta x_{ij}| = |x_{ij}(B) - x_{ij}(A_1, A_2, A_3)|. \tag{5.33}$$

The logic of feature choice is that the value of an effective feature at the current line and its analogues must be as close to each other as possible. A large discrepancy in the value means the feature does not define the output value Y; i.e., it is ineffective. The criterion CV_j is calculated as the difference of feature values of the line, and the analogues averaged over the sample columns.

$$CV_j = \frac{1}{N}\sum_{i=1}^{N} |\Delta x_{ij}| \rightarrow \min . \tag{5.34}$$

Analogues are searched to find the matrix Y. At least one component of Y must be measured continuously and accurately for a unique analogue. However, if the analogue is not unique as defined, then the two components of a target function, which are derived from the Karhunen-Loeve algorithm, are added to the vector Y.

The schematic explanation to the algorithm is exhibited in Figure 5.21. Here "a" is the analogue choice, "b" is the calculation of the partial cross-validation criterion $CV_j \rightarrow \min$ for

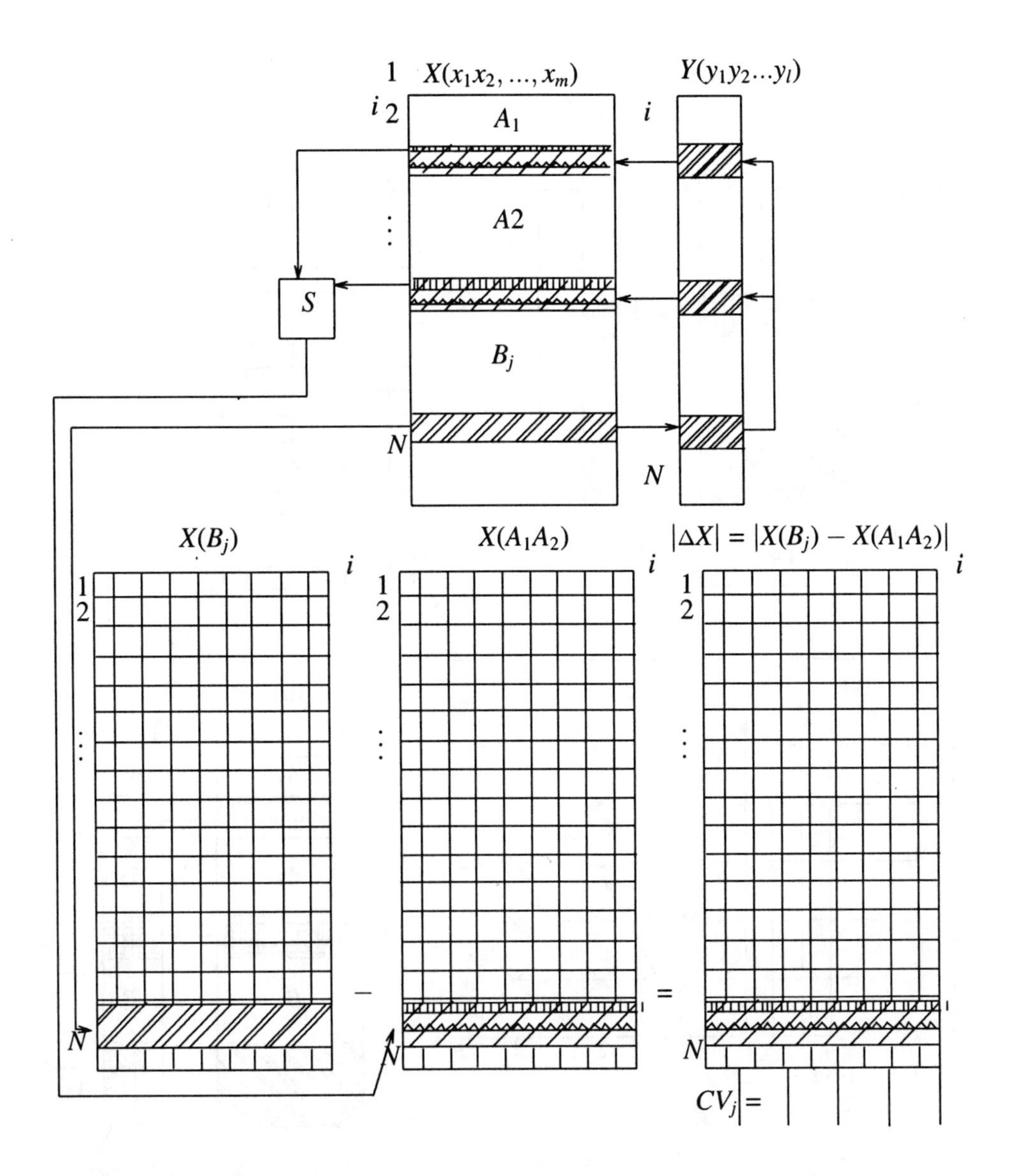

Figure 5.20. Schematic flow of the algorithm corresponding to events forecasts for calculating the cross-validation criterion CV_j when two analogues complexed as per the occurring events, where B-current position of sliding window, and S-spline; the criteria evaluated are $CV \rightarrow \min$, $CV_{\min} \rightarrow \min$, and $\sum CV = \left(\frac{CV}{CV_{\max}} + \frac{CV_{\min}}{(CV_{\min})_{\max}}\right) \longrightarrow \max$.

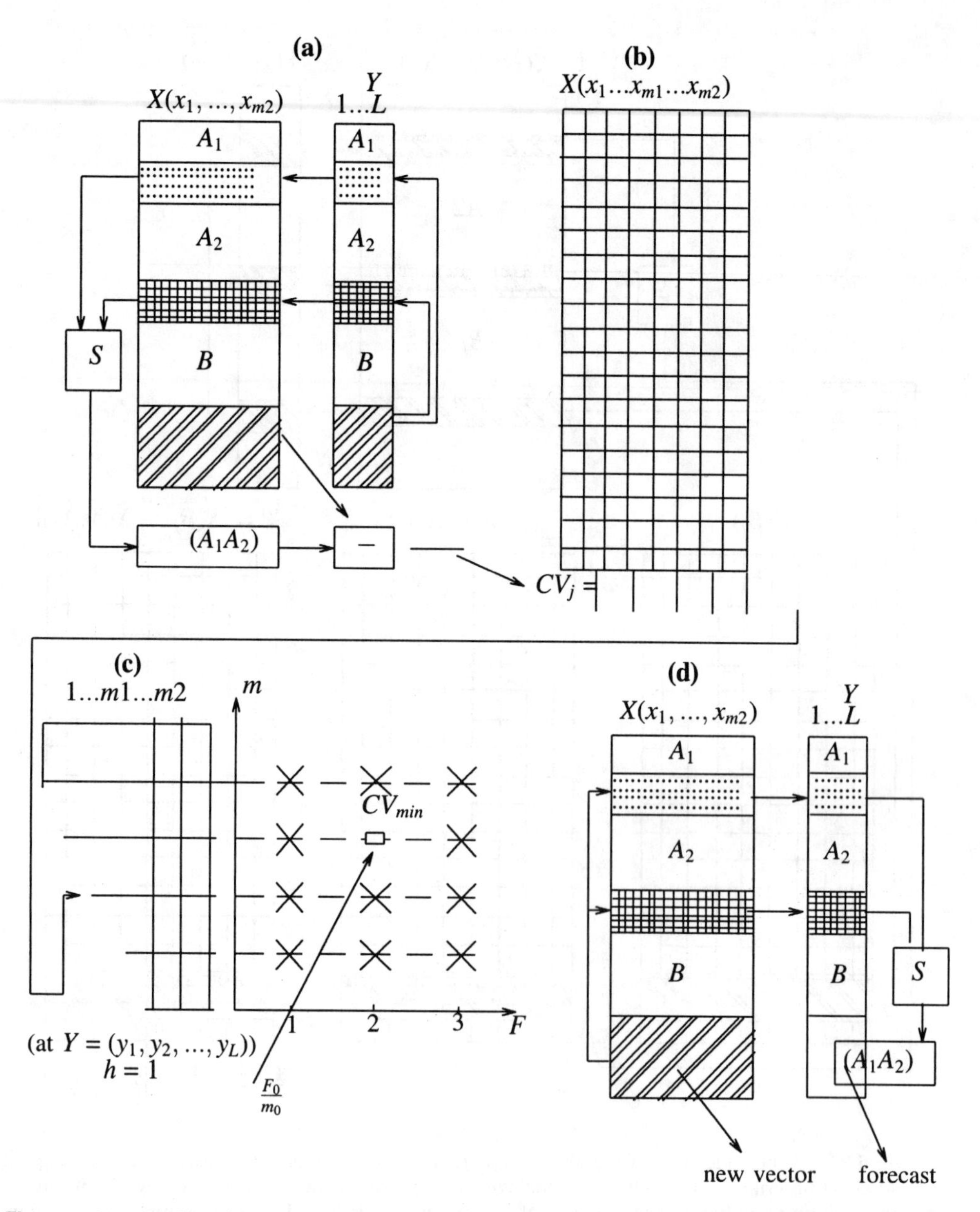

Figure 5.21. Modes of operation of an events forecast algorithm when two analogues A_1 and A_2 complexed (a) choice of analogue, (b) calculation of partial cross-validation criterion CV_j, (c) arranging on the plane to obtain optimal point, and (d) the second stage of the event recognition/ forecast

ordering features, "c" is the calculation of values of the complete cross-validation criterion $\sum CV = \left(\frac{CV}{CV_{\max}} + \frac{CV_{\min}}{(CV_{\min})_{\max}}\right) \longrightarrow \max$. with the purpose of defining the optimal values of F_{opt} and m_{opt}. r is the forecast of the event corresponding to the $(N+1)$st sample line under optimal algorithm parameter values.

Note that matrices X and Y are used in one direction (anti-clockwise) at the optimization stage, and in the opposite one (clockwise) at the forecast stage.

Other features

Use of convolution for an analogue choice in sorting out the vector components of Y. One can use a convolution of components in the target function instead of calculating the analogues in the multidimensional space. This helps the modeler to include components which lead to more accurate forecast. The analogues will be the same, but the calculations are simpler. The target function Y must have a continuous scale for a unique definition of the analogues. Thus, when at least one of the components of Y has such a reading scale, it is recommended that the convolution of the normalized component values $Y = \sqrt{(y_1^2 + y_2^2 + \cdots + y_l^2)}$ for analogue searching be used. If all components are binary variables (equal to 0 or 1), it is necessary to expand the component set by introducing one or two components of the orthogonal Karhunen-Loeve transform (for the joint sample XY).

$$Y = \sqrt{(y_1^2 + y_2^2 + \cdots + y_l^2 + z_1^2 + z_2^2)}, \tag{5.35}$$

where z_1 and z_2 are components of the artificial target function [137]. Sorting out of the target function is meant for excluding some items from the expression.

The complete sorting of variants of criterion values $CV \to \min$ is carried out in a three-dimensional space of (F, m, l) as $h = 1$, where F is the number of complex analogues, m is the number of feature sets, and l is the number of components in the target function.

Correlation measure of distances between points and "Wroslaw taxonomy." The simplest measure to calculate the distance between the points of the multidimensional feature space is the Euclidean distance for continuous features and Hamming distance for binary ones. If the data are nonstationary, for example, values will show an increasing or decreasing trend. The trend is then defined either as an averaged sum of normalized values of the variables or each variable trend is found separately (by a regression line in the form of polynomial of second- or third-order). Deviation of the variable from its trend is read out individually. The correlation coefficient of the deviation of each of the two measured points serves as a correlation measure of distance between them.

When the distance correlation measure is used, it is logical to apply the "Wroslaw taxonomy" algorithm for feature-ordering according to their efficiency. This algorithm is based on the partial cross-validation criterion $CV_j \to \min$ and makes it possible to order features according to their efficiency, and then excludes them one by one in the optimization process of the events-forecasting procedure to find the optimal feature set and the optimal number of complexed analogues.

The "Wroslaw taxonomy" algorithm is applicable only when the target function is defined in the problem. For this reason it is useful only in event forecasting, but not in the process forecasting.

Once the system is trained for a specific problem of event forecasting, it can be considered as the algorithm for recognition of new images. Thus, the event forecasting algorithm is treated as a particular case of the more general problem of image recognition; i.e., when recognizing the $(N+1)$st vector of the target function Y is necessary.

Chapter 6
Applications

Inductive learning algorithms, called Group Method of Data Handling (GMDH) were developed using the principles of self-organization modeling. Self-organization of modeling is the process of finding the optimal structure of mathematical description of a complex object by sorting many variants according to a certain ensemble of external criteria. Unlike the traditional modeling approaches which are deductive in character, the inductive learning algorithms are based on the sorting of models according to the external criteria agreed on by experts. The inductive approach does not eliminate the experts or take them away from the computer, but rather assigns them a special position. Experts indicate the selection criteria of a very general form and interpret the chosen models given by the criteria as the best. They can influence the result of modeling by formulating new criteria. The overall approach of the expert becomes that of an objective referee in resolving scientific controversies. The improvement of man-machine (ergatic) systems is based on the gradual reduction of the human involvement in the process. The automated systems become more mechanized; i.e., they are not only automated but are fully automatic as well. The human element often involves errors and undesired decisions. One of the examples is the process of specifying the objectives, or determining the set of criteria. Future development might lead to sequential decision-making algorithms that include an automatic setting of selection criteria, their sequential determination, freedom-of-choice, and so on—almost without having to involve experts to solve important problems. This means that the involvement of human element will be reduced.

At the present state of development, experts specify external criteria and consider results without interfering with the optimization processing. In case of disagreement, the experts can utilize higher levels of criteria (noise stability criteria), so that controversies are quickly resolved. The problems of nonlinear identification and of long-range forecasting of complex processes are solved by the computer.

Two-level prediction schemes of learning allow one to leave the choice of the model to the computer, significantly increasing the lead time of forecasting. The objective character of the modeling speeds up the understanding of the object, allowing us to avoid false subjective selection.

The following features of inductive approach allow us to improve the quality of self-organization modeling and to give the procedure an objective character [32].

1. The minimum-bias criterion agrees that models obtained with the use of two different sets of data be identical. Such a model leads to a physical model that is isomorphic to the mechanism of the object under consideration. In self-organization modeling, a

physical model is used only to determine the composition of the set of output variables subject to forecasting. Knowing this, we proceed to self-organization of nonphysical models.

Nonphysical models differ from physical models in the composition of the arguments used and in the external criteria. The step-by-step prediction criterion assumes great importance. Nonphysical models can be constructed in various mathematical languages that differ in the degree of prediction accuracy which is measurable by the correlation time [44]. The nonphysical models can use two- or three-dimensional time readout. All this is important because the limit of informativeness of a forecast is determined by the degree of blurredness of the modeling language.

Nonphysical models are used in the self-organization modeling of two-level schematic predictions—for example, annual and seasonal predictions for which the balance-of-predictions criterion allows the increase in the lead time of detailed (seasonal) predictions. The usage of an auxiliary criterion like prediction criterion is necessary to curtail the volume of sorting to the reasonable number of annual and seasonal models and to ensure the uniqueness of the choice of predictions according to the prediction-balance criterion. The balance criterion serves as the choice of both short-term as well as long-term forecasts.

2. The class of the equations and the form of support functions are selected by sorting many variants of models according to the selection criteria. For example, a system of finite-difference equations are formed for use in self-organization modeling of ecological systems. In traditional approaches, the physical equations are considered and their discrete analogues with variable coefficients are formed to represent the object, but ultimately it ends up with poor predictive characteristics.
3. The third feature of the inductive approach is the selection of the set of output and input variables, and of the "leading" variable among them. The objective system analysis allows us to obtain the least biased system of equations according to the system criterion of minimum bias.

   The inductive algorithm for OSA allows us to sort out all possible systems of equations consisting of one, two, three, and so on equations and to select the most unbiased model of system of equations. This determines the structure of the object and the set of output variables. The best output variable which forecasts better than all others is called as the "leading" variable.

   The variables that are interested may not enter into the set of output variables during OSA and have to be predicted as a supplement in terms of the functions of the output variables according to the two-level prediction scheme in time.
4. The self-organization modeling allows us to obtain models with optimal complexity even in case of incomplete information; i.e., without having the data of many important arguments. This is an antithesis of the idea of increasing the information basis up to some universal measure. The inductive learning approach, however, demonstrates the success of such modeling; for example, for predicting winter wheat harvest more than 50 arguments are required to have—the use of fertilizer, method of tillage, periods of irrigation, and so on. All available arguments are very important. Nevertheless, in self-organization modeling, only two to three arguments participate for sufficient accuracy of forecasts (this is further explained in the given examples). Although the connections among the arguments made during the self-organization modeling remain unknown to us, these may be taken into account for an accurate forecasts.
5. The self-organization modeling is possible with noisy data. The inductive learning algorithms obey the laws valid in the communication theory—particularly, Shannon's

second-limit theorem for transmission of noisy signals.

The inductive algorithms allow us to restore the physical model of the object under study even in the case of noisy data exceeding three to four times the regular signal [63]. The methodologies used in the communication theory and the pattern recognition theory allow us to raise the noise immunity of the algorithms. This means that the accurate models can be obtained with an incomplete information basis and noisy data in the same way as the accurate signal is restored under the noisy conditions and distortions of various kinds.

The overall modeling is object-oriented because all problems are solved according to the agreed selection criteria. The final results of modeling may not coincide with the ideas of the modeler about the object being modeled.

1 FIELD OF APPLICATION

The inductive method is an empirical method and is intended for self-organization of mathematical models based on measured data. The object of the modeling is identification and prediction of the object. The usual methods of regularization (for example, regression analysis) are mathematically elegant but inexpedient external additions.

In self-organization theory, an entire series of more apt criteria are proposed—regularity, minimum bias, prediction and others oriented towards satisfaction of the practical users of the models. These criteria are applied sequentially one after another, to eliminate the difficulties of normalization (choice of the coefficients) of the criteria. Multicriterial choice of the model is one of the foundations of noise immunity of the inductive algorithms.

Gödel's incompleteness theory is fulfilled by the problem being concerned with the choice of the ensemble of external complements, and its composition and sequence of application at different levels is solved by sifting a number of variants. The upper level criteria is based on:

1. criterion for noise stability: $\Theta_1 = \frac{\text{noise}}{\text{signal}} \rightarrow \max$,
2. criterion for maximizing the lead time of predictions: $\Theta_2 = \frac{\text{lead time}}{\text{observation time}} \rightarrow \max$, and
3. criterion for minimizing the amount of computational time: Θ_3 = computer operational time $\rightarrow$ min.

The ensemble of criteria is to better ensure the required value of the pair of criteria Θ_1 and Θ_2. The objectivity is based on the choice of the set of criteria.

The field of application of the inductive learning algorithms is shown in Figure 6.1. This is widely spread on the plane with the coordinating values of Θ_1 and Θ_2, and $\Theta_3 \geq 3$ hours of operational time (usually on minicomputers).

The modeling of optical systems requires application of single-level models with a usual time reference, accurate initial measured data, and the complete information basis. Two-dimensional time readout (in terms of seasons and years) in case of noisy data is required in predicting certain agricultural productions, river flows, and so on. Two-level forecasting with two-dimensional time readout with incomplete information basis is used for econometric models, modeling of climatic changes, and ecological systems. Some of the practical examples corresponding to these are given in the preceding chapters. This chapter is extended further for more specific examples in the areas of weather modeling, ecosystem studies, economical systems modeling, agricultural system studies, and solar activity.

We hope that the reader will get an overall idea of how to approach to this type of model-

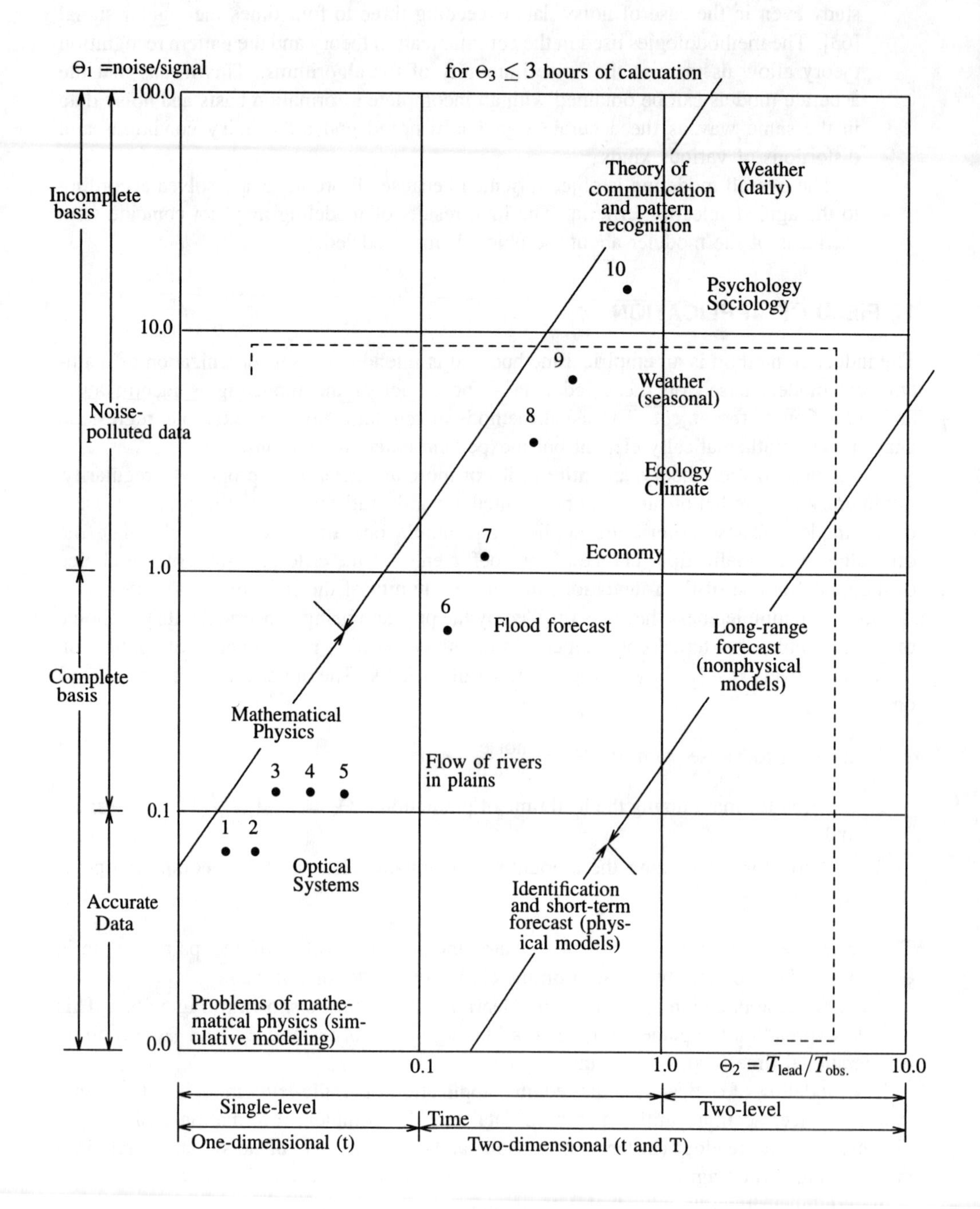

Figure 6.1. Various fields of application using inductive algorithms

ing. However, the reader must keep in mind that more complex modeling problems with very high noise levels require development of new algorithms using special mathematical formulas, new external criteria, and criteria with canonical forms. The field of self-organization modeling is open to receiving and providing further technological achievements.

2 WEATHER MODELING

Self-organization modeling requires the presence of two components: a generator of variety of models (combinatorial or multilayer) and some sensibly chosen ensemble of external criteria to evaluate these models. A relatively short data sample is needed to estimate the parameters of the models and to compute the values of the criteria. A general description of the equations comprising the system to be formed may be known to the modeler. So, a composite approach is frequently the optimum approach, whereby a general description specified *a priori* as a reference by the human author and remaining analysis is done by computer sorting in accordance to various criteria.

2.1 Prediction balance with time- and space-averaging

The problem of identification of complex objects based on the emperical data is treated as an ill-posed problem, in as much as a unique solution always requires the use of some external information or an external supplement according to Gödel. This means that it is in principle impossible to obtain a unique model in optimal complexity without regularization. In addition to the basic criteria like regularity and minimum-bias, the following are some criteria are convenient to use for cylindrical polar coordinates of meteorological field.

Interpolation balance criterion

The difference equation pattern has only six nodes (Figures 6.2 and 6.3). During the training process the patterns are moved vertically upwards in the time axis at one-day steps. Each position of the pattern yields one conditional equation. All models are trained according to the inductive algorithm. The expression is

$$q_{ij}^{t} = \frac{1}{4}(q_{i+1j}^{t} + q_{i-1j}^{t} + q_{ij+1}^{t} + q_{ij-1}^{t}), \tag{6.1}$$

which, from the theory of difference equations, can be used to construct the criterion of interpolation balance. If N is the number of pattern positions in the interpolation region, then the criterion is written as

$$b^2 = \sum_{i=1}^{N}[q_{ij}^{t} - \frac{1}{4}(q_{i+1j}^{t} + q_{i-1j}^{t} + q_{ij+1}^{t} + q_{ij-1}^{t})]^2. \tag{6.2}$$

Balance criteria

Balance-of-variables and balance-of-predictions criteria are used when several variables are predicted simultaneousy. The former requires that some relationship that exists between the variables at a given time should also exist in the future, whereas the latter should be used in an ensemble with the predicting models.

As given before, the prediction balance criterion with time averaging of variables (averaging over a season and over a year) which is used mainly for cyclic processes can be

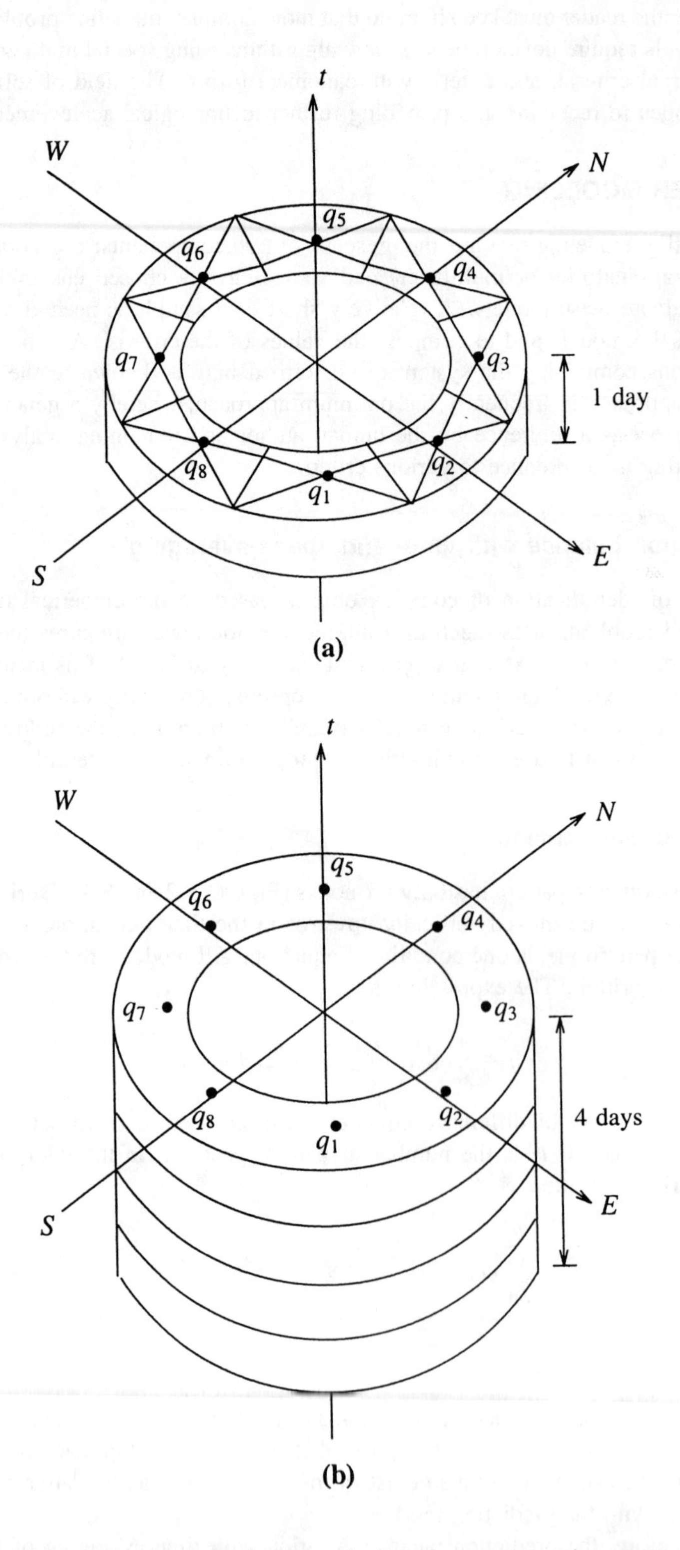

Figure 6.2. Application of interpolation balance criteria with averaging a) in space (average of eight predictions is balanced to the average over entire area of the ring), and b) in time and space (average of $8 \times 4 = 32$ predictions is balanced to the prediction of the average over entire volume of the tube

written as

$$b^2 = [\hat{q}_{year} - \frac{1}{4}(\hat{q}_w + \hat{q}_{sp} + \hat{q}_{su} + \hat{q}_f)]^2,$$
$$B^2 = \sum_{i \in N} b_i^2 \rightarrow \min, \quad (6.3)$$

where $\hat{q}_w, \hat{q}_{sp}, \hat{q}_{su},$ and $\hat{q}_f$ are predicted average seasonal values of the winter, spring, summer and fall variables correspondingly and $\hat{q}_{year}$ is the predicted average annual value. N is the number of years covered by the prediction.

Prediction balance criterion with space averaging of variables

The prediction balance criterion can be applied in a similar way with the averaging over the area of a ring, which can be written as

$$b^2 = [\frac{1}{8}(\hat{q}_1 + \hat{q}_2 + \cdots + \hat{q}_8) - \hat{q}_0]^2,$$
$$B^2 = \sum_{i \in N} b_i^2 \rightarrow \min, \quad (6.4)$$

where $\hat{q}_1, \hat{q}_2, \cdots, \hat{q}_8$ are predictions obtained for eight patterns which form a ring around the axis of a cylinder (Figure 6.2), $\hat{q}_0$ is a prediction of the variable averaged over the entire area of the ring, and N is the number of steps at which the prediction is checked on the time axis.

Prediction balance criterion with time and space averaging

Here time intervals are, for example, days and 4-day period and space intervals are one pattern and the area of the ring; this can be written as

$$b^2 = [\frac{1}{8}\{\frac{1}{4}(\hat{q}_1 + \hat{q}_2 + \hat{q}_3 + \hat{q}_4)_1 + \frac{1}{4}(\hat{q}_1 + \hat{q}_2 + \hat{q}_3 + \hat{q}_4)_2 + \cdots$$
$$\cdots + \frac{1}{4}(\hat{q}_1 + \hat{q}_2 + \hat{q}_3 + \hat{q}_4)_8\} - \hat{q}_0]^2,$$

$$B^2 = \sum_{i \in N} b_i^2 \rightarrow \min, \quad (6.5)$$

where $\hat{q}_0$ is the predicted value of the variable averaged over the entire volume of the pipe consisting of four rings.

Normalized combined criterion

$$c6^2 = (\frac{\eta_{bs} - \eta_{bs_{\min}}}{\eta_{bs_{\max}} - \eta_{bs_{\min}}})^2 + (\frac{\Delta(C) - \Delta(C)_{\min}}{\Delta(C)_{\max} - \Delta(C)_{\min}})^2 + (\frac{b - b_{\min}}{b_{\max} - b_{\min}})^2 \rightarrow \min \quad (6.6)$$

Alternatively, normalization can be avoided by using the criteria $\eta_{bs}^2 \rightarrow \min$, $\Delta^2(C) \rightarrow \min$ and $b^2 \rightarrow \min$ in a sequence one after the other.

To ensure the needed freedom of choice is the goal of every sequential decision making procedure; first F_1 models are selected from total of F_0 models using the first criterion, from this F_2 are selected by the second criterion, and finally the third criterion is used to

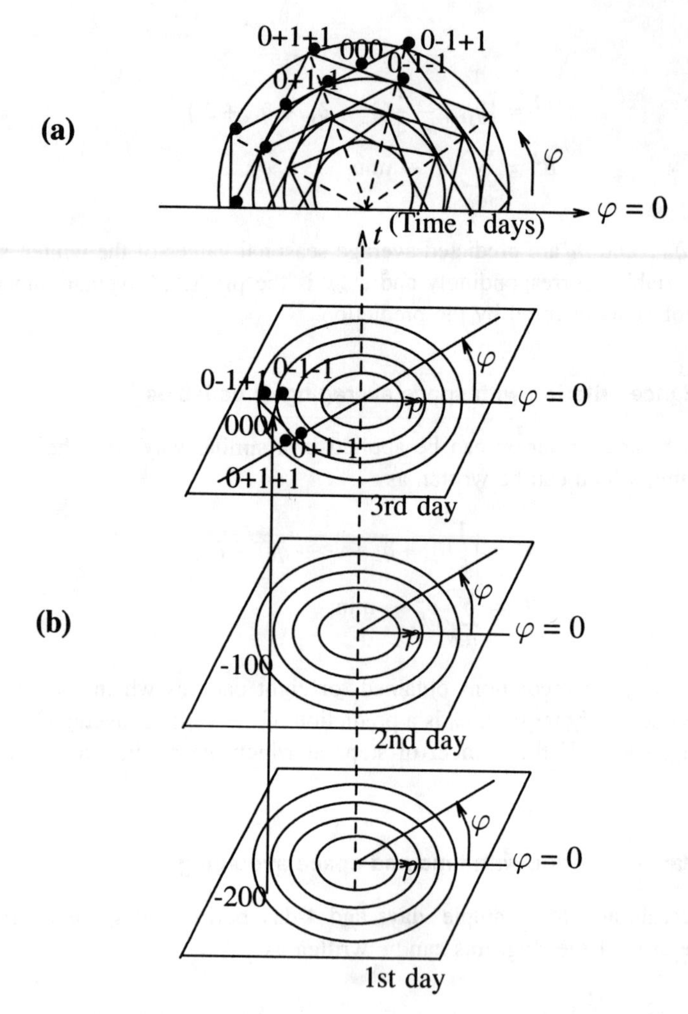

Figure 6.3. Positioning of implicit patterns on three daily charts in cylindrical polar coordinates ρ and φ; a) plan, and b) axonometry

select a single model of optimal complexity $F_0 > F_1 > F_2 > F_3 \geq 1$. The sequence of application is

$$F_0 \rightarrow \eta_{bs}^2(F_1) \rightarrow \Delta^2(C)(F_2) \rightarrow b^2(1). \tag{6.7}$$

To optimize the freedom-of-choice, F_1 and F_2 are chosen to select several identical models on the basis of η_{bs}^2 and $\Delta^2(C)$. A single optimal model is selected from this group using the balance criterion. The number of models tested by the balance criterion can be increased depending on the computer capacity—usually $F_3 \leq 8$.

Sequential application of the criteria does not require the normalization of their values and also there is no need of introducing criterion weighting coeficients.

2.2 Finite difference schemes

Self-organization modeling requires one to indicate the list of variables containing a large access, an appropriate emperical data sample, and a reference function. The computer selects

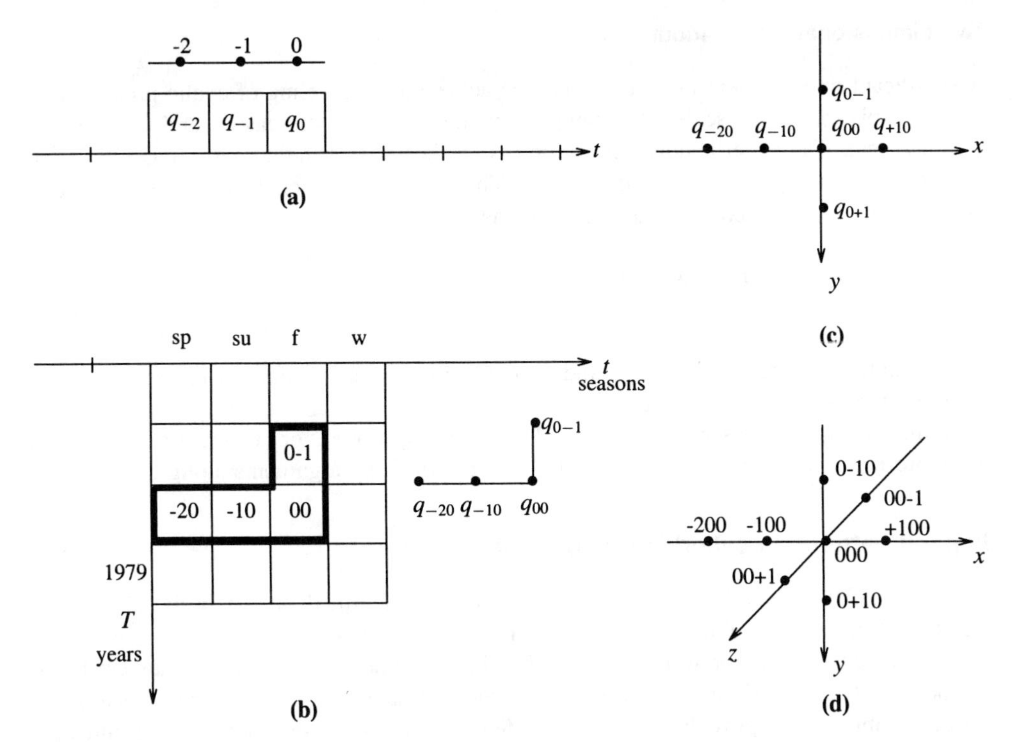

Figure 6.4. Explicit elementary patterns: (a) for the point problem $q(t)$, (b) for the two-dimensional time scale $q(t, T)$, (c) for the problem $q(x, y)$, and (d) for the three-dimensional problem $q(x, y, z)$

the ensemble of most effective arguments, finds the optimum structure of each equation, and estimates the coefficients. The depth of minimum of the selection criterion is the measure of the modeling.

Fields of meteorological parameters (temperature, pressure, humidity, etc.) are usually described by the linear partial differential equations. The linear finite-difference equations, which are discrete analogues of the linear differential equations are suggested as reference functions; these equations can be expanded by introducing nonlinear terms into them.

The following equation is an example of a point problem:

$$\frac{d^2q}{dt^2} + a_1\frac{dq}{dt} + a_0q = f(t). \tag{6.8}$$

By using the forward finite-differences of $\Delta q = (q^t - q^{t-1})$ and $\Delta^2 q = (q^t - q^{t-1}) - (q^{t-1} - q^{t-2})$, this equation has the following finite-difference form:

$$(q^t - 2q^{t-1} + q^{t-2}) + a_1(q^t - q^{t-1}) + a_0q^t = f(t) \tag{6.9}$$

The elementary pattern corresponding to this equation represents the arguments q^{t-1} and q^{t-2} which have an effect on the output q^t.

$$q^t = f_1(t) + f_2(q^{t-1}, q^{t-2}), \tag{6.10}$$

where $f_1(t)$ is the so-called source function or trend function.

To construct the system of conditional equations, the data collected by moving the pattern step-by-step along the time axis (Figure 6.4a) are arranged in the table form with the columns of t, q^t, q^{t-1}, and q^{t-2}. Each position of the pattern corresponds to one row of the data table which is divided into different parts. The prediction is obtained through step-by-step integration by moving the pattern into the region of future time.

Two-dimensional time-readout

A significant improvement in accuracy can be achieved in modeling of cyclic processes by using the bivariate time scale; for example, months and years or seasons and years, hours and days, etc. The pattern shown in Figure 6.4b indicates that the output $q_{t,T}$ is influenced by the arguments $q_{t-1,T}, q_{t-2,T}$, and $q_{t,T-1}$. This pattern can include not only the output variable but also the auxiliary variable like x as

$$q_{t,T} = f_1(t,T) + f_2(q_{t-1,T}, q_{t-2,T}, q_{t,T-1}, q_{t,T-2}, \cdots,$$
$$x_{t,T}, x_{t-1,T}, x_{t-2,T}, x_{t,T-1}, \cdots). \tag{6.11}$$

The patterns related to the (x,y) and (x,t) planes can be identical to the patterns with bivariate time scale (t,T) (Figure 6.4c).

Similarly, the patterns related to the planes $(x,y,t), (x,y,z)$, or (x,y,z,t) (Figure 6.4d) are represented for spatial problems having number of delay arguments along each axis.

Implicit patterns in cylindrical coordinates

In self-organization modeling, the physical fields are identified on the basis of different reference functions; algebraic, harmonic, or finite-difference equations. Usually the finite-difference equations are preferred because they have additional advantages. They are linear in coefficients and nonlinear in variable parameters. But sometimes they may create some serious problems like providing unstable predictions. In such cases addtional measures are taken in achieving the convergence of step-by-step predictions such as

- decreasing the sampling interval of variables;
- simplifying the patterns and functions (not considering the nonlinear terms);
- changing from "explicit" to "implicit" patterns; and so on.

Let us see the concept of *implicit patterns*; they are realized on a closed curve and are moved simultaneously by one step. This yields a simultaneous system of equations. The most promising are the finite-difference models realized by implicit patterns that are constructed in cylindrical-polar coordinates (t, ρ, ϕ) (Figure 6.3). The differential equation of diffusion in this plane contains a linear sum of derivatives of not higher than the second order

$$a_0 \frac{\partial^2 y}{\partial t^2} + a_1 \frac{\partial^2 y}{\partial \rho^2} + a_2 \frac{\partial^2 y}{\partial \phi^2} + a_3 \frac{\partial y}{\partial t} + a_4 \frac{\partial y}{\partial \rho} + a_5 \frac{\partial y}{\partial \phi} = f(t, \rho, \phi), \tag{6.12}$$

where $a_0, a_1, a_2, \cdots, a_5$ are the constants and $f(t, \rho, \phi)$ is the source function.

The following finite-difference analogue can be replaced considering one additive trend equation.

$$y_{ij}^t = f_1(t, \rho, \phi) + f_2(y_{ij}^{t-1}, y_{ij}^{t-2}, y_{i+1j+1}^t, y_{i-1j+1}^t, y_{i-1j-1}^t, y_{i+1j-1}^t,$$
$$z_{ij}^t, z_{ij}^{t-1}, z_{ij}^{t-2}, z_{i+1j+1}^t, z_{i-1j+1}^t, z_{i-1j-1}^t, z_{i+1j-1}^t), \tag{6.13}$$

where $f_1(t, \rho, \phi)$ is the trend function, y is the predicted variable, z are the variables which are correlated with the predicted variable, and i and j are the indices of ρ and ϕ coordinates.

The table of data sample is prepared for each position of the pattern along the closed circle that is achieved by placing the patterns along circular layers of the cylinder in such a way that the adjacent patterns have two common points. The patterns are trained using the part of the data in one region and are used for step-by-step predictions in the other region. The meteorological factors are predicted for the first layer near the central axis of the cylinder making it possible to determine the weather in the second layer, and then the third layer, etc. for the entire region included in the cylinder.

For example, Figure 6.3 shows the placement of eight patterns on one of the three layers of the cylinder. The eight pattern equations can be written for the first layer as

$$a_0 + a_1 q_{ij}^{t-2^k} + a_2 q_{ij}^{t-1^k} + a_3 q_{i+1j+1}^{t^k} + a_4 q_{i+1j-1}^{t^k} + a_5 q_{i-1j+1}^{t^k} + a_6 q_{i-1j-1}^{t^k} = 0 \qquad (6.14)$$

where $k = 1, 2, \cdots, 8$ and the coefficients $a_0, a_1, \cdots, a_6$ are estimated using one of the inductive learning algorithms. While predicting, the values of all $q_{ij}^{t-2^k}$ and $q_{ij}^{t-1^k}$ are known as these were positioned in the past and the values of $q_{i+1j-1}^{t^k}$ and $q_{i-1j-1}^{t^k}$ are determined with a separate prediction because these are located near the cylinder axis. Pattern coupling equations $q_{i+1j-1}^{t^k} = q_{i+1j+1}^{t^{k-1}}$ or $q_{i-1j+1}^{t^k} = q_{i-1j-1}^{t^{k+1}}$ hold for $q_{i+1j+1}^{t^k}$ and $q_{i-1j+1}^{t^k}$. Thus, 16 necessary and sufficient equations are available for determination of 16 variables (eight $q_{i+1j+1}^{t^k}$ and eight $q_{i-1j+1}^{t^k}$).

The equations are obtained for the second and third layers of the cylinder; the only difference is that the values of $q_{i+1j-1}^{t^k}$ and $q_{i-1j-1}^{t^k}$) are determined from the data of the previous layer, not by a separate prediction. This means that a separate prediction is required only in the region located near the cylinder axis.

The self-organization theory offers the additional means for improving the convergence of step-by-step prediction:

1. By using the finite-difference equations with variable coefficients; these have simpler patterns than the equations with constant coefficients.
2. By using the step-by-step prediction criterion (i^2) in the ensemble of the external criteria; this criterion selects from the set of possible models those that have adequate convergence.

2.3 Two fundamental inductive algorithms

Combinatorial algorithm

Since the original differential equation is idealized, the nonlinearities in the system are taken into account by expanding the finite-difference analogue (reference function) with higher ordered terms. For example, for a point problem we can have

$$\begin{aligned} q^t \;=\; & f_1(t) + (a_0 + a_1 q^{t-1} + a_2 q^{t-2} + a_3 q^{t-3} + a_4 q^{t-1} q^{t-2} + a_5 q^{t-1} q^{t-3} \\ & + a_6 q^{t-2} q^{t-3} + a_7 q^{t-1^2} + a_8 q^{t-2^2} + a_9 q^{t-3^2}), \end{aligned} \qquad (6.15)$$

where $f_1(t)$ is the trend or coarse model. This is usually obtained in advance by applying the least squares method through the experimental data or by using an inductive algorithm with the minimum-bias criterion as an external criterion. The purpose of this model is to decrease the number of arguments of the difference part, lumping together some of them into a separate term. In weather forecasting this model can correspond to a climatic forecast averaged over a long time. The remained difference part of the model refines this forecast.

The combinatorial inductive algorithm enables us to evaluate the "structure of functions" obtainable from this equation on the basis of minimum-bias and regularity criteria. There are 10 terms in the difference part of the above equation and the total number of equations to be tested is 2^{10}.

Multilayer algorithm

If the difference part of the above polynomial reference function contains more than 20 terms (varies according to the computer capacity), the multilayer algorithm with linearized terms is applied. The linearized version of the reference function is

$$q^t = f_t + (a_0 + a_1 w_1 + a_2 w_2 + \cdots + a_m w_m), \tag{6.16}$$

where $m \geq 20$ and w represent the terms q.

The multilayer algorithm determines the complete linear polynomial as a superposition of partial polynomials with two-variable of type ($\hat{q}_k = a_{0k} + a_{1k} w_i + a_{2k} w_j$, where k indicates the unit number; $i = 1, 2, \cdots, m$; $j = 1, 2, \cdots, m$; and $i \neq j$). This algorithm realizes the method of incomplete induction by omitting some partial polynomials during sorting and is never tested against the criteria. This is done in a multilayer feedforward network structure. The result of self-organization modeling is the output model with an optimal complexity. The computer selects the structure of the model, its nonlinearity, and the content of its arguments. During the processing the ineffective terms of the reference function are discarded.

2.4 Problem of long-range forecasting

The method of analogues is one of the interesting methods considered in the literature for reliable long-term weather forecasting. The idea of analogues is based on finding the interval whose meteorological characteristics are identical to those presently observed in the measured data and the future of this interval which is measured in the past is the best forecast at the present time. Although the idea is so simple, attempts to apply this idea always produce results that are not very convincing because with a large number of observed variables it is not possible to find the exact analogues in the pre-history data. The self-organization method based on the inductive approach can be interpreted as an improved method of group analogues in which the analogues of the present state of the atmosphere are selected by using special criteria to produce the most probable forecast.

The problem encountered is how to estimate, at least approximately, the achievable prediction time. The maximum achievable prediction time T_p of a one-step forecast is determined by the coherence time τ_c of the autocorrelation function $A_q(\tau)$. The maximum allowed prediction time of a multiple step-by-step forecast is equal to the coherence time multiplied by the number of steps ($T_{p_{\max}} = \tau_c . n$). This means that prediction error increases with each integration step that imposes a definite limit on the step-by-step forecast. This leads to seeking of the maximum capabilities of multiple step-by-step prediction, assuming that they are determined by the coherence time in the same way as they are for one-step prediction. However, studies of autocorrelation functions of meteorological parameters to determine maximum prediction time have not yet been completed. The expected results should be similar to the studies of autocorrelation functions for other complex systems. It turns out that averaging of variables in time increases the coherence time. One has to remember that it also depends on the physical properties of the process being predicted as well as on the quality and characteristics of the mathematical apparatus. This corresponds to the extreme variations from predicting "purely" deterministic objects like motions of planets to "purely" random objects like games of "lotto." The actual physical problems are always located in between these two cases.

The autocorrelation function of a process contains some information on its predictability (the degree of determinancy or randomness). According to studies it is evident that by increasing the averaging interval of variables in time or space shift the process from the region of unpredictability into the region of long-term calculability—i.e., centennial averages and global averages of pressure, temperature, or humidity can be predicted a thousand years in advance. At the same time the predictions based on the daily avearges cannot be valid for more than 15 days. Nevertheless, it is possible to overcome the predictability limit by the following suggested possibilities that are applicable in predicting some weather variables like temperature, pressure, etc. at the surface layer.

2.5 Improving the limit of predictability

Here we discuss the possibilities of increasing the time of weather forecasts in the self-organization method.

1. The first potential contribution of the self-organization method in improving the predictability is in the mathematical apparatus and objective synthesis of the system of equations; the structures of the equations are selected by the minimum-bias criterion.

   The proposed equations are to be valid not only in the training region of the data sample, but also in the testing region. This is precisely what is done by the minimum-bias criterion. This means that the computer sorts out the number of equations for each specified output variable separately and finds out a system of equations that is invariant in time. This is the objectively evaluated system of equations based on the empirical data.
2. The second potential contribution of the self-organization theory in improving predictability is the composite use of different averaging of variables in time and space with the help of the prediction balance criteria. The balance criterion provides a reference point in the future, allows one to perturb the divergence of solutions which is a typical property of the hydrodynamic equations, and from there increases the entropy of predictions.

   In predicting the average daily values, the reference point in the future can be the forecast of average monthly values. Similarly, in predicting the variable of average over an area, the reference point of a forecast at a specified point on the surface of the earth can be the forecast of a sum of variables at several points. This means that when the space averaging is used, the forecast ability of variables averaged over large areas becomes higher. The forecast validity time based on short averages is pulled up toward the validity time on a large regions.
3. The third possibility provided by the self-organization theory is a significant widening of the complete set of arguments (input variables). It is followed by selecting the most effective of these arguments. It is reasonable to include the connected patterns that realize the long-range effect inserted with appropriate delays in time in the use of moving average sums of variables, which are analogous to the integral terms of hydrodynamic equations. If the proposed arguments turn out to be ineffective, these will not be included in the ensemble of predictor arguments.

Example 1. Self-organization modeling of air pressure and temperature at a point located on the cylindrical axes [58].

The possibilities mentioned for improving the limit of predictability in the self-organization method, which uses implicit patterns in cylindrical coordinates, allow one to compute future values of the meteorological parameters. These parameters are already predicted one-step (day) ahead at nodes of all patterns located on the cyldrical axes. The problem requires the one-step prediction of the parameters at one point on the surface of the earth. This can be obtained by using the multilayer algorithm which is demonstrated in this example.

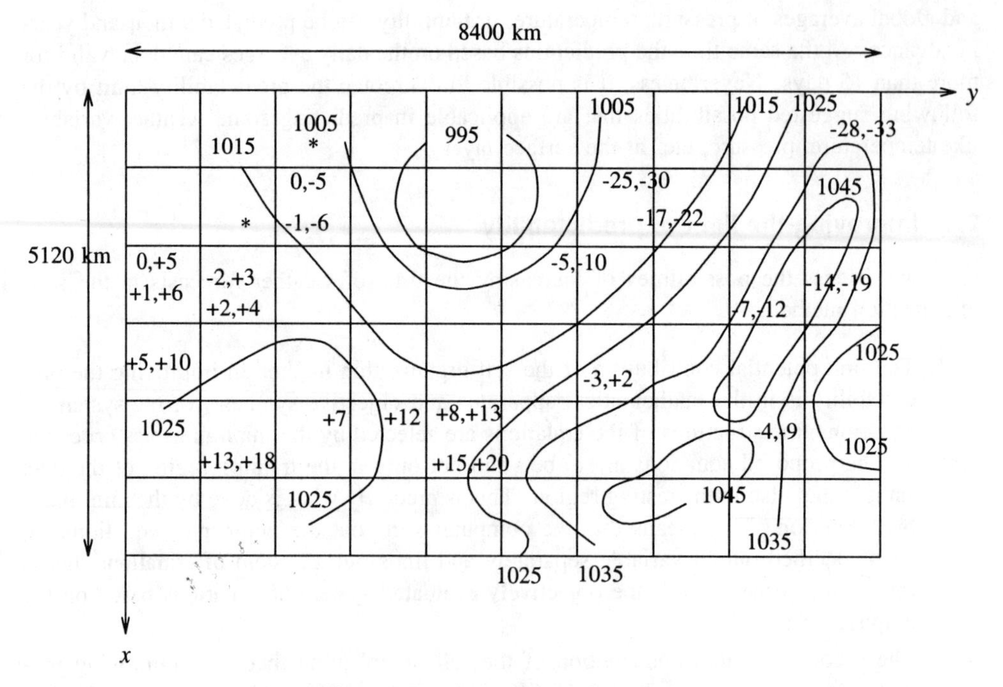

Figure 6.5. A sample weather chart

The example shows how one can find the optimum structure of equations which express the variation of temperature and pressure fields as a function of time. To determine the models used for predicting pressure (P) and temperature (T), the meteorological data charts (Figure 6.5) are divided into elementary cubes and average daily values of the meteorological parameters are prepared in each cube. These charts are stacked on time axis as a parallelopiped-space of given observations as shown in Figure 6.6. This consists of $6 \times 10 \times 10$ unit cubes with the units of measure 5120/6 km along the x-axis, 8400/10 km along the y-axis, and 1 day along the time axis.

Six shapes of even patterns which are tested in the example are given in Figure 6.7 and the corresponding finite-difference models in otimal complexity with the estimates of their accuracy are listed in the Table 6.1. (Here $\bar{P}$ and $\bar{T}$ are the average values of the arguments for the four nodes.) The equations of the complete pattern have the following general forms:

$$\begin{aligned} P_{ij}^{t+1} &= f_1(t,x,y) + f_2(P_{ij}^t, P_{ij}^{t-1}, P_{ij}^{t-2}, \cdots, T_{ij}^t, T_{ij}^{t-1}, T_{ij}^{t-2}, \cdots) \\ T_{ij}^{t+1} &= f_3(t,x,y) + f_4(T_{ij}^t, T_{ij}^{t-1}, T_{ij}^{t-2}, \cdots, P_{ij}^t, P_{ij}^{t-1}, P_{ij}^{t-2}, \cdots). \end{aligned} \tag{6.17}$$

The data are collected according to each pattern from the parallelopiped-space of the observed data and arranged in the tables in relation to the output and input variables; the data are normalized. The data are divided into sequences of A, B, and C, where $W = A \cup B = 80\%$, and $C = 20\%$. The data sequence C is used as a separate sequence for checking the models. The system of equations corresponding to each pattern are selected on the basis of the combined criterion ($c3^2 = \eta_{bs}^2 + \Delta^2(C)$). All equations are compared with each other on the basis of the relative error on the sequence C. The results show that the system of equations corresponding to the fifth pattern (Figure 6.7e) is the most accurate. The relative error is measured as

$$\delta_P = \sqrt{[}\frac{1}{N_C}\sum_t\sum_x\sum y(P-\hat{P})^2]/(P_{\max}-P_{\min}) \rightarrow \min$$

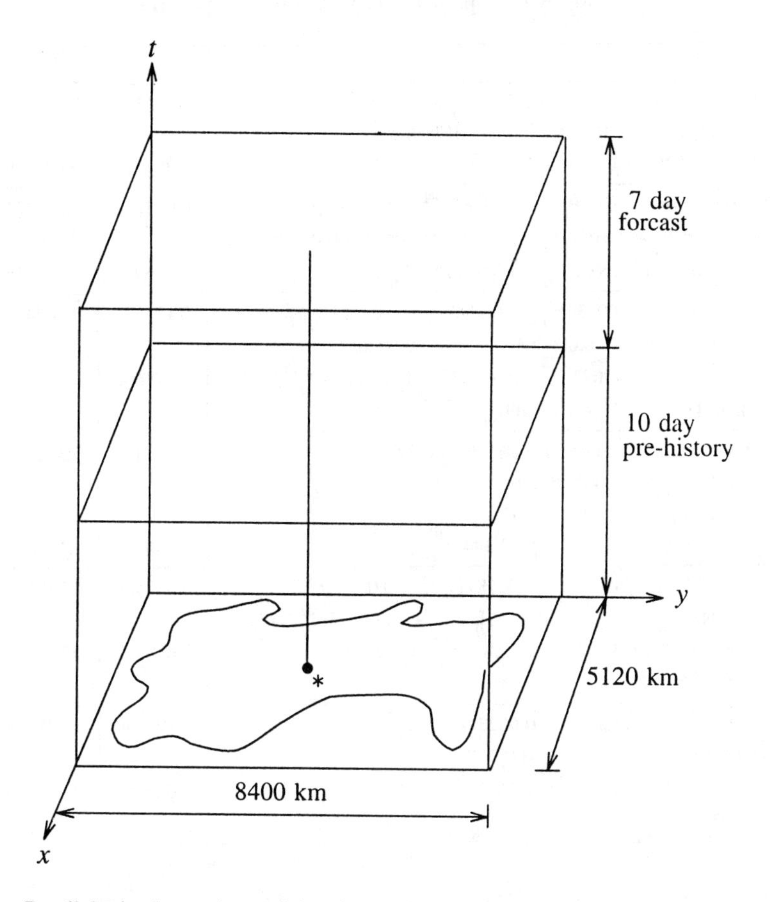

Figure 6.6. Parallelopiped experimental data for ten days (training and testing data) and of predictions for seven days; '*' indicates the point of interest, a location on the earth

$$\delta_T = \sqrt{[\frac{1}{N_C}\sum\sum_t\sum_x y(T-\hat{T})^2]}/(T_{\max}-T_{\min}) \rightarrow \min, \tag{6.18}$$

where P and T are the actual values, $\hat{P}$ and $\hat{T}$ are the predicted values, $P_{\max}$ and $T_{\max}$ are the maximum values, and $P_{\min}$ and $T_{\min}$ are the minimum values of the pressure and temperature, correspondingly; N_C is the number of data points in the sequence C.

For $N_C = 7$ (, i.e., for a seven-day forecast,) the relative prediction errors on pressure and temperature are $\delta_P = 0.3709$ and $\delta_T = 0.3043$, correspondingly. Figure 6.8 shows the curves of predicted temperature and pressure for the axial point shown in the parallelopiped-space of data. The system of equations corresponding to the pattern V (Table 6.1) is used for this purpose. The system is adaptive—i.e., the forecast is updated everyday as new data are received.

The example given above is conducted on an experimental basis. In this example, neither prediction balance nor an objective synthesis of system of equations are used. This means that the possibilities of improving the accuracy of the forecasts have not been exhausted.

Table 6.1. Optimal models for different patterns

No.	System of equations	η_{bs}	$\Delta(C)$	$c3$
I	$P_{ij}^{t+1} = -1.0083 - 0.1099t + 0.3920T_{ij}^{t} + 0.8809\bar{P}_{ij}^{t} +0.0047P_{ij}^{t}T_{ij}^{t} + 0.0121\bar{P}_{ij}^{t}\bar{P}_{ij}^{t} + 0.000177\bar{P}_{ij}^{t}\bar{T}_{ij}^{t}$	0.021138	0.276223	0.277031
	$T_{ij}^{t+1} = -1.7269 + 0.1411t + 0.9834\bar{T}_{ij}^{t}$	0.201362	0.290929	0.353817
II	$P_{ij}^{t+1} = 6.0649 + 0.2366\bar{P}_{ij}^{t} - 0.00348T_{ij}^{t-1} + 0.0059P_{ij}^{t}P_{ij}^{t-1} +0.0047P_{ij}^{t}\bar{P}_{ij}^{t} + 0.01059P_{ij}^{t-1}T_{ij}^{t} + 0.00503\bar{P}_{ij}^{t}\bar{P}_{ij}^{t} + 0.00405\bar{P}_{ij}^{t}\bar{T}_{ij}^{t}$	0.022273	0.289461	0.290317
	$T_{ij}^{t+1} = -0.6725 + 0.9838\bar{T}_{ij}^{t} + 0.00172P_{ij}^{t-1}T_{ij}^{t-1}$	0.176932	0.352216	0.394159
III	$P_{ij}^{t+1} = 4.0505 - 0.00734P_{ij}^{t-2} + 0.4682\bar{P}_{ij}^{t} + 0.00745P_{ij}^{t}P_{ij}^{t-1} +0.0124P_{ij}^{t-2}P_{ij}^{t-2} + 0.00566P_{ij}^{t-2}T_{ij}^{t-1} - 0.00265P_{ij}^{t-2}\bar{T}_{ij}^{t}$	0.121926	0.362449	0.382407
	$T_{ij}^{t+1} = 1.6342 + 0.4908T_{ij}^{t-2} + 0.5019\bar{T}_{ij}^{t} - 0.0000256P_{ij}^{t-1}P_{ij}^{t-2} -0.00311P_{ij}^{t-2}P_{ij}^{t-2} - 0.00309T_{ij}^{t-2}T_{ij}^{t-2}$	0.079602	0.233970	0.247141
IV	$P_{ij}^{t+1} = -3.6239 + 0.6961t + 0.8436P_{ij}^{t} + 0.2970P_{i+1j}^{t} +0.2335P_{ij-1}^{t} + 0.00815P_{i+1j}^{t}P_{i+1j}^{t} - 0.000075P_{i+1j}^{t}T_{ij-1}^{t} +0.00238P_{ij-1}^{t}P_{i-1j}^{t} + 0.0112P_{ij-1}^{t}T_{ij-1}^{t}$	0.057442	0.352762	0.357409
	$T_{ij}^{t+1} = 1.28603 + 0.4706T_{i+1j}^{t} + 0.3842T_{ij-1}^{t}$	0.186287	0.197606	0.271572
V	$P_{ij}^{t+1} = 5.06 + 0.3648P_{ij-1}^{t} + 0.00787T_{ij-1}^{t} - 0.00538P_{ij}^{t}T_{i-1j}^{t} +0.01848P_{i+1j}^{t}P_{ij-1}^{t} - 0.005722T_{ij}^{t}T_{i-1j}^{t} + 0.0000252P_{ij}^{t-1}P_{ij}^{t}$	0.080550	0.235598	0.248988
	$T_{ij}^{t+1} = -0.8886 + 0.2417T_{ij}^{t} + 0.2834T_{i+1j}^{t} + 0.2712T_{ij-1}^{t} +0.00961P_{i+1j}^{t}T_{ij}^{t-1}$	0.038617	0.235150	0.238300
VI	$P_{ij}^{t+1} = 5.056 + 0.572P_{ij+1}^{t} - 0.00234P_{ij}^{t-2}T_{ij}^{t-1} +0.0108P_{ij}^{t-1}T_{i+1j}^{t} + 0.006951P_{ij}^{t-1}T_{ij}^{t}$	0.190290	0.198452	0.279023
	$T_{ij}^{t+1} = 3.9901 + 0.642T_{i+1j}^{t} + 0.2215T_{ij}^{t-1} +0.003696P_{i+1j}^{t}T_{ij}^{t-1} - 0.000228T_{ij}^{t}T_{ij}^{t-2}$	0.046587	0.298915	0.302524

2.6 Alternate approaches to weather modeling

Here are some other suggestions on how to use the inductive approach in solving the weather forecasting [60].

The self-organization method is of a heuristic nature. Its main idea is to generate a large variety of variables and functions connecting them, and to choose the best structure in optimal complexity according to an external criterion. The following proposals are made for better predictions:

- an autonomous system of homogeneous difference equations (for short-range predictions) is proposed to describe the change in instantaneous as well as the averaged values of the variables and to include any source function and external disturbances;
- the use of two-level predictions on the basis of several balances (for example, year-season, and year-month);
- a two-level algorithm (for medium-range predictions) to use with several balance criteria;
- the use of correlational models for predicting weather in movable coordinates;
- the use of ecological variables in a combined system of weather-climate equations to increase prediction accuracy and prediction time; and

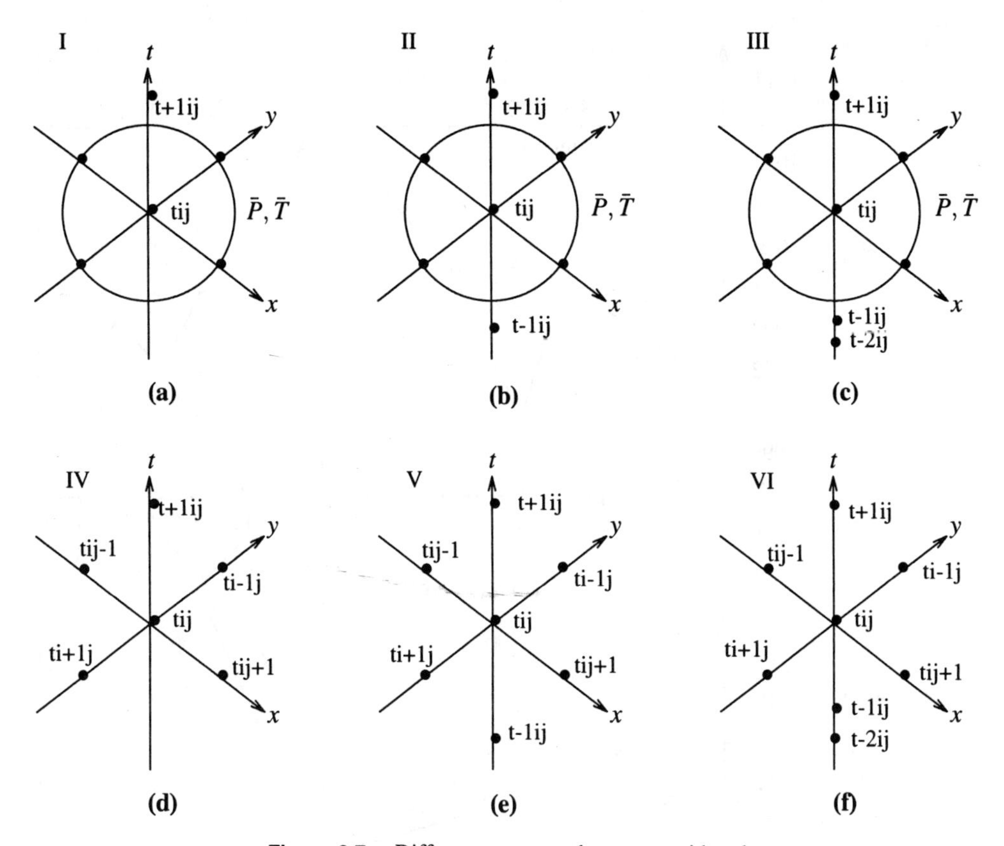

Figure 6.7. Different pattern schemes considered

- the use of a method of group analogues (for long-range predictions) that is based on objective clustering of the weather into number of clusters (not specified in advance) with averaging of the predictions of the variables one by one; the set of significant variables can be set up objectively by the objective system analysis algorithm.

The meteorological variables that determine the weather (air pressure, temperature, wind force, humidity, etc.) oscillate continuously around a mean climatic value (or a trend) in a random manner. The trend is usually known and can be predicted rather accurately. This means that the problem of weather forecasting is reduced to predicting the random deviations of the variables from the trend. These deviations are called the "remainder." One should note that all of the variables referred below correspond to such meteorological variables.

The suggested approaches are described below.

Weather modeling in fixed coordinates

The first approach suggests including the variables of external disturbing influences with the averaged, delayed, and higher-ordered arguments into the reference functions under consideration. The candidate variables usually used are in the original equations; that is, $q_1 = u$, the projection of the wind velocity onto the north-south axis, $q_2 = v$, the same onto the east-west axis, $q_3 = \rho$, the air density, and $q_4 = h$, the air humidity are introduced along with the variables $q_5 = P$, the pressure, $q_6 = T$, the temperature, and $t, x,$ and y are

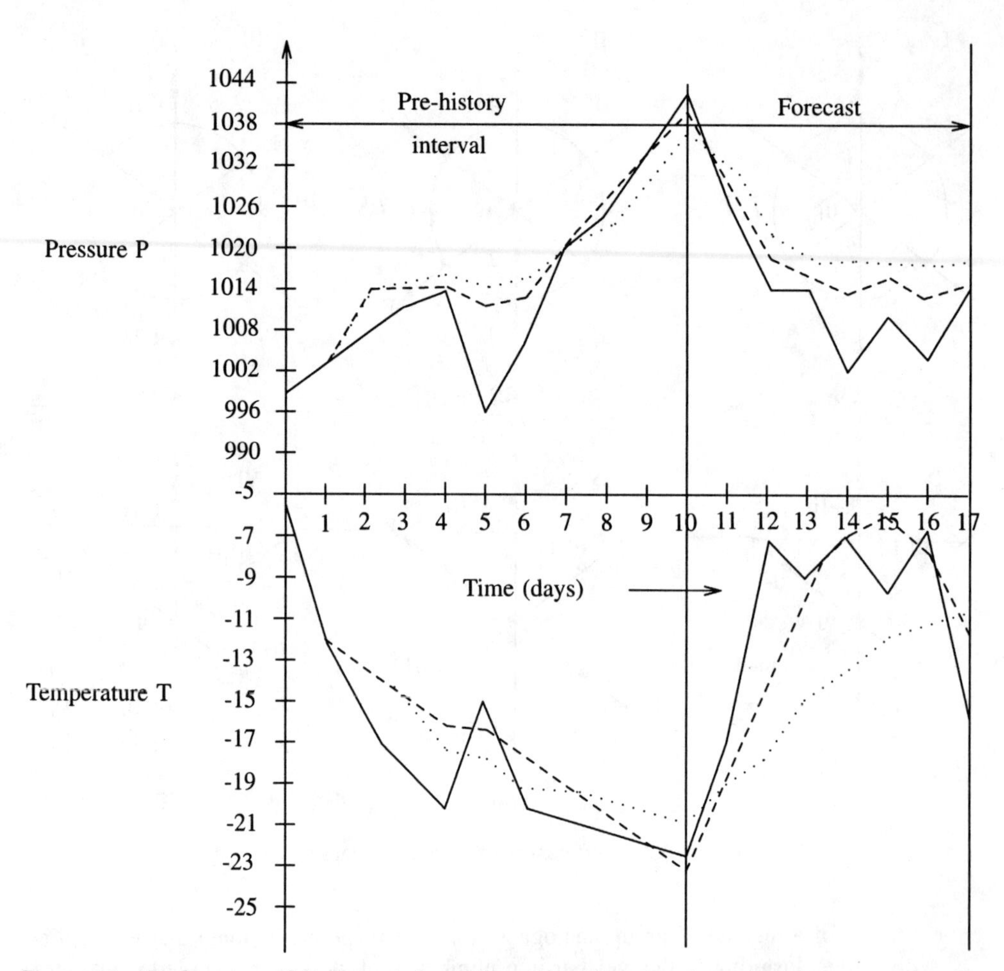

Figure 6.8. Performance of the system of equations trained based on the optimal pattern structure for pressure and temperature for pressure and temperature at the point of interest; the full line indicates the actual measured data, the broken line indicates the pattern training based on the measured data from the point of interest, and the dotted line indicates the pattern training based on the whole parallelopiped of data

the time-space coordinates. The variables of external influences can be the parameters of distant points of the geosphere; for example, the temperature or pressure of air at the centers of the Atlantic ocean (minimum and maximum values), etc.

Using the implicit form of the pattern (Figure 6.9), the following equations can be formed for each of the variables listed.

(a) autoregression equations ($k = 1, 2, \cdots, 6$):

$$q^t_{k_{ij}} = f_1(q^{t-1}_{k_{ij}}, q^{t-2}_{k_{ij}}, q^{t-3}_{k_{ij}}, q^t_{k_{i-1j}}, q^t_{k_{i-2j}}, q^t_{k_{ij-1}}, q^t_{k_{ij-2}}); \tag{6.19}$$

(b) multivariate equations ($k = 1, 2, \cdots, 6;\ l = 1, 2, \cdots, 6;\ k \neq l$):

$$\begin{aligned} q^t_{k_{ij}} &= f_1(q^{t-1}_{k_{ij}}, q^{t-2}_{k_{ij}}, q^{t-3}_{k_{ij}}, q^t_{k_{i-1j}}, q^t_{k_{i-2j}}, q^t_{k_{ij-1}}, q^t_{k_{ij-2}}) \\ &\quad + f_2(q^t_{l_{ij}}, q^{t-1}_{l_{ij}}, q^{t-2}_{l_{ij}}, q^{t-3}_{l_{ij}}, q^t_{l_{i-1j}}, q^t_{l_{i-2j}}, q^t_{l_{ij-1}}, q^t_{l_{ij-2}}); \end{aligned} \tag{6.20}$$

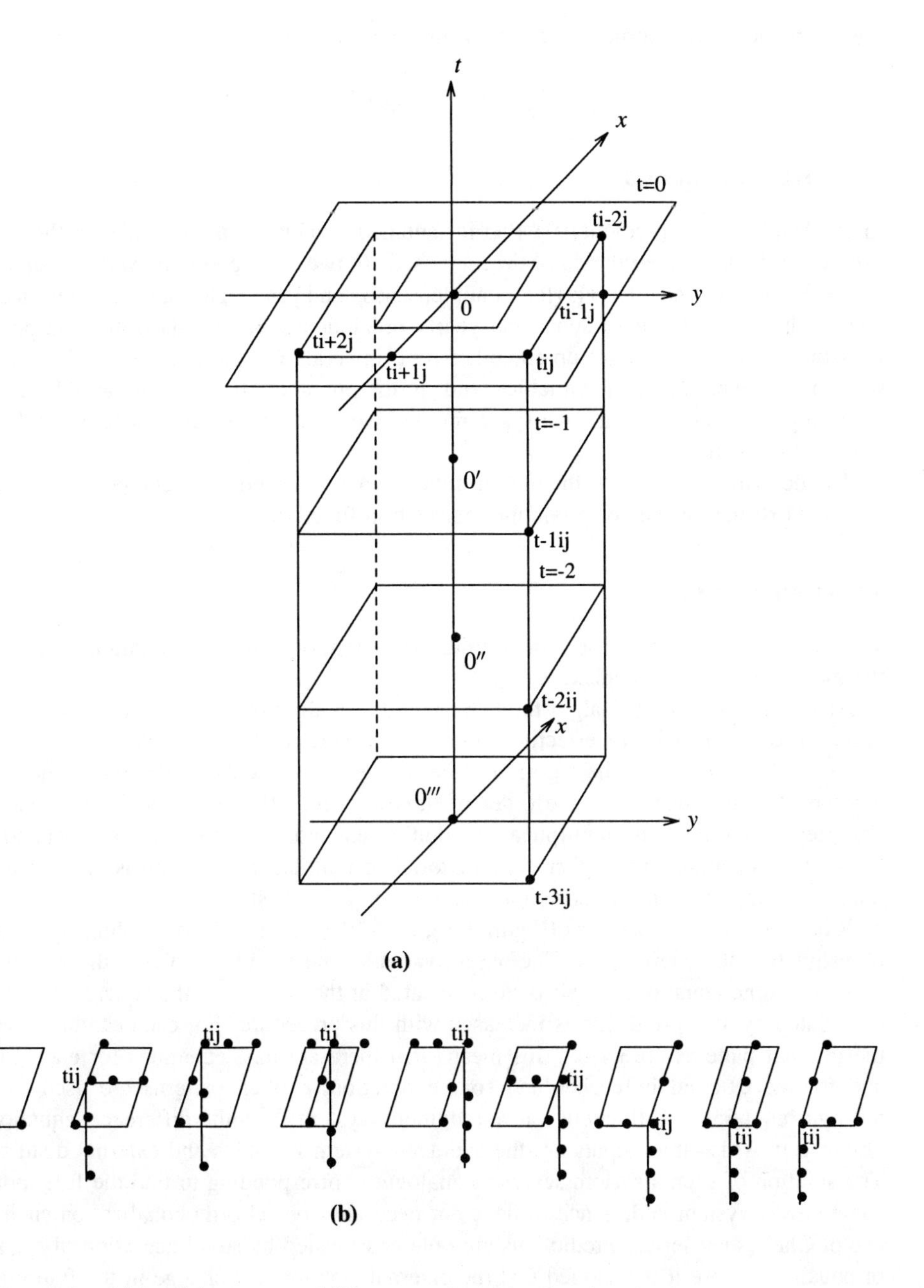

Figure 6.9. (a) implicit pattern and (b) patterns formation for nine simultaneous equations

(c) equations with source functions ($k = 1, 2, \cdots, 6$):

$$q^t_{k_{ij}} = f_1 + f_2 + f_3(t, x, y); \quad (6.21)$$

(d) equations considering the external influences ($k = 1, 2, \cdots, 6$):

$$q^t_{k_{ij}} = f_1 + f_2 + f_3 + f_4(u_1, u_2, \cdots). \quad (6.22)$$

Two-level predictions

In one-level weather predictions, either instantaneous values of the variables or the variables averaged over the same interval of time are used. In two-level predictions, these calculations are made for two time intervals (for example, season and year). The balance criterion enables one to choose a pair of seasonal and yearly predictions. For regularization purposes one uses the balance criteria set up for other time intervals or space regions. The prediction time in the averaging of the variables tends to the prediction time of the variables averaged on a long time interval or a large space region. We recall that balance criteria can be either temporal or spatial.

The descritization step of the data in time is chosen for both the components in relation with the extrema of the corresponding correlation functions.

Two-step algorithm

The two-step algorithm is used for solving the system of difference equations and to obtain the model in optimal complexity.

At the first step of the algorithm, the multilayer algorithm is used with the regularity criterion to obtain only the effective candidate variables and their estimates.

At the second step of the algorithm, the reference function includes the higher ordered arguments (product terms to the order of three) for the effective variables selected in the first step. This uses the multiplicative-additive and non-linear functions. Depending on the number of arguments, either combinatorial or multilayer algorithm is applied with the minimum-bias or regularity criterion to obtain the optimal structure.

With the "implicit" patterns (Figure 6.9), each of them is trained up along the time axis by estimating the coefficients. The output variables are found by solving the simultaneous system of nine equations for all patterns located at the corners of the square. The stability of the step-by-step prediction is increased with this procedure. For each candidate variable the implicit patterns are used. This means that there are nine equations for each variable, and for the external influences and source function, explicit patterns are used to obtain a single equation. Ultimately, an autonomous system of finite-difference equations are obtained in which the outputs are the averaged system variables and external disturbances. The solution of such a system serves as analogues corresponding to find the free motion of some closed system and, hence, it does not need any special orthogonalization such as the use of Chebyshev series. Predictions are obtained by step-by-step integration of the system of equations. Here it is assumed that the external disturbances change in the future as they have in the past.

Weather forecasting in movable coordinates

In meteorological problems, the equations of motion of a cyclone or anticyclone are uncoupled into an equation of motion of the center or a system of equations describing the motion around the center. The center is defined as the point corresponding to the two-dimensional correlation function formulation. The most accurate prediction is given by a method using the weather satellites' data. The modeling is done by using synoptics and that is why it is subjective and why it depends on the experience of the modeler. For example, in determining the position of the center of a cyclone (about 20 to 40 km/hr) for a prediction time of 24 hrs, the average error in determining the speed of displacement is $\delta_s = 300km/hr$ and the average error in determining the direction of motion is $\delta_\phi = 12°$. This leads to an average error of $\delta_s = 200km$ for the position of the center. This is because the pictures taken from the satellites do not show the wind at high levels when there is no cloudiness. This leads to unexpected motions of the formations [81].

In place of subjective forecasting, one has to use mathematical modeling, particularly the inductive approach which is based on objective reasoning.

Atmospheric formations like cyclones exist for only a short time. This means that only few points of observations are available within that short interval of time. Under such conditions, the inductive algorithms work very efficiently to study the situation. For example, the model that can be obtained from the above two-step algorithm consists of a system of two finite-difference equations (one equation for each coordinate). This can be integrated for step-by-step predictions of several steps ahead.

Prediction of the change in the atmospheric formations

It is convenient to locate the coordinate origin at the center of an atmospheric formation such as the center of a cyclone. This means that the problem is predicting change in the shape of the formation around the coordinate origin. The two-step algorithm is based on the use of finite-difference equations with the implicit form of the patterns. The pressure at a definite point in the x, y- plane which moves together with the center of the cyclone over the surface of the earth, can be predicted using its delayed values and the pressures at the neighboring points located at the corners of the square (Figure 6.9).

$$\begin{aligned} P_{ij}^t &= a_0 + a_1 P_{ij}^{t-1} + a_2 P_{ij}^{t-2} + a_3 P_{ij}^{t-3} + \cdots + b_1 P_{i-1j}^t + b_2 P_{i+1j}^t \\ &= f_1(P_{ij}^{t-1}, P_{ij}^{t-2}, P_{ij}^{t-3}, \cdots, P_{i-1j}^t, P_{i+1j}^t), \end{aligned} \tag{6.23}$$

where the time axis t is located at the center of the square as shown.

Use of correlational models

The isobars shown on the meteorological charts can be considered as random functions in the space coordinates and time. They can also be represented as a two-dimensional spectrum and a two- dimensional correlation function of the surface.

We assume that the self-organizing correlation function (its numerator) is stable; i.e., it holds the same characteristic in the prediction region as in the interpolation region. This condition enables us to obtain an optimal nonphysical nonlinear model according to the combined criterion of minimum-bias plus prediction.

The advantage of the correlation models is that they can become multifactor models in a simple manner and that they take into account several meteorological variables and their delayed arguments.

Use of graphs with binary transformation [55]

The problem of predicting the shape of a cyclone (in mobile coordinates with its center) can be simplified by considering the shape of isobar curve as a representative of binary discretization of the pressure (for example, for P = 1000 mm, the pressure in the region close to the center (Figure 6.9) is taken equal to $P = -1$ and outside the isobar it is $P = +1$). Such a binary approach decreases the requirements on the number of observations and increases the prediction accuracy and its time.

The algorithm consists of three parts: (i) training on the graphs of prehistory, (ii) adaptation of the graphs as per the predictions (ensuring the stability of step-by-step predictions), and (iii) selection of the best graph according to the criteria.

The field is partitioned into small squares in which the pressure is either +1 or −1. Four adjacent squares provide a single input for the graph. The training is conducted by calculating the number of transitions of the output variables to +1 and −1 (Figure 6.10). The adaptation is done by changing the graphs with each step in the prediction such that the number of transitions on the entire prediction interval is equal to their number in the observation interval; accordingly, the number of transitions in the graph is decreased by one at each step.

The graphs can be considered with the preceding states (the last value or the last two or three values of the pressure in time). All the graphs are used for predictions and the best one is chosen according to the combined criterion.

Use of ecological variables

Usually, prediction of climate involves the prediction of variables measured with a large siding time averaging or moving average interval. Many ecological variables result from averaging different influences activating in the process—first, meteorological variables; hence, there is considerable correlation between the ecological and the meteorological variables. One can refer to the work of Lebow et al. [80] for such practical examples.

The possibility of using ecological variables for predicting weather is debated. The objective system analysis algorithm often yields a set of significant variables that include both ecological and meteorological variables. For example, a model obtained for the ecosystem of lake Baykal includes the variables (yearly average values): q_1—the water transparency, q_3—the biomass of the plant life in the water, q_4—the biomass of zooplankton, and u_1—the water temperature at the surface layer. These are effective output variables in studying the ecosystem. Corresponding models are obtained for each of these variables. They are used for step-by-step predictions to the year 2000 to study the changes in the system. This means that one meteorological variable like water temperature can be predicted by using the ecological data.

The equation obtained for the temperature in the system is not a physical law, but is merely a tool for predicting the temperature. For extrapolation or prediction of a variable we can use the selected model, whether it be a finite-difference equation or an algebraic equation treated in the same way. For example, Figure 6.11 shows the autocorrelation function of temperature and the cross-correlation function of temperature and biomass of zooplankton. It shows the presence of considerably higher frequencies in the temperature than in the process of interaction between the temperature and the biomass of zooplankton. The correlation time of the remainder of the first process is less than that of the second. Hence, the prediction interval of the variable temperature when using an ecological variable is greater because the limiting attainable prediction interval is proportional to the correlational interval. This means that the ecological variables help in the process of predicting meteorological variables by increasing accuracy and the prediction interval.

In this example, the difference equations used for the ecological variables in predicting the temperature are treated as approximations of their variation. Similarly, the difference equation used for temperature in predicting the ecological variables should be treated as an approximation of its variation with time.

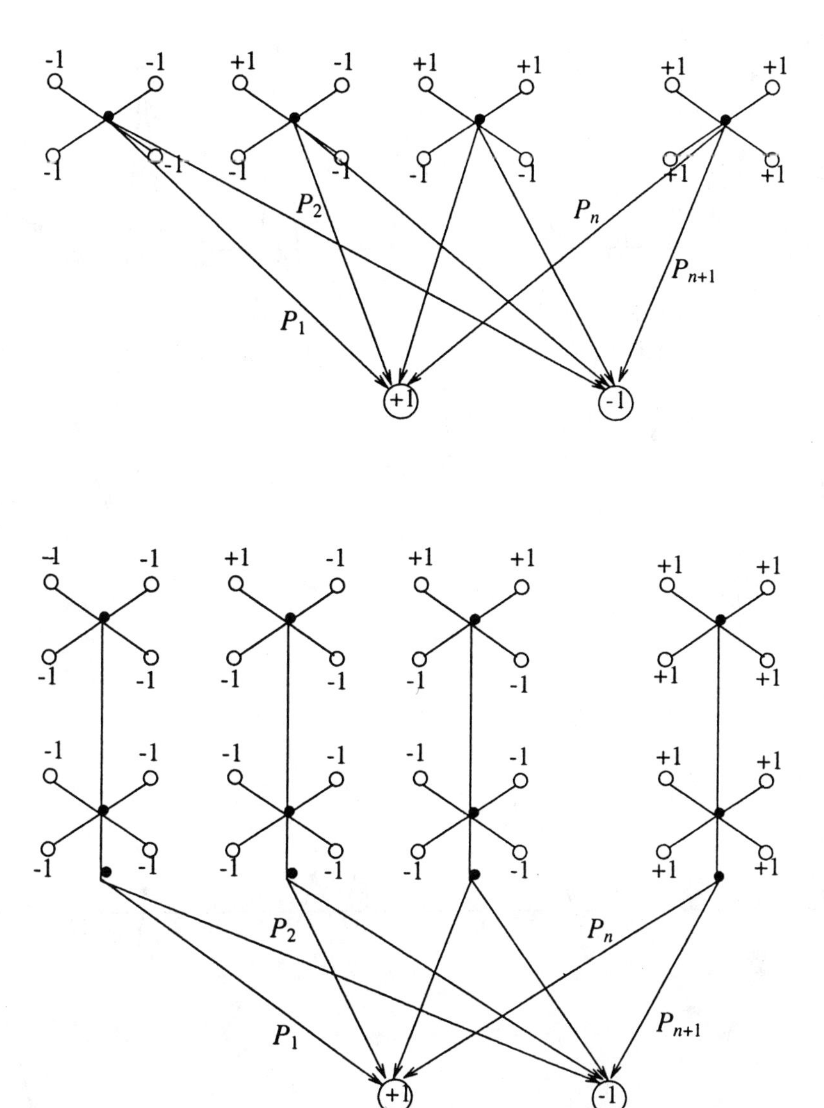

Figure 6.10. Graphs of the number of transitions: upper graph with allowance only for the last state of the field and the lower graph with allowance for the last several states of the field

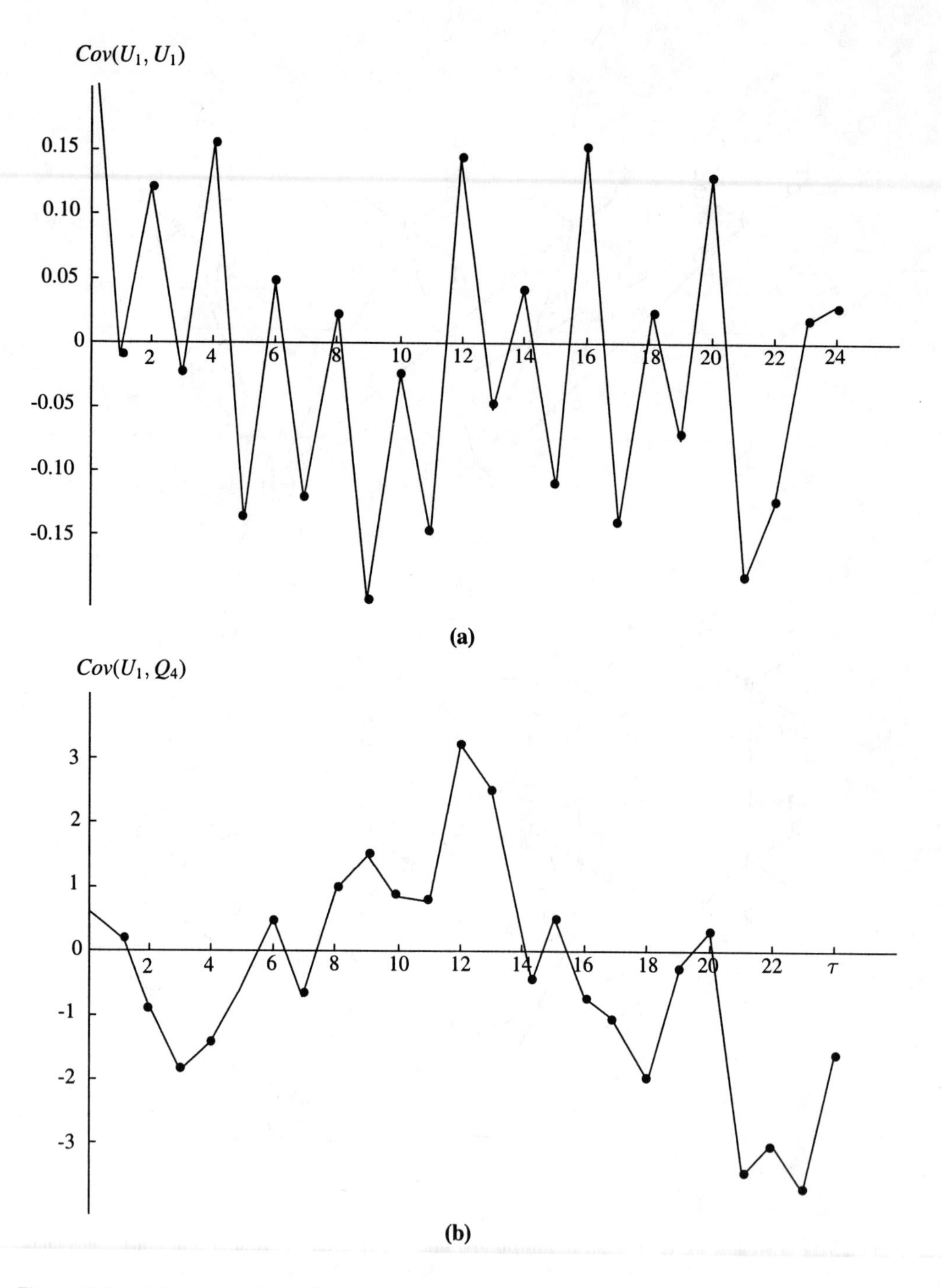

Figure 6.11. (a) autocorrelation function for the temperature U_1 and (b) cross-correlation function of temperature U_1 and biomass Q_4

Use of objective clustering

Let us suppose that the computer has selected four out of six variables to use in the system of autonomous equations listed for characterizing the weather. Any point in the four-dimensional coordinate space represents the weather at a given time. Similar points constitute a cluster of "good" or "bad," "dry" or "rainy" weather. The number of clusters can be found by self-organization clustering with an algorithm of objective clustering of multidimensional space into an unknown number of clusters. A more accurate and detailed representation of the input information helps to find more weather clusters. Here the threshold values for the clusters are not necessarily specified.

Method of group analogues

According to the method of group analogues, when using prehistory data one has to find several situations of the weather (in some interval of time) that is very similar to the present weather conditions under consideration. The duration of a situation is equal to the correlation interval of the anamoly. Then, prediction of the weather can be obtained by averaging all similar situations. This is the foundation for contemporary long-range prediction.

The selection of an analogue for a given synoptic situation or pattern is a common operation in synoptic practice, particularly in long-range weather predictions. To avoid the subjective choice of thresholds for distinguishing situations, one can entrust this operation to the objective clustering algorithm. Here the more extensive prehistory data the computer can examine, the closer to each other will be the situations associated with a single cluster. However, there is a limit to the length of the data sample. Beyond that limit the number of clusters does not increase. Thus, the inductive approach can help in comparing many candidate cases and in finding the optimal number of clusters which can help predict the near future. Usually, the computer chooses several noncontradictory clusterings differing from each other in the number of clusters and in the set of variables. It is expedient to choose from these a single clustering which corresponds to the longest correlation interval. This can be done by further regularization as described in the previous chapter.

Long-range weather prediction

There are at least five different types of atmospheric circulations known in the northern hemisphere. If one type of circulation exists more often than the others, that is called the prevailing circulation. The change from one type of circulation to another is a purely random process and is not subject to prediction (like the result of flipping a coin—"head" or "tail"). However, the change of prevailing circulation from one type to another does lend itself to prediction like any averaged variable, because the correlation function of the process of change of prevailing circulation must be rather broad. Each cluster of the non-contradictory clustering enables us to obtain a long-range weather prediction.

3 ECOLOGICAL SYSTEM STUDIES

Here the performance of the multilevel algorithm along with the objective systems analysis described in the second chapter will be demonstrated in the study of ecological systems.

Let us briefly discuss various stages of this algorithm.

The *first stage* is to divide the set of variables into three subsets: the output variables, the input variables, and variables which have no substantial effect on the subsets. This is the first level of the multilevel algorithm (also called as objective system analysis) used in detecting the relationships among the variables.

The next *two stages* belong to the two-level analysis of the algorithm; the purpose of the *first level* is to divide the set of predictions of the average annual values of the variables (those not discarded during the first stage) into "good," "satisfactory," and "unsatisfactory" predictions and to select the best predictions (one for each variable). The purpose of the *second level* is to predict the average seasonal values of the output variables on the basis of a series of sets of seasonal models.

3.1 Example—ecosystem modeling

Example 2. Self-organization modeling in the Lake Baykal ecological system.

Lake Baykal was thought to be exposed to so-called anthropogenic perturbations because of industrial waste, tourism, etc. The views of scientists on this are diversified. Finding an objective method for predicting the condition of the lake is desirable. The inductive learning methods based on the principle of self-organization are good candidates for exploring the objective characteristics of the system.

The list of possible variables is given by biologists. The seasonal and annual values of the following parameters for a 23-year period are used in this example;

q_1 – the transparency of the water in meters,
q_2 – the biomass of the phytoplankton in mg/m^3,
q_3 – the biomass of the small plants(*Melosira*)in mg/m^3,
q_4 – the biomass of the zooplankton in mg/m^3,
q_5 – the biomass of the *epischura* in g/m^2,
u_1 – the surface water temperature in$^\circ C$,
u_2 – the water level in meters, and
u_3 – the number of hours of sunlight.

Small letters denote the seasonal values and big letters denote the mean annual values.

Here the problem identifies a point physical model that represents the ecosystem. This is solved by using the multilevel iterative algorithm which has the levels of the objective system analysis to identify the characteristic variables of the system and the two-level scheme to select two nonphysical models (annual and seasonal) for long-range quantitative predictions.

Objective system analysis

This level is used to synthesize a model in the form of systems of from one to five equations. The primary variables are used to form the polynomials in the form of finite-difference equations.

First layer. The finite-difference models of the form given below are used for each of the five variables $(q_1 - q_5)$.

$$\begin{aligned} q_i^t = {} & a_0 + a_1 u_1^t + a_2 u_1^{t-1} + a_3 u_1^{t-2} + a_4 u_1^{t-3} + a_5 u_1^{t-4} \\ & + a_6 u_2^t + a_7 u_2^{t-1} + a_8 u_2^{t-2} + a_9 u_2^{t-3} + a_{10} u_2^{t-4} \\ & + a_{11} u_3^t + a_{12} u_3^{t-1} + a_{13} u_3^{t-2} + a_{14} u_3^{t-3} + a_{15} u_3^{t-4} \\ & + a_{16} q_i^{t-1} + a_{17} q_i^{t-2} + a_{18} q_i^{t-3} + a_{19} q_i^{t-4}. \end{aligned} \tag{6.24}$$

Second layer. At the second layer, there are $C_5^2(=10)$ systems of two equations of the form as given below:

$$\begin{aligned} q_i^t = & a_0 + a_1 u_1^t + a_2 u_1^{t-1} + a_3 u_1^{t-2} + a_4 u_1^{t-3} + a_5 u_1^{t-4} \\ & + a_6 u_2^t + a_7 u_2^{t-1} + a_8 u_2^{t-2} + a_9 u_2^{t-3} + a_{10} u_2^{t-4} \\ & + a_{11} u_3^t + a_{12} u_3^{t-1} + a_{13} u_3^{t-2} + a_{14} u_3^{t-3} + a_{15} u_3^{t-4} \\ & + a_{16} q_i^{t-1} + a_{17} q_i^{t-2} + a_{18} q_i^{t-3} + a_{19} q_i^{t-4} \\ & + a_{20} q_j^t + a_{21} q_j^{t-1} + a_{22} q_j^{t-2} + a_{23} q_j^{t-3} + a_{24} q_j^{t-4}. \end{aligned} \tag{6.25}$$

In the *third and fourth layers*, $C_5^3(=10)$ systems of three equations and $C_5^4(=5)$ systems of four equations are used correspondingly. In the *fifth layer*, a single system of five equations which contain 40 terms is used;

$$\begin{aligned} q_i^t = & a_0 + a_1 u_1^t + a_2 u_1^{t-1} + a_3 u_1^{t-2} + a_4 u_1^{t-3} + a_5 u_1^{t-4} \\ & + a_6 u_2^t + a_7 u_2^{t-1} + a_8 u_2^{t-2} + a_9 u_2^{t-3} + a_{10} u_2^{t-4} \\ & + a_{11} u_3^t + a_{12} u_3^{t-1} + a_{13} u_3^{t-2} + a_{14} u_3^{t-3} + a_{15} u_3^{t-4} \\ & + a_{16} q_i^{t-1} + a_{17} q_i^{t-2} + a_{18} q_i^{t-3} + a_{19} q_i^{t-4} \\ & + a_{20} q_j^t + a_{21} q_j^{t-1} + a_{22} q_j^{t-2} + a_{23} q_j^{t-3} + a_{24} q_j^{t-4} + \\ & \ldots \\ & + a_{35} q_l^t + a_{36} q_l^{t-1} + a_{37} q_l^{t-2} + a_{38} q_l^{t-3} + a_{39} q_l^{t-4}. \end{aligned} \tag{6.26}$$

The computational volume can be reduced at the higher levels by discarding the terms that are not effective at the preceding layers.

The F best models are selected at each layer by using the system criterion of the minimum bias. In this selection, a system with at least one equation having $\eta_{bs} \geq 0.01$ is eliminated from the sorting. The optimal model is selected according to the step-by-step integrated prediction accuracy of the prediction criterion on the total points N.

From the above analysis, the following system of equations containing the variables q_3 and q_4 is obtained as the optimal one with the limit of $\eta_{bs_s} \leq 0.005$.

$$\begin{aligned} q_3^t = & 22.4042 - 10.0977 u_1^{t-3} + 1.8842 u_2^{t-3} \\ & - 2.4647 u_2^{t-3} - 0.1673 q_4^{t-1} \\ q_4^t = & 58.9093 + 0.0233 u_1^{t-2} - 0.1382 u_2^{t-1} \\ & - 0.0425 u_2^{t-2} - 0.0325 u_2^{t-4} - 0.0132 u_3^{t-1} \\ & - 0.0386 u_3^{t-3} - 0.0062 q_1^{t-4} + 0.0006 q_4^{t-4}. \end{aligned} \tag{6.27}$$

The characteristic vector of the system includes the variable q_1 along with the variables q_3 and q_4.

Two-stage scheme

There are different ways for solving the two-stage scheme of long-range quantitative predictions of this problem [48], [50]. The latter work is proven to be the best heuristic approach for this scheme. We present here both approaches for giving an idea of using different heuristics.

First approach: Examining the above system of equations, one can easily establish that the vector of output variables consists of q_3 and q_4, and the vector of inputs is q_1, u_1, u_2 and u_3. The variables q_2 and q_5 are excluded from further consideration.

The two-stage scheme comprises of identifying the average annual models and the system of seasonal models and the optimal system of equations to be selected from the both using the prediction balance criterion.

In the *first stage*, for identifying the average annual models for the variables Q_3, Q_4, Q_1, U_1, U_2 and U_3, the harmonical and the single layered combinatorial algorithms are used. The harmonical algorithm is recommended when a large number of data points are available, while the combinatorial algorithm is recommended for a small number of input variables. The finite-difference scheme with the delayed arguments considered in the combinatorial algorithm is

$$Q_{i_{(T)}} = a_0 + a_1 Q_{i_{(T-1)}} + a_2 Q_{i_{(T-2)}} + \cdots + a_m Q_{i_{(T-m)}}, \tag{6.28}$$

where T denotes the number of the year.

Harmonic models have shown better performance than the finite-difference models obtained from the combinatorial algorithm for the variables Q_1, Q_4, and U_1 with five, ten, and eight harmonic components in the trends, correspondingly. The predictive models obtained for the variables Q_3, U_2, and U_3 are insufficiently accurate; variables U_2 and U_3 are excluded from the future consideration. Along the four variables Q_3, Q_4, Q_1, and U_1 considered, the variable Q_4 is called the leading variable because of its better annual predictions.

In the *second stage*, the combinatorial algorithm is used to obtain the seasonal models for the variables q_3 and q_4 having the following type reference functions:

$$\begin{aligned} x_{1_{(t,T)}} &= a_0 + a_1 x_{1_{(t-1,T)}} + a_2 x_{1_{(t-2,T)}} + a_3 x_{1_{(t-3,T)}} + a_4 x_{1_{(t,T-1)}} \\ &+ a_5 x_{2_{(t,T)}} + a_6 x_{2_{(t-1,T)}} + a_7 x_{2_{(t-2,T)}} + a_8 x_{2_{(t-3,T)}} + a_9 x_{2_{(t,T-1)}} \\ &+ a_{10} X_{2_{(T)}} + a_{11} X_{2_{(T-1)}} + a_{12} X_{3_{(T)}} \\ &+ a_{13} X_{3_{(T-1)}} + a_{14} U_{1_{(T)}} + a_{15} U_{1_{(T-1)}}, \end{aligned} \tag{6.29}$$

where t and T denote the season and year; $x_1 = q_3$, $x_2 = q_4$; $X_{2_{(T)}} = Q_{4_{(T)}}$, $X_{3_{(T)}} = Q_{1_{(T)}}$ and $U_{1_{(T)}}$ are the average annual values at the year T.

Similarly, the reference function for the variable x_2 is considered. The reference functions can be expanded further with the trend equation of two-dimensional time read-out and with the variables of $x_3 = q_1$ and u_1 and with their delayed values according to the data points and allotted computer time. The equations containing the variables $x_{1_{(t,T)}}$ and $x_{2_{(t,T)}}$ means that during the step-by-step predictions both equations must be integrated jointly—the system of two equations use their estimated values.

Here, for each season five best models are selected for the leading variable x_2 and one model for the variable x_1 according to the criteria minimum-bias and prediction. Using the prediction balance, the optimal system of equations is selected for each season. This is done by using the seasonal models of the output variables one after the other in step-by-step predictions.

The balance-of-predictions criterion is used for selecting the system of seasonal equations for variable x_2 on the total data sample.

$$\begin{aligned} b_i &= \frac{1}{4}(x_{2_{(w)}} + x_{2_{(sp)}} + x_{2_{(su)}} + x_{2_{(f)}})_i - X_{i2_{(yr)}} \\ c_i &= \frac{1}{4}(x_{2_{(w)}} + x_{2_{(sp)}} + x_{2_{(su)}} + x_{2_{(f)}})_i + X_{i2_{(yr)}} \end{aligned}$$

$$B = \sum_{i=1}^{N} b_i^2 / \sum_{i=1}^{N} c_i^2 \rightarrow \min, \tag{6.30}$$

where $x_{2(w)}, x_{2(sp)}, x_{2(su)}$, and $x_{2(f)}$ are the predicted values of the variable x_2 for winter, spring, summer, and autumn, correspondingly; and $X_{i2(yr)}$ is the predicted average annual value of the variable X_2. There are a total of 5^4 (= 625) formations of system of equations evaluated for their predictions. The optimal system which is found to be better has the value of the criterion $B_{x_2 min} = 0.06$.

The balance of prediction criterion for both the variables x_1 and x_2 is evaluated using the criterion as a system criterion

$$B^* = \sqrt{(B_{x_1}^2 + B_{x_2}^2)}, \tag{6.31}$$

where B_{x_1} is computed the same way as the variable x_2, but only on the interpolation interval.

Second approach: In the *first stage*, an algorithm similar to the objective systems analysis is used for sorting the systems of equations only for those variables Q_1, Q_3, and Q_4 that appear in the characteristic vector. The trend component is included into the equations. The complexity of the models is increased by replacing the addition of polynomials in the right side with multiplication. This means that it indicates switching to nonlinear equations or equations with variable coefficients.

Three layers are necessary for comparison because of the three variables. The multilayer algorithm is used to select the best models at each layer because of the large number of input variables.

The first layer consists of three equations with $9 \times 5 = 45$ terms in the right side of each equation.

$$\begin{aligned} Q_{i(T)} = {} & (a_0 + a_1 T + a_2 T^2 + a_3 U_{1(T)} + a_4 U_{1(T-1)} \\ & a_5 U_{2(T)} + a_6 U_{2(T-1)} + a_7 U_{3(T)} + a_8 U_{3(T-1)}) \\ & .(1 + a_9 Q_{i(T-1)} + a_{10} Q_{i(T-2)} + a_{11} Q_{i(T-3)} + a_{12} Q_{i(T-4)}). \end{aligned} \tag{6.32}$$

The second layer consists of three systems of two equations containing the $8 \times 3 \times 3 = 72$ terms in the left side of each equation;

$$\begin{aligned} Q_{i(T)} = {} & (a_0 + a_1 T + a_2 U_{1(T)} + a_3 U_{1(T-1)} \\ & a_4 U_{2(T)} + a_5 U_{2(T-1)} + a_6 U_{3(T)} + a_7 U_{3(T-1)}) \\ & .(1 + a_8 Q_{i(T-1)} + a_9 Q_{i(T-2)}) \\ & .(a_{10} Q_{j(T)} + a_{11} Q_{j(T-1)} + a_{12} Q_{j(T-2)}). \end{aligned} \tag{6.33}$$

The third layer is a single system of three equations, each of which contains $8\times3\times3\times3 = 216$ terms on its right side;

$$\begin{aligned} Q_{i(T)} = {} & (a_0 + a_1 T + a_2 U_{1(T)} + a_3 U_{1(T-1)} \\ & a_4 U_{2(T)} + a_5 U_{2(T-1)} + a_6 U_{3(T)} + a_7 U_{3(T-1)}) \\ & .(1 + a_8 Q_{i(T-1)} + a_9 Q_{i(T-2)}) \\ & .(a_{10} Q_{j(T)} + a_{11} Q_{j(T-1)} + a_{12} Q_{j(T-2)}) \\ & .(a_{13} Q_{l(T)} + a_{14} Q_{l(T-1)} + a_{15} Q_{l(T-2)}). \end{aligned} \tag{6.34}$$

If the total number of data points permits, one can introduce the moving average terms as inputs into the equations. Here another difference from the objective systems analysis is that the optimal model is chosen using the system criterion of prediction based on its step-by-step prediction accuracy. If the prediction error of any equation of a system is above 5%, then that system is eliminated from further sorting.

The optimal system obtained has two equations for the variables $Q_{3_{(T)}}$ and $Q_{4_{(T)}}$.

The predictions of the annual values of the variables Q_3 and Q_4 are obtained by step-by-step integration of the optimal system of equations obtained above. Stability of the integration is ensured by the explicit form of the pattern used; this means that the delayed arguments identical to the output variable used in the equations. In doing the step-by-step predictions, the future values of the external influences U_1, U_2, and U_3 are needed. Harmonical algorithm is used successfully to obtain the optimal harmonical trends of these variables. The predictions using the harmonical trends which have the prediction errors ≤ 0.05, are used for step-by-step integration of the system of equations. If the harmonical trends of the external influences are poorly predicted (> 0.05), then they are treated along with the variables Q_3 and Q_4 and the order of the systems of equations is increased with those external influences.

In this example U_1 has achieved better predictions with the harmonical model with the norm (≤ 0.05). The other two disturbances U_2 and U_3 could not achieve the norm. The system of difference equations in the output variables are $Q_{3_{(T)}}, Q_{4_{(T)}}, U_{2_{(T)}}$, and $U_{3_{(T)}}$. Harmonic predictions of U_1 and step-by- step predictions of U_2 and U_3 helped in predicting the Q_3 and Q_4 to the year 2000. The variable Q_4 is chosen as the leading variable because of its better performance than the other two output variables and the best F =10 models for the leading variable are selected by using the prediction criterion.

When we are sure that the statistical characteristics of the predictions are stable, we can use the statistical criteria of stability of moments or the stability of correlations to select the best F =3 models out of the 10 selected models of the leading variable.

At the *second stage* the nonphysical seasonal predictive models are identified. Here the two-dimensional time-read out is used in constructing the models and the arguments of the models include both seasonal and annual values. The objective systems analysis is used with a general form of the additive type descriptions of the three output variables q_1, q_3, and q_4. The two-dimensional time trend is considered in the descriptions.

In the *first layer*, there are three first-order systems, one for each output variable with 18 terms on the right side of the quation,

$$\begin{aligned}
q_{i_{(t,T)}} &= a_0 + a_1 t + a_2 T + a_3 t^2 + a_4 T^2 + a_5 u_{1_{(t,T)}} + a_6 U_{1_{(T)}} \\
&+ a_7 u_{2_{(t,T)}} + a_8 U_{2_{(T)}} + a_9 u_{3_{(t,T)}} + a_{10} U_{3_{(T)}} \\
&+ a_{11} Q_{i_{(T)}} + a_{12} q_{i_{(t-1,T)}} + a_{13} q_{i_{(t-2,T)}} + a_{14} Q_{i_{(T-1)}} \\
&a_{15} q_{i_{(t,T-1)}} + a_{16} q_{i_{(t,T-2)}} + a_{17} Q_{i_{(T-2)}},
\end{aligned} \tag{6.35}$$

where i = 1, 3, and 4; and t and T denote the season and year, correspondingly.

In the *second layer*, there are three systems of second-order equations with 26 terms on the right side of each equation as shown below:

$$\begin{aligned}
q_{i_{(t,T)}} &= a_0 + a_1 t + a_2 T + a_3 t^2 + a_4 T^2 + a_5 u_{1_{(t,T)}} + a_6 U_{1_{(T)}} \\
&+ a_7 u_{2_{(t,T)}} + a_8 U_{2_{(T)}} + a_9 u_{3_{(t,T)}} + a_{10} U_{3_{(T)}} \\
&+ a_{11} Q_{i_{(T)}} + a_{12} q_{i_{(t-1,T)}} + a_{13} q_{i_{(t-2,T)}} + a_{14} Q_{i_{(T-1)}} \\
&a_{15} q_{i_{(t,T-1)}} + a_{16} q_{i_{(t,T-2)}} + a_{17} Q_{i_{(T-2)}} \\
&+ a_{18} q_{j_{(t,T)}} + a_{19} Q_{j_{(T)}} + a_{20} q_{j_{(t-1,T)}} + u_{21} q_{j_{(t-2,T)}} + u_{22} Q_{j_{(T-1)}} \\
&a_{23} q_{j_{(t,T-1)}} + a_{24} q_{j_{(t,T-2)}} + a_{25} Q_{j_{(T-2)}},
\end{aligned} \tag{6.36}$$

where $i,j = 1,3$, and 4; $i \neq j$. Similarly, one can write another equation of the system for the output variable $q_{j_{(t,T)}}$.

Consecutively, the third layer has a single system of three equations with 34 terms in each equation.

The best models are selected for each season separately in the form system of equations by using the prediction criterion. The seasonal predictive models for the external disturbances are identified as it was in the case of annual models. Using the prediction criterion, the best F =10 systems of equations are selected for the leading variable q_4.

Next, the complete sorting of the annual and seasonal models are made and evaluated by the balance-of-predictions criterion,

$$b_i^2 = [Q_{(yr)} - \frac{1}{4}(q_{(w)} + q_{(sp)} + q_{(su)} + q_{((f))})]_i^2;$$

$$B^2 = \sum_{i=N_1}^{N_2} b_i^2 \rightarrow \min, \tag{6.37}$$

where b_i is the balance relation for the ith year and B is the total balance on the interval N_1 to N_2.

The overall sorting on the selected models from the annual as well as seasonal models involves 3×10 predictions. The optimal system of equations for the leading variable q_4 is selected according to the balance of predictions criterion.

A measure of success of the modeling is the global minimum achieved on the balance criterion; the error must not exceed 5%. If it exceeds, the freedom-of-choice of models at both the stages needs to be broadened to increase the volume of sorting and the divergence can also be reduced by including more input variables in the descriptions.

3.2 Example—ecosystem modeling using rank correlations

Example 3. Modeling of ecosystems of Kakhovka and Kremenchug reservoirs using the rank correlations [62], [66].

The objective systems analysis (OSA) used in the multilevel iterative algorithm which is described in the above example serves for singling out the least biased systems of equations by establishing the relations of the modeled object. This determines the set of characteristic variables of the object and its simplest physical model which is suitable for a short-range forecast. This set is used to find a nonlinear model of the object which improves the forecast. In case of a large amount of data, it would be easy to take into account more delayed values of the arguments, and to obtain dynamic and nonlinear models.

The best structures of the individual equations are obtained by using the single-layered combinatorial algorithm. Although the inductive learning algorithms are developed for use with a very short data samples, there are limitations in using the single-layered combinatorial algorithm. Experiments spread over time and experiments spread over space are assumed to be equivalent to increase the number of conditional equations. This means that the observations that are spread in space (over the stations) and the observations that differ in season number possess equal validity. The variables of the object are selected from the general set of variables indicated at the start of the investigation of the hydrobiological experimentations and are usually very large in number. If all the variables are considered along with their delayed arguments in space and time, it increases the computational volume of the combinatorial algorithm. A rank correlation method is used to classify the initial variables. A modified version of the OSA algorithm allows one to eliminate the arguments whose rank correlation is of modulus unity. The equations with high collinearity between the arguments are eliminated and the ill-conditioned matrices are avoided.

The *modified objective systems analysis* was used independently for analysing the ecological systems of Kakhovka and Kremenchug reservoirs. Here we present briefly the general procedure used in the analysis. The averaged seasonal values are calculated from the data obtained at six stations in the middle and lower part of the reservoirs. The initial data obtained on the avearge seasonal values of 37 state variables and four external influences in spring, summer and fall of a year has of 18 years. The following notations are used for some variables;

x_2 – chromaticity (degrees),
x_6 – oxygen content (mg/liter),
x_8 – bichromatic oxidizability (mg/liter),
x_9 – suspended particles (mg/liter),
x_{11} – nitrates (mg/liter),
x_{12} – nitrites (mg/liter),
x_{16} – soluble iron (mg/liter),
x_{17} – total phosphorus (mg/liter),
x_{21} – zinc (μg/liter),
x_{25} – organic phosphorus (mg/liter),
x_{27} – biomass of blue-green algae (mg/liter),
x_{30} – number of phytoplankton (millions of cells/liter),
x_{31} – number of blue-green algae (millions of cells/liter),
u_1 – water temparature (°C), and
u_2 – effective sun's energy

For calculating the rank correlation, all seasonal mean observations are given three values of rank (1 for the minimum value, 2 for the medium value, and 3 for the maximum of the three values). The paired rank correlation coefficients are determined from the Spearman's formula [72],

$$\rho_{x_j,x_k} = 1 - \frac{6S(d_i^2)}{N^3 - N}, \tag{6.38}$$

where $S(d_i^2) = \sum_{i=1}^{N} d_i^2$, d_i is the difference of ranks of the corresponding pairs of observations of the variables x_j and x_k, and N is the number of observations. All variables can be divided into three groups (classes I, II and III) with respect to the existing three points (spring, summer, and fall).

The delayed values of the arguments are not taken into account because of the small amount of data. According to the modified OSA, the complete equations from which the sorting of systems of equations is generated belong to one of the following forms:

$$\begin{aligned}
x^I &= a_0 + a_1x^{II} + a_2x^{III} + a_3u_1 + a_4u_2 \\
x^{II} &= b_0 + b_1x^I + b_2x^{III} + b_3u_1 + b_4u_2 \\
x^{III} &= c_0 + c_1x^I + c_2x^{II} + c_3u_1 + c_4u_2,
\end{aligned} \tag{6.39}$$

where x^I, x^{II}, and x^{III} are the variables from the first, second, and third classes, respectively.

The single-layered combinatorial algorithm is used with the symmetrical criteria of η_{bs} and $\Delta(C)$ to select the set of possible system of equations with various composition of the output variables. For example, the following optimum system of equations is obtained in case of Kremenchug reservoir.

$$\begin{aligned}
x_2 &= 78.0 - 2.57u_1, \\
x_6 &= 17.0 - 0.4u_1, \\
x_{30} &= 12400 - 166x_2; \quad \eta_{bs} = 0.006,
\end{aligned} \tag{6.40}$$

where $x_2, x_6,$ and x_{30} belong to the classes I, II, and III respectively. We use the notation $x^I = x_2,\ x^{II} = x_6,$ and $x^{III} = x_{30}$ further.

Based on the prediction errors, the variable X^{III} is chosen as the leading variable which is predicted better than the other two.

The two-stage scheme. In constructing the forecating model of mean annual values, the following linear form of the complete equation is used;

$$\begin{aligned} X^I_{(T)} &= a_0 + a_1X^I_{(T-1)} + a_2X^I_{(T-2)} + a_3X^{II}_{(T)} + a_4X^{II}_{(T-1)} + a_5X^{II}_{(T-2)} \\ &\quad + a_6X^{III}_{(T)} + a_7X^{III}_{(T-1)} + a_8X^{III}_{(T-2)} + a_9U_{1_{(T)}} + a_{10}U_{2_{(T)}} + a_{11}T, \end{aligned} \tag{6.41}$$

where upper-case letters denote the mean annual variables; T is time in years; and the equations for $X^{II}_{(T)}$ and $X^{III}_{(T)}$ are analogous.

These three equations are integrated step-by-step jointly to obtain the yearly forecast.

In obtaining the seasonal average forecasting models at the second stage, the following complete descriptions are used.

$$\begin{aligned} x^I_{(t,T)} &= a_0 + a_1x^I_{(t-1,T)} + a_2x^I_{(t-2,T)} + a_3x^I_{(t,T-1)} + a_4X^I_{(T)} + a_5x^{II}_{(t-1,T)} \\ &\quad + a_6x^{II}_{(t-2,T)} + a_7x^{II}_{(t,T)} + a_8x^{II}_{(t,T-1)} + a_9X^{II}_{(T)} + a_{10}x^{III}_{(t-1,T)} + a_{11}x^{III}_{(t-2,T)} \\ &\quad + a_{12}x^{III}_{(t,T)} + a_{13}x^{III}_{(t,T-1)} + a_{14}X^{III}_{(T)} + a_{15}u_1 + a_{16}u_2, \end{aligned} \tag{6.42}$$

where t and T denote the season and year, correspondingly; equations for $x^{II}_{(t,T)}$ and $x^{III}_{(t,T)}$ are written analogously.

The system of mean seasonal equations are integrated season by season to obtain the forecast.

For example, the following system of equations for annual model of Kremenchug reservoir is obtained as one of the best:

$$\begin{aligned} X^I_{(T)} &= -18.0 + 0.00147X^{III}_{(T)} + 5.58U_1, \\ X^{II}_{(T)} &= 7.78 + 0.284X^{II}_{(T-1)} - 0.00052X^{III}_{(T-1)}, \\ X^{III}_{(T)} &= -817 - 67.5X^{II}_{(T)} + 214U_1 + 0.811T. \end{aligned} \tag{6.43}$$

Similarly four more sytems of equations are selected for future evaluation; i.e., $F = 5$. The values of the external disturbances are taken according to the given scenario.

By the combinatorial algorithm, the seasonal forecasting models for the reservoirs are found; there are three equations in the system for each season. A total of five systems of equations ($F = 5$) are chosen for each season. The best one among them is given below for the Kremenchug reservoir:

(i) winter:

$$\begin{aligned} x^I_{(t,T)} &= 0.00926x^{III}_{(t-1,T)} - 2.59x^{II}_{(t-1,T)} + 0.729x^I_{(t,T-1)} + 1.15X^I_{(T)} - 26.1u_1, \\ x^{II}_{(t,T)} &= 1.06X^{II}_{(T)} + 0.706u_1, \\ x^{III}_{(t,T)} &= -0.0155x^I_{(t,T)} + 0.108x^{II}_{(t,T)} - 0.185x^{II}_{(t-1,T)} \\ &\quad -0.0223x^{III}_{(t,T-1)} - 0.00002X^{III}_{(T)} + 0.769u_2; \end{aligned}$$

(ii) spring:

$$\begin{aligned} x^I_{(t,T)} &= -4.12x^{II}_{(t,T)} - 0.0839x^{III}_{(t,T)} + 0.243x^I_{(t-1,T)} + 1.15X^I_{(T)} + 2.61u_2 + 7.06, \\ x^{II}_{(t,T)} &= 2.02x^{III}_{(t-1,T)} + 0.834X^{II}_{(T)}, \\ x^{III}_{(t,T)} &= -0.501x^I_{(t-1,T)} - 1.11x^{II}_{(t-1,T)} - 2.75x^{III}_{(t-1,T)} \\ &\quad -0.0432x^{III}_{(t,T-1)} - 0.0543X^{III}_{(T)} + 23.1u_1 - 10.1u_2; \end{aligned}$$

(iii) summer:

$$\begin{aligned}
x^{I}_{(t,T)} &= -2.35x^{II}_{(t-1,T)} + 0.0267x^{III}_{(t-1,T)} + 0.584x^{I}_{(t,T-1)} + 1.04X^{I}_{(T)},\\
x^{II}_{(t,T)} &= 0.0669x^{I}_{(t-1,T)} + 0.482X^{II}_{(T)},\\
x^{III}_{(t,T)} &= -13.8x^{I}_{(t,T)} - 3.32x^{I}_{(t-1,T)} + 2.93x^{III}_{(t-1,T)}\\
&\quad -0.0906x^{III}_{(t,T-1)} + 2.68X^{III}_{(T)} + 48.8u_1 - 13.1u_2;
\end{aligned}$$

(iv) fall:

$$\begin{aligned}
x^{I}_{(t,T)} &= -0.00205x^{III}_{(t,T)} + 0.834X^{I}_{(T)},\\
x^{II}_{(t,T)} &= 0.00228x^{III}_{(t-1,T)} + 0.391X^{II}_{(T)},\\
x^{III}_{(t,T)} &= 4.95x^{I}_{(t,T)} + 15.6x^{I}_{(t-1,T)} + 62.0x^{II}_{(t-1,T)}\\
&\quad +0.32x^{III}_{(t-1,T)} - 0.333x^{III}_{(t,T-1)} + 0.892X^{III}_{(T)} + 298.0u_2.
\end{aligned} \tag{6.44}$$

In the next step, the balance criterion is used in sorting the forecasts of the leading variable X^{III}.

$$b_i = X_i - \frac{1}{4}(x_{(w)} + x_{(sp)} + x_{(su)} + x_{(f)})_i; \; B^2 = \sum_{i=1}^{Ta} b_i^2 \rightarrow \min, \tag{6.45}$$

where T_a is the anticipated time.

It is necessary to use different schemes as applied above in obtaining the annual and seasonal forecasts; i.e., the annual forecasts are obtained by a one-dimensional linear scheme, whereas the seasonal forecasts are obtained with a two-dimensional time readout.

In evaluating the balance criterion, a total of 25 variants of forecasts (5 seasonal and 5 annual) are carried out and the optimal annual and seasonal models are obtained for the leading variable.

Overall we can say that the calculation of the rank correlation coefficient of data helps in eliminating the collinear factors and thus allows the reduction of computational volume substantially.

4 MODELING OF ECONOMICAL SYSTEM

In economical system modeling, variables are divided into exogenous or external; i.e., those specified outside the model and introduced into it, and endogenous; i.e., those obtained within the model. This means that the input variables are exogenous, while outputs and state variables are endogenous. Sometimes, one variable may be exogenous to a particular model and endogenous to another. In the beginning of the experimentation, it is often not known which variables will be selected for inclusion in the equations; and it is worth adopting a technique to distinguish the varaibles. The ratio of the variables exogenous to endogenous and their participation in the model determines the extent to which the model is open or closed.

The objective systems analysis (OSA) makes it possible to find an autonomous system of linear algebraic or finite-difference equations that is optimal for the given objective criterion assuming that all variables are the system (internal) variables, the "status quo" scenario. The exogenous variables are not indicated *a priori* at the beginning of the analysis. This technique avoids the *a priori* resolution of the difficult and the controversial question: what extent is a variable exogenous. Specialists select the exogenous variables from the set of variables that figured out in the system of equations picked by the computer.

Step-by-step integration of such an autonomous system of equations yields short-range predictions without any special control of the system. This first level scenario often proves useful for showing the use of control to get better processes in the object being modeled, or to get a better estimate for the near future. To proceed with other scenarios, the variables need to be divided into endogenic and exogenic variables or output and input variables. Furthermore, the input variables are to be divided into external disturbances and control variables. This separation of variables is usually done on the basis of physical considerations, which contain the element of subjectivity.

Here various examples of economical system modeling which represent different scenarios in obtaining the optimal systems of equations are presented. They correspond to the modeling of British economy and USA economy based on the studies conducted by Ivakhnenko and his coworkers [40], [41], [38], and Klein, Mueller and Ivakhnenko [76].

In the first example (Example 4, modeling of British economy) the descriptions are considered with linear static elements described by algebraic equations. It was proven that the possibility of changing the control actions is severely limited in such systems.

The second example (Example 5) is the continuation of the first one with an introduction of finite-difference equations with one delayed argument. It is proven that the modified systems analysis is useful in identifying the macroeconomic variables for long-range predictions of up to 10 to 15 years. The predictions of these variables can serve as a basis for estimating other economic indices used in macroeconomic modeling of a country. A solution for the problem of control action promises to design more efficient systems based on the control criteria and control actions.

The third example (Example 6) is meant for briefing the idea of modeling US economy by extending the delayed arguments in the descriptions. The selected system of equations is used to measure the prediction accuracy on the average quarterly values.

The fourth example (Example 7) is the result of extended studies on British economy to adopt a special procedure that measures the cause-effect relationships among the variables. The degree of exogenicity of the variables is defined on the basis of a harmonic criterion. Here the external influences are included in the descriptions for studying the possible changes in the system.

4.1 Examples—modeling of British and US economies

Example 4. The OSA at the level of trends for modeling the British economy.

Usually, a stationary process is represented as the sum of low-frequency and high-frequency components:

$$q(t) = f(t) + \sum_{i=1}^{m} (A_i \cos w_i t + B_i \sin w_i t), \tag{6.46}$$

where the low-frequency part $f(t)$ is called the quasistatic part or the trend, and the high-frequency part is called the dynamic remainder. The former is represented as polynomials or sums of exponentials, whereas the latter is expressed as a finite-difference equation or a harmonical trend. The structural complexity of these functions is uniquely determined by the inductive learning algorithms. One should understand the uniqueness of the selected trend as a single completely defined optimal structure of the equation of the trend and that this uniquely determines the dynamic remainder.

In the sense of objective systems analysis, a system of algebraic equations describing the trends of the output variables is used at the level of trends, and a system of finite-difference equations is used at the level of the dynamics. These both results of analyses are different because they are related to different mechanisms of statics and dynamics. The coefficients of the former give the interrelationships among the variables, whereas the latter give the interrelationships among their differences.

Frequently, the object itself suggests the type of analysis required. For example, the activity of an aquatic ecosystem is not determined by the level of just one variable like the biomass of phytoplankton, but rather by the rate of its change under the action of external influences. This means that the analysis of dynamic interrelationships is important for ecosystems. For economical systems, it is more important in analyzing the coarse and interaction of the variables.

In general, the operation of the basic inductive learning algorithms is very similar in obtaining the optimal trends and the dynamic remainders. In a given set of multivariables, a nonautonomous physical model in a state space that does not contain the time coordinate is obtained with any algorithm. In case of multilevel algorithm, objective system analysis is used as a first step to obtain a set of output variables characterizing the object and then the two-stage approach is used in time space to obtain an autonomous nonphysical predicting model of the trend or remainder.

The general scheme of the multilevel algorithm at the level of trends is discussed below: (i) the table of initial data sample is supplied, (ii) objective systems analysis with linear algebraic equations (not including the time coordinates) is conducted and a characteristic set of output variables is determined, (iii) by the objective systems analysis, the best F_1 mean annual models are identified using the systems of nonlinear algebraic equations including the time as an argument $(T, T^2, \cdots)$ for the output variables, (iv) the best $F_2 (< F_1)$ systems of equations are selected from the above annual models using the stability criterion of multiple correlations, (v) by the objective systems analysis, the best F_3 quarterly systems of nonlinear equations including the time coordinates $(t, T, t^2, T^2, \cdots)$ are identified for the output variables, (vi) the balance criterion of predictions is evaluated by sorting the predictions of the F_2 annual models and the F_3 quarterly models, and (vii) the system of equations with optimal complexity is given with their predictions.

The notations x and u denote the quarterly values, X and U denote the annual values of the variables and of the external influences, and t and T denote the quarterly and annual time coordinate values. The data table should at least contain the data of one external influence variable to maintain the static stability of the system of linear equations. This rigid necessity does not arise in case of nonlinear equations. The data is separated into three sets $A, B,$ and C.

The single layer combinatorial algorithm is used to select the best structure of each equation at each layer of the objective systems analysis according to the minimum-bias η_{bs} and regularity $\Delta(B)$; and the system criterion of minimum- bias $\eta_{s_{(bs)}}$ and prediction criterion $\Delta(C)$ are used in further analysis of the equations. Keeping in view the drawbacks of the modeling such as multiple values of the predictions and the presence of noise in the data, a confidence level D is fixed so that $\eta_{s_{(bs)}} < D$. Usually the confidence level is kept below the value of 10^{-4}. This explains why the algorithm chooses an optimal system with the least noise having the most complete informational basis with the necessary state variables and the external disturbances. In an ideal data (without noise) the algorithm indicates that all systems of equations with complete information are of equal value.

Identifying the process of inflation in case of British economy

The following mean annual values of 26 variables are:

X_1 – the national product in million pound-sterlings,
X_2 – the energy consumption in million tons of conventional coal,
X_3 – the steel production in millions of tons,
X_4 – the automobile production in thousands,
X_5 – the companies' revenues in million pound-sterlings,
X_6 – the individuals revenues in million pound-sterlings,
X_7 – the savings in million pound-sterlings,
X_8 – the capital investment in million pound-sterlings,
X_9 – the capital investment in the public sector in million pound-sterlings,
X_{10} – the capital construction in percentages,
X_{11} – the index of industrial production in percentages,
X_{12} – the wholesale costs for materials and fuel in percentages,
X_{13} – the volume of retail trade in percentages,
X_{14} – the index of retail costs in percentages,
X_{15} – the registration of new automobiles in thousands,
X_{16} – the purchasing value of the pound in percentages,
X_{17} – the avearge wages in percentages,
X_{18} – the number of unemployed in thousands,
X_{19} – the number of employed persons in millions,
X_{20} – the number of employement vacancies in thousands of working places,
X_{21} – the labor productivity in percentages,
X_{22} – the exports in million pound-sterlings,
X_{23} – the imports in million pound-sterlings,
X_{24} – the current balance in million pound-sterlings,
X_{25} – the money supply (group 1) in million pound-sterlings,
X_{26} – the money supply (group 2) in million pound-sterlings,
U_1 – the tax rate on the company's revenues,
U_2 – the tax rate on the individual's incomes,
U_3 – the government expenditure in millions of pound-sterlings, and
U_4 – the cost of oil in percentages.

Data that covered 15 years (1964 to 1978) were used from the "*Economic Trends: Annual Supplement of 1980 Edition,* London." The data is divided into three sets as $A + B + C = 6 + 6 + 3$.

The results of the OSA and the best systems of equations obtained as a result of the chosen confidence level are given below:

The system of equation at the first layer is

(I)

$$X_{14} = 33.11 - 63.88U_2 + 1.575U_4; \qquad \eta_{s_{(bs)}} = 0.0002534 \tag{6.47}$$

The other number of systems obtained below the confidence level are:

(II)

$$\begin{aligned} X_6 &= -22.28 + 3.906U_1 + 1.198X_{14}, \\ X_{14} &= 18.7 - 3.437U_1 + 0.8334X_6, \\ X_{22} &= -1.89 + 0.2259X_6; \\ \eta_{s_{(bs)}} &= 0.0000225; \end{aligned} \tag{6.48}$$

(III)

$$
\begin{aligned}
X_6 &= -22.28 + 3.906U_1 + 1.198X_{14}, \\
X_8 &= 4.729 + 1.967U_1 + 0.4364X_6 - 0.2653X_{14}, \\
X_{14} &= 18.7 - 3.437U_1 + 0.8334X_6; \\
\eta_{s_{(bs)}} &= 0.00003024.
\end{aligned}
\tag{6.49}
$$

The system IV consists of the equations for X_6 and X_{14} as they are in the above systems. The systems II, III, and IV have achieved the limits of confidence level in this analysis; the system IV has attained the global minimum. The drawback of using this system is that it is statically unstable and it is admitted in the later works that the selection of control actions applied in the system to make it statically stable is not quite correct. Further developments of the modeling of British economy is described in the Examples 5 and 7.

Example 5. Modeling of the British economy using one delayed argument and without specifying the external influences.

This example illustrates the extension the use of the OSA in modeling the British economy with the following features:

(i) It is assumed that the study consists not only of linear static elements described by algebraic equations, but also of links with one delayed argument, which is also called the first-order linear link.

This can be more generalized by considering the second- and higher-ordered links which can be replaced by a sequential combination of first-order links.

(ii) The exogenous varaibles are not indicated *a priori* at the beginning of the analysis.

(iii) The delayed arguments in the equations enable the selection of equations on the basis of the prediction criterion i or $\Delta(C)$, where i indicates the step-by-step prediction integration and $\Delta(C)$ indicates the error on an examining set C:

$$
\Delta^2(C) = \sum_{i=1}^{N_C} \frac{(x - \hat{x})_i^2}{(x - \bar{x})_i^2} \leq 1.0, \tag{6.50}
$$

where $x, \hat{x}$, and $\bar{x}$ indicate the actual, estimated and average values of the variable x.

The selected equations satisfy the objective characteristics of both the criteria of minimum-bias and prediction; this means that the system criteria are used wherever applicable in the analysis of the systems of equations.

The data and the data separation are the same as in the previous example.

The *OSA algorithm* is given as below:

The algorithm consists of several layers with gradual increase in complexity of the equations. The single-layered combinatorial algorithm with the two criteria described above is used at all layers. During the selection, the equations with $\Delta(C) > 1.0$ are excluded from the search as they contain false information.

The following general form of the polynomials are considered:

(i) the *first layer*, the equations with not more than three terms of the form

$$
X_{i_{(T)}} = a_0 + a_1 X_{i_{(T-1)}} + a_2 T, \tag{6.51}
$$

where T is the time coordinate in years;

(ii) the *second layer* , the equations with not more than five terms of the form

$$
X_{i_{(T)}} = a_0 + a_1 X_{i_{(T-1)}} + a_2 T + a_3 X_{j_{(T)}} + a_4 X_{j_{(T-1)}}; \tag{6.52}
$$

(iii) the *third layer*, the quations of not more than seven terms of the form

$$X_{i_{(T)}} = a_0 + a_1 X_{i_{(T-1)}} + a_2 T + a_3 X_{j_{(T)}} + a_4 X_{j_{(T-1)}} + a_5 X_{k_{(T)}} + a_6 X_{k_{(T-1)}}, \qquad (6.53)$$

and so on until the number of terms not exceeding 18 to 20 as the limit of the combinatorial algorithm or the layers are extended while there is at least one equation among all in the last layer satisfying the condition $\eta_{bs} < D$, where D is a certain "confidence level" chosen on the basis of experience.

Systems of equations for prediction and control

The above OSA is applied in modeling the British economy. In addition to the 26 variables used in the previous example, an additional four economic indices are used as input to the algorithm. The indices picked by the computer are

X_{27} – the budget deficit in million pound-sterlings,
X_{28} – the trade balance in million pound-sterlings,
X_{29} – the current balance in million pound-sterlings,
X_{30} – the overall government expenditure in million pound- sterlings,
X_{31} – the tax on private income in million pound-sterlings, and
X_{32} – the corporate tax in million pound-sterlings.

At the first layer, one equation below the confidence level ($D \leq 10^{-4}$) is selected,

$$\begin{aligned} X_{22_{(T)}} &= -1.017 + 1.293 X_{22_{(T-1)}}, \\ \eta_{bs} &= 0.00004062. \end{aligned} \qquad (6.54)$$

At the second layer, ten equations are selected and the best among them is

$$\begin{aligned} X_{22_{(T)}} &= -1.032 + 1.297 X_{22_{(T-1)}} - 0.003908 X_{29_{(T)}}, \\ \eta_{bs} &= 0.00002226. \end{aligned} \qquad (6.55)$$

At the third layer, eight equations are selected and the best among them is

$$\begin{aligned} X_{14_{(T)}} &= 18.64 + 1.064 X_{6_{(T)}} - 0.2526 X_{6_{(T-1)}} + 0.0653 X_{22_{(T)}}, \\ \eta_{bs} &= 0.0000376. \end{aligned} \qquad (6.56)$$

The selection is terminated further because the minimum-bias begins to increase from the fourth layer onward.

The important feature of the algorithm is that the variables selected at the previous layer are refined by supplying additional output variables and arguments at the successive layer.

In the example given here, the most representative output variables X_1, X_8, X_{11}, X_{14}, X_{17}, and X_{22} are selected at the first and second layers. At the third layer, variables X_6, X_{23}, and X_{30} are added to this list. The best equations with least minimum-bias from the three levels are obtained as below:

$$\begin{aligned} X_{1_{(T)}} &= -10.99 + 0.145 X_{1_{(T-1)}} + 0.002955 X_{25_{(T)}} + 0.001323 X_{26_{(T-1)}}; \\ \eta_{bs} &= 0.8902E - 5, \\ X_{6_{(T)}} &= -16.76 + 1.033 X_{14_{(T)}} + 0.4645 X_{23_{(T)}}; \\ \eta_{bs} &= 0.1.524E - 5, \\ X_{8_{(T)}} &= -8.381 + 0.002038 X_{25_{(T-1)}}; \end{aligned}$$

$$
\begin{aligned}
&\eta_{bs} = 2.78E-5,\\
X_{11_{(T)}} &= 25.87 + 0.7407 X_{11_{(T-1)}};\\
&\eta_{bs} = 6.936E-4,\\
X_{14_{(T)}} &= 18.64 + 1.061 X_{6_{(T)}} - 0.2526 X_{6_{(T-1)}} + 0.06531 X_{28_{(T)}};\\
&\eta_{bs} = 0.376E-5,\\
X_{17_{(T)}} &= -5.045 - 1.518T + 2.159 X_{1_{(T)}} - 0.02077 X_{27_{(T)}}\\
&\quad +0.06109 X_{27_{(T-1)}};\\
&\eta_{bs} = 2.568E-5,\\
X_{22_{(T)}} &= -1.032 + 1.297 X_{22_{(T-1)}} - 0.003908 X_{29_{(T)}};\\
&\eta_{bs} = 2.226E-5,\\
X_{23_{(T)}} &= -3.392 - 0.007218 X_{23_{(T-1)}} + 1.013 X_{22_{(T)}} - 0.09933 X_{28_{(T)}};\\
&\eta_{bs} = 2.21E-5,\\
X_{30_{(T)}} &= -5.6 + 0.001638 X_{26_{(T-1)}};\\
&\eta_{bs} = 4.253E-5.
\end{aligned} \tag{6.57}
$$

The algorithm has yielded a system of nine equations with 14 variables. Values of five variables $X_{25}, X_{26}, X_{27}, X_{28}$, and X_{29} are to be selected based on the nature of the problem or by using auxiliary equations. The time variables in some of the equations, primarily of those selected from a set of arguments are assigned by specialists for prediction of norms. The equations are tested for their stability; the roots of the characteristic matrix equation lie within the limits of the unit circle. Step-wise predictions reveal that the processes converge rapidly with the trends exhibited by them. This means that the obtained system of equations can be used for identifying the system structures and for short-term predictions.

Here we are not presenting the two-stage approach used for long-term quantitative predictions as used before, but the solution of a control problem specified in the current example. Usually in solving the economic control problem, experts indicate not only the variables related to control actions, but also their number and interrelations in the form of objective criteria.

Solution of a control problem

Control variables are chosen by the specialists either from the set of characteristic variables from the OSA or from the variables that correlate fairly with the characteristic variables. This is checked according to the values of the external criteria for the auxiliary equations formed for these variables. It is also necessary to check that the selected control variables are adequate to satisfy the controllability conditions (according to Kalman). These conditions might require expansion of the control variables or a change in the control objective.

For example, let us assume that the variable X_{27} (budget deficit) is chosen by experts as the control action and the control objective is defined as maintaining the ratio of the retail price index to the average wages (X_{14}/X_{17}) at the year 1978 level. The following basic system's equations are used to build up a closed system with the selected control variable X_{27} as one of the arguments.

$$
\begin{aligned}
X_{1_{(T)}} &= -10.99 + 0.145 X_{1_{(T-1)}} + 0.002955 X_{25_{(T)}} + 0.001323 X_{26_{(T-1)}};\\
&\eta_{bs} = 0.8902E-5,\\
X_{17_{(T)}} &= -5.045 - 1.518T + 2.159 X_{1_{(T)}} - 0.02077 X_{27_{(T)}} + 0.06109 X_{27_{(T-1)}};\\
&\eta_{bs} = 2.568E-5,
\end{aligned}
$$

$$X_{14_{(T)}} = 0.6X_{17_{(T)}}, \quad \text{and} \quad \Delta X_{14_{(T)}} = 0.6\Delta X_{17_{(T)}}. \tag{6.58}$$

The ratio is given as 0.6; that does not mean that wages rise faster than prices. The fact is that the British economic statistics take the year 1975 as the base year for computing the average wage. If the same base year were to be used for the control problem, then the ratio would be 1.56; this means that the price rise outstrips the average wage as compared to 1976.

The system of auxiliary equations is obtained using the combinatorial algorithm with the minimum-bias criterion and the prediction criterion. The reference function considered in the algorithm takes into account all the variables with one delayed argument.

$$X_{i_{(T)}} = a_0 + a_1T + a_2T^2 + a_3X_{i_{(T-1)}} + \sum_{j\,(j \neq i)} (b_jX_{j_{(T)}} + c_jX_{j_{(T-1)}}), \tag{6.59}$$

where $i = 25, 26, 27$; and $j = 1, 11, 14, 17, 23, 26, 27$.

The system of equations identified are

$$\begin{aligned}
X_{25_{(T)}} &= 2729 + 0.4277X_{25_{(T-1)}} + 131.4X_{1_{(T)}} - 47.79X_{14_{(T)}} + 58.27X_{17_{(T)}} \\
&\quad +0.1648X_{26_{(T)}} - 0.4907X_{26_{(T-1)}}; \quad \eta_{bs} = 1.258E-3, \\
X_{26_{(T)}} &= 6024 + 0.9666X_{26_{(T-1)}} + 566.7X_{1_{(T-1)}} - 310.3X_{14_{(T)}} \\
&\quad -611.2X_{14_{(T-1)}} + 2.315X_{25_{(T)}} + 24.77X_{27_{(T-1)}}; \quad \eta_{bs} = 9.29E-4, \\
X_{27_{(T)}} &= 88.29 + 1.188T^2 - 1.61X_{17_{(T)}} + 1.304X_{17_{(T-1)}} - 0.013X_{25_{(T)}} \\
&\quad +0.009885X_{25_{(T-1)}} - 0.004399X_{26_{(T)}}; \quad \eta_{bs} = 4.306E-1.
\end{aligned} \tag{6.60}$$

The resulting closed system contains six equations with six variables. Kalman's controllability conditions are satisfied. Step-wise integration of the system yields the predictions to achieve the ratio (X_{14}/X_{17} = constant) through variation of budget deficit X_{27} (Figure 6.12).

Prediction of variables other than the characteristic variables.

Out of forty variables fed into the computer while modeling the British economy, only 14 were selected as the characteristic variables. All other variables were rejected as unsuitable for modeling continuous consistent features, both as output variables and as constituents of a set of arguments. This does not mean that the rejected variables are impossible to identify using the inductive learning algorithms; this can be done as a supplementary analysis. They can be predicted as a function of time and the variables selected by computer. For example, the following full polynomial can be used in predicting the number of unemployed:

$$\begin{aligned}
X_{18_{(T)}} &= a_0 + a_1T + a_2T^2 + a_3X_{18_{(T-1)}} + a_4X_{18_{(T-2)}} \\
&\quad + \sum_j (b_jX_{j_{(T)}} + c_jX_{j_{(T-1)}} + d_jX_{j_{(T-2)}}); \quad j = 1, 11, 14, 17.
\end{aligned} \tag{6.61}$$

The combinatorial algorithm with the criteria minimum-bias and prediction selected the following equation:

$$\begin{aligned}
X_{18_{(T)}} &= 238.9 - 0.5211X_{18_{(T-2)}} - 10.18X_{1_{(T)}} + 26.55X_{1_{(T-1)}}; \\
&\quad \eta_{bs} = 0.0989.
\end{aligned} \tag{6.62}$$

The suggested aproach of OSA described in the example affords a novel approach to select a system of indices used at various layers of aggregation. Conventional models

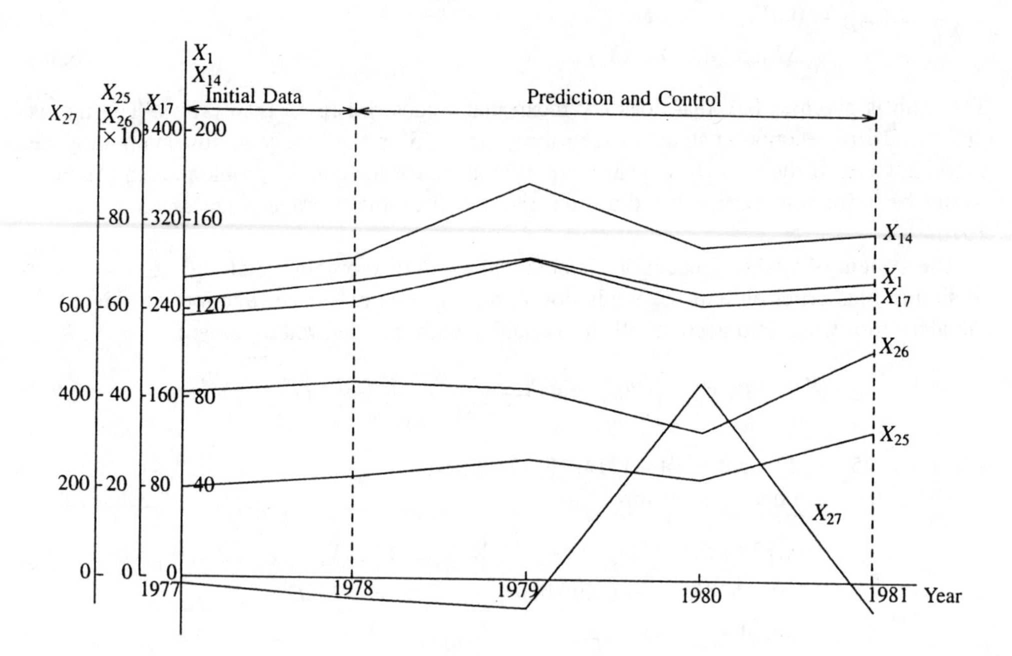

Figure 6.12. Prediction and control action in achieving X_{14}/X_{17} = const. through variation of budget deficit X_{27}

normally use many variables and equations describing their interrelations. The system of indices in a national economic model usually is included with 17 to 130 variables depending on the extent of process elaboration. For instance, in the project headed by L.P. Klein (University of Pennsylvania, USA), the model for British economy contains 226 equations, 206 for the USA economy, and 183 for Canadian economy.

Case studies of the design of the control system have shown that the method described here can be used to design more efficient systems based on other control criteria and control actions. This opens up a vast area of research. It would also be of interest to stimulate control criteria of the form $X_{14}/X_{17} = 0.6 + K(T - 15)$, where T is the time coordinate in years ($T_{(1978)} = 15$), and assigning different values for the coefficient $K = \cdots, -0.2, -0.1, 0, 0.1, 0.2, \cdots$ to determine the effect of economic processes of relative rise or drop in prices.

One can note that in the proposed method of OSA, prediction and control is also applicable in studying other complex systems.

Example 6. Self-organization modeling of US economy.

The objective systems analysis is used in obtaining the system of finite- difference equations. The multilayer algorithm is used to identify each equation. The variables are chosen using the system criterion of minimum-bias and the number of equations is chosen with the prediction criterion using all data. The data used in the algorithm is quarterly data from 1969 to 1974 [76] for the following variables:

x_1 – national product,

x_2 – real national product,

x_3 – national income,

x_4 – personal income,
x_5 – deflation of prices of the national product,
x_6 – deflation of personal prices,
x_7 – consumer price index,
x_8 – whole sale price index,
x_9 – private production in man-hour,
x_{10} – earnings per man-hour,
x_{11} – rate of unemployement,
x_{12} – net running export,
x_{13} – money supply M1,
x_{14} – money supply M2,
x_{15} – rate of 3-month treasury notes,
x_{16} – usage rate of corporate funds,
x_{17} – rate of 6-month commercial paper,
x_{18} – rate of general growth of corporate funds,
x_{19} – rate of personal savings,
x_{20} – pre-tax corporate earnings, and
x_{21} – federal surplus.

The data is separated as even and odd points for the roles of sequences A and B and used in calculating the minimum-bias criterion.

The OSA algorithm is realized in the example below.

At the *first step* of the algorithm, the finite-difference models of all variables are synthesized using the minimum-bias criterion.

$$x_i^t = a_0 + \sum_{j=1}^{8} a_j x_i^{t-j} + \sum_{k=1(k\neq i)}^{21} \sum_{j=0}^{8} b_{kj} x_k^{t-j}. \tag{6.63}$$

All variables are measured with eight delayed values. Each equation is synthesized using the multilayer algorithm. In the first layer of the multilayer algorithm 780 partial models are generated, of which 30 of the most unbiased models are allowed to proceed to the next layer. The selection continues (from 435 models, 30 models are selected) until the minimum-bias criterion ceases to decrease. The number of layers never exceeds ten for any model.

At the *second step*, further selection of system of equations continues from among the obtained models of first step. Here the models that yield a good prediction have priority in the selection process; i.e., the models are selected according to the step-by-step prediction criterion. From among the models obtained from the first step, eight models corresponding to the output variables $x_1, x_3, x_4, x_5, x_6, x_{10}, x_{13}$, and x_{14} are selected and considered as the sought system, the model of the US economy.

The selection threshold is established such that the obtained system of equations would be autonomous and convenient for step-wise predictions. After the selection, the estimates of the equations are adapted using the whole data sample. The system of equations are given below.

$$x_1^t = -114.86 + 0.894 x_1^{t-1} + 0.05625 x_4^{t-5} - 2.3476 x_5^{t-3} + 23.587 x_5^{t-6}$$

$$
\begin{aligned}
& -18.275x_5^{t-7} + 5.376x_5^{t-8} - 5.5217x_6^{t-4} - 0.8283x_{13}^{t-7};\\
x_3^t &= -451.37 + 0.2563x_4^{t-3} - 1.3502x_5^{t-7} - 3.231x_6^{t-6} + 8.1061x_6^{t-8}\\
& -140.92x_{10}^{t-2} - 236.96x_{10}^{t-3} + 87.045x_{10}^{t-7} + 3.5427x_{13}^{t-1}\\
& +2.8526x_{13}^{t-2} + 1.2499x_{13}^{t-5} + 1.1724x_{13}^{t-6};\\
x_4^t &= -238.64 - 84.163x_{10}^{t-5} + 82.507x_{10}^{t-7} + 3.7134x_{13}^{t-2} + 0.6916x_{14}^{t-2};\\
x_5^t &= 11.895 + 1.088x_5^{t-1} - 0.00551x_1^{t-8} + 0.293x_3^{t-6} - 0.2725x_6^{t-3}\\
& -0.086x_{13}^{t-7} + 0.03914x_{14}^{t-7} - 0.00515x_{14}^{t-8};\\
x_6^t &= 5.585 + 0.1306x_6^{t-2} - 0.01343x_3^{t-7} - 0.00243x_4^{t-3} + 1.2604x_5^{t-1}\\
& -0.2224x_5^{t-2} - 0.04754x_{13}^{t-5} - 0.2118x_{13}^{t-6} + 0.0876x_{14}^{t-6};\\
x_{10}^t &= 0.19596 + 0.6092x_{10}^{t-2} + 0.3788x_{10}^{t-4} + 0.1746x_{10}^{t-7} - 0.34245x_{10}^{t-8}\\
& -0.00001x_3^{t-8} + 0.000145x_{13}^{t-1} - 0.00242x_{13}^{t-5}\\
& -0.000319x_{14}^{t-6} + 0.003319x_{14}^{t-7};\\
x_{13}^t &= 67.893 + 0.0645x_1^{t-1} + 0.0465x_4^{t-3} - 1.0195x_5^{t-4} + 1.1876x_5^{t-6}\\
& -0.6898x_5^{t-7} + 35.405x_{10}^{t-5} + 19.755x_{10}^{t-7} - 0.4934x_{13}^{t-6}\\
& +0.18827x_{13}^{t-7} - 0.15198x_{13}^{t-8};\\
x_{14}^t &= 4.3623 + 1.0298x_{14}^{t-1} - 18.541x_{10}^{t-1} + 15.657x_{10}^{t-3} + 45.187x_{10}^{t-7}\\
& -0.4316x_{13}^{t-1} - 0.4448x_{13}^{t-4} + 0.1071x_{13}^{t-6}. \qquad (6.64)
\end{aligned}
$$

The prediction accuracy is checked for each equation in the system on the data of the years 1975 and 1976. The range of the residual sum of squares vary from 0.28% to 6.87% for one year and from 1.23% to 11.44% for two years.

Example 7. Modeling of the British economy for restoration of the governing laws in the object.

The variables participating in the modeling of a complex object are assumed to have some degree of exogenicity. In this example, the degree of exogenicity of the variables is defined on the basis of a special criterion that is used to find more objective ways of dividing the variables. The objective system-analysis algorithm makes it possible to find an autonomous or closed system of algebraic or finite-difference equations that is optimal for a given criterion by assuming that all the variables in the equations are endogenous or system variables. If we remove those equations from the closed autonomous system whose output variables have proven to be exogenic with the greatest degree, then a governing principle is obtained in the form of an underdetermined system of equations or an approximation of such governing principle, which is suitable for short-range predictions (a "status quo" scenario). Further studies on the system yields a number of other scenarios that are useful for analyzing the object. Obviously, the corresponding equations are not physical laws reflecting the mechanism of objective modeling, but they make it possible to study the possible changes by introducing the external influences with respect to an objective control criterion.

The *OSA* algorithm is given below:

The choice of equations at each layer is made on the basis of two criteria; the minimum-bias (η_{bs}) and prediction (i). The equations with the criterion value of $i^2 > 1.0$ are eliminated from the preliminary selection as the equations providing disinformation.

First layer: Equations of the following form containing not more than four terms on the right side are evaluated.

$$X_{i_{(T)}} = a_0 + a_1 X_{i_{(T-1)}} + a_2 X_{i_{(T-2)}} + a_3 X_{i_{(T-3)}}. \qquad (6.65)$$

Second layer: Systems of two equations of the following form whose right sides contain not more than eight terms are evaluated.

$$\begin{aligned} X_{i_{(T)}} &= a_0 + a_1 X_{i_{(T-1)}} + a_2 X_{i_{(T-2)}} + a_3 X_{i_{(T-3)}} \\ &+ a_4 X_{j_{(T)}} + a_5 X_{j_{(T-1)}} + a_6 X_{j_{(T-2)}} + a_7 X_{j_{(T-3)}}. \end{aligned} \tag{6.66}$$

Third layer: Systems of three equations of the following form containing not more than 12 terms are evaluated.

$$\begin{aligned} X_{i_{(T)}} &= a_0 + a_1 X_{i_{(T-1)}} + a_2 X_{i_{(T-2)}} + a_3 X_{i_{(T-3)}} + a_4 X_{j_{(T)}} + a_5 X_{j_{(T-1)}} + a_6 X_{j_{(T-2)}} \\ &+ a_7 X_{j_{(T-3)}} + a_8 X_{k_{(T)}} + a_9 X_{k_{(T-1)}} + a_{10} X_{k_{(T-2)}} + a_{11} X_{k_{(T-3)}}. \end{aligned} \tag{6.67}$$

All the systems of equations obtained are autonomous; i.e., the number of variables in them is the same as the number of equations. This enables us to make step-by-step predictions in the system. The layer by layer procedure continues until it reaches the limits of the self-organization modeling according to the basic algorithms used in them. The end result is that one chooses some of the non-contradictory systems of equations, which are attained below the confidence level set for this purpose.

The minimum-bias criterion as a criterion of exogenicity of the variables:

The original data of the variables exhibit information regarding the changes in them. One considers two competitive hypotheses to decide which of two chosen variables X_i and X_j is cause and effect: hypothesis H_1 says that X_i is effect and X_j is cause, and hypothesis H_2 says that X_i is cause and X_j is effect.

These hypotheses can be tested using the single-layered combinatorial algorithm with the criteria of minimum-bias and prediction; one finds two dynamic optimal models using the following complete descriptions:

for H_1,

$$\begin{aligned} X_{i_{(T)}} &= a_0 + a_1 X_{i_{(T-1)}} + a_2 X_{i_{(T-2)}} + a_3 X_{i_{(T-3)}} \\ &+ a_4 X_{j_{(T)}} + a_5 X_{j_{(T-1)}} + a_6 X_{j_{(T-2)}} + a_7 X_{j_{(T-3)}}; \end{aligned}$$

for H_2,

$$\begin{aligned} X_{j_{(T)}} &= b_0 + b_1 X_{j_{(T-1)}} + b_2 X_{j_{(T-2)}} + b_3 X_{j_{(T-3)}} \\ &+ b_4 X_{i_{(T)}} + b_5 X_{i_{(T-1)}} + b_6 X_{i_{(T-2)}} + b_7 X_{i_{(T-3)}}. \end{aligned} \tag{6.68}$$

This means that for each hypothesis one finds a model of optimal complexity. The hypothesis is true for finding which minimum-bias criterion is deeper than the other. This procedure is suitable only for the variables of dynamical systems—it works without an error with a sufficiently large sample of experimental data.

A harmonic criterion of exogenicity of variables

Considering each variable in turn as an output variable, let us look at the above dynamic models. By compiling tables of data approximating the change in the two variables, one can make expert evaluations on the models obtained earlier. The problem is reduced to considering one pair of variables as cause and effect, or the other way around.

The period of the lowest-frequency component of each variable [11], [73] is found using

$$S(w) = \frac{2}{N}\left[\sum_{t=1}^{N} X_k^2(t) + 2\sum_{\nu=1}^{N-1}\sum_{t=1}^{N-\nu}(X_k(t) - \bar{X}_k)(X_k(t+\nu) - \bar{X}_k)\cos(\nu w)\right], \tag{6.69}$$

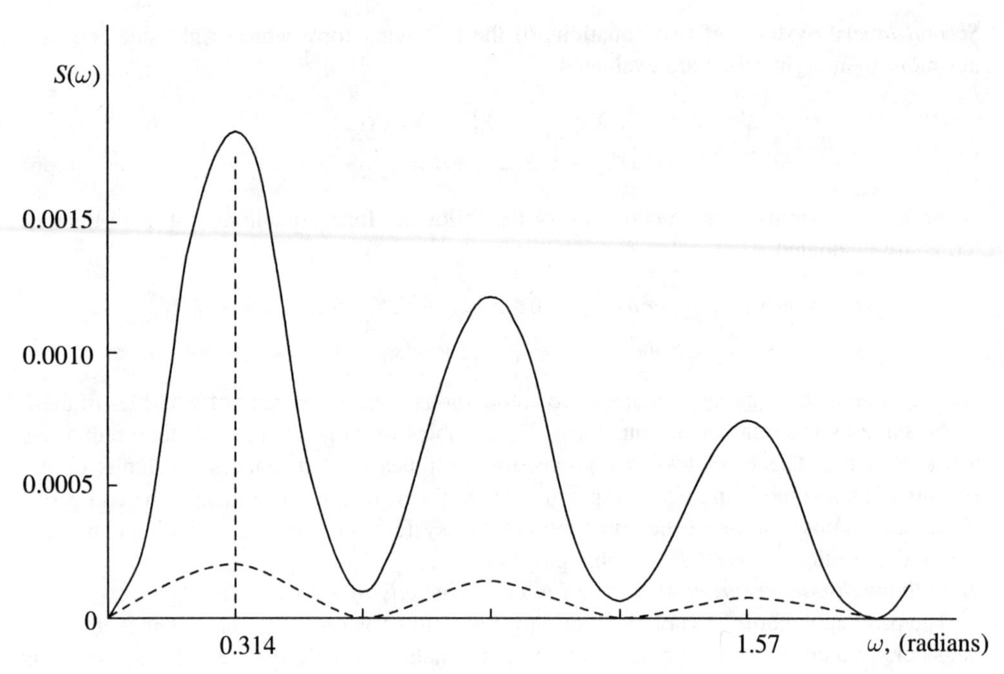

Figure 6.13. Spectrograms of the variables x_1 (solid curve) and x_2 (dashed curve) for determining the lowest frequency

where N is the number of measured data points of the variables $X_k(t)$, $k = 1, 2; \bar{X}_k$ is the average of $X_k(t)$ on N points.

Assigning a sequence of frequencies $0 \leq w \leq \pi$, spectrogram $S(w)$ is graphed until the first local maximum is obtained. This gives us the period T and the frequency w_M of the lowest harmonic. The period of this harmonic should be the same for the two variables (Figure 6.13). The phase of each of the harmonic components is then identified by using the regression equations of the following form:

$$X_i = a_0 + (a_1 \sin w_M t + a_2 \cos w_M t) = a_0 + A \sin(w_M t + \theta_i);$$
$$X_j = b_0 + (b_1 \sin w_M t + b_2 \cos w_M t) = b_0 + B \sin(w_M t + \theta_j);$$

$$\theta_i = \arctan \frac{a_1}{a_2}; \quad \theta_j = \arctan \frac{b_1}{b_2}, \tag{6.70}$$

where w_M is the frequency corresponding to the first local maximum of the spectrogram. The criterion of exogenicity for the two variables is given as $\Delta E = \theta_i - \theta_j$. In dynamical systems, the cause cannot overtake the effect in regard to phase, and consequently, the hypothesis H_1 is chosen if $\Delta E > 0$ (the cause is X_i and the effect is X_j), and H_2 if $\Delta E < 0$ (the cause is X_j and the effect is X_i).

This type of analysis of cause and effect makes it possible to eliminate those equations formed with the output variables that are formally causes rather than effects. Ultimately, the determinate system of equations becomes underdetermined, because the number variables exceeds the number of equations.

The results of modeling of the British economy

The above described OSA algorithm is applied for modeling the British economy. All 26 variables participating in the algorithm, both the endogenic variables X and the four exogenic external disturbances U are treated as equally significant. For the description of the variables and the data, refer to Example 1. The following autonomous system of equations and the values of criteria are obtained as below; the values of the prediction criterion i belong to the set C:

$$
\begin{aligned}
X1_{(T)} &= -0.5739 + 0.5197X1_{(T-1)} + 2.766X8_{(T)} - 15.17U2_{(T-3)}, \\
&\quad \eta_{bs}^2 = 1.389E-3 \text{ and } i = 2.821E-1; \\
X6_{(T)} &= -0.5937 + 2.518X8_{(T)} + 3.104X8_{(T-1)} + 38.74U1_{(T-1)} \\
&\quad -5.379U1_{(T-3)} - 64.54U2_{(T-1)}, \\
&\quad \eta_{bs}^2 = 1.590E-3 \text{ and } i = 2.331E-1; \\
X8_{(T)} &= 0.4487 + 0.02953X1_{(T-1)} + 0.2508X6_{(T)} - 0.555X8_{(T-2)} \\
&\quad -4.606U1_{(T-1)} + 12.25U2_{(T-3)}, \\
&\quad \eta_{bs}^2 = 4.176E-4 \text{ and } i = 5.838E-2; \\
X14_{(T)} &= 4.076 + 5.229X8_{(T-1)} + 22.94U1_{(T-1)} - 33.22U1_{(T-2)} \\
&\quad +86.97U2_{(T-2)} - 54.13U2_{(T-3)}, \\
&\quad \eta_{bs}^2 = 4.694E-4 \text{ and } i = 3.014E-1; \\
X22_{(T)} &= -5.769 + 0.03826X1_{(T-2)} - 0.2461X6_{(T)} + 2.814X8_{(T-1)} \\
&\quad -17.69U1_{(T-2)} + 7.553U1_{(T-3)}, \\
&\quad \eta_{bs}^2 = 2.902E-2 \text{ and } i = 8.91E-2; \\
U1_{(T)} &= 0.2981 - 0.01838X1_{(T)} + 0.01304X6_{(T)} + 0.006959X6_{(T-2)} \\
&\quad -0.8259U2_{(T-2)}, \\
&\quad \eta_{bs}^2 = 1.893E-2 \text{ and } i = 2.123E-3; \\
U2_{(T)} &= 0.1425 - 0.01253X1_{(T-1)} + 0.03903X8_{(T-1)} - 0.00358X8_{(T-2)} \\
&\quad +0.005462X14_{(T-1)} - 0.2325U1_{(T-1)}, \\
&\quad \eta_{bs}^2 = 3.442E-3 \text{ and } i = 8.210E-4.
\end{aligned}
\tag{6.71}
$$

The step-by-step predictions of the variables are checked to the year 1981 without taking into account any hypothesis to control the system for improving it.

The degree of exogenicity

The degree of exogenicity is used in the search for boundaries in dividing a significant variable into endogenic and exogenic variables. This is measured for each variable by $E_i = \sum_{j=1}^{7} \Delta E_{ij},\ i = \overline{1,7},\ i \neq j$. The harmonic criterion of exogenicity is analyzed for the variables of the above system and the results are tabulated in the Table 6.2.

According to the rank of the degree of exogenicity, the variables U_1 and U_2 proved to be the most exogenic. By removing the corresponding equations from the system, it becomes an underdetermined system of equations or governing principle of the system under the requirements that: (i) it is necessary that the selected candidate models be the result of true reference functions of the system and (ii) it is necessary that the initial data be sufficiently accurate. If the first requirement is not true, then the system of difference equations obtained is only an optimal approximation of the governing laws under consideration. With reference to the second requirement, it should be noted that with the increase of noise in the initial data, the OSA algorithm chooses systems with fewer equations. Thus, the physical model corresponding to the object can be obtained with the minimum-bias criterion only

Table 6.2. Results of the analysis on harmonic criterion of exogenicity

Varia-	ΔE							E	Rank
ble	X_1	X_6	X_8	X_{14}	X_{22}	U_1	U_2		
X_1	-	0	0.123	0.033	0.19	0.2	-0.097	0.449	5
X_6	0	-	-0.036	0	-0.38	0.092	0.0066	0.0245	4
X_8	-0.123	0.036	-	-0.22	-0.05	0.83	-0.081	0.492	6
X_{14}	-0.033	0	0.22	-	0	1.04	0.3	1.527	7
X_{22}	-0.19	0.38	0.05	0	-	-1.51	0.031	-0.234	3
U_1	-0.2	-0.092	-0.83	-1.04	0.51	-	0.247	-1.405	1
U_2	0.097	-0.0066	0.082	-0.3	-0.031	-0.247	-	-0.4056	2

in exact data. In the presence of noise in the data, the algorithm gives only the optimum approximation of these laws under the given conditions.

The underdetermined system of equations representing the governing laws or their optimal approximation can be used for constructing other scenarios for short-range normative predictions that are of interest to us.

Principles involved in short-range and long-range predictions

In the above analysis, when the term "governing law" is used, it refers to some mathematical description ensuring a sufficiently accurate short-range predictions of one to three steps ahead. This means that it is interpreted as an approximation of the characteristics of an object in a narrow sense. To restore such a governing law, the inductive approach is used with the minimum-bias criterion.

To restore a governing law in the broad sense and ensure exact long-range prediction with the inductive approach, one uses the minimum-bias and balance-of-predictions criteria. The originality of the multilevel approach is indicated first of all by the fact that these criteria are used at different levels of the analysis. No one mathematical language can exactly express the true physical or other governing laws that are suitable for long-range predictions. A single language can be used only for obtaining a physical model using exact data for short-range predictions. This is substantiated by the existence of a limiting attainable prediction time for all individually accepted mathematical languages. Only multilevel algorithms based on the balance criterion ensure long-range predictions; i.e., these can be applied for obtaining the governing laws of an object in the broad sense.

The governing laws corresponding to an economical system should be sought in the form of a consistent system of annual and quarterly models that ensures an exact qualitative long-range prediction.

5 AGRICULTURAL SYSTEM STUDIES

Various studies [65], [74] on agricultural systems reveal that fulfilling agricultural production forecasts, particularly in large irrigation systems remains a difficult problem. Difficulties arise which seemingly cannot be overcome by conventional modeling techniques.

Problems concerning the control and planning of irrigated forms are divided into three basic groups:

The *first group of problems* is related to the selection of varieties and hybrids of farm products that can be effectively cultivated in soil irrigated with a controlled quantity of water, and the optimum planning of the structure of irrigated areas for the entire crop rotation cycle.

A crop or a hybrid is assumed to be superior if it provides a maximum average yield ($\bar{y}$) on an observable series of meteorological factors $\{X = (x_1, x_2, \cdots, x_n)\}$ for standard values of the vector of control variables $\{U = (u_1, u_2, \cdots, u_k\}$ and a specified limit (δ) imposed on the standard deviation (s) of the yield:

$$\begin{aligned} \bar{y} &= \tfrac{1}{T} \textstyle\sum_{i=1}^{T} y_i(X, U) \rightarrow \max, \\ s &= \tfrac{1}{T} \textstyle\sum_{i=1}^{T} \sqrt{(y_i - \bar{y})^2} \leq \delta, \end{aligned} \tag{6.72}$$

where T is the number of years in which the meteorological factors are observed and y_i is the crop yield calculated from the meteorological factors of the ith year.

The *second group* is of problems dealing with crop rotation cycle on each field for maximizing the production with the given limitations of water and fertilizers. The task of this group is to maximize the gross yield of a crop-rotation area in hundred-kilograms of grain (y_m) for a known structure of the sowing area:

$$y_m = \sum_{j=1}^{m} y_j(X, U) \rightarrow \max, \tag{6.73}$$

where m is the number of fields included in the crop-rotation area; y_j is the yield of crop j, and $X(x_1, x_2, \cdots, x_n)$ are predicted values of the meteorological factors. The structure of the optimal model contains the selected arguments from the vector of controlled variables (U) for specified crop-yield planning intervals.

The *third one* encompasses operative production control/ planning of the irrigation-crop rotation system with the effective use of fertilizer, water, climatic, and technical resources. Here the goal is to determine the optimum distribution of water resources, fertilizers, and other yield control factors (U_i) that can be used during the ith interval:

$$y_i = \sum_{j=1}^{m} y_{ji}(X_i, U_i) \rightarrow \max, \tag{6.74}$$

where $i = 1, 2, 3$ is the control interval number (intervals between watering); $j = 1, 2, \cdots, m$ is the field number included in the crop-rotation area; and y_{ji} is the yield model of crop j during control interval i.

On each control interval, a separate yield model is used and the vector X consists of a combination of measured and predicted climatic factors. The control intervals are selected to be compatible with the phases of development of various crops, but are not longer than the period between the waterings.

Each of these problem groups requires different arguments of the yield model, control intervals, and allowed accuracy of solutions. These requirements are usually noncontradictory, requiring one to synthesize a complex of yield models for each group of problems. Solving these problems on the basis of the experimental field data and by using trial and error methods yields no guaranteed solutions to the problem of effective use of irrigation systems for a given period. But these problems can be adequately solved by using the production models synthesized on the basis of experimental data obtained from soil and climatic conditions of a specific irrigation system. This is adressed in the work of Khomovnenko and Kolomiets [74] by the use of inductive learning algorithms. In this work, the self-organization modeling is described for modeling of winter wheat productivity. Various problems are organized independently to model them with partial models using small groups of arguments. The experimental data used in this study is collected from one of the agricultural experimental stations located in Ukraine during 1967 to 1975.

5.1 Winter wheat modeling using partial summation functions

Example 8. Modeling of winter wheat productivity. Agricultural production depends on the number of natural factors and on the agro-technology used in the process; i.e., on the possibility of selecting the control factors. Natural factors can be divided into soil and climate factors. Soil factors are mechanical composition and depth of profile (which characterize mechanical strength, speed of infiltration, water conductivity of unsaturated soil, and the characteristic of water retention), aeration, fertility, salinity (which influences structure, weight potential, and toxicity), temperature, and water levels of the soil. Climatic factors are temperature and humidity of air, sedimentation, wind, light intensity, length of the day, and length of the growing period. Some of these factors are independent and the changes in some make changes in others. These factors influence the soil-climate zone and determine the possible selection of varieties and hybrids of plants in the irrigated plant cycle. They have various effects on the production of agricultural products depending on the period of vegetation.

Three main situations in irrigated soil management are pointed out below:

(i) photosynthesis is limited by factors not related to water delivery, but the water is limited by water reserves. An increase in water use due to sprinkling either does not increase the harvest much or has no effect on it;

(ii) if both the photosynthesis and water consumption are limited by water supplies, then the harvest increases with an increase of water use, and sprinkling is most effective; and

(iii) photosynthesis is affected by factors other than water supply; the water consumption is not affected by the water supply in the soil. In this case, it is necessary to use other methods of controling crop production; for example, by increasing the mineral feeding.

In this example of modeling winter wheat, the following identifiable phenophases of development are considered:

1 – from planting to sprouting,
2 – from sprouting to the beginning of tilling,
3 – period of wintering,
4 – from the end of tilling to stem formation,
5 – head formation,
6 – milk formation,
7 – waxy milk formation,
8 – waxy formation, and
9 – complete ripeness.

The model of winter wheat harvest is represented in a general form of output and participating input variables as

$$\begin{aligned} y = f(&B, Q, H, N, P, K, T^{\circ}_{min}, h, l, C, \\ &W_1, \cdots, W_9, t_1, \cdots, t_9, T^{\circ}_1, \cdots, T^{\circ}_9, t^{\circ}_1, \cdots, t^{\circ}_9, \\ &S_1, \cdots, S_9, E_1, \cdots, E_9, R_1, \cdots, R_9, N_{r_1}, \cdots, N_{r_9}, N_{h_1}, \cdots, N_{h_9}, \\ &t_{N_1}, \cdots, t_{N_9}, t_{P_1}, \cdots, t_{P_9}, t_{K_1}, \cdots, t_{K_9}), \end{aligned} \quad (6.75)$$

where y is the wheat harvest in 100 kg-units (220.462 lbs)/ hectare; B is the index of the predecessor; Q is the soil fertility; W is the water in the soil; H is the amount of seed;

C is the period of planting; N, P, K are the amount of nitrogen, potassium, and calcium introduced into the soil; T°_{min} is the sum of minimal temperatures below $-10°C$ during wintering; h is the depth of frost in the soil; l is the amount of snow cover; t_i is the length of the phenophase; T°_i is the sum of effective midday tempeartures during the phenophase; t°_i is the average daily air temperature; S_i is the number of sunshine hours; E_i is the sum of evaporation from the water surface during the phenophase; R_i is the amount of rainfall; N_{r_i} is the number of rain days; N_{e_i} is the number of days having a relative air humidity of less than 30%; $t_{N_i}, t_{P_i}, t_{K_i}$ are the periods of fertilizing; and $i = 1, 2, \cdots, 9$ are the index numbers for indicating the phenophases.

The factors considered in the general form are regularly observed variables at the farm's experimental station and the standard observations made by the national meteorological network. One can also include other factors into the model [64]. The above general form is analyzed for various aspects of the modeling given below. All models are obtained using the single-layered combinatorial algorithm.

A special aspect of this modeling is that the described model is replaced by a selection of partial models, each of which is either linear or a quadratic polynomial with small number of arguments. The action of the remaining factors is averaged and used in representing the precision of the model. The importance of this approach is indicated by the demands of the problems under study and also of the availablity of measuring factors in determining the productivity of crops. The major advantage is that it is simple and considered carrying experiments in accordance with the theory of planning experiments.

Modeling of wheat harvest (y) as a function of the time of planting (C) and the rate of seeding (H)

The data used in this modeling is obtained during 1967 to 1971 for five years of 525 experiments. The seeding rate was maintained as $H = 1, 2, \cdots, 7$ million seeds per hectare and the planting time C was from September 25th to October 25th with three variants of water use; without sprinkling, watering of 1200 m^3/ hectare, and watering plus vegetational sprinkling assuring soil moisture not lower than 80% of the least capacity.

The combinatorial algorithm is used with the combined criterion of minimum-bias plus regularity. All 525 data points are used in the analysis. The best model obtained is given below:

$$y = 24.7 + 4.9H - 0.58H^2 + 0.03HC - 0.0056C^2 \tag{6.76}$$

The quadratic character of the model indicates that the model achieves the maximum value of the optimal norm of planting and periods of planting guaranteeing a maximum harvest. However, the large bias term and the mean squared error (ε) of 33% on all points indicate the influence of other factors like the water regime and the climatic factors that are not part of the model.

To determine the degree of the effect of the water regime, the experimental data points are divided into three sets of 175 points each and partial models are identified for three separate water systems:

$$\begin{aligned} y_1 &= 6.32H - 0.51H^2 + 6.96C - 0.17HC - 1.09C^2, \ \varepsilon_1 = 41\%, \\ y_2 &= 6.89H - 0.70H^2 + 15.01C - 0.22HC - 2.46C^2, \ \varepsilon_2 = 31\%, \\ y_3 &= 11.88H - 0.98H^2 + 12.37C - 0.49HC - 1.83C^2, \ \varepsilon_3 = 27\%, \end{aligned} \tag{6.77}$$

where the model for y_1 is constructed using the data on without sprinkling; the model for y_2 is with one watering in the fall; and the model for y_3 is under conditions with optimal water supply as suggested by agrotechnology. One can see that the accuracy of models increases from 41% to 27%.

The optimal norm of seeding H and periods of planting C have achieved the above model for maximizing the harvest with

$$\begin{aligned} H_1 &= 5.73; \quad C_1 = 2.67; \\ H_2 &= 5.16; \quad C_2 = 3.43; \\ H_3 &= 5.37; \quad C_3 = 2.64. \end{aligned} \tag{6.78}$$

Here the rate of seeding is given in millions of seeds per hectare and the time of planting in days counting from September 1st multiplied by 0.1. In this way, the optimal period of planting for wheat without sprinkling and with an optimal water supply is September 26 to 27, and for wheat with one initial fall watering is October 4 to 5. The maximum harvests of winter wheat for different watering systems are

$$y_{1max} = 27.66; \; y_{2max} = 41.13; \; \text{and} \; y_{3max} = 48.29. \tag{6.79}$$

Using these values of maximum harvests for various watering systems, one can construct a function $y_{max} = f(W)$, where W is the quantity of water in thousands of m^3 per hectare lost in sprinkling. The following water losses are accepted for each case as (i) without sprinkling, 0.1 (moisture supply to assure germination); (ii) one initial watering, 1.2 thousand m^3 per hectare; and (iii) total losses to maintain soil dampness not lower than 80% of the moisture capacity, 3.0 thousand m^3 per hectare. The second ordered function is considered and the estimated model is obtained as

$$y_{max} = 26.02 + 16.14W - 2.90W^2. \tag{6.80}$$

This model is useful in planning the yield depending on the water supply of the sprinkling system, and also in determining the specific losses in obtaining the quantity of agricultural production.

Modeling of wheat harvest (y) as a function of the rate of fertilizing (N, P) and the water supply (W)

The experimental data observed during the years 1969 to 1972 are used; altogether 168 experiments with 14 different mineral feedings and three types of water supplies were conducted. The mineral feeding consists of nitrogen (N = 30, 60, 90, 120, 150 acting units), and phosphorus (P = 30, 60, 90, 120 acting units).

The data points are divided into three sets; each set consists of 56 points for each water supply. The full description of second-order is used for each water system and the best models are selected according to the combined criterion of "minimum-bias plus regularity" of the form:

$$\begin{aligned} y_1 &= 18.1 + 1.2P - 0.069P^2 + 0.98N, \; c2 = 49.3\%, \\ y_2 &= 23.8 + 1.76P - 0.105P^2 + 1.12N, \; c2 = 9.8\%, \\ y_3 &= 27.0 + 2.24P - 0.115P^2 + 1.18N, \; c2 = 12.2\%. \end{aligned} \tag{6.81}$$

The lower accuracy of the model of y_1 with experiments without watering indicates a strong influence of other factors which are not taken into consideration in the model. Among them are the fluctuation of natural soil moisture and other controlling and disturbing actions whose influence is reduced because of improved water delivery in the other models. The increase in the values of the coefficients in response to the improved water delivery indicates the increase of intensiveness of the mineral feedings introduced into the soil.

Considering that the harvest models for three watering systems have the same structure of the form,

$$y_i = a_{i0} + a_{i1}P - a_{i2}P^2 + a_{i3}N, \quad i = 1,2,3, \tag{6.82}$$

and that the effect of water supply on the production is nonlinear, a second-ordered function is selected for constructing the dependence of coefficients of the above model on the water supply. The general form of such a dependence is given as

$$a_{ij} = b_{ij} + c_{ij}W + d_{ij}W^2, \quad i = 1,2,3; \quad j = 0,1,2,3. \tag{6.83}$$

After finding the dependencies of all coefficients a_{ij} as $f(W)$, the general model of the wheat production taking into account the water supply and fertilizers used is found as

$$\begin{aligned} y = {} & (18.1 + 6.7W - 1.17W^2) + (1.13 + 0.62W - 0.84W^2)P \\ & -(0.065 - 0.05W - 0.011W^2)P^2 + (0.96 + 0.17W - 0.032W^2)N. \end{aligned} \tag{6.84}$$

This model can be used for predicting the harvest depending the nitrogen and phosphorus additives and also on the water delivery system. One can study the effectiveness of using water resources for all periods of the watering season under the conditions of deficient water supplies using the analogous models for crops in plant rotation as functions of fertilizers.

Modeling of wheat harvest as a function of various climatic factors

The data corresponding to the 60 field experiments conducted during 1973 to 1975 are used; the experiments are aimed to define the effects on the harvest of the amount of seed planted and the quantity of nitrogen fertilizers used with an optimal water supply. The meteorological data are obtained from the meteorological station located within a distance of one kilometer from the experimental fields.

(i) *Models based on the duration of phenophases.* To determine critical phases of development in winter wheat and dominating meteorological factors, the models are constructed by taking into account the dynamics of separate meteorological factors in the main phases of vegetation.

Keeping in view the limited quantity of data points, including those obtained under conditions of good wintering, the models for phenophases are synthesized. Linear models which characterize the dependence of the main quantitative characteristics of the harvest are considered on the duration of development phases in days:

$$\begin{aligned} y_1 &= 4.75t_5 - 1.96t_6 + 1.29t_7 - 2.83t_8 - 2.78t_9, \quad \varepsilon_1 = 15.6\%, \\ y_2 &= 75.8 + 93.9t_7 - 25.6t_8 - 45.1t_9, \quad \varepsilon_2 = 9.1\%, \\ y_3 &= 37.5 - 0.076t_9, \quad \varepsilon_3 = 4.8\%, \end{aligned} \tag{6.85}$$

where y_1 is the harvest in 100 kg/ hectare; y_2 is the weight of thousand seeds in grams; and y_3 is the number of seeds in a head.

The selected models serve only for characterizing the tendency toward a decrease in wheat harvest (y_1) with an increase of the length of the milk (t_6), waxy (t_8), and complete ripeness (t_9), at the expense of a decrease in kernel weight (y_2). Similarly, it is better to mention that an increase of harvest (y_1) with the milky wax degree of ripeness (t_7) occurs at the expense of an increase in kernel weight (y_2). This is because of the positive coefficients of t_7 in the first two models for y_1 and y_2.

(ii) *Models based on the number of sunshine hours.* The purpose of extending the above modeling is to find the climatic factors that exert more influence than the duration of the phenophase on the production. Linear models are synthesized depending on the amount of sunshine S during the different phases. The optimal models are obtained as below:

$$\begin{aligned} y_1 &= 3.83S_5 - 0.313S_6 - 3.65S_7 + 0.88S_9, \quad \varepsilon_1 = 10.5\%, \\ y_2 &= 42.8 - 0.313S_5 - 0.685S_7 + 1.114S_9, \quad \varepsilon_2 = 3.2\%, \\ y_3 &= 1.51S_5 + 0.056S_6 + 0.474S_9, \quad \varepsilon_3 = 11.1\%. \end{aligned} \tag{6.86}$$

According to the length of the phases and the amount of sunshine from the above models, one can state that the amount of sunshine is the essential element in the harvest. Sunshine hours during wheat head formation S_5 affect the quantity of kernels in the head y_3. It is also correct to state that an increase in S_5 means a slight decrease in the weight of the 1000 kernels y_2.

(iii) *Models based on the temperature.* Models are constructed for the sum of effective temperatures for the phases in the winter wheat development. Additional arguments are included into the models for determining the effect of the early phases of development in wheat production. The resulting models obtained for y_1 and y_2 have the form

$$\begin{aligned} y_1 &= 6.35L_5 + 1.76t_5^\circ - 0.07T_6^\circ + 0.88T_9^\circ, \quad \varepsilon_1 = 14.6\%, \\ y_2 &= 2.92t_5^\circ - 0.08T_5^\circ + 0.445T_7^\circ, \quad \varepsilon_2 = 2.5\%, \end{aligned} \tag{6.87}$$

where t_i° and T_i° are the average daily air temperature and the sum of effective midday temperatures during the ith phenophase, correspondingly; L_5 is the degree of development of leafy surface during the phase of head formation (for the April 9th).

These models take the second place to the models depending on the duration of sunshine hours. Here also the strong dependence of productivity on the conditions of head formation is evident.

(iv) *Models based on the water evaporation.* The evaporation capacity of the atmosphere is an important indicator of the conditions of an agricultural production. One can use various meteorological indicators such as the relative and absolute humidity of the air, its temperature, atmospheric pressure, air dryness, evaporation from the water surface, and various combinations of these and other indicators in constructing the models (also unified into one model with the variable coefficients using the inductive learning algorithms) for climate zones with appropriate adaptation for each irrigated field.

Here the variables concerning the total and midday evaporation from the water surface of the standard evaporation tank are used in models for each phase of winter wheat development. The following models are obtained:

$$\begin{aligned} y_1 &= 12.3E_5 - 8.71E_7 - 8.83E_9, \quad \varepsilon_1 = 10.1\%, \\ y_2 &= 113.5 - 0.464E_5 - 3.09E_6 - 10.3E_9, \quad \varepsilon_2 = 2.1\%, \\ y_3 &= 3.54E_5 - 0.368E_9, \quad \varepsilon_3 = 9.3\%. \end{aligned} \tag{6.88}$$

According to the structures of these models, these are analogous to the models that represent the dependence of the harvest using the number of sunshine hours. These models also trace the relationship with E_5 during the head formation. One can draw the conclusion that the number of kernels is established during the period of head formation, and that an increase in evaporation causes an increase in the number of kernels and better harvest.

Models with respect to the end of the head formation phase.

From the above modeling results, it is clear that climatic factors affect the wheat harvest and its characteristics mainly prior to the stage of head formation. This makes it possible to build up a good prediction model for the wheat harvests for the end of the head formation phase using the amount of evaporation from the water sutrface (E_5), the number of sunshine hours (S_5), the duration of the phenophase (t_5), and the relative air humidity (e_5).

The linear models obtained are as follows:

$$\begin{aligned} y_1 &= 5.91S_5 - 2.06t_5 + 11.13e_5 - 3.65E_5, \ \varepsilon_1 = 10.1\%, \\ y_2 &= -0.56S_5 + 0.12t_5 + 16.19e_5, \ \varepsilon_2 = 2.6\%, \\ y_3 &= 0.51t_5 + 6.26e_5, \ \varepsilon_3 = 5.1\%. \end{aligned} \tag{6.89}$$

The models can be used for the evaluation of agricultiral-climatic resources of a specific irrigation system for determining the possibilities of raising a particular crop for which the model is identified. In this case, the observed data can be averaged to the available series of years. The same models can be used for predicting the future harvest with known factors of crop development for the end of the head formation phase or earlier—for example, after wintering with a favorable prediction of the sunshine, evaporation of water surface, and length of the head formation phase.

It is necessary to supplement with certain appropriate biological indicators for a more objective prediction of wheat harvest after the beginning of spring growth. The variables such as the degree of development of the leafy surface L and the quantity of dry matter V at the beginning and end of the head formation are used in constructing the second-order polynomial models; the optimal models have the form

$$\begin{aligned} y_1 &= 17.6L_4 + 0.39V_4 - 1.29L_4^2, \ \varepsilon_1 = 6.7\%, \\ y_2 &= 77.5 + 0.006V_4^2 + 3.19L_4 - 5.85L_4V_4, \ \varepsilon_2 = 4.7\%, \\ y_3 &= 37.6 - 0.00066V_4^2 + 0.0065L_4^2, \ \varepsilon_3 = 4.7\%; \end{aligned} \tag{6.90}$$

$$\begin{aligned} y_1 &= 40.55 - 0.0023V_5 + 0.086V_5L_5 - 0.325L_5^2, \ \varepsilon_1 = 13.6\%, \\ y_2 &= 43.17 - 0.06V_5 + 1.12L_5, \ \varepsilon_2 = 6.7\%, \\ y_3 &= 37.69 - 0.00003V_5^2 + 0.0014V_5L_5, \ \varepsilon_3 = 4.9\%. \end{aligned} \tag{6.91}$$

The latter which is a more precise model of the harvest specifies the dependence of wheat production on the degree of development of the leafy surface at the end of the stalk formation phase. The dependence of the production on the quantity of dry matter and the amount of leaf surface at the end of the head phase decreases because of biological changes occurring in the head stage.

This means that it is sufficient to indicate the amount of leaf surface and the amount of dry matter at the end of the the stalk formation phase for building up the prediction models. The linear models that consider the climatic factors during the head stage have the form

$$\begin{aligned} y_1 &= -77.5 + 2.89L_4 + 3.07S_5 - 0.032t_5 - 0.139e_5 + 17.19E_5, \ \varepsilon_1 = 6.8\%, \\ y_2 &= 0.048V_4 - 0.503S_5 + 0.079L_5 + 16.05E_5, \ \varepsilon_2 = 2.5\%, \\ y_3 &= 32.94 - 0.061V_4 + 0.563e_5, \ \varepsilon_3 = 5.1\%. \end{aligned} \tag{6.92}$$

The inclusion of the variables of biological indicators of development at the end of stalk formation permit a higher degree of accuracy in the models. The variable corresponding to the amount of dry matter is discarded by the computer in the model of wheat harvest (y_1) and the variable corresponding to the surface of green foliage is left in the model.

The above selected models can be used effectively for predicting the wheat harvest since the measurement of foliage area can be carried out with high precision by indirect methods—for example, with the aid of aerial photography.

The studies on agricultural productions with irrigation are related to a large number of meteorological and agro-technical factors. The majority of the factors are cross-correlated and the construction of one universal model for studying various aspects of agricultural production is inadequate. This study indicates that as a result of directed selection of partial models constructed for a large group of arguments, it is possible to get models of production corresponding to certain practical considerations.

Example 9. Adaptation of yield models to crop regionalization.

The principal sources of model error in the above example are the limited usage of the variables. Here is a trial to further extend modeling of the yield of agricultural crops in adapting to regionalization. The organization of the modeling process objectively takes into account the expansion of input variables. The modeling errors depend on the noise in the experimental data, assumptions made in the design of the complete model (i.e., the maximum complexity of the model and correspondence to the physical process), the method used to divide the experimental data into sets, the choice of control intervals (averaging periods of input variables), the criterion of model selection, etc.

A particular feature of crop rotation-structure planning to solve the different groups of problems in adapting yield models to crop regionalization is addressed in the work of Khomovnenko [75] with the exclusive use of predicted input variables. The intermediate yield models are developed with the use of calender (monthly, seasonal, or average of several months) values of climatic factors.

For example, considering the modeling of wheat harvest as $y = f(H, C)$, 20 more variables which characterize the climatic conditions are supplemented to the original list of variables H and C; average daily temperatures $t_{(i-j)}$, precipitation $R_{(i-j)}$, sunshine hours $S_{(i-j)}$, and lack of air humidity $e_{(i-j)}$ are summed over the period from ith to jth month and used as an input in the model (when $i = j$, only one subscript is used).

The meteorological data used in modeling is collected during 1945 to 1974 by one of the agricultural weather stations that is located about 100 km away from the experimental farm fields. This a situation in which time and spatial extrapolation of input variables could be used.

The data is divided into four sets: training (A), testing (B), and two examin sets (C_1, C_2, and $C = C_1 \cup C_2$). The second examin set is produced from the sequence of meteorological data, measured in the region for which the yield model is to be used. The regularity criterion $\Delta(A \cup B)$ is the squared error measured on the training and test sets, $\Delta^2(A \cup B) = \sum_{i \in A \cup B}(y - \hat{y})_i^2 < 0.02$; $\Delta(C_1)$ is the squared error measured on the first examin set, and $\Delta(C_2)$ is the squared error measured on the second examin set. The model accuracy is improved further by an appropriate choice of averaging intervals of the input variables—for example, by making these intervals equal to the development phases of the crop, and the model adaptive with respect to the climatic conditions of the irrigation system. To adapt the model, the step-by-step correction of equations is used by excluding the extremum test data points. The algorithm with orthogonalized complete description (a generalized algorithm) is used [112] (also refer to Chapter 2).

In the first case, the criteria used in estimating the potential effectiveness are the average annual yield ($\bar{y}$) and the mean square deviation $\Delta(A \cup B)$. A complete sifting of all intermediate models yields the model

$$y = 53.41 - 0.000203e_{(5)}\, t^{\circ}_{(6)} - 4588.8\frac{1}{H\, e_{(3-5)}} + 0.048R_{(3-5)}, \tag{6.93}$$

which is the optimal yield model for winter wheat; the average annual yield computed from this model is $\bar{y}$ = 23.25 100- kg./hectare and the error $\Delta(C) = 0.004$. In computing the criterion error, the variable y is replaced by $\bar{y}$.

Similarly, in the second case the yield models used for annual planning of water distribution have different lists of input variables that characterize watering and fertilization schedules and climatic factors. The models obtained are as:

$$\begin{aligned} y_1 &= 69.1 - 0.0000023\frac{S^2_{(5)}}{t^{\circ}_{(10)}} + 551.1\frac{N}{t^{\circ}_{(4)}} - 0.0023\frac{t^{\circ}_{(5)}\, t^{\circ}_{(3-4)}}{P\, t^{\circ}_{(4)}} - 6.0\frac{S_{(4)}}{t^{\circ}_{(4)}}, \\ y_2 &= 30.9 + 0.0000079R_{(12-2)}\, N\, S_{(4)}\, t^{\circ}_{(5)} - 57.7\frac{1}{N\, S_{(3)}} - 0.94\frac{1}{R_{(11-2)}}, \\ y_3 &= 31.91 + 0.007N\, S_{(4)} + 0.33N\, R_{(3-5)} - 0.022\frac{1}{N^2\, R_{(3-5)}} + 1.72\frac{1}{N^2\, S_{(3)}}. \end{aligned} \tag{6.94}$$

The models for y_2 and y_3 are obtained as optimal and acceptable solutions, but the intermediate models for y_1 do not yield any acceptable solutions due to the limited nature of the experimental series. However, in order to illustrate the potential effectiveness of the above selected models for y_1, y_2, and y_3, which characterize different irrigation schedules, the yields are simulated for a period of 30 years (1945 to 1974).

The results of the investigation of the models prove to be physically sound and agree with the experimental results conducted by various scientists. A further increase in accuracy can be achieved by optimizing the average intervals of the input variables.

6 MODELING OF SOLAR ACTIVITY

Model as a sum of trend and remainder

The random processes being modeled can be represented as the sum of trend and a remainder.

$$y(t) = Q(t) + q(t), \tag{6.95}$$

where $Q(t) = \frac{1}{T_m}\sum_{t=1}^{T_m} y(t)$ is the moving average about the center of the averaging interval and $q(t)$ is the remainder (called "anamoly" in meteorology). Both are modeled using one of the iductive learning algorithms. The process averaging interval T_m gradually increases until the sum of the prediction errors of the trend and of the remainder decreases. The global minimum of the sum is the optimal value of T_m.

Additive-multiplicative trend

Alternatively, more complex is the additive-multiplicative trend of the form

$$y(t) = Q_1(t) + Q_2(t)q_1(t) + Q_3(t)q_3(t), \tag{6.96}$$

where Q_1, Q_2, and Q_3 are polynomials in time with the dimensions of the moments of a random function—the mathematical expectation, the variance, and the third moment, respectively. Such trends are first of its kind in the literature to be used.

The limiting admissible prediction time with respect to a trend is estimated on the basis of the averaging interval; $T_{tr} \leq T_m$, where T_{tr} is the averaging interval of trend. Usually this interval is sufficiently large. For autoregression single factor models of the form $q_t = f(q_{t-1}, q_{t-2}, \cdots)$ the limiting admissible prediction time of the remainder in the case of a single-level prediction does not exceed the correlation interval $T_{rem} \leq T_k$, where T_{rem} is the averaging interval used in the remainder part.

To determine the limiting admissible prediction time for multifactor models of the form $q_t = f(q_{t-1}, q_{t-2}, \cdots, u_t, u_{t-1}, \cdots)$, it is necesssary to choose according to the larger of the autocorrelation and cross-correlation intervals.

Multistep prediction as a transient process

Long-range multistep prediction has many features of transient process. In the prediction region of a random process, the predicting model shifts from a steady state regime of continuous step-by-step renewal of information (regime of observation) into another steady-state regime for which no new information of the object is being fed in (regime of prediction). As in the theory of tracking systems of servomechanisms, the predicting model can be represented as the sum of two components—the trend and remainder.

In the interpolation interval, the prediction differs only slightly from the actual data since it is based on the minimization of the mean-square error of the residuals. In turn, in the prediction interval, it is convenient to represent the remainder in the form as a series

$$q(t) = q_1(t) + q_2(t) + q_3(t) + q_4(t), \tag{6.97}$$

where $q_1(t)$ is the exponential component of the remainder, $q_2(t)$ is the attenuating transient error, $q_3(t)$ is the nonattenuating component of the remainder, and $q_4(t)$ is the constant component.

After singling out the trend $Q(t)$ from the actual data, it is expected that harmonic components exist in the remainder with the same frequencies as in the trend because only nonattenuating oscillations of given frequencies are singled out in the trend. The presence of nonattenuating component $q_3(t)$ in the remainder is the result of imperfection of the algorithm for singling out the harmonic trend; in an ideal case $q_3(t) = 0$. In long-range predictions, there is a small steady state angular tracking error ($\theta =$ const) by which the prediction differs from the trend. As in the servosystems, this can be determined without integrating the differential equations of the model, but by substituting the steady state forcing function such as $Q(t) = \sin \omega t$ and the response, $y(t) = A \sin(\theta + \omega t)$ for determining the angle θ.

The analysis of steady-state regime is simple in a single-frequency trend and complicated in several frequencies. In particular, one needs to solve nonlinear equations to determine the tracking error without integrating the equations. But it is simpler to first integrate the equations of the predicting model and, thus, find the tracking error and all other components of the prediction. Once the angular tracking error is determined, then it is easier to determine the quality factor of the predicting model as a ratio of the angular frequency of the trend to the tracking error.

Correlation interval of the transient component

Correlation function of a typical non-steady-state process can be represented as a sum of steady-state and transient components:

$$y(t) = y_{ss}(t) + y_{tr}(t), \tag{6.98}$$

where

$$y_{ss}(t) = Q(t) + q_3(t) + q_4(t) \text{ and } y_{tr}(t) = q_1(t) + q_2(t).$$

The corresponding autocorrelation functions can be found for the three terms in the above equation as

$$A_{yy}(\tau) = A_{ss}(\tau) + A_{tr}(\tau). \tag{6.99}$$

The basic part of the steady-state component is the trend and the remainder is the transient component.

If we construct a correlation function of steady-state process, we will notice that the prediction time of a steady-state process does not exceed the length of its correlation interval; $T_{pr} \leq \tau_c$, where τ_c is the coherence time (refer chapter 2).

The time interval throughout which the correlation function exceeds the value of the delta function, $\delta = 0.05$. The value of $A_{yy} = 0.05$ is taken from the experiments of "tossing a coin." The correlation interval for a purely random process is given as less than unity—$\tau_c \leq 1$; this means that the prediction is impossible from observations. From these experiments, we can conclude that the steady-state component can achieve infinite prediction time and the transient component can achieve the prediction time of less than or equal to unity with the help of the best inductive learning algorithms.

An example of modeling solar activity

The basic steps in the algorithm for predicting oscillatory processes are listed as follows:

1. spectral analysis of the process,
2. singling out the harmonic trend and the remainder,
3. calculating the steady-state and transient components,
4. obtaining the optimal difference equation of the transient component using an inductive learning algorithm,
5. predicting the process as the sum of the steady-state and transient components, and
6. determining the accuracy of the prediction in case of prediction time equal to the correlation interval of transient and evaluating the results.

Here the problem of predicting solar activity characterized by Wolf numbers is considered. The data are taken for the years 1700 to 1978.

1. *Spectral analysis of the process.* The spectral analysis on the series of data reveals that it contains a sharp harmonic component with period T_0 = 11.2 years, multiple harmonics with periods $T_1 = (1/2)(11.2)$, $T_2 = (1/3)(11.2)$, $T_3 = (1/4)(11.2)$, and $T_4 = (1/5)(11.2)$ years, and also low frequency harmonics with periods $T_{-1} = 2(11.2)$, $T_{-2} = 3(11.2)$, $T_{-3} = 4(11.2)$, and $T_{-4} = 5(11.2)$ years.
2. *Singling out harmonic trend and the remainder.* The optimal harmonic trend is obtained by gradually increasing harmonic components until it leads to the lowering of the approximation error on the remainder. In this example, the trend obtained using all initial data has only a single harmonic with period T_0 = 11.2 years (frequency $\omega_0 = 2\pi/T_0 = 0.56$ radians/ year).

$$Q(t) = 49.9 - 8.3 \sin \frac{2\pi}{11.2} t + 25.4 \cos \frac{2\pi}{11.2} t \tag{6.100}$$

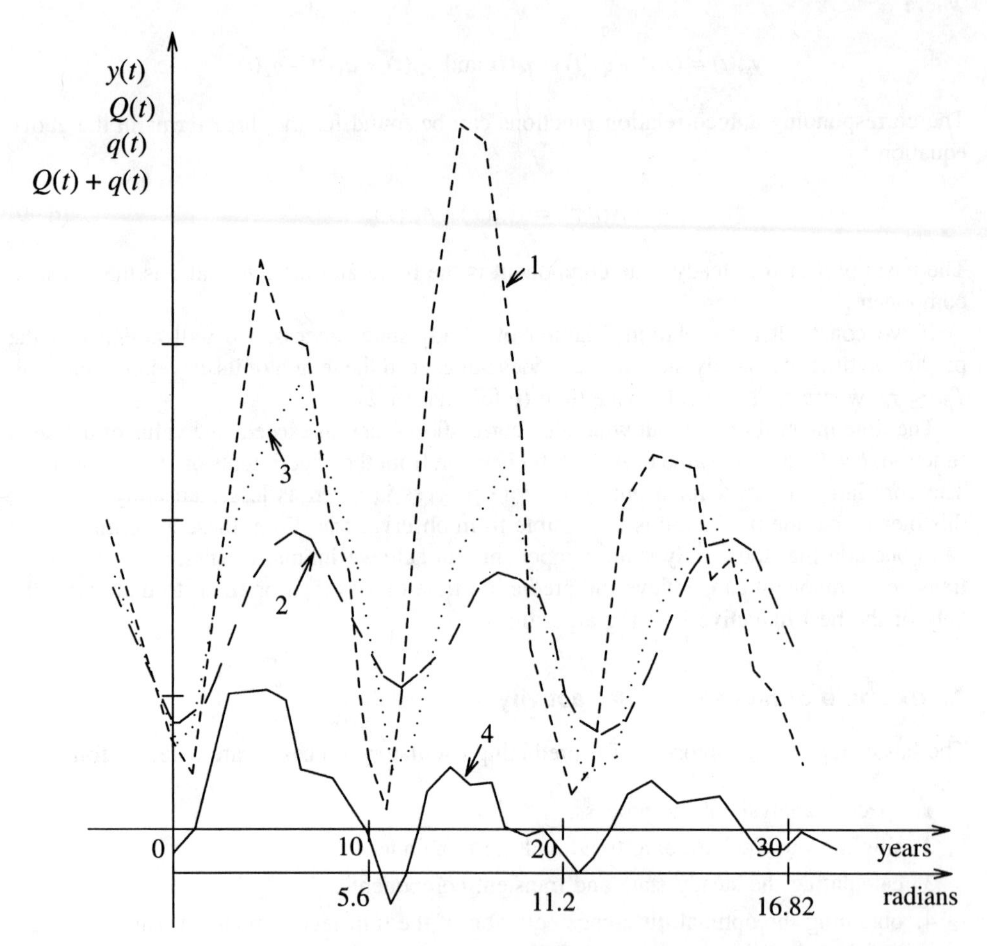

Figure 6.14. Processes being investigated. (1) the actual data of Wolf numbers, (2) the harmonic trend $Q(t)$, (3) the prediction with the trend and the estimated remainder $Q(t)+q(t)$, and (4) the output of the difference equation obtained with the inductive learning algorithm for the remainder data

It is aimed at preserving the possibility of checking the accuracy of long-range prediction over the course of 40 years with the starting year of 1943; i.e., where $t = 0$.

The residual data, which yields the remainder is calculated as the difference between the actual data and the trend. Figure 6.14 illustrates the trend $Q(t)$ and the remainder $q(t)$. As it is expected, the remainder is of an attenuating nature. The same frequency $\omega_0 = 2\pi/11.2$ radians/ year is found as it is used in the trend.

3. *Obtaining a difference model of the remainder.* Here combinatorial algorithm is used to obtain the optimal difference equation; the complete polynomial has 15 lagging arguments and the model is chosen in the plane of two criteria $\eta_{bs}^2 - i^2(N)$ with the constraint of $i^2 \leq 1.0$. The following model is obtained:

$$q(t) = 0.153q_{t-3} - 0.144q_{t-5} + 0.526q_{t-11} - 0.225q_{t-15}. \tag{6.101}$$

Figure 6.14 exhibits the approximation of the remainder by step-by-step integration of this equation.

4. *Singling out the nonattenuating harmonic part of the remainder.* It is proceeded by the singling out of the harmonic trend and obtained the following harmonic trend, as in step 2,

$$q_3(t) = 1.184 \sin \frac{2\pi}{11.2}t + 5.046 \cos \frac{2\pi}{11.2}t, \tag{6.102}$$

the amplitude of which is $A_3 = \sqrt(1.184^2 + 5.046^2) = 5.183$. A_1 and A_2 are the amplitudes of the harmonic part of the trend and the remainder, correspondingly. This enables us to find the angular tracking error $\theta = 0.061$ radians. The quality factor of the predicting model is computed as $G = \omega/\theta = 0.56/0.061 = 9$ year^{-1}.

5. *Determining the constant component of the remainder.* This is estimated as $q_4(t) = 2.329$ = const by using the least squares technique.

6. *Predicting the solar activity.* The equations are obtained for the trend $Q(t)$ and the remainder $q(t)$ as well as for all four of its components $q_1(t), q_2(t), q_3(t)$, and $q_4(t)$. The difference equation for the remainder gives the step-by-step predictions for it. The sum of the trend and the remainder gives the single-level prediction of solar activity without improving the prediction result.

   In short-range predictions with the prediction interval of $T_{pr} = 11$ years, the accuracy of prediction is 0.64 (without adaptation of the degree of stability) and in long-range predictions with $T_{pr} = 33$ years, it is 1.54.

Correlation function of the transient component of the remainder

The transient component of the remainder is defined as $y_{tr} = q_1(t) + q_2(t)$ and its correlation function is calculated from

$$A_{tr}(\tau) = \sum_{i=1}^{N-\tau} q_i q_{i+\tau}, \quad \tau = 1, 2, \cdots, N. \tag{6.103}$$

All values of $A_{tr}(\tau)$ are normalized with respect to the maximum value and construct the correlational graph. The coherence time obtained is nine years, confirming the prediction interval on which it is possible to increase the accuracy considerably by optimal choice of the degree of the predicting model.

Further analysis on the problem is based on the theory of servosystems with the analysis of the dynamic equations of a tracking system [52], and we leave it to the reader. However, in long-range predictions it makes no sense to increase the time of transient process of the remainder beyond the correlation interval which is approximately three time constants of τ_c, according to the concepts of the information theory.

Two-level algorithm

Alternatively, the modeling of solar activity is conducted through the two-level algorithm of multilevel objective analysis using the balance criterion. The above examined optimization (structure and estimates of the predicting models) is performed for both levels as seasonal and annual predictions. With regard to the compatibility of the predictions, the balance criterion makes it possible to raise the prediction time of seasonal predictions to the prediction time of annual predictions.

Chapter 7
Inductive and Deductive Networks

Mathematical literature reveals that the number of neural network structures, concepts, methods, and their applications have been well known in neural modeling literature for sometime. It started with the work of McCulloch and Pitts [93], who considered the brain a computer consisting of well-defined computing elements, the neurons. Systems theoretic approaches to brain functioning are discussed in various disciplines like cybernetics, pattern recognition, artificial intelligence, biophysics, theoretical biology, mathematical psychology, control system sciences, and others. The concept of neural networks have been adopted to problem-solving studies related to various applied sciences and to studies on computer hardware implementations for parallel distributed processing and structures of non-von Neuman design.

In 1958 Rosenblatt gave the theoretical concept of "perceptron" based on the neural functioning [105]. The adaptive linear neuron element (adaline), which is based on the perceptron theory, was developed by Widrow and Hopf for pattern recognition at the start of the sixties [131]. It is popular for its use in various applications in signal processing and communications. The inductive learning technique called group method of data handling (GMDH) and which is based on the perceptron theory, was developed by Ivakhnenko during the sixties for system identification, modeling, and predictions of complex systems. Modified versions of these algorithms are used in several modeling applications. Since then, one will find the studies and developments on perceptron-based works in the United States as well as in other parts of the world [3], [26], [82].

There is rapid development in artificial neural network modeling, mainly in the direction of connectionism among the neural units in network structures and in adaptations of "learning" mechanisms. The techniques differ according to the mechanisms adapted in the networks. They are distinguished for making successive adjustments in connection strengths until the network performs a desired computation with certain accuracy. The least mean-square (LMS) technique that is used in adaline is one of the important contributions to the development of the perceptron theory. The back propagation learning technique has become well known during this decade [107]. It became very popular through the works of the PDP group who used it in the multilayered feed-forward networks for various problem-solving.

1 SELF-ORGANIZATION MECHANISM IN THE NETWORKS

Any artificial neural network consists of processing units. They can be of three types: input, output, and hidden or associative. The associative units are the communication links between input and output units. The main task of the network is to make a set of associations of the input patterns x with the output patterns y. When a new input pattern is added to the configuration, the association must be able to identify its output pattern. The units are connected to each other through connection weights; usually negative values are called inhibitory and positive ones, excitatory.

A process is said to undergo self-organization when identification or recognition categories emerge through the system's environment. The self-organization of knowledge is mainly formed in adaptation of the learning mechanism in the network structure [5], [8]. Self-organization in the network is considered while building up the connections among the processing units in the layers to represent discrete input and output items. Adaptive processes (interactions between state variables) are considered within the units.

Linear or nonlinear threshold functions are applied on the units for an additional activation of their outputs. A standard threshold function is a linear transfer function that is used for binary categorization of feature patterns. Nonlinear transfer functions such as sigmoid functions are used to transform the unit outputs. Threshold objective functions are used in the inductive networks as a special case to measure the objectivity of the unit and to decide whether to make the unit go "on" or "off." The strategy is that the units compete with each other and win the race. In the former case the output of the unit is transformed according to the threshold function and fed forward; whereas in the latter, the output of the unit is fed forward directly if it is "on" according to the threshold objective function. A state function is used to compute the capacity of each unit. Each unit is analyzed independently of the others. The next level of interaction comes from mutual connections between the units; the collective phenomenon is considered from loops of the network. Because of such connections, each unit depends on the state of many other units. Such a network structure can be switched over to self-organizing mode by using a statistical learning law. A learning law is used to connect a specific form of acquired change through the synaptic weights—one that connects present to past behavior in an adaptive fashion so that positive or negative outcomes of events serve as signals for something else. This law could be a mathematical function, such as an energy function that dissipates energy into the network or an error function that measures the output residual error.

A learning method follows a procedure that evaluates this function to make pseudorandom changes in the weight values, retaining those changes that result in improvements to obtain the optimum output response. Several different procedures have been developed based on the minimization of the average squared error of the unit output (least squares technique is the simplest and the most popular).

$$\varepsilon = \frac{1}{2}\sum_{j,i}(\hat{y} - y)_{j,i}{}^2, \tag{7.1}$$

where $\hat{y}_j$ is the estimated output of jth unit depending on a relationship, and y_j is the desired output of the ith example. Each unit has a continuous state function of their total input and the error measure is minimized by starting with any set of weights and updating each weight w by an amount proportional to $\partial\varepsilon/\partial w$ as $\delta w_{ij} = -\alpha\,\partial\varepsilon/\partial w_{ij}$, where α is a learning rate constant.

The ultimate goal of any learning procedure is to sweep through the whole set of associations and obtain a final set of weights in the direction that reduces the error function. This is realized in different forms of the networks [29], [77], [107], [131].

The statistical mechanism built in the network enables it to adapt itself to the examples of what it should be doing and to organize information within itself and, thereby, to learn. The collective computation of the overall process of self-organization helps in obtaining the optimum output response.

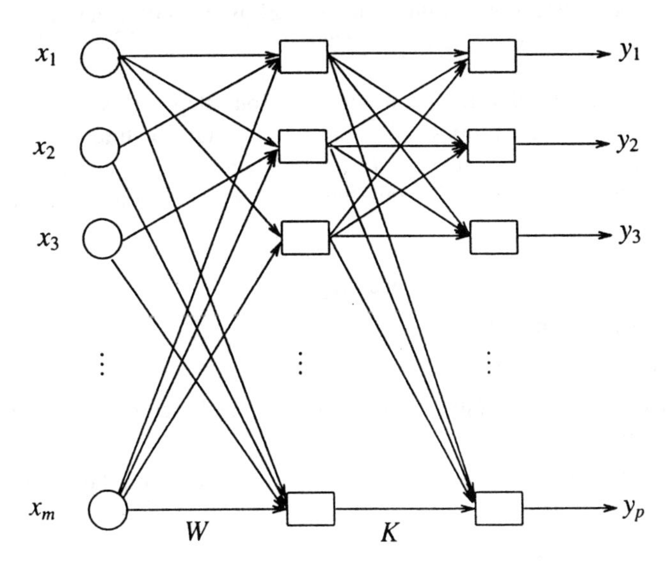

Figure 7.1. Unbounded feedforward network where X and Y are input/output vectors and W and K are weight matrices

This chapter presents differences and commonalities among inductive-based learning algorithms, deductive-based adaline, and backpropagation techniques. Multilayered inductive algorithm, adaline, backpropagation, and self-organization boolean logic techniques are considered here because of their commonality as parallel optimization algorithms in minimizing the output residual error and for their inductive and deductive approaches in dealing with the state functions. Self-organizing processes and criteria that help in obtaining the optimum output responses in the algorithms are explained through the collective computational approaches of these networks. The differences in empirical analyzing capabilities of the processing units are described. The relevance of local minima depends on various activating laws and heuristics used in the networks and knowledge embedded in the algorithms. This comparison study would be helpful in understanding the inductive learning mechanism compared with the standard neural techniques and in designing better and faster mechanisms for modeling and predictions of complex systems.

1.1 Some concepts, definitions, and tools

Let us consider a two-layered feedforward unbounded network with the matrices of connected weights of W at first layer and K at output layer (Figure 7.1). The functional algorithm is as follows:

Step 1. Initialize with random weights. Apply set of inputs and compute resulting outputs at each unit.

Step 2. Compare these outputs with the desired outputs. Find out the difference, square it, sum all of the squares. The object of training is to minimize this difference.

Step 3. Adjust each weight by a small random amount. If the adjustment helps in minimizing the differences, retain it; otherwise, return the weight to its previous value.

Step 4. Repeat from step 2 onward until the network is trained to the desired degree of minimization.

Any statistical learning algorithm follows these four steps. In working with such self-organization networks, one has to specify and build certain features of the network such as type of "input-output" processing, state function, threshold transfer function (decision function), and adopting technique. Overall, the networks can be comprised according to the following blocks:

1. "Black box" or "input-output" processing
   - batch processing
   - iterative processing
   - deductive approach (summation functions are based on the unbounded form of the network)
   - inductive approach (summation functions are based on the bounded form of the network)
   - multi-input single output
   - multi-input multi-output
2. Considering state functions
   - linear
   - nonlinear [29], [103], [132]
   - boolean logic
   - parallel
   - sequential
3. Activating with threshold transfer functions
   - linear threshold logic unit (TLU)
   - nonlinear or sigmoid
   - objective function (competitive threshold without transformations)
4. Adapting techniques
   - minimization of mean square error function (simplest case)
   - backpropagation of the output errors
   - minimizing an objective function ("simulated annealing")
   - front propagation of the output errors.

Some of the terminology given above are meant mainly for comparing self-organization networks. The term "deductive approach" is used for the network with unbounded connections and a full form of state function by including all input variables—contrary to the inductive approach that considers the randomly selected partial forms.

State functions

Unbounded structure considers the summation function with all input variables at each node:

$$s_j = \sum_{i=1}^{n} w_{ji}x_i + w_{j0}, \tag{7.2}$$

where n is the total number of input variables; s_j is the output of the node; x_i are the input terms; w_{j0} is the biased term, and w_{ji} are the connection weights.

Bounded structure considers the summation function with a partial list (r) of input variables:

$$s_j = \sum_{i=n1}^{n2} w_{ji} x_i + w_{j0}, \tag{7.3}$$

where $(n2 - n1) = r$ and $r+1$ is the number of the partial list of variables. A network with an unbounded/bounded structure with threshold logic function is called *deductive* because of its *apriori* fixedness. A network with a bounded structure and a threshold objective function is *inductive* because of its competitiveness among the units with randomly connected partial sets of inputs.

Parallel function is defined as the state function with the inputs from the previous layer or iteration $'j'$; whereas, the *sequential form* depends on the terms from the previous iteration and the past ones of the same iteration:

$$s_j = \sum_{k=1}^{r} w_{j,j-k}\, s_{j-k} + \sum_{i=n1}^{n2} w_{ji}\, x_i + w_{j0}. \tag{7.4}$$

The computationally sequential one takes more time and can be replaced by a parallel one if we appropriately choose input terms from the previous layer.

Transfer functions

These are used in the TLUs for activating the units. Various forms of transfer functions are used by scientists in various applications. The analytical characteristics of linear type TLUs are extensively studied by the group of Fokas [19]. Here is a brief listing of linear and nonlinear TLUs for an interested reader.

Linear type TLUs or discrete-event transformations. The following are widely used threshold logic functions in perceptron and other structures.

(i) *Majority rule*:

$$F(u) = \begin{cases} 1 & \text{if } u > 0 \\ 0 & \text{if } u \le 0; \end{cases}$$

(ii) *Signum function*:

$$F(u) = \begin{cases} 1 & \text{if } u > 0 \\ -1 & \text{if } u \le 0; \end{cases}$$

(iii) *Piecewise linear function*:

$$F(u) = \begin{cases} u & \text{if } u > 0 \\ 0 & \text{if } u \le 0; \end{cases}$$

(iv) *Signum-0-function*:

$$F(u) = \begin{cases} 1 & \text{if } u > 0 \\ 0 & \text{if } u = 0 \\ -1 & \text{if } u < 0; \quad \text{and} \end{cases}$$

(v) *Parity rule*:

$$F(u) = 1 \quad \text{if} \quad u \text{ is even}$$
$$\quad\; 0 \quad \text{if} \quad u \text{ is zero or odd .} \tag{7.5}$$

This is used in cellular automata and soliton automata [19]. In all the cases u is unit output.

Nonlinear or discrete analogue transformations

(i) Here are some forms of sigmoid function ($F(u) = \tanh u$) often used in various applications. They provide continuous monotonic mapping of the input; some map into the range of -1 and 1, and some into the range of 0 and 1:

$$F(u) = (1 + e^{-u'})^{-1};$$
$$F(u) = (e^{u'} - e^{-u'}) * (e^{u'} + e^{-u'})^{-1}, \quad F'(u) = (1 + F) * (1 - F);$$
$$F(u) = (1 - e^{-2u}) * (1 + e^{-2u})^{-1};$$
$$F(u) = 2 * (1 + e^{-2u})^{-1} - 1;$$
$$F(u) = \frac{1}{2}(1 + \tanh\, u'); \quad \text{and}$$
$$F(u) = (e^{u} - 1) * (e^{u} + 1)^{-1}, \tag{7.6}$$

where $u' = u * g$, in which g is the gain width. In all the nonlinear cases the curve has a characteristic sigmoidal shape that is symmetrical around the origin. For example, take the last one. When u is positive, the exponential exceeds unity and the function is positive, implying preference for growth. When u is negative, the exponential is less than unity and the function is negative, reflecting a tendency to retract. When u is zero, the function is zero, corresponding to a 50–50 chance of growth or retraction. For large positive values of u, the exponentials dominate each term and the expression approaches unity, corresponding to certain growth. For large negative values of u, the exponentials vanish and the expression approaches -1, corresponding to certain retraction. Here are some other types of transformations:

(ii) *Sine function*:

$$F(u) = \sin(u').$$

The use of this function leads to a generalized Fourier analysis.

(iii) *Parametric exponential function*:

$$F(u) = a + be^{-u'},$$

where a and b are the parameters;

(iv) *Gaussian function*:

$$F(u) = e^{\sum_i -\frac{1}{2}\frac{(u_i - \mu_i)^2}{\sigma_i^2}},$$

where μ is the mean value and σ is the covariance term; and

(v) *Green function*:

$$F(u) = \sum_{\alpha=1}^{n} c_\alpha G(u; t_\alpha), \tag{7.7}$$

where c_α are coefficients which are unknown, and t_α are parameters which are called centers in the radial case [101].

Threshold objective functions. There are various forms of threshold objective functions such as regularity, minimum-bias, balance-of-variables, and prediction criterion, used mainly in inductive networks. These are built up based on objectives like regularization, forecasting, finding physical law, obtained minimum biased model or the combination of two or three objectives which might vary from problem to problem.

2 NETWORK TECHNIQUES

The focus here is on the presentation of emperical analyzing capabilities of the networks; i.e., multilayered inductive technique, adaline, backpropagation, and self-organization boolean logic technique, to represent the input-output behavior of a system. The aspects considered are: basic functioning at unit-level based on these approaches connectivity of units for recognition and prediction type of problems.

2.1 Inductive technique

Suppose we have a sample of N observations, a set of input-output pairs (I_1, o_1), $(I_2, o_2), \cdots$, $(I_N, o_N)\,\epsilon N$, where N is a domain of certain data observations, and we have to train the network using these input-output pairs to solve an identification problem. For the given input $I_j(1 \leq j \leq N)$ of variables x corrupted by some noise is expected to reproduce the output o_j and to identify the physical laws, if any, embedded in the system. The prediction problem concerns the given input I_{N+1} that is expected to predict exactly the output o_{N+1} from a model of the domain that it has learned during the training.

In the inductive approaches, a general form of summation function is considered Kolmogorov-Gabor polynomial which is a discrete form of Volterra functional series [21]:

$$\begin{aligned} \hat{y} &= a_0 + \sum_{i=1}^{m} a_i x_i + \sum_{i=1}^{m}\sum_{j=1}^{m} a_{ij} x_i x_j + \sum_{i=1}^{m}\sum_{j=1}^{m}\sum_{k=1}^{m} a_{ijk} x_i x_j x_k + \cdots \\ &= a_0 + a_1 x_1 + a_2 x_2 + \cdots + a_{11} x_1^2 + a_{12} x_1 x_2 + \cdots + a_{111} x_1^3 + a_{112} x_1^2 x_2 + \cdots \\ &= a_0 + a_1 x_1 + a_2 x_2 + \cdots + a_{11} x_{m+1} + a_{12} x_{m+2} + \cdots + a_{mm} x_{m1}, \end{aligned} \tag{7.8}$$

where the estimated output is designated by $\hat{y}$, the external input vector x by $(x_1, x_2, \cdots, x_{m1})$, and a are the weights or coefficients. This is linear in parameters a and nonlinear in x. The nonlinear type functions were first introduced by the school of Widrow [132]. The input variables x could be independent variables or functional terms or finite difference terms; i.e., the function is either an algebraic equation, a finite difference equation, or an equation with mixed terms. The partial form of this function as a state functional is developed at each simulated unit and activated in parallel to build up the complexity.

Let us see the function at the unit level. Assume that unit n receives input variables; for instance, $(x_2, x_5) \subset x$—i.e., the state function of the unit is a partial function in a finite form of (7.8):

$$s_n = w_{n0} + w_{n1} x_2 + w_{n2}, x_5, \tag{7.9}$$

where w are the connection weights to the unit n. If there are $m1$ input variables and two of them are randomly fed at each unit, the network needs $C_{m1}^2 (= m1(m1-1)/2)$ units at first layer to generate such partial forms. If we denote y^p as the actual value and s_n^p as the estimated value of the output for the function being considered for pth observation, the output error is given by

$$e_n^p = s_n^p - y^p \qquad (p \epsilon N). \tag{7.10}$$

The total squared error at unit n is:

$$\varepsilon = \sum_{p\epsilon N} (e_n^p)^2. \tag{7.11}$$

This corresponds to the minimization of the averaged error ε in estimating the weights w. This is the least squares technique. The weights are computed using a specific training set at all units that are represented with different input arguments of $m1$. This is realized at each unit of the layered network structure.

Multilayered structure is a parallel bounded structure built up based on the connectionistic approach; information flows forward only. One of the important functions built into the structure is the ability to solve implicitly defined relational functionals. The units are determined as independent elements of the partial functionals; all values in the domain of the variables which satisfy the conditions expressed as equations are comprised of possible solutions [15], [29]. Each layer contains a group of units that are interconnected to the units in the next layer. The weights of the state functions generated at the units are estimated using a training set A which is a part of N. A threshold objective function is used to activate the units "on" or "off" in comparison with a testing set B which is another part of N. The unit outputs are fed forward as inputs to the next layer; i.e., the output of nth unit if it is in the domain of local threshold measure would become input to some other units in the next level. The process continues layer after layer. The estimated weights of the connected units are memorized in the local memory. A global minimum of the objective function would be achieved in a particular layer; this is guaranteed because of steepest descent in the output error with respect to the connection weights in the solution space, in which it is searched according to a specific objective by cross-validating the weights.

2.2 Adaline

Adaline is a single element structure with the threshold logic unit and variable connection strengths. It computes a weighted sum of activities of the inputs times the synaptic weights, including a bias element. It takes +1 or -1 as inputs. If the sum of the state function is greater than zero, output becomes +1, and if it is equal to or less than zero, output is -1; this is the threshold linear function. Recent literature reveals the use of sigmoid functions in these networks [98]. The complexity of the network is increased by adding the number of adalines, called "madaline," in parallel. For simplicity, the functions of the adaline are described here.

Function at Single Element

Let us consider adaline with m input units, whose output is designated by y and with external inputs $x_k (k = 1, \cdots, m)$. Denote the corresponding weights in the interconnections by w_k. Output is given by a general formula in the form of a summation function:

$$s = w_0 + \sum_k w_k x_k, \tag{7.12}$$

where w_0 is a bias term and the activation level of the unit output is

$$\mathcal{S} = f(s). \tag{7.13}$$

Given a specific input pattern x^p and the corresponding desired value of the output y^p, the output error is given by

$$e^p = s^p - y^p \qquad (p\epsilon N), \tag{7.14}$$

where N indicates the sample size. The total squared error on the sample is

$$\varepsilon = \sum_{p \epsilon N} (e^p)^2. \tag{7.15}$$

The problem corresponds to minimizing the averaged error ε for obtaining the optimum weights. This is computed for a specific sample of training set. This is realized in the iterative least mean-square (LMS) algorithm.

LMS algorithm or Widrow-Hopf delta rule

At each iteration the weight vector is updated as

$$w^{p+1} = w^p + \frac{\alpha}{|x^p|^2} e^p x^p, \tag{7.16}$$

where w^{p+1} is the next value of the weight vector; w^p is the present value of the weight vector; x^p is present pattern vector; e^p is the present error according to Equation (7.14) and $|x^p|^2$ equals the number of weights.

pth iteration:

$$\begin{aligned} e^p &= y^p - {x^p}^T w^p \\ \delta e^p &= \delta(y^p - {x^p}^T w^p) = -{x^p}^T \delta w^p, \end{aligned} \tag{7.17}$$

where T indicates transpose. From Equation (7.16) we can write

$$\delta w^p = w^{p+1} - w^p = \frac{\alpha}{|x^p|^2} e^p x^p. \tag{7.18}$$

This can be substituted in Equation (7.17) to deduce the following:

$$\begin{aligned} \delta e^p &= -{x^p}^T \frac{\alpha}{|x^p|^2} e^p x^p \\ &= -{x^p}^T x^p \frac{\alpha}{|x^p|^2} e^p \\ &= -\alpha e^p. \end{aligned} \tag{7.19}$$

The error is reduced by a factor of α as the weights are changed while holding the input pattern fixed. Adding a new input pattern starts the next adapt cycle. The next error is reduced by a factor α, and the process continues. The choice of α controls stability and speed of convergence. Stability requires that $2 < \alpha < 0$. A practical range for α is given as $1.0 > \alpha > 0.1$.

2.3 Back Propogation

Suppose we want to store a set of pattern vectors $x^p, p = 1, 2, \cdots, N$ by choosing the weights w in such a way that when we present the network with a new pattern vector x^i it will respond by producing one of the stored patterns which it resembles most closely. The general nature of the task of the feed-forward network is to make a set of associations of the input patterns x_k^p with the output patterns y_l^p. When the input layer units are put in the configuration x_k^p the output units should produce the corresponding y_l^p. S_i are denoted as activations of output units based on the threshold sigmoid function and z_j^p are those of the intermediate or hidden layer units.

(i) For a *2-layer net*, unit output is given by:

$$S_i^p = f(\sum_k w_{ik} x_k^p); \tag{7.20}$$

(ii) For a *3-layer net:*

$$S_i^p = f(\sum_j w_{ij} z_j^p) = f(\sum_j w_{ij} f(\sum_k w_{jk} f_{j.k}^p)). \tag{7.21}$$

In either case the connection weights w's are chosen so that $S_i^p = y_i^p$. This corresponds to the gradient minimization of the average of ε (7.22) for estimating the weights. The computational power of such a network depends on how many layers of units it has. If it has only two, it is quite limited; the reason is that it must discriminate solely on the basis of the linear combination of its inputs [95].

Learning by Evaluating Delta Rule

A way to iteratively compute the weights is based on gradually changing them so that the total squared-error decreases at each step:

$$\varepsilon = \frac{1}{2} \sum_{i,p} (S_i^p - y_i^p)^2. \tag{7.22}$$

This can be guaranteed by making the change in w proportional to the negative gradient ε with respect to w (sliding down hill in w space on the error surface ε).

$$\delta w_{ij} = -\alpha \frac{\partial \varepsilon}{\partial w_{ij}}, \tag{7.23}$$

where α is a learning rate constant of proportionality. This implies a gradient descent of the total error ε for the entire set p. This can be computed from Equations (7.20) or (7.21).

For a *2-layer net:*

$$\begin{aligned} \delta w_{ik} &= -\alpha \frac{\partial \varepsilon}{\partial w_{ik}} = -\alpha (\sum_{i,p} \frac{\partial \varepsilon}{\partial S_i} \frac{dS_i}{dx_i} \frac{\partial x_i}{\partial w_{ik}}) \\ &= \alpha \sum_{i,p} [y_i^p - f(s_i^p)] f'(s_i^p) x_k^p \equiv \alpha \sum_{i,p} \delta_i^p x_k^p, \end{aligned} \tag{7.24}$$

where $s_i^p = \sum_k w_{ik} x_k^p$ is the state function and $f'()$ is the derivative of the activation function $f()$ at the output unit i. This is called a generalized delta rule.

For a *3-layer net:* input patterns are replaced by z_j^p of the intermediate units.

$$\delta w_{ij} = \alpha \sum_p \delta_i^p z_j^p. \tag{7.25}$$

By using the chain rule the derivative of (7.21) is evaluated:

$$\delta w_{jk} = \alpha \sum_{i,p} \delta_i^p w_{ij} f'(s_j^p) x_k^p \equiv \sum_p \delta_i^p x_k^p. \tag{7.26}$$

This can be generalized to more layers. All the changes are simply expressed in terms of the auxiliary quantities δ_i^p, δ_j^p, $\cdots$ and the δ's for one layer are computed by simple recursions from those of the subsequent layer. This provides a training algorithm where the responses are fed forward and the errors are propagated back to compute the weight changes of layers from the output to the previous layers.

2.4 Self-organization boolean logic

In the context of principle of self-organization, it is interesting to look at a network of boolean operators (gates) which performs a task via learning by example scheme based on the work of Patarnello and Carnevali [99].

The *general problem of modeling* the boolean operator network is formulated as below. The system is considered for a boolean function like addition between two binary operands, each of L bits, which gives a result of the same length. It is provided with a number of examples of input values and the actual results. The system organizes its connections in order to minimize the mean-squared error on these examples between the actual and network results. Global optimization is achieved using simulated annealing based on the methods of statistical mechanics.

The *overall system* is formalized as follows. The network is configured by N_G gates and connections, where each gate has two inputs, an arbitrary number of outputs, and realizes one of the 16 possible boolean functions of two variables. The array $\Lambda_i\,(i = 1, 2, \cdots, N_G)$ with integer values between 1 and 16 indicates the operation implemented by ith gate. The experiments performed are chosen to organize the network in such a way that a gate can take input either from the input bits or from one of the preceding gates (the feedback is not allowed in the circuit). This means that $X_{i,j}^{(l,r)} = 0$ when $i \geq j$. The incidence matrices $X_{i,j}^{(l)}$ and $X_{i,j}^{(r)}$ represent the connections whose elements are zero except when gate j takes its left input from output gate i; then $X_{i,j}^{(l)} = 1$ and $X_{i,j}^{(r)} = 1$ is for right input. The output bits are connected randomly to any gate in the network.

The *training* is performed by identifying and correcting, for each example, a small subset of network connections which are considered responsible for the error. The problem is treated as a global optimization problem, without assigning adhoc rules to back propagate corrections on some nodes. The optimization is performed as a Monte Carlo procedure toward zero temperature (simulated annealing), where the energy or "cost" function ε of the system is the difference between the actual result and the calculated circuit output, averaged over the number of examples N_A fed to the system (chosen randomly at the beginning and kept fixed during the annealing).

$$\varepsilon(\Lambda, X) = \sum_{l=1}^{L} \varepsilon_l = \sum_{l=1}^{L} \frac{1}{N_A} \sum_{k=1}^{N_A} (q_{lk} - \hat{q}_{lk})^2, \tag{7.27}$$

where q_{lk} is the actual result of the lth bit in the kth example, $\hat{q}_{lk}\,(= f(\Lambda, X))$ is the estimated output of the circuit. Thus, ε is the average number of wrong bits for the examples used in the training for a random network of $\varepsilon_l \sim 1/2$.

The search for the optimal circuit is done over the possible choice for X by choosing Λ randomly at the beginning and keeping it fixed during the annealing procedure and performing the average. The optimization procedure proceeds to change the input connection of a gate according to the resulting energy change $\Delta\varepsilon$. If $\Delta\varepsilon < 0$, the change is accepted; otherwise, it is accepted with the probability $\exp(-\Delta\varepsilon/T)$, where T is the temperature—a control parameter which is slowly decreased to zero according to some suitable "annealing schedule." The "partition" function for the problem is considered as

$$Z = \sum_{(X)} \exp(-\varepsilon/T). \tag{7.28}$$

The *testing part* of the system is straight forward; given the optimal circuit obtained after the training procedure, its correctness is tested by evaluating the average error over the exhaustive set of the operations, in the specific case all possible additions of $2L$-bit integers, of which there are $N_B = 2^L.2^L$ $(= 2^{2L})$.

$$\Delta(B) = \sum_{l=1}^{L} \Delta_l(B) = \sum_{l=1}^{L} \frac{1}{N_B} \sum_{k=1}^{N_B} (q_{lk} - \hat{q}_{lk})^2, \tag{7.29}$$

where the quantities q_{lk} and $\hat{q}_{lk}$ are the same as those in the above formula.

The performance of the boolean network is understood from the quantities ε and $\Delta(B)$; the low values of the ε mean that the system is trained very well and the small values of $\Delta(B)$ mean that the system is able to generalize properly. So, usually one expects the existence of two regimes (discrimination and generalization) between which possibly a state of "confusion" takes place.

Experiments are shown [100] for different values N_G and N_A with $L = 8$. It is found that a typical learning procedure requires an annealing schedule with approximately 3.10^6 Monte Carlo steps per temperature, with temperature ranging from $T \sim O(1)$ down to $T \sim O(10^{-6})$ (roughly 70 temperatures for a total of $\sim$ 200 million steps). The schedule was slow enough to obtain correct results when N_G is large, and is redundantly long when N_G is small. The system achieved zero errors ($\Delta(B) = 0$ as well as $\varepsilon = 0$; i.e., it finds a rule for the addition) in some cases considered ($N_G = 160, N_A = 224$ or 480). In these cases, as not all possible two-input operators process information, one can consider the number of "effective" circuits, which turn out to be approximately 40.

According to the annealing schedule, reaching $T \sim 0$ implies that learning takes place as an ordering phenomenon. The studies conducted on small systems are promising. Knowing Z exactly, the thermodynamics of these systems are analyzed using the "specific heat," which is defined as

$$C_v = \frac{\partial \varepsilon}{\partial T}. \tag{7.30}$$

The "specific heat" C_v is a response function of the system and a differential quantity that indicates the amount of heat a system releases when the temperature is infinitesimally lowered. The interesting features of these studies are given below:

- for each problem there is a characteristic temperature such that C_v has a maximum value;
- the harder problem, the lower its characteristic temperature; and
- the sharpness of the maximum indicates the difficulty of the problem, and in very hard problems, the peak remains one of the singularities in large critical systems.

In these networks, the complexity of a given problem for generalization is architecture-dependent and can be measured by how many networks solve that problem from the trained circuits with a reasonably high probability. The occurrence of generalization and learning of a problem is an entropic effect and is directly related to the implementation of many different networks.

3 GENERALIZATION

Studies have shown that any unbounded network could be replaced by a bounded network according to the capacities and energy dissipations in their architectures [18]. Here two types of bounded network structures are considered.

One of the important functions built into the feedforward structure is the ability to solve implicitly defined relational functionals—the units of which are determined as independent elements of the partial functionals. All values in the domain of the variables that satisfy the conditions, expressed as equations are comprised of possible solutions.

3.1 Bounded with transformations

Let us assume that unit k receives variables. For instance, $(x_k, x_{k+1}) \subset X$; that is, the state function of the unit is a partial function in a finite form of (7.8):

$$s_k = w_{k0} + w_{k1}x_k + w_{k2}x_{k+1} = f(x_k, x_{k+1}), \tag{7.31}$$

where w are the connection weights to the unit k. There are n input variables and two of them are consecutively fed at each unit. There are n units at each layer. If we denote y^p as the actual value and s_k^p as the estimated value of the output for the function being considered for the pth observation, the output error is given by

$$e_k^p = s_k^p - y^p \qquad (p \epsilon O). \tag{7.32}$$

The total squared-error at unit k is:

$$\varepsilon^2 = \sum_{p \epsilon O} (e_k^p)^2. \tag{7.33}$$

This corresponds to the minimization of the averaged error ε in estimating the weights w. The output s_k is activated by a transfer function such as a sigmoid function $F(\,)$:

$$x_k' = F(s_k), \tag{7.34}$$

where x_k' is the activated output fed forward as an input to the next layer.

The schematic functional flow of the structure can be given as follows. Let us assume that there are n input variables of x including nonlinear terms fed in pairs at each unit of the first layer (Figure 7.2). There are n units at each layer. The state functions at the first layer are:

$$\begin{aligned} s_j &= w_{j0} + w_{j1}x_j + w_{j2}x_{j+1} && 1 \le j < n \\ &= w_{j0} + w_{j1}x_j + w_{j2}x_1 && j = n. \end{aligned} \tag{7.35}$$

These are formed in a fixed order of cyclic rotation. The outputs $s_j, (j = 1, 2, \cdots, n)$ are activated by a sigmoid function and fed forward to the second layer:

$$\begin{aligned} s_j' &= w_{j0}' + w_{j1}'x_j' + w_{j2}'x_{j+1}' && 1 \le j < n \\ &= w_{j0}' + w_{j1}'x_j' + w_{j2}'x_1' && j = n, \end{aligned} \tag{7.36}$$

where $x_j' = F(s_j), (j = 1, 2, \cdots, n)$ are the activated outputs of first layer and s_j' are the outputs of the second layer. The process is repeated at the third layer:

$$\begin{aligned} s_j'' &= w_{j0}'' + w_{j1}''x_j'' + w_{j2}''x_{j+1}'' && 1 \le j < n \\ &= w_{j0}'' + w_{j1}''x_j'' + w_{j2}''x_1'' && j = n, \end{aligned} \tag{7.37}$$

where $x_j'' = F(s_j'), (j = 1, 2, \cdots, n)$ are the activated outputs of the second layer fed forward to the third layer; s_j'' are the outputs; and x_j''' are the activated outputs of the third layer. The process goes on repetitively as the complexity of the state function increases as given

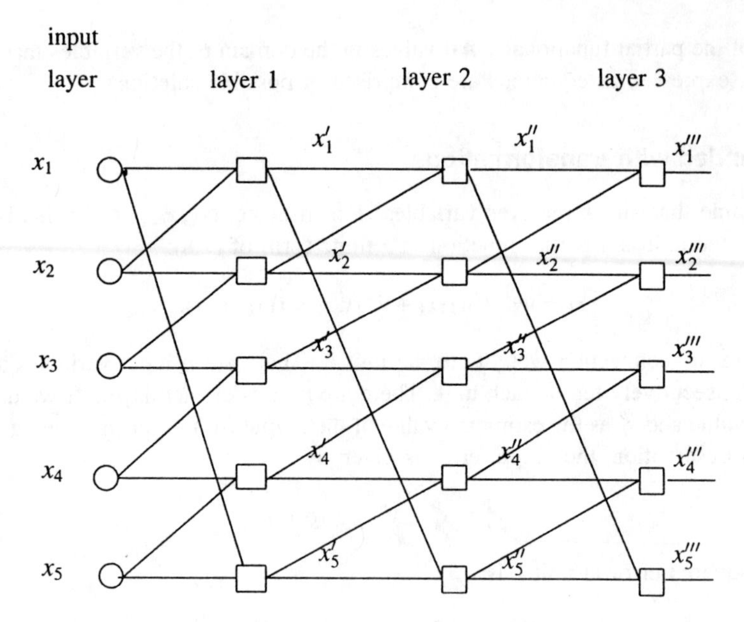

Figure 7.2. Bounded network structure with five input terms using a sigmoid function

below. For example, the state function at the unit k of the third layer with the activated output of x'''_k is described as:

$$\begin{aligned} x'''_k &= F(w''_{k0} + w''_{k1}x''_k + w''_{k2}x''_{k+1}) = F(f(x''_k, x''_{k+1})) \\ &= F(s''_k) \\ &= F(F(s'_k), F(s'_{k+1})) \\ &= F(F(F(s_k), F(s_{k+1})), F(F(s_{k+1}), F(s_{k+2}))) \qquad (7.38) \\ &= F(F(F(f(x_k, x_{k+1})), F(f(x_{k+1}, x_{k+2}))), F(F(f(x_{k+1}, x_{k+2})), F(f(x_{k+2}, x_{k+3})))), \end{aligned}$$

where s_k, s_{k+1}, s_{k+2} are the unit outputs at the first layer evaluated from the input variables of $(x_k, x_{k+1}, x_{k+2}, x_{k+3}) \subset X$. The optimal response according to the transformations is obtained through the connecting weights and is measured by using the standard average residual sum of squared error. This converges because of the gradient descent of the error by least-squares minimization and reduction in the energy dissipations of the network that is achieved by nonlinear mapping of the unit outputs through the threshold function, such as the sigmoid function.

3.2 Bounded with objective functions

Let us assume that unit j at the first layer receives variables. For instance, $(x_2, x_5) \subset X$; i.e., the state function of the unit is a partial function in a finite form of (7.8):

$$s_j = w_{j0} + w_{j1}x_2 + w_{j2}x_5 = f(x_2, x_5), \qquad (7.39)$$

where w are the connection weights to the unit j. If there are $m1$ input variables and two of them are randomly fed at each unit, the network needs $C^2_{m1} (= m1(m1 - 1)/2)$ units at the first layer to generate such partial forms. If we denote y^p as the actual value and s^p_j as the estimated value of the output for the function being considered for pth observation, the output error is given by (7.28). The total squared error at unit j is computed as in (7.29).

This corresponds to the minimization of the averaged error ε in estimating the weights w. Each layer contains a group of units, which are interconnected to the units in the next layer. The weights of the state functions generated at the units are estimated using a training set A which is a part of N. An objective function as a threshold is used to activate the units "on" or "off" in comparison with a testing set B which is another part of N. The unit outputs are fed forward as inputs to the next layer; i.e., the output of jth unit—in the domain of local threshold measure—would become input to some other units in the next level. The process continues layer after layer. The estimated weights of the connected units are memorized in the local memory. A global minimum of the objective function would be achieved in a particular layer; this is guaranteed because of steepest descent in the output error with respect to the connection weights in the solution space, in which it is searched according to a specific objective by cross-validating the weights.

The schematic functional flow of the structure can be described as follows. Let us assume that there are $m1$ input variables of x, including nonlinear terms fed in pairs randomly at each unit of the first layer. There are C^2_{m1} units in this layer that use the state functions of the form (7.35):

$$\begin{aligned} x'_n &= f(x_i, x_j) \\ &= w'_{n0} + w'_{n1}x_i + w'_{n2}x_j, \end{aligned} \tag{7.40}$$

where x'_n is the estimated output of unit n, $n = 1, 2, \cdots, C^2_{m1}$; $i, j = 1, 2, \cdots, m1$; $i \neq j$; and w' are the connecting weights. Outputs of $m2\,(\leq C^2_{m1})$ units are made "on" by the threshold function to pass on to the second layer as inputs. There are C^2_{m2} units in the second layer and state functions of the form (7.35) are considered:

$$\begin{aligned} x''_n &= f(x'_i, x'_j) \\ &= w''_{n0} + w''_{n1}x'_i + w''_{n2}x'_j, \end{aligned} \tag{7.41}$$

where x''_n is the estimated output, $n = 1, 2, \cdots, C^2_{m2}$; $i, j = 1, 2, \cdots, m2$; $i \neq j$; and w'' are the connecting weights. Outputs of $m3\,(\leq C^2_{m2})$ units are passed on to the third layer according to the threshold function. In the third layer C^2_{m3} units are used with the state functions of the form (7.35):

$$\begin{aligned} x'''_n &= f(x''_i, x''_j) \\ &= w'''_{n0} + w'''_{n1}x''_i + w'''_{n2}x''_j, \end{aligned} \tag{7.42}$$

where x'''_n is the estimated output, $n = 1, 2, \cdots, C^2_{m3}$; $i, j = 1, 2, \cdots, m3$; $i \neq j$; and w''' are the connecting weights. This provides an inductive learning algorithm which continues layer after layer and is stopped when one of the units achieves a global minimum on the objective measure. The state function of a unit in the third layer might be equivalent to the function of some original input variables of x:

$$\begin{aligned} x'''_n &= f(x''_i, x''_j) \\ &\equiv f(f(x'_g, x'_h), f(x'_k, x'_l)) \\ &\equiv f(f(f(x_p, x_q), f(x_p, x_r)), f(f(x_q, x_r), f(x_u, x_v))) \\ &\equiv f(x_p, x_q, x_r, x_u, x_v), \end{aligned} \tag{7.43}$$

where $(x''_i, x''_j) \subset X''$ and $(x'_g, x'_h, x'_k, x'_l) \subset X'$ are the estimated outputs from the second and first layers, respectively, and $(x_p, x_q, x_r, x_u, x_v) \subset X$ are from the input layer (Figure 7.3). A typical threshold objective function such as regularization is measured for its total squared

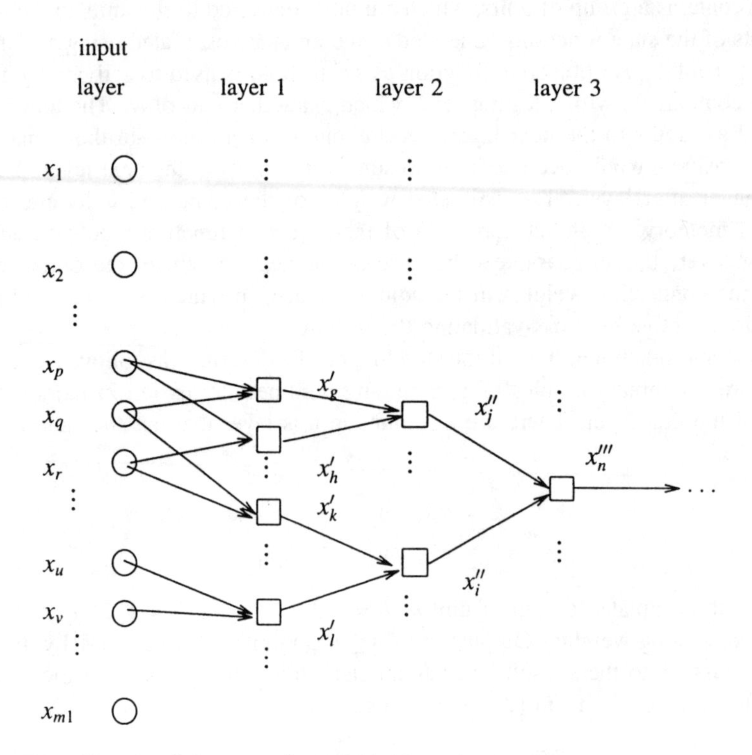

Figure 7.3. Functional flow to unit n of third layer in a multilayered inductive structure

error on testing set B as:

$$\Delta^2(B) = \sum_{s \epsilon B} (x_n''' - y)_s^2, \tag{7.44}$$

where y is the actual output value and x_n''' is the estimated output of unit n of the third layer. The optimal response according to the objective function is obtained through the connecting weights w, which are memorized at the units in the preceding layers [90]. Figure 7.4 illustrates the multilayered feedforward network structure with five input variables and with the selections of five at each layer.

4 COMPARISON AND SIMULATION RESULTS

The major difference among the networks is that the inductive technique uses a bounded network structure with all combinations of input pairs as it is trained and tested by scanning the measure of threshold objective function through the optimal connection weights. This type of structure is directly useful for modeling multi input single-output (MISO) systems, whereas adaline and backpropagation use an unbounded network structure to represent a model of the system as it is trained and tested through the unit transformations for its optimal connection weights. This type of structure is used for modeling multi-input multi-output (MIMO) systems.

Mechanisms shown in the generalized bounded network structures are easily worked out for any type of systems—MISO or MIMO. In adaline and backpropagation, input and

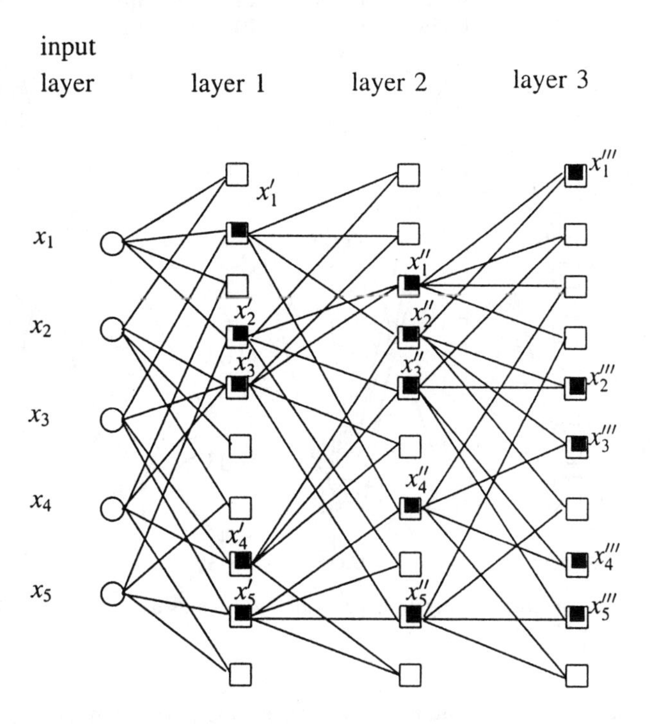

Figure 7.4. Feedforward multilayered inductive structure with $m1 = 5, m2 = 5$, and $m3 = 5$ using threshold objective function

output data are considered either $\{-1, +1\}$ or $\{0, 1\}$. In the inductive approach, input and output data are in discrete analogue form, but one can normalize data between $\{-1, +1\}$ or $\{0, 1\}$. The relevance of local minima depends on the complexity of the task on which the system is trained. The learning adaptations considered in the generalized networks differ in two ways: the way they activate and forward the unit outputs. In backpropagation the unit outputs are transformed and fed forward. The errors at the output layer are propagated back to compute the weight changes in the layers and in the inductive algorithm the outputs are fed forward based on a decision from the threshold function. The backpropagation handles the problem that gradient descent requires infinitesimally small steps to evaluate the output error and manages with one or two hidden layers. The adaline uses the LMS algorithm with its sample size in minimizing the error measure, whereas in the inductive algorithm it is done by using the least squares technique. The parameters within each unit of inductive network are estimated to minimize, on a training set of observations, the sum of squared errors of the fit of the unit to the final desired output.

The batchwise procedure of least squares technique sweeps through all the points of the measured data accumulating $\partial\varepsilon/\partial w$ before changing the weights. It is guaranteed to move in the direction of steepest descent. The online procedure updates the weights for each measured data point separately [131]. Sometimes this increases the total error ε, but by making the weight changes sufficiently small the total change in the weights after a complete sweep through all the measured points can be made to closely and arbitrarily approximate the steepest descent. The use of batchwise procedure in the unbounded networks requires more computer memory, whereas in the bounded networks such as multilayered inductive networks, this problem does not arise.

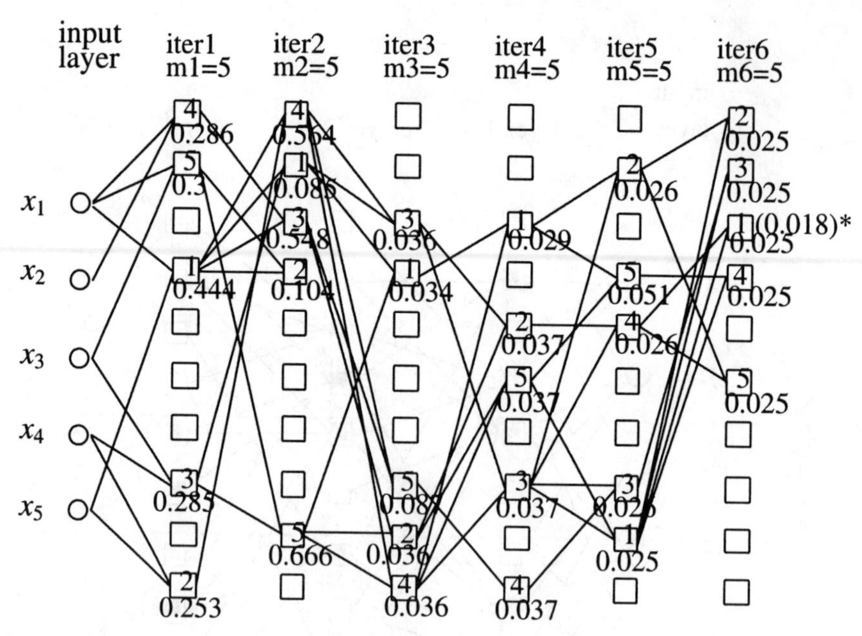

Figure 7.5. Bounded inductive network structure with linear inputs using threshold objective function (only activated links are shown)

Simulation experiments are conducted to compare the performances of inductive versus deductive networks by evaluating the output error as a learning law [91], [92]. Here the above general types of bounded network structures with inputs fed in pairs are considered. One is deductive network with sigmoid transfer function $\tanh(y * u_0)$, where u_0 is the gain factor and another is inductive network with threshold objective function which is a combined criterion ($c2$) of regularity and minimum-bias. As a special case, sinusoidal transformations are used for deductive network in one of the studies. In both the structures, the complexity of state function is increased layer by layer. The batchwise procedure of least squares technique is used in estimating the weights. Various randomly generated data and actual emperical data in the discrete analogue form in the range $\{-1, +1\}$ are used in these experiments. The network structures are unique in that they obtain optimal weights in their performances. Two examples for linear and nonlinear cases and another example on deductive network without any activations are discussed below:

(i) In linear case, the output data is generated from the equation:

$$y = 0.433 - 0.195 x_1 + 0.243 x_2 + 0.015 x_3 - 0.18 x_4 + \epsilon, \qquad (7.45)$$

where $x_1, \cdots, x_4$ are randomly generated input variables, y is the output variable, and ϵ is the noise added to the data.

(a) Five input variables $(x_1, x_2, \cdots, x_5)$ are fed to the inductive network through the input layer. The global measure is obtained at a unit in the sixth layer ($c2 = 0.0247$). The mean-square error of the unit is computed as 0.0183. Figure 7.5 shows the iterations of the self-organization network (not all links are shown for clarity). The values of $c2$ are given at each node.

(b) The same input and output data are used for the deductive network; unit outputs are activated by sigmoid function. It converges to global minimum at a unit in the third layer. The residual mean-square error (MSE) of the unit is 0.101.

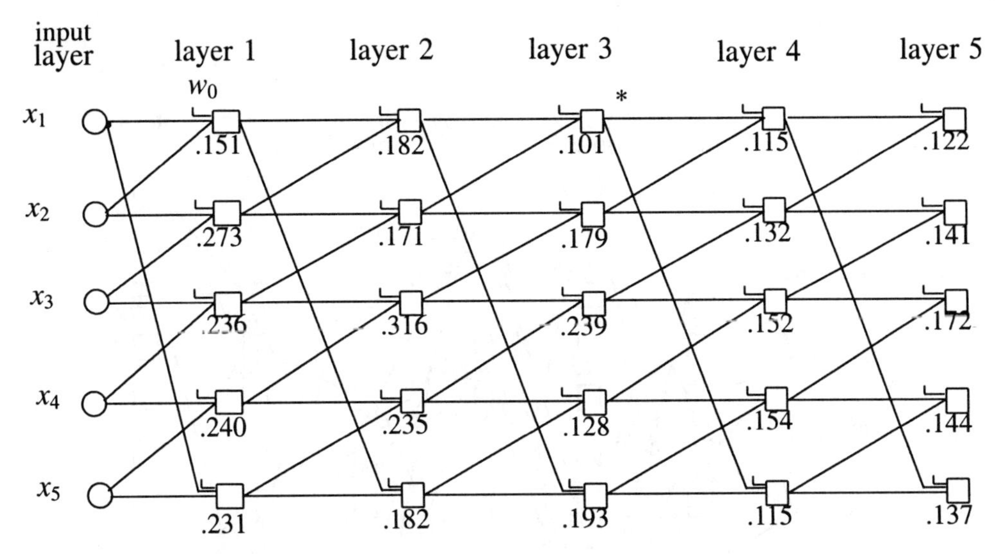

Figure 7.6. Bounded network structure with linear inputs and sigmoid output activations; w_0 is the biased term at each node

Figure 7.6 gives the evolutions of the generation of nodes by the network during the search process and residual MSE at each node is also given. "*" indicates the node which achieved the optimum value in all the networks given.

(ii) In a nonlinear case, the output data is generated from the equation:

$$y = 0.433 - 0.095x_1 + 0.243x_2 + 0.35x_1^2 - 0.18x_1x_2 + \epsilon, \tag{7.46}$$

where x_1, x_2 are randomly generated input variables, y is the output variable, and ϵ is the noise added to the data.

(a) x_1, x_2, x_1^2, x_2^2, x_1x_2 are fed as input variables. In the inductive case the global measure is obtained at a unit in the third layer ($c2 = 0.0453$). The residual MSE of the unit is computed as 0.0406. Figure 7.7 gives the combined measure of all units and residual MSE at the optimum node. Table 7.1 gives the connecting weight values (w_0, w_j, and w_i), the value of the combined criterion, and the residual MSE at each node.

(b) The same input/output data is used for the deductive network; sigmoid function is used for activating the outputs. It is converged to global minimum at a unit in the second layer. The average residual error of the unit is computed as 0.0223 for an optimum adjustment of $g = 1.8$. Figure 7.8 gives the residual MSE at each node. Table 7.2 gives the connecting weight values (w_0, w_j, and w_i), and the residual MSE at each node.

(c) In another case, the deductive network with the same input/output data is activated by the transfer function $F(u) = \sin(u * g)$, where u is the unit output and g is the gain factor. The global minimum is tested for different gain factors of g $(= 1 \pm \tau)$, where τ varies from 0.0 to 1.0. As it varies, optimal units are shifted to earlier layers with a slight change of increase in the minimum. For example, at $\tau = 0.5$ the unit in the third layer achieves the minimum of 0.0188 and at $\tau = 0.8$ the unit in the second layer has the minimum of 0.0199. The global minimum of 0.0163 is achieved at the second unit of the sixth layer for $\tau = 0.0$ (Figure 7.9).

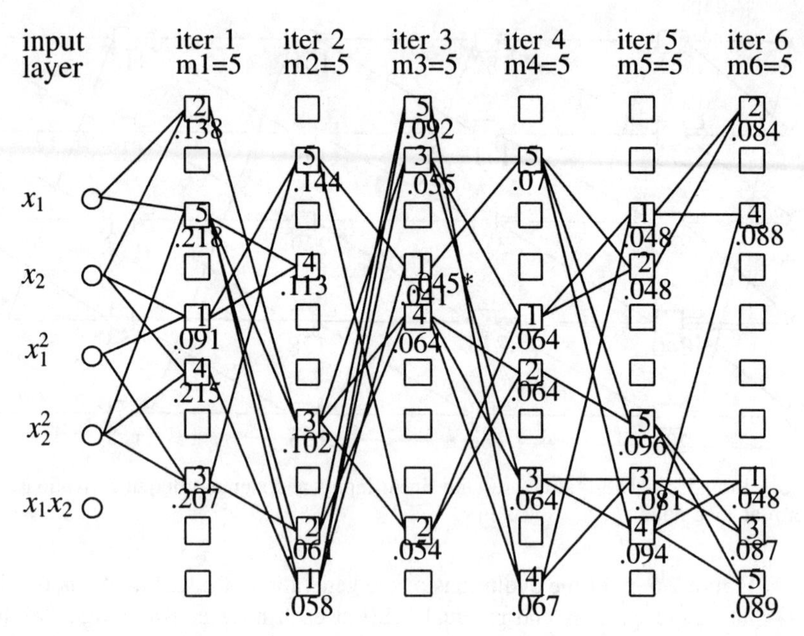

Figure 7.7. Bounded inductive network structure with nonlinear inputs using threshold objective function (only activated links are shown)

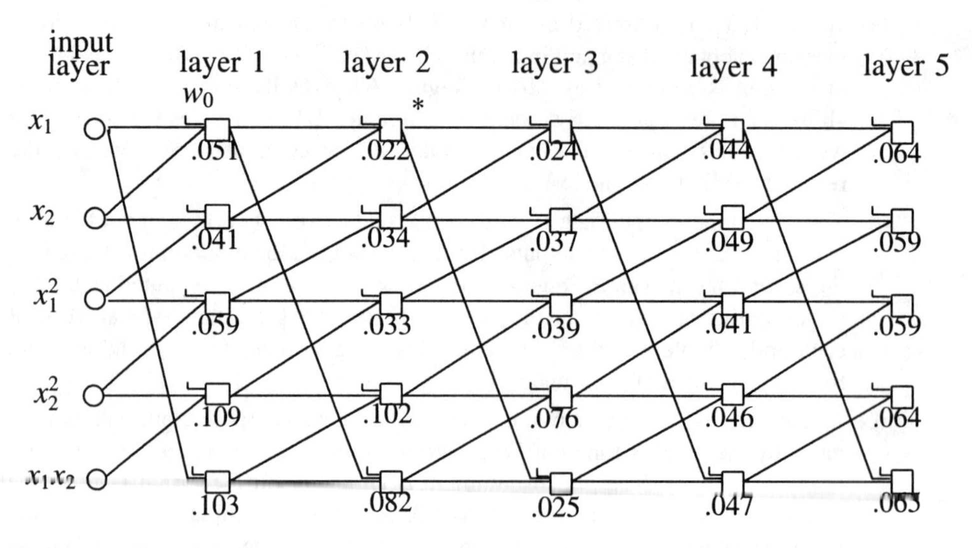

Figure 7.8. Bounded network structure with nonlinear inputs and sigmoid output activations; w_0 is the biased term at each node

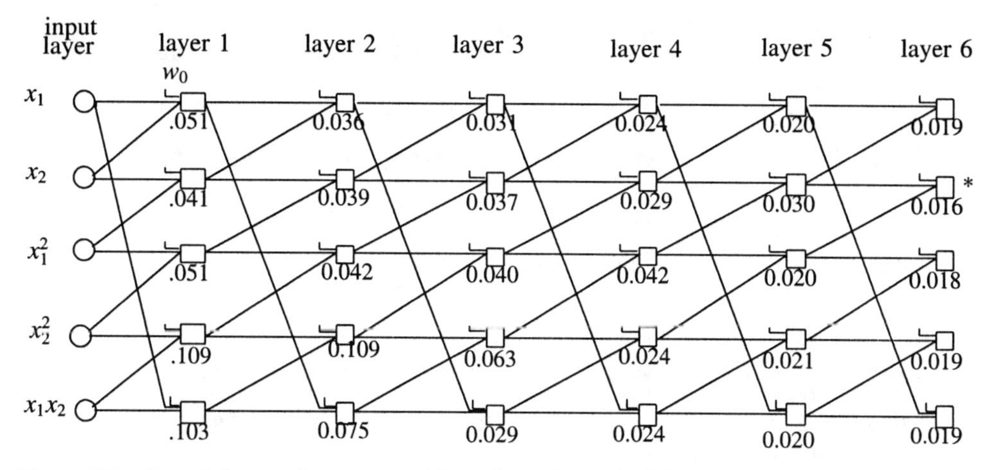

Figure 7.9. Bounded network structure with nonlinear inputs and sinusoidal output transformations; w_0 is the biased term at each node

(iii) Further, the network structures are tested for their performances without any threshold activations at the units; i.e., the unit outputs are directly fed forward to the next layer. Global minimum is not achieved; the residual error is reduced layer-by-layer as it proceeds—ultimately, the network becomes unstable. This shows the importance of the threshold functions in the convergence of these networks.

The resulting robustness in computations of self-organization modeling is one of the features that has made these networks attractive. It is clear that network models have a strong affinity with statistical mechanics. The main purpose of modeling is to obtain a better input-output transfer relationship between the patterns by minimizing the effect of noise in the input variables. This is possible only by providing more knowledge into the network structures; that is, improving the network performance and achieving better computing abilities in problem solving. In the inductive learning approach the threshold objective function plays an important role in providing more informative models for identifying and predicting complex systems. In the deductive case the unit output transformation through the sigmoid function plays an important role when the functional relationship is sigmoid rather than linear. Over all, one can see that the performance of the neural modeling can be improved by adding one's experience and knowledge into the network structure as a self-organization mechanism. It is an integration of various concepts from conventional computing and artificial intelligence techniques.

Table 7.1. Network structure with threshold objective function

LAYER=	1	(m1= 5)		
	J= 1	I= 2		
.411	.186	.147;	c2= .138E+00,	MSE= .513E-01
	J= 1	I= 3		
.454	.145	.134;	c2= .416E+00,	MSE= .122E+00
	J= 1	I= 4		
.425	.213	.120;	c2= .218E+00,	MSE= .657E-01
	J= 1	I= 5		
.455	.069	.268;	c2= .279E+00,	MSE= .103E+00
	J= 2	I= 3		
.434	.155	.179;	c2= .907E-01,	MSE= .406E-01

```
          J= 2      I= 4
   .405    .629     -.376;  c2=  .215E+00,  MSE=  .137E+00
          J= 2      I= 5
   .458    .052      .284;  c2=  .226E+00,  MSE=  .997E-01
          J= 3      I= 4
   .452    .203      .133;  c2=  .207E+00,  MSE=  .589E-01
          J= 3      I= 5
   .465    .073      .260;  c2=  .266E+00,  MSE=  .102E+00
          J= 4      I= 5
   .466    .008      .329;  c2=  .257E+00,  MSE=  .109E+00
        ---------------------
 LAYER=   2        (m2= 5)
          J= 1      I= 2
   .024   1.097     -.151;  c2=  .154E+00,  MSE=  .523E-01
          J= 1      I= 3
   .033   2.313    -1.363;  c2=  .144E+00,  MSE=  .609E-01
          J= 1      I= 4
  -.033    .208      .822;  c2=  .295E+00,  MSE=  .527E-01
          J= 1      I= 5
   .004   -.451     1.423;  c2=  .113E+00,  MSE=  .412E-01
          J= 2      I= 3
  -.079    .186      .933;  c2=  .224E+00,  MSE=  .523E-01
          J= 2      I= 4
  -.054    .076      .989;  c2=  .165E+00,  MSE=  .536E-01
          J= 2      I= 5
   .020   -.099     1.045;  c2=  .102E+00,  MSE=  .443E-01
          J= 3      I= 4
  -.019   -.665     1.664;  c2=  .263E+00,  MSE=  .598E-01
          J= 3      I= 5
   .020   -.437     1.388;  c2=  .613E-01,  MSE=  .381E-01
          J= 4      I= 5
   .023   -.794     1.747;  c2=  .581E-01,  MSE=  .417E-01
        ---------------------
 LAYER=   3        (m3= 5)
          J= 1      I= 2
   .008   1.439     -.472;  c2=  .919E-01,  MSE=  .390E-01
          J= 1      I= 3
   .001    .098      .886;  c2=  .548E-01,  MSE=  .374E-01
          J= 1      I= 4
  -.008    .399      .596;  c2=  .119E+00,  MSE=  .399E-01
          J= 1      I= 5
   .008   4.123    -3.144;  c2=  .453E-01,  MSE=  .406E-01*
          J= 2      I= 3
   .000    .047      .939;  c2=  .642E-01,  MSE=  .374E-01
          J= 2      I= 4
  -.013    .146      .858;  c2=  .111E+00,  MSE=  .404E-01
          J= 2      I= 5
   .003   -.456     1.430;  c2=  .128E+00,  MSE=  .411E-01
          J= 3      I= 4
   .004   1.154     -.174;  c2=  .969E-01,  MSE=  .372E-01
          J= 3      I= 5
   .001    .929      .055;  c2=  .537E-01,  MSE=  .373E-01
          J= 4      I= 5
  -.009    .715      .281;  c2=  .105E+00,  MSE=  .406E-01
        ---------------------
 LAYER=   4        (m4= 5)
          J= 1      I= 2
   .004   -.390     1.372;  c2=  .896E-01,  MSE=  .353E-01
          J= 1      I= 3
   .004   -.400     1.385;  c2=  .699E-01,  MSE=  .353E-01
```

```
         J= 1      I= 4
 -.007    .713      .283;  c2=  .918E-01,  MSE=  .363E-01
         J= 1      I= 5
  .002   -.172     1.156;  c2=  .121E+00,  MSE=  .351E-01
         J= 2      I= 3
  .001    .001      .986;  c2=  .636E-01,  MSE=  .350E-01
         J= 2      I= 4
  .000    .867      .121;  c2=  .636E-01,  MSE=  .350E-01
         J= 2      I= 5
  .002   2.012    -1.025;  c2=  .819E-01,  MSE=  .351E-01
         J= 3      I= 4
  .001    .992     -.005;  c2=  .636E-01,  MSE=  .350E-01
         J= 3      I= 5
  .001   1.118     -.130;  c2=  .716E-01,  MSE=  .350E-01
         J= 4      I= 5
 -.002    .253      .738;  c2=  .669E-01,  MSE=  .351E-01
        ---------------------
LAYER=   5        (m5= 5)
         J= 1      I= 2
  .004   1.419     -.436;  c2=  .971E-01,  MSE=  .352E-01
         J= 1      I= 3
  .003   3.864    -2.879;  c2=  .105E+00,  MSE=  .354E-01
         J= 1      I= 4
  .001    .337      .649;  c2=  .484E-01,  MSE=  .350E-01
         J= 1      I= 5
  .001   -.137     1.123;  c2=  .484E-01,  MSE=  .350E-01
         J= 2      I= 3
  .004   -.585     1.567;  c2=  .113E+00,  MSE=  .351E-01
         J= 2      I= 4
  .004   -.438     1.421;  c2=  .983E-01,  MSE=  .352E-01
         J= 2      I= 5
  .004   -.446     1.429;  c2=  .964E-01,  MSE=  .352E-01
         J= 3      I= 4
  .003  -2.476     3.461;  c2=  .814E-01,  MSE=  .353E-01
         J= 3      I= 5
  .003  -2.602     3.587;  c2=  .935E-01,  MSE=  .353E-01
         J= 4      I= 5
  .001   -.172     1.158;  c2=  .340E+01,  MSE=  .350E-01
        ---------------------
LAYER=   6        (m6= 5)
         J= 1      I= 2
 -.004   -.141     1.132;  c2=  .836E-01,  MSE=  .364E-01
         J= 1      I= 3
  .004   -.555     1.539;  c2=  .899E-01,  MSE=  .353E-01
         J= 1      I= 4
  .004   -.557     1.542;  c2=  .883E-01,  MSE=  .353E-01
         J= 1      I= 5
  .003  -7.773     8.758;  c2=  .983E-01,  MSE=  .352E-01
         J= 2      I= 3
  .004   -.456     1.439;  c2=  .972E-01,  MSE=  .352E-01
         J= 2      I= 4
  .004   -.445     1.428;  c2=  .982E-01,  MSE=  .352E-01
         J= 2      I= 5
 -.004    .666      .323;  c2=  .895E-01,  MSE=  .363E-01
         J= 3      I= 4
  .001    .492      .494;  c2=  .483E-01,  MSE=  .350E-01
         J= 3      I= 5
  .004   1.659     -.675;  c2=  .870E-01,  MSE=  .353E-01
         J= 4      I= 5
  .004   1.677     -.693;  c2=  .888E-01,  MSE=  .354E-01
```

```
---------------------------------------


Table 7.2. Network structure with sigmoid function

 LAYER=    1
           J= 1      I= 2
   .411     .186      .147;     MSE=  .513E-01
           J= 2      I= 3
   .434     .155      .179;     MSE=  .406E-01
           J= 3      I= 4
   .452     .203      .133;     MSE=  .589E-01
           J= 4      I= 5
   .466     .008      .329;     MSE=  .109E+00
           J= 5      I= 1
   .455     .268      .069;     MSE=  .103E+00
          ---------------------
 LAYER=    2
           J= 1      I= 2
  -.500   -1.100     2.489;     MSE=  .223E-01*
           J= 2      I= 3
  -.477    1.803     -.436;     MSE=  .336E-01
           J= 3      I= 4
  -.489    1.989     -.613;     MSE=  .328E-01
           J= 4      I= 5
  -.856     .115     1.709;     MSE=  .102E+00
           J= 5      I= 1
  -.757    1.304      .402;     MSE=  .824E-01
          ---------------------
 LAYER=    3
           J= 1      I= 2
  -.484    1.052      .329;     MSE=  .242E-01
           J= 2      I= 3
  -.456    1.464     -.117;     MSE=  .368E-01
           J= 3      I= 4
  -.614     .960      .577;     MSE=  .393E-01
           J= 4      I= 5
  -.722    1.497      .169;     MSE=  .764E-01
           J= 5      I= 1
  -.488     .158     1.229;     MSE=  .249E-01
          ---------------------
 LAYER=    4
           J= 1      I= 2
  -.458     .905      .454;     MSE=  .438E-01
           J= 2      I= 3
  -.441    1.641     -.304;     MSE=  .492E-01
           J= 3      I= 4
  -.502    1.757     -.349;     MSE=  .410E-01
           J= 4      I= 5
  -.467     .290     1.080;     MSE=  .456E-01
           J= 5      I= 1
  -.436   -4.405     5.736;     MSE=  .465E-01
          ---------------------
 LAYER=    5
           J= 1      I= 2
  -.437    1.088      .253;     MSE=  .643E-01
           J= 2      I= 3
  -.455     .954      .405;     MSE=  .591E-01
           J= 3      I= 4
  -.476     .560      .830;     MSE=  .586E-01
```

```
          J= 4      I= 5
  -.422   19.103  -17.783;    MSE=  .642E-01
          J= 5      I= 1
  -.426    -.459    1.786;    MSE=  .651E-01
         --------------------
LAYER=    6
          J= 1      I= 2
  -.427    1.159     .175;    MSE=  .778E-01
          J= 2      I= 3
  -.428    -.111    1.444;    MSE=  .741E-01
          J= 3      I= 4
  -.441    1.755    -.406;    MSE=  .729E-01
          J= 4      I= 5
  -.413     .121    1.195;    MSE=  .796E-01
          J= 5      I= 1
  -.417    -.037    1.358;    MSE=  .791E-01
       --------------------------------------
```

Chapter 8
Basic Algorithms and Program Listings

The computer listings of the basic inductive network structures for multilayer, combinatorial and harmonical techniques, and their computational aspects are given here. Multilayer algorithm uses a multilayered network structure with linearized input arguments and generates simple partial functionals. Combinatorial algorithm uses a single-layered structure with all combinations of input arguments including the full description. Harmonical algorithm follows the multilayered structure in obtaining the optimal harmonic trend with nonmultiple frequencies for oscillatory processes. One can modify these source listings as per his/her needs. These programs run on microcomputers and SPARC stations of SUN microsystems. To some extent they were also previously given for NORD-100/500 systems [88].

1 COMPUTATIONAL ASPECTS OF MULTILAYERED ALGORITHM

The basic schematic functional flow of the multilayered inductive learning algorithm is given in Chapters 2 and 7.

As the multilayer network procedure is more repetitive in nature, it is important to consider the algorithm in modules and facilitate repetitive characteristics. The most economical way of constructing the algorithm is to provide three main modules: (i) the first module is for computations of common terms in the conditional symmetric matrix of the normal equations for all input variables. This is done at the beginning of each layer with all fresh input variables entering into the layer using the training set, (ii) the second module is for generating the partial functions by forming the symmetric matrices of the normal equations for all pairs of input variables, for estimating their coefficients, for computing the values of the threshold objective functions on the testing set, and for memorizing the information of coefficients and input variables of the best functions (this is done for each layer), and (iii) the third module is for computing the coefficients of the optimal model by recollecting the information from the associated units.

To initiate the program one has to specify the control parameters:

M1	—	no. of input variables
N	—	total no. of data points
PE	—	percentage of points on training and testing sets; $50 < PE < 100$; if $PE = 80$, then $A = 80\%$, $B = 80\%$, and $C = 20\%$

PM	—	no. of layers
ALPHA	—	weightage used in the combined criterion as C = ALPHA*C1 + (1-ALPHA)*C2, where C indicates the combined criterion (c2), C1 indicates the minimum-bias criterion, C2 indicates the regularity criterion, and 0 ≤ ALPHA ≤ 1
CHO(I), I = 1,PM	—	freedom-of-choice at each layer of *PM* layers
FF	—	choice of optimal models at the end (FF ≥ 1)

The values of these parameters are supplied through the file "param.dat." The file "input.dat" supplies the output and input data measurements.

The "input.dat" file is to be supplied according to the specified reference function. If the reference function is a linear function (for example, ($M1 = 6$)), then

$$y_1 = a_0 + a_1x_1 + a_2x_2 + \cdots + a_6x_6, \tag{8.1}$$

where a are the coefficients; $x_1, \cdots, x_6$ are the inputs to the network; and y_1 is the desired output variable. One has to supply the data file with N rows of points as

y_1	x_1	x_2	x_3	x_4	x_5	x_6

If the reference function is a nonlinear function (for example, ($M1 = 5$)), then

$$y_1 = a_0 + a_1x_1 + a_2x_2 + a_3x_1^2 + a_4x_2^2 + a_5x_1x_2, \tag{8.2}$$

where a are the coefficients; $x_1, x_2, x_1^2, x_2^2, x_1x_2$ are the inputs to the network, and y_1 is the desired output variable. One has to supply the data file with N rows of points as

y_1	x_1	x_2	x_1^2	x_2^2	x_1x_2

The higher-ordered terms are to be calculated and supplied in the file. Data sets A and B are separated according to the dispersion analysis.

In the first module, common terms in the conditional matrix XH is computed using the P2 input variables and the output variable Y. P1 and PU indicate the number of functions to be selected at the first layer and number of the layer, correspondingly.

In the second module, it forms the matrices (HM1, HM2, HM3) of normal equations for each pair of input variables J and I, and estimates the weights or coefficients (KO1, KO2, KO3) using the data sets A, B, and W (=A∪B), correspondingly. All partial functions are evaluated by the combined criterion. It stores the information on coefficients (KOE) and input variables (NK) of the best P1 nodes. Subroutine RANG is used to arrange all values in ascending order. Standard subroutine GAUSS is used to estimate the coefficients of each partial function.

Futhermore, the estimated outputs (YY) of P1 functions are calculated to send it to the next layer. To repeat the above two modules, we have to convert the outputs (YY) as inputs (XX) and initialize with fresh control parameters of the layer—the number of the layer PU is updated as PU+1, the number of input arguments P2 is equated to P1, and the number of functions to be selected (freedom-of-choice) is taken from CHO(PU) as specified at the beginning. This procedure is repeated until PU becomes the number of specified layers (PM).

Modules 1 and 2 with the subroutine NM, help in forming normal equations for each pair in a more economical of utilizing computer time.

In the third module, it recollects the information for the function that has achieved global minimum or FF functions. The parameter PDM is calculated in advance as an indicator of the number of original input arguments u activating in the function at a particular layer—in the first layer PDM = 2 and in consecutive layers PDM = PDM*2. The coefficients and number of input arguments of the optimal function are computed using the stored information from KOE and NK.

The program listing and the sample output for a chosen example are given below.

1.1 Program listing

```
C
C**********************************************
C THIS FORTRAN VERSION IS DEVELOPED BY H. MADALA
C**********************************************
C  MULTILAYER INDUCTIVE LEARNING ALGORITHM
C
C     MAIN PROGRAM
C
        INTEGER N,M,M1,PE,PM,N1,I,J,K,S,P,R,T,GG,PN,
     1          FF,SH,PU,YP,P1,BM,P2,NI,PDM,
     2          PL,NL,EG,SS,MH,MH1,MH2,IFAIL
        REAL XS,XM,OSH,TL,TX,YB,C,C1,C2,YM,AL,OL,H21,H22,Y3,Y11,
     1          Y22,CTROO
        REAL CML (30,10),X(15,200),Y(1,200),KX(15),AX(200),
     1       XX(15,200),KO1(15),KO2(15),KO3(15),KO4(15),CM(30),
     2       HM1(15,16),HM2(15,16),HM3(15,16),CMM(30,10),
     3       KOE(30,10,20),CT(15),CTRO(15),D2(15),AY(200),
     4       XH(15,10,10),YY(20,200),SK(20),A(256),AD (256),
     5       D22(200)
        INTEGER NPP (200),NP1(200),NP2(200),NO1(200),NO2(200),
     1          CHO(10),NK(30,10,20),NC(30),ND(15),ST(20,5),
     2          NDD(200),AN (256),AND (256),OB(200,5)
C
         OPEN(1,FILE='param.dat')
         OPEN(8,FILE='input.dat')
         OPEN(3,FILE='output.dat')
C****************
C INITIALIZATION
C****************
         READ(1,*)M1,N,PE,PM,ALPHA
         READ(1,*)(CHO(I),I=1,PM), FF
         XS =PE*N
         PE =INT(XS/100.)
C*******************************************************
C  M1 - NO. OF INPUT VARIABLES
C  N  - NO. OF DATA OBSERVATIONS
C  PE - PERCENTAGE OF TOTAL PTS. ON TRAIN AND TESTING SETS
C  PM - NO. OF LAYERS
C  (CHO(I), I =1,PM) - CHOICE OF MODELS AT EACH LAYER
C  FF - CHOICE OF OPTIMAL MODELS AT THE END
C*******************************************************
         M=1
         DO 91 I=1,N
         READ(8,*)Y(1,I),(X(J,I),J=1,M1)
   91 CONTINUE
C
   92  FORMAT (2X,'CONTROL PARAMS:'/2X,'--------------'/)
   95  FORMAT (3x,'NO.OF INPUT VARIABLES (M1) ',I2)
   97  FORMAT (3x,'NO.OF DATA POINTS (N) ',I3)
   99  FORMAT (3X,'PERCENTAGE OF TRAIN AND TEST POINTS (PE) ',I2)
```

```
 100   FORMAT (3X,'NO.OF LAYERS (PM) ',I2)
 102   FORMAT (3X,'WEIGHTAGE VALUE IN COMBINED CRIT (ALPHA) ',F3.1)
 104   FORMAT (3X,'FREEDOM-OF-CHOICE AT EACH LAYER(CHO) ',10I3)
 106   FORMAT (3X,'NO.OF OPTIMAL MODELS (FF) ',I2)
 108   FORMAT (3X,'NO.OF OUTPUT VARIABLES (M) ',I2)
 110   FORMAT (//)
 120   FORMAT (2X,10E10.3)
 125   FORMAT (1X,'PERFORMANCE OF THE NET:'/1X,'------------------'/)
 130   FORMAT (2X,'EQUATION NUMBER= ',I2/)
 140   FORMAT (3X,'LAYER=',I4,2X,'SELECTED DESCRIPTION=',I5)
 150   FORMAT (5X,'ERROR GAUSS='I4)
 160   FORMAT (5X,'COMBINED ERROR BEST= ',E10.3,4X,'WORST= ',E10.3)
 165   FORMAT (5x,'RESIDUAL MSE= ',E10.3,'AT THE BEST COMBINED NODE')
 170   FORMAT (5X,'RESIDUAL MSE   BEST= ',E10.3,4X,'WORST= ',E10.3)
 175   FORMAT (1X,'OPTIMAL MODELS:'/1X,'--------------'/)
 180   FORMAT (2X,'MODEL',I3,1X,'(LAYER ',I2,3X,'COMBINED=',E10.3,1X,
     1                  'MIN BIAS=',E10.3,1X,'MSE=',E10.3,1X,')')
 190   FORMAT (2X,'COEFFICIENTS=',/2X,E12.3)
 200   FORMAT (/(2X,10I10))
 210   FORMAT (2X,10E10.3)
 220   FORMAT (7X,' ----------------------------------------')
 230   FORMAT (10X,'---------------------')
 240   FORMAT(2X,'Y=')
 250   FORMAT(2X,'X=')
 260   FORMAT(/13X,'MULTI   L A Y E R E D   ALGORITHM'//)
C
       WRITE(3,260)
       WRITE (3,92)
       WRITE(3,95)M1
       WRITE(3,97)N
       WRITE(3,99)PE
       WRITE(3,100)PM
       WRITE(3,102)ALPHA
       WRITE(3,104) (CHO(I),I=1,PM)
       WRITE(3,106)FF
       WRITE(3,108)M
C
         PN=0
         S=M1+PN
                P=S
         N1=N
C
                P=M1
                S=M1
                CHO(0)=M1
                 WRITE(3,240)
                DO 71 J=1,M
                WRITE (3,120) (Y(J,I),I=1,N1)
  71            CONTINUE
               WRITE (3,250)
               WRITE (3,120) ((X(I,J), I=1,M1),J=1,N1)
C*********************************************************
C NORMALIZATION AND RANGE OF DATA AS PER DISPERSION ANALYSIS
C*********************************************************
         DO 5 J=1,S
         DO 3 I=1,N1
  3      AX(I)=ABS(X(J,I))
         CALL NORM(AX,N1,XS)
         KX(J)=XS
         DO 4 I=1,N1
```

```
 4      X(J,I)=X(J,I)/XS
 5      CONTINUE
        DO 6 I=1,N1
 6      Y(1,I)=Y(1,I)/KX(1)
            NI=0
            BM=CHO(0)
               DO 7 I=1,PM
               IF (BM.LT.CHO(I))BM=CHO(I)
 7      CONTINUE
               YP=1
 8      P2=CHO(0)
             P1=CHO(1)
            DO 9 I=1,BM
             DO 9 J=1,PM
 9      NK (I,1,J)=0
               WRITE (3,110)
       WRITE(3,125)
               WRITE (3,130)YP
        IF(P2.EQ.P)THEN
        DO 10 I=1,P
10      ND(I)=P-I+1
        GOTO 13
        ENDIF
               DO 12 J=1,P
               D2 (J)=0.0
               DO 11 I=1,N1
11       D2 (J)=D2(J)+X(J,I)*Y(YP,I)
12       D2(J)=ABS (D2(J))
               CALL RANG (D2,ND,P)
13       CONTINUE
               DO 14 J=1,P2
               DO 14 I=1,N1
            I1=P-J+1
               MH1=ND(I1)
14       XX(J,I)=X(MH1,I)
               PU=1
               PDM=2
C****************************************************************
C FIRST MODULE TO CALCULATE COMMON TERMS IN CONDITIONAL MATRICES
C****************************************************************
15             DO 16 I=1,N1
               D22(I)=0.0
               DO 16 J=1,P2
16       D22(I)=D22(I)+XX(J,I)**2
               CALL RANG(D22,NDD,N1)
               DO 17 I=1,PE
               NP1 (I)=NDD(I)
            I1=N1-I+1
17       NP2 (I)=NDD(I1)
               CALL OPE (NP1,NO1,PE,N1)
               CALL OPE (NP2,NO2,PE,N1)
            EG=0
               K=0
               DO 18 I=1,PE
               SH=NP1 (I)
               DO 18 J=1,PE
         IF(SH.EQ.NP2(J))THEN
         K=K+1
         NPP(K)=SH
         GOTO 18
```

```
      ENDIF
 18    CONTINUE
      IF(PU.EQ.1)THEN
      DO 19 I=1,N1
 19   AY(I)=ABS(Y(YP,I))
      CALL FMAX(AY,N1,YM,I)
      ENDIF
             R=N1-PE
             Y3=0.0
             Y22=0.0
             DO 74 K=1,N1
             Y3=Y3+Y(YP,K)
             Y22=Y22+Y(YP,K)**2
 74        CONTINUE
       Y22 =SQRT(Y22)
             DO 20 J=1,R
             OB(J,1)=NO1(J)
 20    OB(J,2)=NO2(J)
             DO 21 J=1,P2
             XH (J,1,3)=0.0
             XH (J,2,3)=0.0
             XH (J,3,3)=0.0
             DO 75 K=1,N1
             XH (J,1,3)=XH(J,1,3)+XX(J,K)
             XH (J,2,3)=XH(J,2,3)+XX(J,K)**2
             XH (J,3,3)=XH(J,3,3)+XX(J,K)*Y(YP,K)
 75    CONTINUE
             DO 21 T=1,2
             XH(J,1,T)=0.0
             XH(J,2,T)=0.0
             XH(J,3,T)=0.0
             DO 76 K=1,R
             MH=OB(K,T)
             XH(J,1,T)=XH(J,1,T)+XX(J,MH)
             XH(J,2,T)=XH(J,2,T)+XX(J,MH)**2
             XH(J,3,T)=XH(J,3,T)+XX(J,MH)*Y(YP,MH)
 76   CONTINUE
 21    CONTINUE
             XS=0.0
             XM=0.0
             DO 22 I=1,R
             MH1=NO1(I)
             MH2=NO2(I)
             XS=XS+Y(YP,MH1)
 22    XM=XM+Y(YP,MH2)
C*************************************************************
C SECOND MODULE FOR FORMING THE CONDITIONAL MATRICES FOR EACH
C    PARTIAL FUNCTION
C*************************************************************
             SH=1
             J=0
 23    J=J+1
             I=J+1
 24    HM1(1,1)=R
             HM2(1,1)=R
             HM1(1,4)=XS
             HM2(1,4)=XM
             H21=0.0
             H22=0.0
             DO 77 K=1,R
```

```
               MH1=NO1(K)
               MH2=NO2(K)
               H21=H21+XX(J,MH1)*XX(I,MH1)
               H22=H22+XX(J,MH2)*XX(I,MH2)
  77      CONTINUE
               HM1(2,3)=H21
               HM1(3,2)=H21
               HM2(2,3)=H22
               HM2(3,2)=H22
               HM2(1,4)=XM
               CALL NM (HM1,XH,1,J,I).
               CALL NM (HM2,XH,2,J,I)
               DO 25 K=1,3
               DO 25 S=1,4
 25      HM3(K,S)=HM1(K,S)+HM2(K,S)
C*****************************
C     ESTIMATING COEFFICIENTS
C*****************************
               CALL GAUSS(HM1,3,4,KO1,IFAIL)
               IF (IFAIL.EQ.0)GO TO 29
               CALL GAUSS(HM2,3,4,KO2,IFAIL)
               IF(IFAIL.EQ.0)GO TO 29
               CALL GAUSS (HM3,3,4,KO3,IFAIL)
               IF (IFAIL.EQ.0)GO TO 29
C*******************************************
C     COMPUTING THE VALUES OF EXTERNAL CRITERIA
C*******************************************
               C1=0.0
               C2=0.0
C***************************************************
C  C1 - MEAN SQUARED MINIMUM BIAS ERROR ON TOTAL POINTS
C  C2 - MEAN SQUARED RESIDUAL ERROR ON EXAMIN SET
C  C  - ROOT MEAN COMBINED ERROR  OF (C1 + C2)
C***************************************************
               DO 78 S=1,N1
         C1=C1+(KO1(1)-KO2(1)+(KO1(2)-KO2(2))*XX(J,S)+(KO1(3)-
     1    KO2(3))*XX(I,S))**2
 78     CONTINUE
               C1=C1/(Y22**2)
        Y11 =0.0
               MH1=2*PE-N1
               DO 79 S=1,MH1
               MH=NPP(S)
        C2=C2+(Y(YP,MH)-KO3(1)-KO3(2)*XX(J,MH)-KO3(3)*XX(I,MH))**2
        Y11 =Y11+Y(YP,MH)**2
 79    CONTINUE
               C2=C2/Y11
        C = SQRT( ALPHA*C1 + (1-ALPHA)*C2)
C
        CALL NM(HM3,XH,3,J,I)
               HM3(1,1)=N1
             HM3(1,4)=Y3
               HM3(2,3)=0.0
               DO 80 K=1,N1
               HM3(2,3)=HM3(2,3)+XX(J,K)*XX(I,K)
 80     CONTINUE
               HM3(3,2)=HM3(2,3)
               CALL GAUSS(HM3,3,4,KO4,IFAIL)
               IF(IFAIL.EQ.0)GO TO 29
               IF(SH.GT.P1)GO TO 27
```

```
            CM(SH)=C
            DO 26 K=1,3
 26      KOE(SH,K,PU)=KO4(K)
            CMM(SH,1)=C1
            CMM(SH,2)=C2
            NK(SH,2,PU)=J
            NK(SH,3,PU)=I
            IF(SH.EQ.P1)CALL RANG(CM,NC,P1)
            SH=SH+1
            GO TO 30
 27     MH1=NC(P1)
            IF(C.GT.CM(MH1))GO TO 30
            GG=NC(P1)
            CMM(GG,1)=C1
            CMM(GG,2)=C2
            CM(MH1)=C
            DO 28 K=1,3
 28      KOE(MH1,K,PU)=KO4(K)
            NK(MH1,2,PU)=J
            NK(MH1,3,PU)=I
            CALL RANG(CM,NC,P1)
            GO TO 30
 29      EG=EG+1
 30      I=I+1
            IF(I.LE.P2)GO TO 24
            IF(J.LT.P2-1)GO TO 23
            DO 33 S=1,P1
            OSH=0.0
            DO 32 J=1,N1
            YB=KOE(S,1,PU)
            DO 31 I=2,3
            MH1=NK(S,I,PU)
 31      YB=YB+KOE(S,I,PU)*XX(MH1,J)
 32      OSH=OSH+(Y(YP,J)-YB)**2
C           OSH=SQRT(OSH/N1)/Y22
        OSH =SQRT(OSH)/Y22
        IF(S.EQ.1)THEN
        TX=OSH
        TL=OSH
        ENDIF
        IF(NC(1).EQ.S)THEN
        AL=OSH
        IF(PU.EQ.1)THEN
        OL=OSH
        PL=1
        NL=S
        ENDIF
        IF(OL.GE.OSH)THEN
        OL=OSH
        PL=PU
        NL=S
        ENDIF
        ENDIF
        IF(OSH.LT.TL)TL=OSH
        IF(OSH.GT.TX)TX=OSH
 33             CONTINUE
C*****************************************************
C PRINTING THE PERFORMANCE OF THE NETWORK AT EACH LAYER
C*****************************************************
            MH1=NC(1)
```

```
              MH2=NC(P1)
              WRITE(3,140)PU,P1
              WRITE(3,150)EG
              WRITE(3,160)CM(MH1),CM(MH2)
              WRITE(3,170)TL,TX
              WRITE(3,165)AL
              WRITE(3,230)
 34        PDM=2*PDM
        DO 35 J=1,P1
        DO 35 I=1,N1
        YY(J,I)=KOE(J,1,PU)
        DO 35 S=2,3
 35     YY(J,I)=YY(J,I)+XX(NK(J,S,PU),I)*KOE(J,S,PU)
        DO 36 J=1,P1
        DO 36 I=1,N1
 36     XX(J,I)=YY(J,I)
        IF(PU.EQ.1)THEN
        DO 39 I=1,FF
        IF(I.LE.P1)THEN
        CML(I,1)=CM(NC(I))
        CML(I,2)=PU
        CML(I,3)=NC(I)
        DO 38 J=1,2
 38     CML(I,J+3)=CMM(NC(I),J)
        ELSE
        CML(I,1)=10000.
        ENDIF
 39     CONTINUE
        ELSE
        K=1
 40     I=1
        C=CML(1,1)
        DO 41 J=2,FF
        IF(CML(J,1).GT.C)THEN
        C=CML(J,1)
        I=J
        ENDIF
 41     CONTINUE
        IF(C.LE.CM(NC(K)))GOTO 43
        CML(I,2)=PU
        CML(I,1)=CM(NC(K))
        DO 42 J=1,2
 42     CML(I,J+3)=CMM(NC(K),J)
        CML(I,3)=NC(K)
        K=K+1
        IF(K.LE.P1)GOTO 40
 43     ENDIF
              IF(PU.EQ.PM)GO TO 44
              PU=PU+1
              P2=P1
              P1=CHO(PU)
              GO TO 15
C********************************************
C THIRD MODULE TO RECOLLECT THE OPTIMAL MODELS
C********************************************
 44     WRITE(3,110)
        WRITE(3,175)
        SS=0
        PDM=PDM-1
 45     SS=SS+1
```

```
 46      PU=CML(SS,2)
         NCDGE=CML(SS,3)
         K=0
         DO 47 I=1,10
         ST(I,1)=0
         ST(I,2)=0
         SK(I)=0
 47      CONTINUE
       WRITE(3,180)SS,INT(CML(SS,2)),CML(SS,1),
      1SQRT(CML(SS,4)),SQRT(CML(SS,5))
         K=0
         DO  48 I=0,P
         CTRO(I)=0.0
 48      CT(I)=0.0
                DO 49 I=1,PDM
                A(I)=0.0
                AD(I)=0.0
                AN(I)=0
 49      AND(I)=0
                DO 50 I=1,3
                A(I)=KOE(NCDGE,I,PU)
 50      AN(I)=NK(NCDGE,I,PU)
                IF(PU.EQ.1)GO TO 55
                AD(1)=A(1)
 88      SH=1
         DO 53 I=2,PDM
         IF(A(I).NE.0)THEN
         IF(AN(I).EQ.0)THEN
         SH=SH+1
         AD(SH)=A(I)
         AND(SH)=0
         ELSE
         DO 86 S=1,3
         AD(SH+S)=A(I)*KOE(AN(I),S,PU-1)
         AND(SH+S)=NK(AN(I),S,PU-1)
 86      CONTINUE
         SH=SH+3
         ENDIF
         ENDIF
 53      CONTINUE
                DO 54 I=2,PDM
                A(I)=AD(I)
                AN(I)=AND(I)
 54      CONTINUE
         PU=PU-1
         IF(PU.GT.1)GOTO 88
 55      CONTINUE
                DO 56 I=1,PDM
                S=AN(I)
                CT(S)=CT(S)+A(I)
 56      CONTINUE
                DO 57 I=1,P
               IP1=P-I+1
                MH=ND(IP1)
                CTRO(MH)=CT(I)
 57      CONTINUE
                CTRO(0) =CT(0)*KX(YP)
                CTROO =CTRO(0)
                WRITE (3,190)CTROO
             MPN=M1+PN
```

```
              DO 60 J=1,MPN
              IF(CTRO(J).NE.0.0)THEN
              CTRO(J)=CTRO(J)*KX(YP)/KX(J)
              K=K+1
              ST(K,1)=J
              SK(K)=CTRO(J)
                 IF(K.EQ.10)THEN
               WRITE(3,200)(ST(K,1),K=1,10)
               WRITE(3,210)(SK(K),K=1,10)
               DO 61 K=1,10
              ST(K,1)=0
               SK(K)=0
 61         CONTINUE
               K=0
               ENDIF
              ENDIF
 60          CONTINUE
              IF(K.NE.0)THEN
              WRITE(3,200)(ST(I,1),I=1,K)
              WRITE(3,210)(SK(I),I=1,K)
             ENDIF
           WRITE(3,220)
              IF(SS.LT.FF)GO TO 45
              YP=YP+1
              IF(YP.LE.M)GO TO 8
              close(3)
              close(8)
              close(1)
              STOP
              END
```

Subroutines used

```
C
              SUBROUTINE FMAX(X,N,XM,K)
              DIMENSION X(200)
              REAL XM
              INTEGER N,K,I
              XM=X(1)
              K=1
              DO 1 I=2,N
              IF (XM.GE.X(I))GOTO 1
              XM=X(I)
              K=I
  1       CONTINUE
              RETURN
              END
C
C
              SUBROUTINE NORM(XN,N,P)
              DIMENSION XN(200)
              INTEGER N,K
              REAL P,XM
              CALL FMAX (XN,N,XM,K)
              P=1.0
  1       P=P*10
              IF(P.GT.XM)GO TO 2
              GO TO 1
  2       P=P/10
```

```
      IF(P.LT.XM)GO TO 3
      GO TO 2
 3    P=P*10
      RETURN
      END
C
C
      SUBROUTINE RANG(X,NP,N)
      DIMENSION X(200),XD(200)
      INTEGER NP(200),ND(200)
      INTEGER N,K,I,N1
      REAL XM
      DO 1 I=1,N
      XD(I)=X(I)
 1    ND(I)=I
      N1=N
 2    CALL FMAX(XD,N1,XM,K)
      NP(N1)=ND(K)
      K1=K+1
      DO 3 I=K1,N1
      XD(I-1)=XD(I)
 3    ND(I-1)=ND(I)
      N1=N1-1
      IF(N1.GE.2)GO TO 2
      NP(1)=ND(1)
      RETURN
      END
C
C
      SUBROUTINE NM(HM,XH,T,J,I)
      INTEGER T,S,R
      DIMENSION XH(15,10,10),HM(15,16)
      S=2
      R=J
 1    HM(1,S)=XH(R,1,T)
      HM(S,1)=HM(1,S)
      HM(S,S)=XH(R,2,T)
      HM(S,4)=XH(R,3,T)
      S=S+1
      R=I
      IF(S.EQ.3)GO TO 1
      RETURN
      END
C
C
      SUBROUTINE OPE(NP,NO,PE,N1)
      INTEGER I,J,Z,PE
      INTEGER NP(200),NO(200)
      Z=0
      I=1
 1    DO 2 J=1,PE
      IF(I.EQ.NP(J))GO TO 3
 2    CONTINUE
      Z=Z+1
      NO(Z)=I
 3    I=I+1
      IF (I.LE.N1) GO TO 1
      RETURN
      END
C
```

```
C
            FUNCTION RND(S2)
              R1=(S2+3.14159)*5.04
         R1=R1-INT(R1)
         S2=R1
               RND=R1
             RETURN
            END
C
C
         SUBROUTINE GAUSS(A,N,L,X,IF)
         DIMENSION A(15,16),X(15)
         IF=1
         NN=N-1
         DO 99 K=1,NN
         J=K
         KK=K+1
         DO 100 I=KK,N
         IF(ABS(A(J,K)).LT.ABS(A(I,K)))J=I
  100    CONTINUE
         IF(J.EQ.K)GOTO 11
         DO 300 I=1,L
         T=A(K,I)
         A(K,I)=A(J,I)
         A(J,I)=T
  300    CONTINUE
   11    DO 88 J=KK,N
         IF(A(K,K).EQ.0.)GOTO 13
         D=-A(J,K)/A(K,K)
         DO 400 I=1,L
         A(J,I)=A(J,I)+D*A(K,I)
  400    CONTINUE
   88    CONTINUE
   99    CONTINUE
         IF(A(N,N).EQ.0.)GOTO 13
         X(N)=A(N,L)/A(N,N)
         NN=N-1
         DO 500 J=1,NN
         K=N-J
         SUM=0.0
         NNN=N-K
         DO 200 JJ=1,NNN
         M=K+JJ
         SUM=SUM+A(K,M)*X(M)
  200    CONTINUE
         IF(A(K,K).EQ.0.)GOTO 13
         X(K)=(A(K,L)-SUM)/A(K,K)
  500    CONTINUE
         GOTO 14
   13    IF=0
   14    RETURN
         END
C
```

1.2 Sample output

Example. The output data is generated from the equation:

$$y = 0.433 - 0.095\,x_1 + 0.243\,x_2 + 0.35\,x_1^2 - 0.18\,x_1 x_2 + \epsilon,$$

where x_1, x_2 are randomly generated input variables, y is the output variable computed from the above equation, and ϵ is the noise added to the data. The data file "input.dat" is prepared correspondingly.

The control parameters are supplied in the file "param.dat"

```
5  100  75  7  0.5
10  10  10  10  10  10  10  8
```

The parameters take the values as M1 =5, N =100, PE =75, PM =7, ALPHA =0.5, CHO(1) =10, CHO(2) =10, ..., CHO(7) =10, and FF =8.

The program creates the output file "output.dat" with the results.

The results are given first with the control parameters, then the performance of the network at each layer that include the values of the combined criterion for the best and the worst models, the values of the residual mean-square error (MSE) for the best and the worst models, and the residual MSE value for the best model according to the combined criterion. The value of ERROR GAUSS indicates the number of singular nodes, if any in the layer, and the SELECTED DESCRIPTION is the freedom-of-choice at each layer. The EQUATION NUMBER indicates the number of the output variable. It is fixed as one (M = 1) because it is dealt with as a single output equation. This can be changed to a number of output equations and the program is modified accordingly.

The coefficient values of optimal models as a number specified for FF are displayed with the constant term and the numbers of input variables with the layer number and the values of the criteria. The second model in the list, obtained at the seventh layer, is the best among all according to the combined criterion; this is read as

$$y = 0.433 - 0.0948x_1 + 0.248x_2 + 0.340x_1^2 - 0.00593x_2^2 - 0.167x_1x_2. \tag{8.3}$$

The output is written in the file "output.dat" as below:

```
                M U L T I     L   A   Y   E   R      ALGORITHM


 CONTROL PARAMS:
 --------------


  NO.OF INPUT VARIABLES (M1)   5
  NO.OF DATA POINTS (N) 100
  PERCENTAGE OF TRAIN AND TEST POINTS (PE) 75
  NO.OF LAYERS (PM)   7
  WEIGHTAGE VALUE IN COMBINED CRIT (ALPHA) 0.5
  FREEDOM-OF-CHOICE AT EACH LAYER(CHO)   10 10 10 10 10 10 10
  NO.OF OPTIMAL MODELS (FF)   8
  NO.OF OUTPUT VARIABLES (M)   1



PERFORMANCE OF THE NET:
-------------------


 EQUATION NUMBER=   1

  LAYER=   1  SELECTED DESCRIPTION=    10
    ERROR GAUSS=   0
    COMBINED ERROR BEST=   0.644E-01     WORST=   0.275E+00
    RESIDUAL MSE   BEST=   0.304E-01     WORST=   0.961E-01
    RESIDUAL MSE=  0.304E-01 AT THE BEST COMBINED NODE
```

```
        --------------------
 LAYER=    2  SELECTED DESCRIPTION=    10
   ERROR GAUSS=    0
   COMBINED ERROR BEST=  0.202E-01    WORST=  0.538E-01
   RESIDUAL MSE   BEST=  0.160E-01    WORST=  0.334E-01
   RESIDUAL MSE=  0.177E-01 AT THE BEST COMBINED NODE
        --------------------
 LAYER=    3  SELECTED DESCRIPTION=    10
   ERROR GAUSS=    0
   COMBINED ERROR BEST=  0.196E-01    WORST=  0.223E-01
   RESIDUAL MSE   BEST=  0.113E-01    WORST=  0.194E-01
   RESIDUAL MSE=  0.173E-01 AT THE BEST COMBINED NODE
        --------------------
 LAYER=    4  SELECTED DESCRIPTION=    10
   ERROR GAUSS=    0
   COMBINED ERROR BEST=  0.108E-01    WORST=  0.162E-01
   RESIDUAL MSE   BEST=  0.592E-02    WORST=  0.117E-01
   RESIDUAL MSE=  0.608E-02 AT THE BEST COMBINED NODE
        --------------------
 LAYER=    5  SELECTED DESCRIPTION=    10
   ERROR GAUSS=    0
   COMBINED ERROR BEST=  0.614E-02    WORST=  0.127E-01
   RESIDUAL MSE   BEST=  0.470E-02    WORST=  0.878E-02
   RESIDUAL MSE=  0.509E-02 AT THE BEST COMBINED NODE
        --------------------
 LAYER=    6  SELECTED DESCRIPTION=    10
   ERROR GAUSS=    0
   COMBINED ERROR BEST=  0.593E-02    WORST=  0.861E-02
   RESIDUAL MSE   BEST=  0.392E-02    WORST=  0.509E-02
   RESIDUAL MSE=  0.418E-02 AT THE BEST COMBINED NODE
        --------------------
 LAYER=    7  SELECTED DESCRIPTION=    10
   ERROR GAUSS=    0
   COMBINED ERROR BEST=  0.496E-02    WORST=  0.664E-02
   RESIDUAL MSE   BEST=  0.349E-02    WORST=  0.418E-02
   RESIDUAL MSE=  0.362E-02 AT THE BEST COMBINED NODE
        --------------------


OPTIMAL MODELS:
--------------

 MODEL  1 ( LAYER  7   COMBINED= 0.599E-02 MIN BIAS= 0.713E-02
                            MSE= 0.457E-02 )
 COEFFICIENTS=
    0.431E+00

          1         2         3         4         5
 -0.813E-01 0.245E+00 0.326E+00-0.614E-02-0.161E+00
       --------------------------------------
 MODEL  2 ( LAYER  7   COMBINED= 0.496E-02 MIN BIAS= 0.566E-02
                            MSE= 0.415E-02 )
 COEFFICIENTS=
    0.433E+00

          1         2         3         4         5
 -0.948E-01 0.248E+00 0.340E+00-0.593E-02-0.167E+00
       --------------------------------------
 MODEL  3 ( LAYER  7   COMBINED= 0.550E-02 MIN BIAS= 0.654E-02
```

```
                           MSE= 0.420E-02 )
COEFFICIENTS=
   0.433E+00

          1          2          3          4          5
-0.941E-01 0.250E+00 0.339E+00-0.749E-02-0.168E+00
      --------------------------------------
MODEL  4 ( LAYER  7   COMBINED= 0.570E-02 MIN BIAS= 0.685E-02
                           MSE= 0.423E-02 )
COEFFICIENTS=
   0.432E+00

          1          2          3          4          5
-0.937E-01 0.250E+00 0.339E+00-0.795E-02-0.168E+00
      --------------------------------------
MODEL  5 ( LAYER  7   COMBINED= 0.580E-02 MIN BIAS= 0.663E-02
                           MSE= 0.483E-02 )
COEFFICIENTS=
   0.431E+00

          1          2          3          4          5
-0.813E-01 0.245E+00 0.326E+00-0.619E-02-0.161E+00
      --------------------------------------
MODEL  6 ( LAYER  6   COMBINED= 0.593E-02 MIN BIAS= 0.702E-02
                           MSE= 0.458E-02 )
COEFFICIENTS=
   0.431E+00

          1          2          3          4          5
-0.812E-01 0.245E+00 0.326E+00-0.617E-02-0.161E+00
      --------------------------------------
MODEL  7 ( LAYER  7   COMBINED= 0.576E-02 MIN BIAS= 0.696E-02
                           MSE= 0.421E-02 )
COEFFICIENTS=
   0.432E+00

          1          2          3          4          5
-0.923E-01 0.251E+00 0.338E+00-0.828E-02-0.169E+00
      --------------------------------------
MODEL  8 ( LAYER  7   COMBINED= 0.578E-02 MIN BIAS= 0.700E-02
                           MSE= 0.423E-02 )
COEFFICIENTS=
   0.432E+00

          1          2          3          4          5
-0.915E-01 0.251E+00 0.338E+00-0.863E-02-0.169E+00
      --------------------------------------
```

2 COMPUTATIONAL ASPECTS OF COMBINATORIAL ALGORITHM

The algorithm given is for a single-layered structure. The mathematical description of a system is represented as a reference function in the form of discrete Volterra series in multivariate data and finite-difference equations in time series data.

$$y = a_0 + \sum_{i=1}^{l} a_i x_i + \sum_{i=1}^{l}\sum_{j=1}^{l} a_{ij} x_i x_j + \sum_{i=1}^{l}\sum_{j=1}^{l}\sum_{k=1}^{l} a_{ijk} x_i x_j x_k + \cdots$$

$$y_t = a_0 + a_1 y_{t-1} + a_2 y_{t-2} + \cdots, \tag{8.4}$$

where y and x_i are the desired and input variables in the first polynomial; l is the number of input variables; y_t is the desired output at the time t; $y_{t-1}, y_{t-2}, \cdots$ are the delayed arguments of the output as inputs in the finite-difference scheme.

The combinatorial algorithm frames all combinations of partial functions from the given reference function. If the reference function is a linear function; for example,

$$y = f(x_1, x_2) = a_0 + a_1 x_1 + a_2 x_2, \tag{8.5}$$

then it generates

$$y = a_0,\ y = a_1 x_1,\ y = a_2 x_2,\ y = a_0 + a_1 x_1,$$

$$y = a_0 + a_2 x_2,\ y = a_1 x_1 + a_2 x_2,\ \text{and}\ y = a_0 + a_1 x_1 + a_2 x_2. \tag{8.6}$$

Suppose there are $m(= 3)$ parameters in the reference function, then the total combinations are $2^m - 1(= 7)$. The "structure of functions" is used to generate these partial models.

a_2	a_1	a_0
0	0	1
0	1	0
1	0	0
0	1	1
1	0	1
1	1	0
1	1	1

where each row indicates a partial function with its parameters represented by "1," the number of rows indicates the total number of units, and the number of columns indicates total number of parameters in the full description. This matrix is referred further in forming the normal equations.

The weights are estimated for each partial equation by using the least squares technique with a training data set at each unit and computed at its threshold measure according to the external criterion using the test set. Then the unit errors are compared with each other and the better functions are selected for their output responses and evaluated further.

For simplicity, the external criteria used in this algorithm are the minimum-bias, regularity, and combined criterion of minimum-bias and regularity.

Three ways of splitting data are used here: sequential, alternative, and dispersion analysis. The user can choose one of them or experiment with them for different types of splittings.

The program works for time series data as well as multivariate data. If it is time series data, the user has to specify the number of autoregressive terms in the finite-difference function and supply the "input.dat" file with the time series data. If it is multivariate data, one has to specify the number of input variables and supply the "input.dat" file with the rows of the data points for output and input variables.

The program listing and an example with the sample output are given below.

2.1 Program listing

```
C
C*********************************************************************
C THIS PROGRAM IS THE RESULT OF EFFORTS FROM VARIOUS GRADUATE STUDENTS
C  AND RESEARCH PROFESSIONALS AT THE COMBINED CONTROL SYSTEMS GROUP OF
C  INSTITUTE OF CYBERNETICS, KIEV (UKRAINE)
```

```
C*****************************************************************************
C     SINGLE LAYER COMBINATORIAL INDUCTIVE ALGORITHM
C
C       M  - TOTAL NO.OF DATA POINTS
C       MP - NO.OF POINTS IN TEST SET
C       MA1 - NO.OF POINTS IN EXAMIN SET
C       IT - ORDER OF THE MODEL
C       L  - NO.OF INPUT VARIABLES
C       NB - FREEDOM OF CHOICE (NO.OF BEST MODELS AT THE OUTPUT)
C       IH - NO.OF DISCRETE POINTS IN SIGNAL DATA
C       G(IH)- DISCRETE SIGNAL DATA
C       LM - SELECTION CRITERION NO.
C       IS = -1 - DATA IS SPLITTED ON THE BASIS OF STD.DEVIATIONS
C          =  0 - DATA IS SPLITTED ALTERNATELY
C          =  1 - DATA IS SPLITTED SEQUENTIALLY
C       Y1(M) - DEPENDENT VARIABLE (OUTPUT VECTOR)
C       X1(M,L) - INDEPENDENT VARIABLES (INPUT MATRIX)
C       Y(M) - OUTPUT VECTOR AFTER SEPARATION OF DATA
C       X(M,L) - INPUT MATRIX AFTER SEPARATION OF DATA
C
C SUBROUTINE DATA - WHICH SUPPLIES THE DISCRETE SIGNAL
C                   DATA G(IH)
C SUBROUTINE FORM - WHICH FORMS THE OUTPUT VECTOR Y1(M)AND
C                   THE INPUT MATRIX  X1(M,L)FROM THE
C                         DISCRETE SIGNAL G(IH). THIS IS MAINLY
C                         FOR FORMING FINITE-DIFFERENCE
C                         EQUATIONS
C
C****************************************************************************
C
C    MAIN PROGRAM
C
           DIMENSION D(100)
           INTEGER NP(9)
           COMMON /GAMA/G(100)
           COMMON /X1Y1/X1(100,15),Y1(100)
           COMMON /XYUD/X(100,15),Y(100),UD(100,15)
           COMMON /PS/NB,N,PS(15,16)
           COMMON /INIT/M,MP,MA1,L
C
             OPEN(3,FILE='results.dat')
             OPEN(8,FILE='innl.dat')
C
        WRITE(3,12)
  12    FORMAT(8X,'SINGLE  L A Y E R E D  COMBINATORIAL ALGORITHM'///)
             WRITE (*,230)
 230         FORMAT(2X,'GIVE TOTAL DISCRETE POINTS')
        READ(*,*)IH
C
         WRITE(*,235)
 235     FORMAT(2X, 'TIME SERIES (1)/ MULTIVARIATE DATA (2)??')
         READ(*,301)IIH
         IF (IIH.EQ.2) GOTO 350
C
        CALL DATA(IH)
        WRITE(3,240)
 240    FORMAT(2X,'DATA:')
             WRITE (3,100) (G(I),I=1,IH)
 100         FORMAT (3X,5F12.2)
        WRITE(3,303)
```

```
 303    FORMAT(//)
        WRITE(*,300)
 300    FORMAT(//3X,'Give No of AR terms in model= ')
        READ(*,301)L
 301    FORMAT(I2)
            CALL FORM(IH,M,L)
            IF (IIH.EQ.1) GOTO 355
C
 350      M =IH
          WRITE(*,245)
 245      FORMAT(2X, 'GIVE NO.OF INPUT VARIABLES??')
          READ(*,301)L
          DO 91 I =1,M
          READ(8,*) Y1(I), (X1(I,J), J=1,L)
  91      CONTINUE
C
 355    WRITE(3,250)M
        WRITE(*,250)M
 250    FORMAT(//2X,'TOTAL NO.OF DATA PTS. =',I3//)
        WRITE(*,280)
 280    FORMAT(2X,'GIVE NO.OF TRAINING PTS??')
        READ(*,290)ME
 290    FORMAT(I2)
        WRITE(*,260)
 260    FORMAT(2X,'GIVE NO.OF TESTING PTS??')
        READ(*,270)MP
 270    FORMAT(I2)
        MA1=M-(MP+ME)
        IF(MA1.LE.0)MA1=0
        WRITE(*,999)
 999    FORMAT(1H$,'DATA SETS BY (-1 DISP, 0 ALTER, 1 SEQUEN)?')
        READ(*,220)IS
 220    FORMAT(I2)
C
             YM=0.0
             DO 5 I=1,M
             YM=YM+Y1(I)
   5         CONTINUE
             YM=YM/M
             IF(IS)15,16,17
  15         DO 7 I=1,M
   7         Y1(I)=(Y1(I)-YM)/YM
             DO 8 I=1,L
             XM=0.0
             DO 9 J=1,M
   9         XM=XM+X1(J,I)
             XM=XM/M
             DO 10 J=1,M
  10         X1(J,I)=(X1(J,I)-XM)/XM
   8         CONTINUE
             DO 11 I=1,M
             D(I)=Y1(I)**2
             DO 13 J=1,L
  13         D(I)=D(I)+X1(I,J)**2
             D(I)=D(I)/(L+1)
  11         CONTINUE
             CALL RANG (D,NP,M)
             DO 14 I=1,M
             I2=M-I+1
             I1=NP(I2)
```

```
          Y(I)=Y1(I1)
          DO 14 J=1,L
          X(I,J)=X1(I1,J)
  14      CONTINUE
          GO TO 3
  16      I1=0
          DO 18 L1=1,2
          DO 18 I=L1,M,2
          I1=I1+1
          Y(I1)=Y1(I)
          DO 18 J=1,L
          X(I1,J)=X1(I,J)
  18      CONTINUE
          GO TO 3
  17      DO 19 I=1,M
          Y(I)=Y1(I)
          DO 19 J=1,L
          X(I,J)=X1(I,J)
  19      CONTINUE
   3      CONTINUE
          CALL COMBI
      NOB=NB
          STOP
          END
C
```

Subroutines used

```
      SUBROUTINE DATA(IH)
      COMMON /GAMA/G(100)
      DO 300 I=1,IH
      READ(8,100)G(I)
 100   format(f12.6)
 300  CONTINUE
      RETURN
      END
C
          SUBROUTINE FORM(IH,M,L)
          COMMON /GAMA/G(100)
          COMMON /X1Y1/X(100,15),Y(100)
          M1=0
          L1=L+1
          DO 2 I=L1,IH
          M1=M1+1
          Y(M1)=G(I)
          DO 1 J=1,L
          IJ=I-J
   1      X(M1,J)=G(IJ)
   2      CONTINUE
          M=M1
          RETURN
          END
C
          SUBROUTINE COMBI
          REAL KCH,IQ
      DIMENSION OS(16),OA(16),FS(15,16),FS1(15,16),
     1    ID(15),P(15),P1(15),IA(15),IP(15)
      COMMON /XYUD/X(100,15),Y(100),UD(100,15)
      COMMON /PS/NB,N,PS(15,16)
```

```
       COMMON /INIT/M,MP,MA1,L
65      FORMAT(/2X,'MODEL ORDER (IT)=',I3/2X,'NO INPUT VAR.(L)=',
     1 I3/2X,'TOTAL NO.PTS.(M)=',I3/2X,'N0.PTS.TESTSET(MP)=',I3/2X,
     2 'NO.PTS.EXAM.SET (MA1)=',I3/)
           WRITE(*,64)
64       FORMAT(2X,'GIVE ORDER OF THE MODEL??')
           READ(*,*)IT
           WRITE (3,65)IT,L,M,MP,MA1
           N=1
           DO 38 J1=1,L
38         N=N*(IT+J1)/J1
           KCH=2.**N-1
           WRITE(3,50)N,KCH
           WRITE(*,50)N,KCH
50      FORMAT (/4X, 'NO.TERMS IN FULL MODEL=',I3/4X,
     1    'NO.PARTIAL MODELS=',F12.0/)
       WRITE(*,320)
320    FORMAT(///2X,'NO OF OPTIMAL MODELS (NB)??')
       READ(*,330)NB
330    FORMAT(I2)
       WRITE(3,321)NB
321    FORMAT(//2X,'NO OF OPTIMAL MODELS = ',I2)
C*********************************
C - FORMING CONDITIONAL EQUATIONS
C*********************************
           N1=N+1
           MA=M-MA1
           MO=MA-MP
           MPR=MO+1
C*********************************
C   STRUCTURE OF FULL POLYNOMIAL
C*********************************
           CALL FORD(IT,L,M,N,IP)
            WRITE(*,100)
100    FORMAT(1H$,'GIVE SELECT CRIT(1-REGUL,2-MINBIAS,3-COMBINED)?')
       READ(*,101)LM
101    FORMAT(I2)
C*********************************
C    FORMING NORMAL EQUATIONS
C*********************************
           CALL NOS(N,N1,M,1,MO,FS)
           CALL NOS(N,N1,M,MPR,MA,FS1)
C************************************
C    SORTING OF PARTIAL DESCRIPTIONS
C************************************
           IQ=0.0
C*********************************************
C    CALCULATION OF COEFFICIENTS OF THE MODELS
C*********************************************
41         IQ=IQ+1
           CALL DICH(IQ,ID,N,2)
           KB=0
           DO 60 I4=1,N
60         KB=KB+ID(I4)
           KB1=KB+1
           CALL PAP(ID,N,N1,KB,KB1,FS,P,IA)
           CALL PAP(ID,N,N1,KB,KB1,FS1,P1,IA)
C**********************************************
C    VALIDATION OF MODELS BY SELECTING CRITERION
C**********************************************
```

```
          IF(LM-2)92,93,92
92        OSH=0.0
          DO 54 J=MPR,MA
          Z=0.0
          DO 55 I=1,N
55        Z=Z+P(I)*UD(J,I)
54        OSH=OSH+(Z-Y(J))**2
          OSH1=SQRT(OSH)/MP
          IF (LM-2)51,93,93
93        OSH=0.0
          DO 56 J3=1,MA
          Z=0.0
          AF=0.0
          DO 57 I3=1,N
          Z=Z+P(I3)*UD(J3,I3)
57        AF=AF+P1(I3)*UD(J3,I3)
56        OSH=OSH+(Z-AF)**2
          OSH2=SQRT(OSH)/MA
          IF(LM-2)51,52,53
51        OSH=OSH1
          GO TO 59
52        OSH=OSH2
          GO TO 59
53        OSH=OSH1+OSH2
C*********************************
C    SELECTION OF THE NB BEST MODELS
C*********************************
59        IF (IQ-NB)42,42,43
42        JF=IQ
          GO TO 47
43        IF(NB-1)45,44,45
44        R5=OS(1)
          GO TO 49
45        CALL FMAX(OS,NB,R5,JF)
49        IF(OSH-R5)47,41,41
47        OS (JF)=OSH
          DO 48 I5=1,N
48        PS (I5,JF)=P(I5)
          IF (IQ.LT.KCH)GO TO 41
C*****************************************************
C    SELECTION CRITERION FOR SORTING OUT THE BEST MODELS
C*****************************************************
          IF(LM-2)88,89,90
88        WRITE(3,85)
85        FORMAT(/4X,'SORTING OUT BY REGULARITY CRITERION')
          GOTO 91
89        WRITE(3,84)
84        FORMAT (/4X,'SORTING OUT BY MINIMUM-BIAS CRITERION')
          GOTO 91
90        WRITE(3,80)
80        FORMAT(/4X,'SORTING OUT BY COMBINED CRITERION')
91        CONTINUE
          WRITE(3,75)
75        FORMAT(4X,'DEPTH OF THE MINIMUM')
          WRITE(3,68)(OS(K),K=1,NB)
C*********************************
C    ADAPTATION OF THE COEFFICIENTS
C*********************************
          DO 76 K=1,NB
          DO 71 I6=1,N
```

```
          IF (PS(I6,K))72,73,72
73        ID(I6)=0
          GO TO 71
72        ID(I6)=1
71        CONTINUE
          CALL NOS(N,N1,M,1,MA,FS)
          KB=0
          DO 70 I7=1,N
70        KB=KB+ID(I7)
          KB1=KB+1
          CALL PAP(ID,N,N1,KB,KB1,FS,P,IA)
          DO 58 I8=1,N
58        PS(I8,K)=P(I8)
          OSH=0.0
          AF=0.0
          DO 77 J=1,M
          Z=0.0
          DO 78 I=1,N
78        Z=Z+PS(I,K)*UD(J,I)
          OSH=OSH+(Z-Y(J))**2
          AF=AF+Y(J)**2
          IF(J-MA)77,79,77
79        R5=OSH
          R1=AF
77        CONTINUE
          R7=R5/R1
          OS(K)=SQRT(R7)
          IF (MA1)83,83,86
83        OA(K)=0.0
          GO TO 76
86        OA(K)=SQRT((OSH-R5)/(AF-R1))
76        CONTINUE
C***********************************************
C    PRINTING OUT THE PARAMETERS OF BEST MODELS
C***********************************************
          WRITE (3,67)
67        FORMAT(4X,'COEFFICIENTS:')
          DO 94 J=1,NB
94        WRITE(3,69) (PS(I,J),I=1,N)
69        FORMAT (8F10.3)
          WRITE (3,95)
95        FORMAT (4X,'MSE AFTER ADAPTATION')
          WRITE (3,68) (OS(K),K=1,NB)
68        FORMAT (2X,5E12.3)
          WRITE (3,87)
87        FORMAT (4X,'ERROR ON THE EXAMIN SET')
          WRITE (3,68) (OA (K),K=1,NB)
          RETURN
          END
C
C
          SUBROUTINE FMAX(G,JE,C,M)
          DIMENSION G(16)
          C=G(1)
          M=1
          I=2
20        IF (C-G(I))21,22,22
21        C=G(I)
          M=I
22        I=I+1
```

```
      IF(I-JE)20,20,23
23    RETURN
      END
C
      SUBROUTINE PAP(ID,N,N1,IS,IS1,FS,P,IA)
      DIMENSION ID(15),FS(15,16),P(15),IA(15)
      DIMENSION QN(15,16),R(15)
      K=0
      DO 34 I=1,N
      P(I)=0.0
      IF (ID(I)) 35,34,35
35    K=K+1
      IA(K)=I
      QN(K,IS1)=FS(I,N1)
34    CONTINUE
      DO 36 I=1,IS
      DO 36 J=1,IS
      L1=IA(I)
      L2=IA(J)
36    QN(I,J)=FS(L1,L2)
      CALL GAUSS(QN,IS,IS1,R)
      DO 37 K=1,IS
      L3=IA(K)
37    P(L3)=R(K)
      RETURN
      END
C
      SUBROUTINE DICH(JQ,ID,JN,JS)
      DIMENSION ID(15)
      REAL JQ,JL
      JL=JQ
      DO 11 I=1,JN
11    ID(I)=0
      IF(JS-1)15,19,15
15    I=0
      JN1=JN+1
16    I=I+1
      IF(JS-JL)17,17,18
17    JC=JL/JS
      L1=JN1-I
      ID(L1)=JL-JC*JS
      JL=JC
      GO TO 16
18    L2=JN1-I
      ID(L2)=JL
19    RETURN
      END
C
      SUBROUTINE FORD(ICT,L,M,N,IP)
      REAL IC
      DIMENSION IP(15)
      COMMON /XYUD/X(100,15),Y(100),UD(100,15)
      WRITE (3,24)
24    FORMAT(4X,'STRUCTURE OF THE FULL POLYNOMIAL')
      IC=0.0
      JF=0
      ICT1=ICT+1
25    CALL DICH(IC,IP,L,ICT1)
      IC=IC+1
      IS=0
```

```
          DO 26 J1=1,L
26        IS=IS+IP(J1)
          IF(IS-ICT)27,27,25
27        JF=JF+1
28        FORMAT(5X,17I3)
          WRITE(3,28)(IP(J),J=1,L)
          DO 32 I=1,M
          UD(I,JF)=1.0
          IF(JF-1)32,32,81
81        DO 31 J=1,L
          IF(IP(J))31,31,82
82        UD(I,JF)=UD(I,JF)*X(I,J)**IP(J)
31        CONTINUE
32        CONTINUE
          IF(IP(1)-ICT)25,30,30
30        RETURN
          END
C
          SUBROUTINE NOS(N,N1,ML,MB,M1,FS)
          DIMENSION FS(15,16)
          COMMON /XYUD/X(100,15),Y(100),UD(100,15)
          DO 31 I=1,N
          FS(I,N1)=0.0
          DO 31 J=MB,M1
31        FS(I,N1)=FS(I,N1)+UD(J,I)*Y(J)
          DO 32 I1=1,N
          DO 32 J1=1,N
          FS(I1,J1)=0.0
          DO 32 K=MB,M1
32        FS(I1,J1)=FS(I1,J1)+UD(K,I1)*UD(K,J1)
          RETURN
          END
C
          SUBROUTINE RANG(X,NP,N)
          DIMENSION X(100),XD(100)
          INTEGER NP(100),ND(100)
          DO 1 I=1,N
          XD(I)=X(I)
1         ND(I)=I
          N1=N
2         CALL FMAX(XD,N1,XM,K)
          NP(N1)=ND(K)
          K1=K+1
          DO 3 I=K1,N1
          XD(I-1)=XD(I)
3         ND(I-1)=ND(I)
          N1=N1-1
          IF (N1.GE.2)GO TO 2
          NP(1)=ND(1)
          RETURN
          END
C
      SUBROUTINE GAUSS(A,N,L,X)
      DIMENSION A(15,16),X(15)
      L=N+1
      NN=N-1
      DO 88 K=1,NN
      J=K
      KK=K+1
      DO 100 I=KK,N
```

```
      IF(ABS(A(J,K)).LT.ABS(A(I,K)))J=I
100   CONTINUE
      IF(J.EQ.K)GOTO 11
      DO 300 I=1,L
      T=A(K,I)
      A(K,I)=A(J,I)
      A(J,I)=T
300   CONTINUE
 11   DO 88 J=KK,N
      IF(A(K,K).EQ.0.)GOTO 600
      D=-A(J,K)/A(K,K)
      DO 88 I=1,L
      A(J,I)=A(J,I)+D*A(K,I)
 88   CONTINUE
      IF(A(N,N).EQ.0.)GOTO 600
      X(N)=A(N,L)/A(N,N)
      NN=N-1
      DO 500 J=1,NN
      K=N-J
      SUM=0.0
      NNN=N-K
      DO 200 JJ=1,NNN
      M=K+JJ
      SUM=SUM+A(K,M)*X(M)
200   CONTINUE
      IF(A(K,K).EQ.0.)GOTO 600
      X(K)=(A(K,L)-SUM)/A(K,K)
500   CONTINUE
600   RETURN
      END
```

2.2 Sample outputs

Example.

I. Here the case of multivariate data is considered. The output data is generated from the equation:

$$y = 0.433 - 0.095\,x_1 + 0.243\,x_2 + 0.35\,x_1^2 - 0.18\,x_1\,x_2 + \epsilon,$$

where x_1, x_2 are randomly generated input variables y is the output variable, and ϵ is the noise added to the data. The "input.dat" file is arranged for 100 measured points with the values of $y, x_1, x_2, x_1^2, x_2^2, x_1x_2$.

y	x_1	x_2	x_1^2	x_2^2	x_1x_2

The initial control parameters of the program are fed through the terminal as it asks inputting the values, starting with

```
  GIVE TOTAL DISCRETE POINTS
100

  TIME SERIES (1)/MULTIVARIATE DATA (2)??
2

  GIVE NO.OF INPUT VARIABLES??
5
```

```
  GIVE NO.OF TRAINING PTS??
80

  GIVE NO.OF TESTING PTS??
15

  DATA SPLITTING BY (-1 DISP, 0 ALTER, 1 SEQUEN)??
1

  GIVE ORDER OF THE MODEL??
1
```

Then it on the screen displays information to the user on how to feed further information:

```
     NO.OF TERMS IN FULL MODEL = 6
     NO.OF PARTIAL MODELS = 63
```

The user has to feed further data such as the number of optimal models to be selected and the selection criterion to be used.

```
  NO.OF OPTIMAL MODELS (NB)??
8

  GIVE SELECT CRIT (1-REGUL, 2-MINBIAS, 3-COMBINED)?
1
```

The output is written in a file "results.dat" given here:

```
        SINGLE  L A Y E R E D  COMBINATORIAL ALGORITHM

  TOTAL NO.OF DATA PTS. =100

  MODEL ORDER (IT)=  1
  NO INPUT VAR.(L)=  5
  TOTAL NO.PTS.(M)=100
  N0.PTS.TESTSET(MP)= 15
  NO.PTS.EXAM.SET (MA1)=  5

    NO.TERMS IN FULL MODEL=  6
    NO.PARTIAL MODELS=         63.

  NO OF SELECT MODELS =  8
    STRUCTURE OF THE FULL POLYNOMIAL
       0  0  0  0  0
       0  0  0  0  1
       0  0  0  1  0
       0  0  1  0  0
       0  1  0  0  0
       1  0  0  0  0

    SORTING OUT BY REGULARITY CRITERION
    DEPTH OF THE MINIMUM
     0.647E-04    0.652E-04    0.219E-02    0.364E-02    0.352E-02
     0.219E-02    0.394E-02    0.409E-02
```

```
COEFFICIENTS:
 0.434    -0.180    0.000    0.350    0.243    -0.095
 0.434    -0.180    0.000    0.350    0.243    -0.095
 0.417    -0.192    0.005    0.266    0.242     0.000
 0.442     0.000    0.000    0.174    0.161     0.000
 0.437     0.000   -0.030    0.173    0.190     0.000
 0.416    -0.191    0.000    0.265    0.247     0.000
 0.458     0.000   -0.033    0.293    0.196    -0.127
 0.463     0.000    0.000    0.292    0.163    -0.126
MSE AFTER ADAPTATION
 0.469E-03   0.470E-03   0.116E-01   0.306E-01   0.303E-01
 0.116E-01   0.260E-01   0.264E-01
ERROR ON THE EXAMIN SET
 0.516E-03   0.527E-03   0.901E-02   0.268E-01   0.266E-01
 0.900E-02   0.182E-01   0.182E-01
```

The STRUCTURE OF THE FULL POLYNOMIAL helps to read the coefficients in order. For example, the first row indicates the constant term; the second row which contains 1 at the fifth column indicates that the second coefficient corresponds to the fifth variable; similarly, the third row for the fourth variable, and so on until the last row indicates the coefficient of first variable.

The COEFFICIENTS are given for eight optimal models; they are given according to the order of STRUCTURE OF THE FULL POLYNOMIAL as $a_0, a_5, a_4, a_3, a_2,$ and a_1. The DEPTH OF THE MINIMUM for regularity criterion, MSE AFTER ADAPTATION, and ERROR ON THE EXAMIN SET are given for each model in the order. The first model is the best one among all; this is read as

$$y = 0.434 - 0.180x_1x_2 + 0.0x_2^2 + 0.350x_1^2 + 0.243x_2 - 0.095x_1 \tag{8.7}$$

II. The above example can also be solved alternatively by forming the "input.dat" with the variables $y, x_1,$ and x_2 as

$$\boxed{y}\boxed{x_1}\boxed{x_2}.$$

The control parameter values are the same as above, except the number of variables and the value of the order of the model which must be fed as

```
  GIVE NO.OF INPUT VARIABLES??
2

  GIVE ORDER OF THE MODEL??
2
```

Then the output in "results.dat" is shown below:

```
        SINGLE  L A Y E R E D  COMBINATORIAL ALGORITHM

  TOTAL NO.OF DATA PTS. =100

  MODEL ORDER (IT)=  2
  NO INPUT VAR.(L)=  2
  TOTAL NO.PTS.(M)=100
  NO.PTS.TESTSET(MP)= 15
  NO.PTS.EXAM.SET (MA1)=  5

    NO.TERMS IN FULL MODEL=  6
```

```
   NO.PARTIAL MODELS=          63.

 NO OF SELECT MODELS =  8
   STRUCTURE OF THE FULL POLYNOMIAL
      0  0
      0  1
      0  2
      1  0
      1  1
      2  0

   SORTING OUT BY REGULARITY CRITERION
   DEPTH OF THE MINIMUM
    0.364E-02   0.646E-04   0.219E-02   0.651E-04   0.394E-02
    0.352E-02   0.409E-02   0.219E-02
   COEFFICIENTS:
    0.442       0.161       0.000       0.000       0.000       0.174
    0.434       0.243       0.000      -0.095      -0.180       0.350
    0.417       0.242       0.005       0.000      -0.192       0.266
    0.434       0.243       0.000      -0.095      -0.180       0.350
    0.458       0.196      -0.033      -0.127       0.000       0.293
    0.437       0.190      -0.030       0.000       0.000       0.173
    0.463       0.163       0.000      -0.126       0.000       0.292
    0.416       0.247       0.000       0.000      -0.191       0.265
   MSE AFTER ADAPTATION
    0.306E-01   0.469E-03   0.116E-01   0.470E-03   0.260E-01
    0.303E-01   0.264E-01   0.116E-01
   ERROR ON THE EXAMIN SET
    0.268E-01   0.516E-03   0.901E-02   0.527E-03   0.182E-01
    0.266E-01   0.182E-01   0.900E-02
```

Notice the change in the order of the coefficients. The first row of the STRUCTURE OF THE POLYNOMIAL indicates that the first coefficient term is the constant term; the second row indicates that the second coefficient term corresponds to the variable x_2; the third row indicates that the third coefficient term corresponds to the variable x_2^2; the fourth row indicates that the fourth coefficient term corresponds to the variable x_1; the fifth row corresponds to the variable x_1x_2; and the sixth row indicates the variable x_1^2. The second model is the best optimal model among the eight models; this is read as

$$y = 0.434 + 0.243x_2 + 0.0x_2^2 - 0.095x_1 - 0.180x_1x_2 + 0.350x_1^2. \tag{8.8}$$

3 COMPUTATIONAL ASPECTS OF HARMONICAL ALGORITHM

This is used mainly to identify the harmonical trend of oscillatory processes [127]. It is assumed that the effective reference functions of such processes are in the form of a sum of harmonics with nonmultiple frequencies. This means that the harmonical function is formed by several sinusoids with arbitrary frequencies which are not necessarily related.

Let us suppose that function $f(t)$ is the process having a sum of m harmonic components with distinct frequencies $w_1, w_2, \cdots, w_m$.

$$f(t) = \mathcal{A}_0 + \sum_{k=1}^{m}[\mathcal{A}_k \sin(w_k t) + \mathcal{B}_k \cos(w_k t)], \tag{8.9}$$

where $\mathcal{A}_0$ is the constant term; $\mathcal{A}_k$ and $\mathcal{B}_k$ are the coefficients; and $w_i \neq w_j,\ i \neq j,\ 0 < w_i < \pi,\ i = 1, 2, \cdots, m$. The process has discrete data points of interval length of N $(1 \leq t \leq N)$.

A balance relation is derived using the trigonometric properties for a fixed point i and any p;

$$\sum_{p=0}^{m-1} \mu_p \,[f(i+p)+f(i-p)] = f(i+m)+f(i-m), \tag{8.10}$$

where $\mu_0, \mu_1, \cdots, \mu_{m-1}$ are the weighing coefficients. This is considered a balance relation of the process and is used as an objective function

$$b_i = [f(i+m)+f(i-m)] - \sum_{p=0}^{m-1} \mu_p [f(i+p)+f(i-p)]. \tag{8.11}$$

If the process is expressed exactly in terms of a given sum of harmonic components, then $b_i = 0$; i.e., the discrete values of $f(t)$ which are symmetric with respect to a point $i\,(m+1 \leq i \leq N-m)$ satisfy the balance relation. The coefficients μ_p are independent of i. It is possible to determine uniquely the coefficients μ_p, $p = 0, 1, \cdots, m-1$ from the balance relation for $i = m+1, \cdots, N-m$. $(N-m)-(m+1) \geq m-1$; i.e., $N \geq 3m$.

The standard trigonometric relation which is used in deriving the balance relation,

$$\mu_0 + \sum_{k=1}^{m-1} \mu_p \cos(pw_k) = \cos(mw_k) \tag{8.12}$$

helps in obtaining the frequencies w_k. This could be formed as mth degree algebraic equation in $\cos w$:

$$\mathcal{D}_m(\cos w)^m + \mathcal{D}_{m-1}(\cos w)^{m-1} + \cdots + \mathcal{D}_1(\cos w) + \mathcal{D}_0 = 0, \tag{8.13}$$

where $\mathcal{D}_i$, $i = 0, 1, \cdots, m$ are the functions of μ_p.

Substituting the values of μ_p, the above equation can be solved for m frequencies w_k of harmonics by using the standard numerical techniques. Various combinations of the harmonic components are formed with the frequencies w_k. The coefficients $\mathcal{A}_0, \mathcal{A}_k$, and $\mathcal{B}_k$ are estimated for each combination by using the least-squares technique. The best combination as an optimal trend is selected according to the value of the balance criterion.

The algorithm functions as below:

The discrete data is to be supplied as training set A and testing set B; one can allot a separate checking set C for examining the final optimal trend; i.e., $N = N_A + N_B + N_C$. The maximum number of harmonics is chosen as M_{max} $(< N/3)$. The coefficients μ_p are estimated by using the least squares technique by forming the balance equations with the training set. The system of equations has the form:

$$\sum_{p=0}^{m-1} \mu_p [y(i+p) + y(i-p)] = y(i+m) + y(i-m);$$

$$i = m+1, \cdots, N_A - m. \tag{8.14}$$

By substituting the values of μ_p in the above mth order polynomial in $\cos w$, the frequencies are estimated; the m roots of the polynomial uniquely determine the m frequencies w_k. These frequencies are fed through the input layer of multilayer structure where the complete sifting of harmonic trends would take place according to the inductive principle of self-organization. This is done by a successive increase in the number of terms of the harmonic components $m = 1, m = 2, m = 3, \cdots$ until $m = M_{max}$. The linear normal equations are constructed in the first layer for any $1 \leq m \leq M_{max}$ number of harmonics. The coefficients $\mathcal{A}_0, \mathcal{A}_k$, and $\mathcal{B}_k$ are estimated for all the combinations based on the training set using the least squares technique; the balance functions are then evaluated. The best trends are selected. The output error residuals of the best trends are fed forward as inputs to the second layer. This procedure is repeated in all subsequent layers. The complexity of the model increases layer by layer as long as the value of the "imbalance" decreases. The optimal trend is the total combination of the harmonical components obtained from the layers. The performance of the optimal trend is tested on the checking set C.

The program listing and sample outputs for an example are given below.

3.1 Program listing

```
C
C**********************************************************************
C THIS PROGRAM IS THE RESULT OF EFFORTS FROM VARIOUS GRADUATE STUDENTS
C  AND RESEARCH PROFESSIONALS AT THE COMBINED CONTROL SYSTEMS GROUP OF
C  INSTITUTE OF CYBERNETICS, KIEV (UKRAINE)
C**********************************************************************
C
C      HARMONICAL INDUCTIVE LEARNING ALGORITHM
C
C
C        N - NO.OF TRAINING SET POINTS
C        NP - NO.OF TEST SET POINTS
C        NE - NO.OF EXAMIN SET POINTS
C        PT - NO.OF PREDICTION POINTS
C        JFM - MAX NO.OF FREQUENCIES
C        JF - FREEDOM OF CHOICE
C        NRM - NO.OF SERIES IN HARMONICAL TREND
C        NN = N+NP+NE
C        NPT = NN+PT
C        G(NN) - DISCRETE SIGNAL DATA
C        APR(NPT) - HARMONICAL MODEL VALUES
C        MA - NO.OF LAG POINTS FOR SMOOTHING PROCEDURE(MOVING AVERAGE
C              VALUE). IF IT IS ONE, DATA REMAINS SAME
C
C**********************************************************************
C   MAIN PROGRAM
C
        INTEGER PT
        DIMENSION GY(120)
        COMMON /AB/G(120)
C
          OPEN(3,FILE='output.dat')
          OPEN(8,FILE='ts.dat')
C
        WRITE(3,4)
4       FORMAT(5X,'        L A Y E R E D  HARMONICAL ALGORITHM'/)
C
        WRITE(*,110)
110      FORMAT(3X,'GIVE NO.OF TRAIN, TEST & EXAM PTS?')
        READ(*,*)N,NP,NE
        NN=N+NP+NE
        WRITE(*,112)
```

```
112       FORMAT(3X,'GIVE NO.OF PRED PTS??')
          READ(*,*)PT
          NPT=NN+PT
          READ(8,*)(G(I),I=1,NN)
          FAX=G(1)
          DO 5 I=2,NN
          IF(G(I).GT.FAX)FAX=G(I)
  5       CONTINUE
           DO 6 I=1,NN
           G(I)=G(I)/FAX
  6        CONTINUE
          WRITE(*,222)
222       FORMAT(3X,'GIVE MOVING AVERAGE VALUE (=1 or >1)?')
          READ(*,111)MA
111       FORMAT(I2)
          WRITE(*,333)
333       FORMAT(3X,'HOW MANY SERIES?')
          READ(*,111)NRM
          WRITE(*,114)
114       FORMAT(3X,'GIVE MAX NO.OF FREQS(<=15)??')
          READ(*,*)JFM
          JF2=2*JFM+2
          WRITE(*,115)
115       FORMAT(3X,'GIVE FREEDOM OF CHOICE(< MAX FREQS)??')
          READ(*,*)JF
          SMA=0.0
          DO 7 I=1,MA
  7       SMA=SMA+G(I)
          SMA=SMA/MA
          GY(1)=SMA
          IX=1
          MHR=MA+1
          DO 8 I=MHR,NN
          IX=IX+1
          IX1=IX-1
          IMA=I-MA
          GY(IX)=GY(IX1)+(G(I)-G(IMA))/MA
  8       CONTINUE
          DO 9 I=1,IX
          G(I)=GY(I)
  9       CONTINUE
          CALL HARMAN(N,NP,NE,NN,PT,JF,JFM,NRM,0,1,JF2,NPT)
          STOP
          END
C
```

Subroutines used

```
          SUBROUTINE WB(N1,M,M1,IER,KA)
          COMMON /BC/X(160),Y(160),Y1(31),Y2(31),A(31),C(31,32),W(15)
          N=N1-2
           M1=M+1
          NM=N-M
          DO 1 I=1,N
 1        Y(I)=X(I+2)-X(I)
          DO 2 J=1,M1
          Y1(J)=0.0
          DO 2 I=1,M
 2        C(I,J)=0.0
```

```
      DO 3 I=M1,NM
      K=I-M
      R=K
      E=1.0/R
      DO 4 J=1,M1
      I1=I+J-1
      I2=I-J+1
      Y2(J)=Y1(J)+Y(I1)+Y(I2)
      IF(KA-0)4,10,4
10    Y2(J)=Y2(J)-Y1(J)
 4    Y1(J)=Y2(J)
 8    DO 5 K1=1,M
      DO 5 J=K1,M1
      E1=Y2(K1)*Y2(J)
      IF(KA-2)5,11,5
11    E1=E1*E
 5    C(K1,J)=C(K1,J)+E1
      IF(KA-2)3,12,12
12    K=K-1
      IF(K-0)13,3,13
13     DO 7 J=1,M1
      I1=I+J-1-K
      I2=I-J+1-K
 7     Y2(J)=Y2(J)-Y(I1)-Y(I2)
      GOTO 8
 3     CONTINUE
      IF(M-1)14,77,14
14    DO 6 I=2,M
      I1=I-1
      DO 6 J=1,I1
 6    C(I,J)=C(J,I)
77    CALL GAUSS(C,M,M1,A,IER)
      RETURN
      END
C
      SUBROUTINE COEF(M,N,IER)
      COMMON /BC/Y(160),Y1(160),WK(31),B(31),A(31),HM(31,32),W(15)
      K=2*M
      K1=K+1
      DO 1 I=1,K1
      HM1=0.0
      IF(I-K)2,2,3
 2    AI=I
      BI=(AI+1.25)/2.
      II=INT(BI)
      BI=(AI+0.1)/2.
      AI=INT(BI)
      TI=BI-AI
      DO 4 J=I,K
      AJ=J
      BJ=(AJ+1.25)/2.
      JJ=INT(BJ)
      BJ=(AJ+0.1)/2.
      AJ=INT(BJ)
      TJ=BJ-AJ
      W1=W(II)-W(JJ)
      W2=W(II)+W(JJ)
      IF(II-JJ)6,5,6
 5    IF(ABS(TI-TJ)-0.01)8,30,30
30    S1=0.0
```

```
      GOTO 9
 8    S1=N
      GOTO 9
 6    AN=N
      CN=AN*W1/2.
      BN=W1/2.
      S1=SIN(CN)/SIN(BN)
 9    AN=N
      CN=AN*W2/2.
      BN=W2/2.
      S2=SIN(CN)/SIN(BN)
      AN=N+1
      BN=AN*W1/2.
      CN=AN*W2/2.
      CN1=COS(BN)
      CN2=COS(CN)
      SN1=SIN(BN)
      SN2=SIN(CN)
      IF(TI-0.25)11,10,10
10    IF(TJ-0.25)13,12,12
12    HM(I,J)=S1*CN1-S2*CN2
      GOTO 40
13    HM(I,J)=S2*SN2+S1*SN1
      GOTO 40
11    IF(TJ-0.25)15,14,14
14    HM(I,J)=S2*SN2-S1*SN1
      GOTO 40
15    HM(I,J)=S1*CN1+S2*CN2
40    HM(I,J)=0.5*HM(I,J)
 4   CONTINUE
      IF(TI-0.25)17,16,16
16    Y1(1)=SIN(W(II))
      Y1(2)=SIN(2*W(II))
      GOTO 18
17    Y1(1)=COS(W(II))
      Y1(2)=COS(2*W(II))
18    WK1=COS(W(II))
      DO 19 J=3,N
      Y1(J)=2.*WK1*Y1(J-1)-Y1(J-2)
19    HM1=HM1+Y1(J)*Y(J)
      HM(I,K+2)=HM1+Y1(1)*Y(1)+Y1(2)*Y(2)
      IF(TI-0.25)21,20,20
20    AN=N+1
      AN=AN*W(II)/2.
      H1=SIN(AN)
      GOTO 22
21    AN=N+1
      AN=AN*W(II)/2.
      H1=COS(AN)
22    AN=N
      BN=W(II)/2.
      CN=AN*BN
      HM(I,K1)=H1*SIN(CN)/SIN(BN)
      GOTO 24
 3   HM(I,K1)=N
      H1=0.
      DO 23 J=1,N
23    H1=H1+Y(J)
      HM(I,K+2)=H1
24    IF(I-2)1,25,25
```

```
25      I1=I-1
        DO 26 J=1,I1
26      HM(I,J)=HM(J,I)
1      CONTINUE
        K11 = K1+1
        CALL GAUSS(HM,K1,K11,B,IER)
        RETURN
        END
C
        SUBROUTINE WB1(M,M1,IER)
        COMMON /BC/YB(160),AP(160),WK(31),B(31),A(31),C(31,32),W(15)
        M1 = M+1
        DO 1 I=1,M
        DO 1 J=1,M1
        AJ=J-1
        AJ=AJ*W(I)
1       C(I,J)=COS(AJ)
        CALL GAUSS(C,M,M1,A,IER)
        RETURN
        END
C
        SUBROUTINE RANG(N,B)
        DIMENSION B(15)
        DO 1 I=1,N
        I1=I+1
        IF(I1-N)7,7,3
7       DO 1 J=I1,N
        IF(B(I)-B(J))1,1,2
2       R=B(I)
        B(I)=B(J)
        B(J)=R
1       CONTINUE
3       RETURN
        END
C
C
        SUBROUTINE HARMAN(N,NP,NE,NN,PT,F,FM,NRM,KA,IP,F2,NPT)
        INTEGER F,FM,PT,F2
        REAL IB(6)
        DIMENSION IST(6),PA(15,120),PA1(15,120),APR(160)
        COMMON /AB/G(120)
        COMMON /TIN/TIN(15,48)
        COMMON /BC/YB(160),AP(160),WK(31),B(31),A(31),C(31,32),W(15)
C
100     FORMAT(//)
101     FORMAT(5X,'FREEDOM OF CHOICE',I3/)
102     FORMAT(5X,'MAX NO.OF FREQUENCIES',I3/)
103     FORMAT(5X,'MAX.NO.OF SERIES',I3/)
104     FORMAT(5X,'LENGTH OF EXAMINING SET (C)',I4/)
105     FORMAT(5X,'LENGTH OF TESTING SET (B)',I4/)
106     FORMAT(5X,'LENGTH OF TRAINING SET (A)',I4/)
107     FORMAT(5X,'NO.OF PREDICTION POINTS',I4/)
109     FORMAT(/)
110     FORMAT(2X,7F11.3)
111     FORMAT(2X,'TIME SERIES')
112     FORMAT(10X,'OPTIMAL TREND',/,10X,'------- -----')
113     FORMAT(3X,'SERIES',I3)
114     FORMAT(3X,'NO.OF FREQUENCIES',I3)
115     FORMAT(3X,'FREE TERM',F13.5)
116     FORMAT(3X,'FREQ',12X,'COEFFS A',9X,'COEFFS B',8X,'AMPLITUDE')
```

```
117       FORMAT(F10.7,3F17.6)
118       FORMAT(2X,'ACTUAL VALUES:')
119       FORMAT(5F16.6)
120       FORMAT(2X,'ESTIMATED VALUES:')
121       FORMAT(5X,'PREDICTED VALUES:')
122       FORMAT(I8,2F28.5)
123       FORMAT(I8,F53.5)
124       FORMAT(/)
127       FORMAT(11X,'NO CORRECT DECISION')
C
          NK=N+NP+NE
          N1=N+NP
           NKT=NK+PT
          PI=3.1415926535/2.
          WRITE(3,100)
          WRITE(3,106)N
          WRITE(3,105)NP
          WRITE(3,104)NE
          WRITE(3,102)FM
          WRITE(3,101)F
          WRITE(3,107)PT
          WRITE(3,103)NRM
          WRITE(*,109)
          WRITE(*,111)
          WRITE(*,110)(G(I),I=1,NN)
          NR=1
1         IT=0
2         IT=IT+1
          M=0
3         M=M+1
          MP=2*M
          DO 4 I=1,NK
          IF(NR-1)6,6,5
6         YB(I)=G(I)
          GOTO 4
5         YB(I)=PA(IT,I)
4         CONTINUE
          CALL WB(N,M,M1,IER,KA)
          IF(IER)77,998,77
77        CALL FRIQ(M,M1)
          DO 7 J=1,M
          AN=1.-WK(J)**2
          BN=WK(J)/SQRT(AN)
7         W(J)=PI-ATAN(BN)
          CALL WB1(M,M1,IER)
          IF(IER)78,999,78
78        CALL COEF(M,N,IER)
          IF(IER)79,997,79
79        B1=0.
          B2=0.
          B3=0.
          D1=0.
           D2=0.
           D3=0.
          M1=M+1
          NKM=NK-M
          DO 11 I=M1,NKM
          R=0.
          DO 12 J=1,M
          I1=I+J-1
```

```
      I2=I-J+1
12    R=R+A(J)*(YB(I1)+YB(I2)-2*B(MP+1))
      IM=I+M
       MI=I-M
      R=(YB(IM)+YB(MI)-R-2*B(MP+1))**2
      IF(I-(N-M))80,80,13
80    B1=B1+R
      GOTO 11
13    IF(I-(N1-M))81,81,14
81    B2=B2+R
      GOTO 11
14    B3=B3+R
11    CONTINUE
      AN=N-MP
      BN=B1/AN
      IB(1)=SQRT(BN)
      AN=NP
      BN=B2/AN
      IB(2)=SQRT(BN)
      DO 15 I=1,MP
      R=0.0
      DO 16 J=1,M
      AI=I
      D=W(J)*AI
      J2 = 2*J
      J21 = J2-1
16    R=R+B(J21)*SIN(D)+B(J2)*COS(D)
      D1=D1+(YB(I)-B(MP+1)-R)**2
15   AP(I)=R
      DO 17 I=M1,NKM
      I1=I-M
      R=-AP(I1)
      DO 18 J=1,M
      I1=I+J-1
      I2=I-J+1
18    R=R+A(J)*(AP(I1)+AP(I2))
      I2=I+M
      AP(I2)=R
      D=(YB(I2)-R-B(MP+1))**2
      IF(I2-N)82,82,19
82    D1=D1+D
      GOTO 17
19    IF(I2-N1)83,83,20
83    D2=D2+D
      GOTO 17
20    D3=D3+D
17    CONTINUE
      AN=N
      BN=D1/AN
      IB(4)=SQRT(BN)
      AN=NP
      BN=D2/AN
      IB(5)=SQRT(BN)
      IF(NE)21,21,22
21    IB(3)=0.
      IB(6)=0.
      GOTO 23
22    IB(3)=SQRT(B3/NE)
      IB(6)=SQRT(D3/NE)
23    IF(IT-1)25,84,25
```

```
84      IF(M-F)24,24,25
24      KP=(NR-1)*8+1
        IF(NR-1)26,26,27
26      TIN(M,KP)=0.
        GOTO 28
27      TIN(M,KP)=IT
28      TIN(M,KP+1)=M
        DO 29 I=1,6
        KS=KP+1+I
29      TIN(M,KS)=IB(I)
        DO 30 I=1,NK
30      PA1(M,I)=YB(I)-AP(I)-B(MP+1)
        GOTO 34
25      R=0.
        IZ=0
        DO 31 I=1,F
        KP=(NR-1)*8+IP+2
        D=TIN(I,KP)
        IF(R-D)85,85,31
85      R=D
        IZ=I
31      CONTINUE
55      IF(R-IB(IP))34,34,86
86      DO 32 I=1,NK
32      PA1(IZ,I)=YB(I)-AP(I)-B(MP+1)
        KP=(NR-1)*8+1
        DO 33 I=1,6
        KS=KP+1+I
33      TIN(IZ,KS)=IB(I)
        TIN(IZ,KP)=IT
        TIN(IZ,KP+1)=M
        IF(NR-1)34,87,34
87      TIN(IZ,KP)=0.0
34      IF(M-FM)3,88,88
88      IF(NR-1)89,35,89
89      IF(IT-F)2,35,35
35      CALL PRI(NR,IP,F)
        NR=NR+1
        DO 136 J=1,F
        DO 136 I=1,NK
136     PA(J,I)=PA1(J,I)
        IF(NR-NRM)1,1,90
90      WRITE (3,100)
        WRITE (3,112)
        IZ=1
        NR=1
        P1=TIN(1,IP+2)
        DO 36 I=1,NRM
        KS=(I-1)*8+IP+2
        DO 36 J=1,F
        D=TIN(J,KS)
        IF(D-P1)91,36,36
91      NR=I
        P1=D
        IZ=J
36      CONTINUE
        KP=(NR-1)*8+2
        IST(NR)=TIN(IZ,KP)
        I1=NR-1
        IF(I1)92,382,92
```

```
92      CONTINUE
        DO 37 I=1,I1
        I2=NR-I
        KS=I2*8+1
        IZ=TIN(IZ,KS)
        KS=(I2-1)*8+2
37      IST(I2)=TIN(IZ,KS)
382     DO 38 I=1,NKT
        APR(I)=0.0
        IF(I-NK)39,39,40
39      YB(I)=G(I)
        GO TO 38
40      YB(I)=0.0
38      CONTINUE
381     IZ=1
41      M=IST(IZ)
        MP=2*M
        CALL WB(N,M,M+1,IER,KA)
        IF(IER)999,999,42
42      CALL FRIQ(M,M+1)
        DO 43 J=1,M
        AN=1.0-WK(J)**2
        BN=WK(J)/SQRT(AN)
43      W(J)=PI-ATAN(BN)
        CALL RANG(M,W)
        CALL WB1(M,M+1,IER)
        IF(IER)93,998,93
93      CONTINUE
        CALL COEF(M,N,IER)
        IF(IER)94,997,94
94      CONTINUE
        WRITE(3,113)IZ
        WRITE(3,115)B(MP+1)
        WRITE(3,114)M
        WRITE(3,116)
        DO 46 I=1,M
        I2 = I*2
        I21 = I2-1
        BN=B(I21)**2+B(I2)**2
        P1=SQRT(BN)
46      WRITE(3,117)W(I),B(I21),B(I2),P1
        DO 47 I=1,MP
        R=0.0
        DO 48 J=1,M
        AI=I
        D=W(J)*AI
        J2 = J*2
        J21 = J2-1
48      R=R+B(J21)*SIN(D)+B(J2)*COS(D)
        APR(I)=APR(I)+R+B(MP+1)
47      AP(I)=R
        M1=M+1
        NKM=NK+PT-M
        DO 53 I=M1,NKM
        I1=I-M
        R=-AP(I1)
        DO 49 J=1,M
        IJ1=I+J-1
        IJ2=I-J+1
49      R=R+A(J)*(AP(IJ1)+AP(IJ2))
```

```
      I2=I+M
      AP(I2)=R
53    APR(I2)=APR(I2)+AP(I2)+B(MP+1)
      DO 50 I=1,NK
50    YB(I)=YB(I)-AP(I)-B(MP+1)
      IZ=IZ+1
      IF(IZ-NR)41,41,95
95    CONTINUE
      WRITE(3,100)
      WRITE(3,118)
      WRITE(3,110)(G(I),I=1,NN)
      WRITE(3,109)
      WRITE(3,120)
      WRITE(3,110)(APR(I),I=1,NN)
      GM=0.0
      DO 54 IH=1,NN
      GM=GM+G(IH)
54    CONTINUE
      GM=GM/NN
      CN=0.0
      CD=0.0
      DO 10 IH=1,NN
      CK=G(IH)-APR(IH)
      CN=CN+CK**2
10    CD=CD+(G(IH)-GM)**2
      CK=SQRT(CN/CD)
      WRITE(3,133)CK
133   FORMAT(/5X,'RESIDUAL SUM OF SQUARES =',5X,E18.7/)
      WRITE(3,100)
      WRITE(3,121)
      I1=N1+1
      I2=NK+PT
      DO 51 I=I1,I2
      IF(I-NK)96,96,52
96    CONTINUE
      WRITE(3,122)I,G(I),APR(I)
      GO TO 51
52    WRITE(3,123)I,APR(I)
51    CONTINUE
      GO TO 1001
999   WRITE(*,124)
      GO TO 1000
998   WRITE(*,124)
      GO TO 1000
997   WRITE(*,124)
1000  WRITE(*,127)
1001  RETURN
      END
C
C
      SUBROUTINE PRI(NR,IP,F)
      INTEGER F
      DIMENSION SERV(6)
      COMMON /TIN/TIN(15,48)
10    FORMAT(//,1X,'SERIES',I2)
11    FORMAT(2X,'TRNO',2X,'FRNO',4X,'BAL A',
     1 6X,'BAL B',6X,'BAL C',6X,'ERR A',6X,'ERR B',6X,
     2 'ERR C',/)
12    FORMAT(3X,I3,2X,I4,6E11.3)
13    FORMAT(3X,'---------------------------------------------')
```

```
      K=1
      KP=(NR-1)*8+IP+2
      P=TIN(1,KP)
      IF(F-1)7,4,7
7     DO 1 I=2,F
      IF(TIN(I,KP)-P)2,1,1
2     P=TIN(I,KP)
      K=I
1     CONTINUE
4     WRITE(3,10)NR
      WRITE(3,11)
      KP=(NR-1)*8+1
      DO 3 I=1,F
      DO 5 J=1,6
      KS=KP+1+J
5     SERV(J)=TIN(I,KS)
      MT=TIN(I,KP)
      MF=TIN(I,KP+1)
      WRITE(3,12)MT,MF,(SERV(J),J=1,6)
      IF(I-K)3,6,3
6     WRITE(3,13)
3     CONTINUE
      RETURN
      END
C
C
      SUBROUTINE NEW(N1,N,MAX,EPS,EPS1)
      DIMENSION DB(31)
      COMMON /BC/YB(160),AP(160),C(31),AD(31),A(31),FI(31,32),W(15)
      DO 11 I=1,N1
11    DB(I)=AD(I)
      DO 12 I=1,N1
      I1=N1+1-I
12    AD(I1)=DB(I)
      I=N
      J=1
      N2=N1
1     IF(I-1)20,20,2
2     R=1.0
      M=0
      DO 3 I1=1,I
3     DB(I1)=(N2-I1)*AD(I1)
      F2=1.0
4     CALL FUNC(AD,N2,R,F)
      CALL FUNC(DB,I,R,F1)
      IM=M+1
      IF(ABS(F1)-EPS1)7,7,8
8     F2=F1
7     R=R-F/F2
      M=M+1
      IF(M-MAX)10,5,5
10    IF(ABS(F)-EPS)5,5,4
5     C(J)=R
      J=J+1
      DO 6 I1=1,I
6     AD(I1)=AD(I1)+AD(I1-1)*R
      I=I-1
      N2=N2-1
      GO TO 1
20    C(J)=-AD(2)/AD(1)
```

```
      RR=-AD(2)/AD(1)
      RETURN
      END
C
C
      SUBROUTINE FUNC(A,N1,R,F)
      DIMENSION A(31)
      N=N1-1
      F=A(1)
      DO 1 I=1,N
1     F=F*R+A(I+1)
      RETURN
      END
C
      SUBROUTINE FRIQ(M,M1)
      COMMON /BC/YB(160),AP(160),WK(31),CO(31),A(31),FI(31,32),W(15)
      M1 = M+1
      DO 1 I=1,M
      DO 1 J=1,M1
1     FI(I,J)=0.0
      FI(1,2)=1.0
      IF(M-1)11,27,11
11    FI(2,1)=-1.0
      FI(2,3)=2.0
      IF(M-2)2,2,12
12    DO 3 I=3,M
      I1=I+1
      DO 4 J=2,I1
4     FI(I,J)=2*FI(I-1,J-1)-FI(I-2,J)
3     FI(I,1)=-FI(I-2,1)
2     M2=M-1
      DO 5 I=1,M2
      DO 5 J=1,M1
5     FI(M,J)=FI(M,J)-FI(I,J)*A(I+1)
27    FI(M,1)=FI(M,1)-A(1)
      DO 6 I=1,M1
6     CO(I)=FI(M,I)
      EP=0.000001
      EPS2=0.000001
      EPS3=0.0001
      MAX=25
      EPS=0.000001
      EPS1=0.001
      ETA=0.00001
      DO 66 I=1,M
66    WK(I)=0
      CALL NEW(M1,M,MAX,EPS,EPS1)
      RETURN
      END
C
      SUBROUTINE GAUSS(A,N,L,X,KGA)
      DIMENSION A(31,32),X(31)
      KGA = 1
      L = N+1
      NN=N-1
      DO 99 K=1,NN
      J=K
      KK=K+1
      DO 100 I=KK,N
      IF(ABS(A(J,K)).LT.ABS(A(I,K)))J=I
```

```
 100      CONTINUE
          IF(J.EQ.K)GOTO 11
          DO 300 I=1,L
          T=A(K,I)
          A(K,I)=A(J,I)
          A(J,I)=T
 300    CONTINUE
  11      DO 88 J=KK,N
          IF(A(K,K).EQ.0.)GOTO 600
          D=-A(J,K)/A(K,K)
          DO 400 I=1,L
          A(J,I)=A(J,I)+D*A(K,I)
 400      CONTINUE
  88      CONTINUE
  99      CONTINUE
          IF(A(N,N).EQ.0.)GOTO 600
          X(N)=A(N,L)/A(N,N)
          NN=N-1
          DO 500 J=1,NN
          K=N-J
          SUM=0.0
          NNN=N-K
          DO 200 JJ=1,NNN
          M=K+JJ
          SUM=SUM+A(K,M)*X(M)
 200      CONTINUE
          IF(A(K,K).EQ.0.)GOTO 600
          X(K)=(A(K,L)-SUM)/A(K,K)
 500      CONTINUE
          GOTO 800
600       KGA = 2
          WRITE(*,700)
700       FORMAT(5X,' SINGULAR')
800       RETURN
          END
```

3.2 Sample output

Example. The time series data sample is supplied with a file "ts.dat." The data corresponds to the air-temperature data that is collected at an interval of one day. The control parameters are fed as input:

```
  GIVE NO.OF TRAIN, TEST & EXAM PTS?
45 1 1
  GIVE NO.OF PRED PTS??
5
  GIVE MOVING AVERAGE VALUE (=1 or >1)?
1
  HOW MANY SERIES?
3
  GIVE MAX NO.OF FREQS(<=15)??
8
  GIVE FREEDOM OF CHOICE(< MAX FREQS)??
7
```

One can choose the MOVING AVERAGE VALUE to smooth out the noises in the data; if it is 1, then it takes the data as it is. SERIES indicates the number of layers in the algorithm. Usually, one or two layers are sufficient to obtain the optimal trend. Even if the user chooses more number of layers, it selects the optimal trend from the layer where it achieves the global minimum of the balance relation. MAX NO.OF FREQS which has the limit of less than or equal to 15 indicates the maximum number of distinct frequencies M_{max} to be determined. FREEDOM OF CHOICE denotes the number of optimal trends to be selected at each layer.

The performance of the algorithm is given for each layer. The values of the balance function for training, testing, and examining sets (BAL A, BAL B, BAL C) and their error values (ERR A, ERR B, ERR C) are given correspondingly for each selected trend. The best trends or combinations of the freedom-of-choice are shown. The best one among them according to the balance relation on training set (BAL A) is underlined. TRNO indicates the trend number or combination number from the previous layer and FRNO indicates the number of harmonical components in the current trend. For example, the optimum trend underlined for SERIES 1 has seven frequencies (see output below). The best trend underlined for SERIES 2 has also seven (FRNO =7) harmonical components. This is based on the seventh trend or combination (TRNO =7) of the SERIES 1. Similarly, the best trend in SERIES 3 has one frequency (FRNO =1) and is based on the second trend or combination (TRNO = 2) of the SERIES 2.

The OPTIMAL TREND is collected starting from the SERIES, where the global minimum on the balance relation (BAL A) is achieved, to the first layer. For the output given below, the global minimum is achieved at the SERIES 3 with the value of BAL A equal to 0.101E+01; it has one harmonical component. This is the follow up of the second combination (TRNO = 2) of the SERIES 2. The second combination of the SERIES 2 has eight harmonical components and is the follow up of the sixth trend (TRNO = 6) of the SERIES 1. The sixth one in the SERIES 1 has six harmonic components. This means that the recollected information of the optimal trend includes six harmonical components from the SERIES 1, eight from the SERIES 2, and one from the SERIES 3 along with a FREE TERM from each SERIES; the OPTIMAL TREND is printed giving the values of the FREE TERMs, the frequencies (FREQ), and the coefficients (COEFFS A and B) at each layer along with the AMPLITUDE values. This is represented as

$$\hat{y}_t = \sum_{j=1}^{s}[A_{0j} + \sum_{k=1}^{m_j}(A_{jk}\sin(w_{jk}t) + B_{jk}\cos(w_{jk}t))], \tag{8.15}$$

where $\hat{y}_t$ is the estimated output value; s denotes the number of series in the optimal trend; $m_j, j = 1, 2, \cdots, s$ denote the number of harmonic components at each series; A_{0j} is the free term at jth SERIES; A_{jk} and B_{jk} are the estimated coefficients of the kth component of the jth SERIES; and w_{jk} are the corresponding frequency components.

ACTUAL and ESTIMATED VALUES are given for comparison and the RESIDUAL SUM OF SQUARES (RSS) is computed as

$$\text{RSS} = \sum_{i=1}^{N}\frac{(y_i - \hat{y}_i)^2}{(y_i - \bar{y})^2} \leq 1, \tag{8.16}$$

where y and $\hat{y}$ are the actual and estimated values and $\bar{y}$ is the average value of the time series.

The PREDICTED VALUES are given as specified using the optimal trend; this includes the predictions for the points N_C.

The output is written in the file "output.dat" below.

```
L A Y E R E D  HARMONICAL ALGORITHM
```

LENGTH OF TRAINING SET (A) 45

LENGTH OF TESTING SET (B) 1

LENGTH OF EXAMINING SET (C) 1

MAX NO.OF FREQUENCIES 8

FREEDOM OF CHOICE 7

NO.OF PREDICTION POINTS 5

MAX.NO.OF SERIES 3

SERIES 1

TRNO	FRNO	BAL A	BAL B	BAL C	ERR A	ERR B	ERR C
0	1	0.464E+01	0.620E+00	0.709E+01	0.455E+01	0.131E+01	0.365E+01
0	2	0.651E+01	0.381E+01	0.654E+01	0.427E+01	0.304E+01	0.687E+01
0	8	0.408E+01	0.149E+02	0.358E+01	0.271E+01	0.628E+01	0.950E+01
0	4	0.607E+01	0.650E+01	0.555E+01	0.419E+01	0.300E+00	0.462E+01
0	5	0.486E+01	0.994E+01	0.512E+00	0.442E+01	0.278E+01	0.548E+01
0	6	0.373E+01	0.883E+01	0.320E+01	0.354E+01	0.133E+01	0.111E+01
0	7	0.356E+01	0.121E+02	0.463E+01	0.296E+01	0.522E+01	0.588E+01

SERIES 2

TRNO	FRNO	BAL A	BAL B	BAL C	ERR A	ERR B	ERR C
7	7	0.215E+01	0.606E+01	0.360E+01	0.158E+01	0.401E+01	0.454E+01
6	8	0.236E+01	0.575E+01	0.385E+01	0.919E+00	0.443E+00	0.207E+00
7	8	0.254E+01	0.829E+01	0.338E+01	0.101E+01	0.447E+01	0.407E+01
6	7	0.252E+01	0.673E+01	0.588E+01	0.157E+01	0.275E+01	0.152E+01
3	7	0.235E+01	0.885E+01	0.203E+01	0.183E+01	0.681E+01	0.902E+01
3	5	0.261E+01	0.981E+01	0.151E+01	0.190E+01	0.842E+01	0.972E+01
7	6	0.255E+01	0.809E+01	0.313E+01	0.258E+01	0.671E+01	0.732E+01

SERIES 3

TRNO	FRNO	BAL A	BAL B	BAL C	ERR A	ERR B	ERR C
3	3	0.120E+01	0.457E+01	0.164E+01	0.909E+00	0.435E+01	0.443E+01
2	4	0.133E+01	0.236E+01	0.490E+00	0.784E+00	0.170E+00	0.929E+00
3	2	0.133E+01	0.563E+01	0.359E+01	0.971E+00	0.467E+01	0.428E+01
2	3	0.123E+01	0.171E+01	0.226E-01	0.838E+00	0.386E+00	0.150E-01
2	2	0.116E+01	0.159E+01	0.115E+01	0.874E+00	0.101E+01	0.200E+00
3	8	0.116E+01	0.456E+01	0.503E+00	0.537E+00	0.361E+01	0.256E+01
2	1	0.101E+01	0.596E-01	0.132E+01	0.902E+00	0.323E+00	0.389E+00

OPTIMAL TREND

SERIES 1
FREE TERM -0.56199
NO.OF FREQUENCIES 6

FREQ	COEFFS A	COEFFS B	AMPLITUDE
0.2369936	-1.056414	1.915627	2.187610

```
0.7902706        -2.265249        -1.351049        2.637553
1.0355266        -0.320283         1.655817        1.686509
1.8367290        -0.274392        -0.120682        0.299759
2.1455603         1.113026         0.479222        1.211809
2.5376661         0.573313        -0.212797        0.611531
  SERIES  2
  FREE TERM     -0.09219
  NO.OF FREQUENCIES  8
  FREQ            COEFFS A         COEFFS B        AMPLITUDE
0.1195246        -3.281033        -2.040643        3.863858
0.6629882         1.209835        -0.435315        1.285768
0.9145533        -1.877773        -0.696096        2.002644
1.3779728        -0.100550        -0.039555        0.108051
1.8496013        -0.052124        -0.297579        0.302110
2.0773623         0.101575         0.242814        0.263203
2.3273549         0.492773         0.068364        0.497493
2.7066665         0.342581        -0.085725        0.353144
  SERIES  3
  FREE TERM     -0.00055
  NO.OF FREQUENCIES  1
  FREQ            COEFFS A         COEFFS B        AMPLITUDE
1.8217989         0.012065         0.247733        0.248027
```

```
ACTUAL VALUES:
     -5.000   -10.000    -1.000    -1.500    -1.000     2.000    -8.500
    -12.500   -10.000    -9.000    -4.000     0.000    -0.250    -5.000
     -7.500    -8.000    -7.000    -2.000     2.000     1.000     2.000
      2.000     2.500     3.000     1.750     1.000     0.000     1.000
      4.000     8.000     6.000     2.500     1.500    -2.500    -0.250
      3.000     0.000     3.500     3.000    -0.250    -2.000     1.750
     -0.250     1.000     4.000     1.000     3.000

ESTIMATED VALUES:
     -3.638    -8.640    -1.738    -0.811    -0.339     1.102    -7.365
    -12.481   -11.739    -9.539    -4.904     1.292    -0.541    -6.119
     -7.006    -8.557    -6.686    -0.882     0.505     0.457     2.355
      1.528     2.405     4.178     2.263     0.802     1.061     1.741
      3.362     8.525     5.694     1.881     2.656    -3.908    -0.646
      3.155     0.691     3.938     1.978    -0.462    -0.409    -0.207
     -0.136     1.278     2.909     1.323     2.611

   RESIDUAL SUM OF SQUARES =        0.1963205E+00
```

```
PREDICTED VALUES:
    47                   3.00000                 2.61100
    48                                           0.34542
    49                                          -2.28130
    50                                          -0.90668
    51                                          -1.17158
    52                                           1.71091
```

Epilogue

When we solve any problem of mathematical or logical origin we take either the deductive or inductive (combined) path and develop corresponding theories and algorithms. Deduction is the application of a general law to many partial problems. Induction is the synthesis of a general law from many particular observations. Since childhood, we have learned to prefer the deductive way of thinking. The most respected sciences adhere to the mathematics of deductive science. Theorems are proven on the basis of axiomatic theory. Thus, we conceptualize scientific way as being deductive. Any other way of thinking is referred to as "not proven" or "not scientific", or simply "heuristic or a rule of thumb." But both ways are equally heuristic, and constrained. The main heuristic feature of the deductive approach is an axiom based on *a priori* accepted information, whereas the main heuristic for the inductive approach is its choice of the external criteria.

The choice of axiomatic or external criteria belongs to experts. But experts informed about general possible properties of every type of criteria. Two types of external criteria are considered in this book: accuracy and differential types. The most interesting criteria are of the differential type. Some scientists conclude that the differential type of criteria (for example, balance-of-variables) do not work (Ihara J, 1976); this is true only of noiseless data. The inductive approach is realized in the form of multilayered perceptron-like and combinatorial algorithms. Further developments are described in the book. For example, the use of implicit patterns are suggested, and the objective computer clusterization algorithm and the method of analogues are explained.

The ways to avoid a multivalued choice of decisions are called the "art of regularization." Regularization is a very sophisticated, but interesting area of investigation. Authors are inclined to use the general algebraic approach in all the investigations. By the solution of algebraic and difference equations, the selection characteristic is investigated. It expresses the dependence of an external criterion from the noise dispersion when the length of data sample is small and having constant dispersion of noise. The usual approach in the pattern recognition theory which, on the contrary, includes investigation of the dependence of criterion from the length of data sample. Thus, Shannon's second-limit theorem as a displacement of criterion minimum is proven. The primary part of the book covers this idea as it touches on parametric models. The second part of the book presents new developments on nonparametric algorithms, particularly in the chapter "clustering." All the methods, algorithms, and applications demonstrate the variety of possibilities of inductive methods that, are very sophisticated in the learning mode, but very simple in the application mode. They are not simple realizations of trial and error methods, but are based on sophisticated theory. The inductive approach promises very simple decisions for many difficult tasks.

The success of the Hopfield network with symmetric components partially reaches its solution by the constrained optimization (for example, the traveling salesman problem). Inductive algorithms can be easily applied to this type of problems too. The difference is that in continuous-valued input data it is necessary to use the two-dimensional selection type of algorithmic structures—binary-valued data, two one-dimensional selection type of structures. The inductive approach rivals the deductive and always wins inspite of data sample that is short length and noisy.

The problems show how wide the application of the inductive approach is in systems modeling, pattern recognition, and artificial intelligence is. Authors express their hope that this book would stimulate an interest in developing and applying inductive learning algorithms to various complex systems studies.

Bibliography

[1] Akishin, B. A. and Ivakhnenko, A. G., "Extrapolation (Prediction) Using Monotonically Varying Noisy Data," *Soviet Automatic Control*, **8**, 4, (1975), 17–23.

[2] Aksenova, T. I., "Sufficient Convergence Conditions for External Criteria for Model Selection," *Soviet Journal of Automation and Information Sciences*, **22**, 5, (1989) 49–53.

[3] Aleksander, I. (ed), *Neural Computing Architectures—the Design of Brain-Like Machines*, (MIT press, 1989), 401.

[4] Arbib, M. A., *Brains, Machines and Mathematics*, (Springer Verlag, NY, 2nd edition, 1987).

[5] Arbib, M. A. and Amari, S. (eds.), *Dynamic Interactions in Neural Networks: Models and Data*, (Springer Verlag, 1989), p. 280.

[6] Barron, R. L., "Adaptive Transformation Networks for Modeling, Prediction, and Control," *IEEE Systems, Man and Cybernetics Group Annual Symposium Record*, (1977) 254–263.

[7] Barron, A. R. and Barron, R. L., "Statistical Learning Networks: A Unifying View," *Symposium on the Interface: Statistics and Computer Science*, Reston, Virginia (1988).

[8] Basar, E., Flohr, H., Haken, H. and Mandell, J. (eds.), *Synergetics of Brain*, (Springer Verlag 1983).

[9] Beck, M. B., "Modeling of Dissolved Oxygen in a Nontidal Stream," in James, A. (ed.) *The Use of Mathematical Models in Water Pollution Control*, (Wiley, NY, 1976), 1–38.

[10] Beer, S. T., *Cybernetics and Management*, (1963).

[11] Box, G. E. P. and Jenkins, G. M., "*Time Series Analysis; Forecasting and Control*," (Holden-Day, San Francisco, USA, Revised edition, 1976).

[12] Duffy, J. and Franklin, M., "An Identification Algorithm and Its Application to an Environmental System," *IEEE Transactions on Systems, Man, and Cybernetics*, **SMC 5**, 2, (1975) 226–240.

[13] Dyshin, O. A., "Noise Immunity of the Selection Criteria for Regression Models with Correlated Perturbations," *Soviet Journal of Automation and Information Sciences*, **21**, 3, (1988) 16–24.

[14] Dyshin, O. A., "Asymptotic Properties of Noise Immunity of the Criteria of Model Accuracy," *Soviet Journal of Automation and Information Sciences*, **22**, 1, (1989) 91–98.

[15] Edelman, G. M., *Neural Darwinism—the Theory of Neural Group Selection*, (Basic Books, 1987), 371.

[16] Farlow, S. J. (ed.), *Self Organizing Methods in Modeling: GMDH Type Algorithms*, (Marcel Dekker Inc., New York, 1984), 350.

[17] Feigenbaum, E. A. and McCorduck, P., *The Fifth Generation*, (Pan Books, 1983).

[18] Fogelman, Soulie, F., Robert, Y. and Tchuente, M. (eds), *Automata Networks in Comp Sci: Theory and Applications*, (Princeton University press, 1987).

[19] Fokas, A. S., Papadopoulou, E. P. and Saridakis, Y. G., "Soliton Cellular Automata," *Physica D*, **41**, (1990), 297–321.

[20] Forrester, J. W., *World Dynamics*, (Wright-Allen Press, 1971).

[21] Gabor, D., Wildes, W. and Woodcock, R., "A Universal Nonlinear Filter, Predictor and Simulator Which Optimizes Itself by a Learning Process," *IEE Proceedings*, **108B**, (1961), 422–438.

[22] Gabor, D., "Cybernetics and the Future of Industrial Civilization," *Journal of Cybernetics*, **1**, (1971), 1–4.

[23] Heisenberg, W., *The Physical Principles of the Quantum Theory*, Eckart, E. and Hoyt, C. (Trans.), (University of Chicago press, Chicago, IL, 1930), 183.

[24] Hinton, G. E. and Anderson, J. (eds.), *Parallel Models of Associative Memory*, (Hillsdale, New Jersey: Lawrence Erlbaum 1981).

[25] Holland, J. H., Holyoak, K J., Nisbett, R. E. and Thagard, P. R., *Induction: Processes of Inference, Learning, and Discovery*, (MIT Press, Cambridge, Massachusetts 1986).

[26] Hopfield, J. J., "Neural Networks and Physical Systems with Emergent Collective Computational Abilities," *Proceedings Natl Acad Sci USA*, **79**, (1982), 2554–2558.

[27] Ihara, J., "Unique Selection of Model by Balance-of-Variables Criterion—Letter to the Editor and Authors' Reply," *Soviet Automatic Control*, **9**, 1, (1976), 70–72.

[28] Ikeda, S., Ochiai, M. and Sawaragi, Y., "Sequential GMDH Algorithm and Its Application to River Flow Prediction," *IEEE Transactions on Systems, Man and Cybernetics*, **SMC** , (1976), 473–479.

[29] Ivakhnenko, A. G., "Polynomial Theory of Complex Systems," *IEEE transactions on Systems, Man, and Cybernetics*, **SMC-1**, 4, (1971), 364–378.

[30] Ivakhnenko, A. G., "The Group Method of Data Handling in Long-Range Forecasting," *Technological Forecasting and Social Change*, **12**, 2/3, (1978), 213–227.

[31] Ivakhnenko, A. G., "Prediction of the Future: State of the Art and Perspectives," *Soviet Automatic Control*, **13**, 2, (1980), 77–81.

[32] Ivakhnenko, A. G., "Features of the Group Method of Data Handling Realizable in An Algorithm of Two-Level Long-Range Quantitative Forecasting," *Soviet Automatic Control*, **16**, 2, (1983), 1–8.

[33] Ivakhnenko, A. G., "Dialogue Language Generalization as a Method for Reducing the Participation of a Man in Solving Problems of System Analysis," *Soviet Automatic Control*, **16**, 5, (1983), 1–11.

[34] Ivakhnenko, A. G. and Ivakhnenko, N. A., "Nonparametric GMDH Predicting Models Part 2. Indicative Systems for Selective Modeling, Clustering, and Pattern Recognition," *Soviet Journal of Automation and Information Sciences*, **22**, 2, (1989), 1–10.

[35] Ivakhnenko, A. G. and Karpinskiy, A. M., "Computer Self Organization of Models in Terms of General Communication Theory (Information Theory)," *Soviet Automatic Control*, **15**, 4, (1982), 5–22.

[36] Ivakhnenko, A. G. and Kocherga, Yu L., "Theory of Two-level GMDH Algorithms for Long-Range Quantitative Prediction," *Soviet Automatic Control*, **16**, 6, (1983), 7–12.

[37] Ivakhnenko, A. G., Koppa, Yu V., Lantayeva, D. N. and Ivakhnenko, N. A., "The Relationship Between Computer Self Organization of Mathematical Models and Pattern Recognition," *Soviet Automatic Control*, **13**, 3, (1980), 1–9.

[38] Ivakhnenko, A. G., Koppa, Yu V. and Kostenko, Yu V., "Systems Analysis and Long-Range Quantitative Prediction of Quasi-static Systems on the Basis of Self-Organization of Models, Part 3. Separation of Output Variables According to Degree of Exogenicity for Restoration of the Laws Governing the Modeling Object," *Soviet Automatic Control*, **17**, 4, (1984), 7–14.

[39] Ivakhnenko, A. G., Koppa, Yu V., Petukhova, S. A. and Ivakhnenko, M. A., "Use of Self-Organization to Partition a Set of Data into Clusters Whose Number is not Specified in Advance," *Soviet Journal of Automation and Information Sciences*, **18**, 5, (1985), 7–14.

[40] Ivakhnenko, A. G. and Kostenko Yu V., "Systems Analysis and Long-Range Quantitative Prediction of Quasistatic Systems on the Basis of Self-Organization of Models. Part I. Systems Analysis at the Level of Trends," *Soviet Automatic Control*, **15**, 3, (1982), 9–17.

[41] Ivakhnenko, A. G., Kostenko, Yu V. and Goleusov, I. V., "Systems Analysis and Long-Range Quantitative Prediction of Quasistatic Systems on the Basis of Self-Organization of Models. Part 2. Objective Systems Analysis without em a priori Specification of External Influences," *Soviet Automatic Control*, **16**, 3, (1983), 1–8

[42] Ivakhnenko, A. G., Kovalchuk, P. I., Todua, M. M., Shelud'ko, O. I., and Dubrovin, O. F., "Unique Construction of Regression Curve Using a Small Number of Points—Part 2," *Soviet Automatic Control*, **6**, 5, (1973), 29–41.

[43] Ivakhnenko, A. G., Kovalenko, S. D., Kostenko, Yu V. and Krotov G. I., "An Experiment of Self-organization of the Models for Forecasting Radio-Communication Conditions," *Soviet Automatic Control*, **16**, 6, (1983), 1–6.

[44] Ivakhnenko, A. G. and Kozubovskiy, S. F., "The Correlation Interval as a Measure of the Limit of Predictability of a Random Process and Detailization of the Modeling Language," *Soviet Automatic Control*, **14**, 4, (1981), 1–6.

[45] Ivakhnenko, A. G. and Kritskiy, A. P., "Recovery of a Signal or a Physical Model by Extrapolating the Locus of the Minima of the Consistency Criterion," *Soviet Journal of Automation and Information Sciences*, **19**, 3, (1986), 25–31.

[46] Ivakhnenko, A. G. and Krotov, G. I., "Simulation of Environmental Pollution in the Absence of Information about Disturbances," *Soviet Automatic Control*, **10**, 5, (1977), 8–22.

[47] Ivakhnenko, A. G. and Krotov G. I., "Comparative Studies in Self-Organization of Physical Field Models," *Soviet Automatic Control*, **11**, 5, (1978), 42–52.

[48] Ivakhnenko, A. G., Krotov, G. I. and Cheberkus, V. I., "Multilayer Algorithm for Self organization of Long Term Predictions (Illustrated by the Example of the Lake Baikal Ecological System)," *Soviet Automatic Control*, **13**, 4, (1980), 22–38.

[49] Ivakhnenko, A. G., Krotov, G. I. and Stepashko, V., "Harmonic and Exponential Harmonic GMDH Algorithms, Part 2. Multilayer Algorithms with and without Calculation of Remainders," *Soviet Automatic Control*, **16**, 1, (1983), 1–9.

[50] Ivakhnenko, A. G. and Krotov, G. I., "Modeling of a GMDH Algorithm for Identification and Two-Level Long-Range Prediction of the Ecosystem of Lake Baykal," *Soviet Automatic Control*, **16**, 2, (1983), 9–14.

[51] Ivakhnenko, A. G. and Krotov, G. I., "A Multiplicative-additive Nonlinear GMDH Algorithm with Optimization of the Power of Factors," *Soviet Automatic Control*, **17**, 3, (1984), 10–15.

[52] Ivakhnenko, A. G., Krotov, G. I. and Kostenko Yu V., "Optimization of the Stability of the Transient Component of a Long-Range Prediction," *Soviet Journal of Automation and Information Sciences*, **18**, 4, (1985), 1–9.

[53] Ivakhnenko, A. G., Krotov, G. I. and Strokova, T. I., "Self-Organization of Dimensionless Harmonic-exponential and Correlation Predicting Models of Standard Structure," *Soviet Automatic Control*, **17**, 4, (1984), 15–26.

[54] Ivakhnenko, A. G., Krotov, G. I. and Yurachkovskiy, Yu P., "An Exponential-harmonic Algorithm of the Group Method of Data Handling," *Soviet Automatic Control*, **14**, 2, (1981), 21–27.

[55] Ivakhnenko, A. G., Osipenko, V. V. and Strokova, T. I., "Prediction of Two-dimensional Physical Fields Using Inverse Transition Matrix Transformation," *Soviet Automatic Control*, **16**, 4, (1983), 10–15.

[56] Ivakhnenko, A. G., Peka, P. Yu and Koshul'ko, A. I., "Simulation of the Dynamics of the Mineralization Field of Aquifers with Optimization of Porosity Estimate of the Medium," *Soviet Automatic Control*, **9**, 4, (1976), 28–35.

[57] Ivakhnenko, A. G., Peka, P. Yu and Yakovenko, P. I., "Identification of Dynamic Equations of a Complex Plant on the Basis of Experimental Data by Using Self-Organization of Models Part 2. Multidimensional Problems," *Soviet Automatic Control*, **10**, 2, (1977), 31–37.

[58] Ivakhnenko, A. G. and Madala H. R., "Prediction and Extrapolation of Meteorological Fields by Model Self Organization," *Soviet Automatic Control*, **12**, 6, (1979), 13–27.

[59] Ivakhnenko, A. G. and Madala H. R., "Self-Organization GMDH Algorithms for Modeling and Long-Term Prediction of Cyclic Processes such as Tea Crop Production," *Proceedings of International Coneference on Systems Engineering, Coventry Polytechnic, England, UK*, (1980), 580–594.

[60] Ivakhnenko, A. G. and Madala H. R., "Application of the Group Method of Data Handling to the Solution of Meteorological and Climatological Problems," *Soviet Journal of Automation and Information Sciences*, **19**, 1, (1986), 72–80.

[61] Ivakhnenko, A. G., Sarychev, A. P., Zalevskiy, P. I. and Ivakhnenko, N. A., "Experience of Solving the Problem of Predicting Solar Activity with Precise and Robust Approaches," *Soviet Journal of Automation and Information Sciences*, **21**, 3, (1988), 31–42.

[62] Ivakhnenko, A. G., Sirenko, L. A., Denisova, A. I., Ryabov, A. I., Sarychev, A. P. and Svetalskiy, B. K., "Objective Systems Analysis of the Ecosystem of the Kakhovka Reservoir Using the Unbiasedness Criterion," *Soviet Automatic Control*, **16**, 1, (1983), 64–73.

[63] Ivakhnenko, A. G. and Stepashko, V. S., "Numerical Investigation of Noise Stability of Multicriterion Selection of Models," *Soviet Automatic Control*, **15**, 4, (1982), 23–32.

[64] Ivakhnenko, A. G., Stepashko, V. S., Khomovnenko, M. G. and Galyamin, E. P., "Self Organization Models of Growth Dynamics in Agricultural Production for Control of Irrigated Crop Rotation," *Soviet Automatic Control*, **10**, 5, (1977), 23–33.

[65] Ivakhnenko, A. G., Stepashko, V. S., Kostenko, Yu V., Zhitorchuk, Yu V. and Madala H. R., "Self organization of Composite Models for Prediction of Cyclic Processes by using Prediction Balance Criterion," *Soviet Automatic Control*, **12**, 2, (1979), 8–21.

[66] Ivakhnenko, A. G., Svetalskiy, B. K., Sarychev A. P., Denisova A. I., Sirenko, L. A., Nakhshina, E. P. and Ryabov, A. K., "Objective Systems Analysis and Two-Level Long-Range Forecast for the Ecological Systems of Kakhovka and Kremenchug Reservoirs," *Soviet Automatic Control*, **17**, 2, (1984), 26–36.

[67] Ivakhnenko, A. G., Vysotskiy, V. N. and Ivakhnenko, NA, "Principal versions of the Minimum Bias Criterion for a Model and an Investigation of Their Noise Immunity," *Soviet Automatic Control*, **11**, 1, (1978), 27–45.

[68] Ivakhnenko, M. A. and Timchenko, I. K., "Extrapolation and Prediction of Physical Fields Using Discrete Correlation Models," *Soviet Journal of Automation and Information Sciences*, **18**, 4, (1985), 19–26.

[69] Ivakhnenko, N. A., "Investigation of the Criterion of Clusterization Consistency by Computational Experiments," *Soviet Journal of Automation and Information Sciences*, **21**, 4, (1988), 23–26.

[70] Ivakhnenko, S. A., Lu, I., Semina, L. P. and Ivakhnenko, A. G., "Objective Computer Clusterization Part 2. Use of Information about the Goal Function to Reduce the Amount of Search," *Soviet Journal of Automation and Information Sciences*, **20**, 1, (1987), 1–13.

[71] Ivakhnenko, N. A., Semina, L. P. and Chikhradze, T. A., "A Modified Algorithm for Objective Clustering of Data," *Soviet Journal of Automation and Information Sciences*, **19**, 2, (1986), 9–18.

[72] Kendall, M. G., *Rank Correlation Methods*, (C. Griffin, London, 3rd edition, 1962), 199.

[73] Kendall, M. G., *Time Series*, (C. Griffin, London, 1973).

[74] Khomovnenko, M. G. and Kolomiets, N. G., "Self-Organization of a System of Simple Partial Models for Predicting the Wheat Harvest," *Soviet Automatic Control*, **13**, 1, (1980), 22–29.

[75] Khomovnenko, M. G., "Self-Organization of Potentially Efficient Crop Yield Models for an Automatic Irrigation Control System," *Soviet Automatic Control*, **14**, 6, (1981), 54–61.

[76] Klein, L. P., Mueller, I. A. and Ivakhnenko, A. G., "Modeling of the Economics of the USA by Self-Organization of the System of Equations," *Soviet Automatic Control*, **13**, 1, (1980), 1–8.

[77] Kohonen, T., *Self Organization and Associative Memory*, (Springer Verlag, 2nd edition, 1988) 312.

[78] Kondo, J., *Air Pollution*, (Tokyo, Corono Co. 1975).

[79] Kovalchuk, P. I., "Internal Convergence of GMDH Algorithms," *Soviet Automatic Control*, **16**, 2, (1983), 88–91.

[80] Lebow, W. M., Mehra, R. K., Toldalagi, P. M. and Rice H., "Forecasting Applications of GMDH in Agricultural and Meteorological Time Series," in Farlow S. J. (ed), *Self-organization Methods in Modeling: GMDH Type Algorithms*, (Marcel Dekker, NY, 1984), 121–147.

[81] Lerner, E. J., "The Great Weather Network," *IEEE Spectrum* , February, (1982), 50–57.

[82] Lippmann, R. P., "An Introduction to Computing with Neural Nets," *IEEE Acoustics, Speech, and Signal Processing (ASSP) Mag April*, (1987), 4–22.

[83] Lorenz, E. N., "Atmospheric Predictability as Revealed by Naturally Occurring Analogues," *Journal of the Atmospheric Sciences*, **26**, 4, (1969), 636–646.

[84] Lorenz, E. N., "Predictability and Periodicity. A review and Extension," *Third Conference on Probability and Statistics in Atmospheric Sciences*, June 11–22, (1971), 1–4.

[85] Maciejowsky, J. M., *Modeling of systems with Small Observation Sets*, (Lecture Notes in Control and Information Sciences, 10, 6, 1978), 242.

[86] Madala, H. R. and Lantayova, D., "Group Method of Data Handling (GMDH)—A Survey," *Proceedings of the 15th Annual Computer Society of India Convention, Bombay, India*, Part II, (1980), 108–114.

[87] Madala, H. R., "Self-organization GMDH Computer Aided Design for Modeling of Cyclic Processes," *Proceedings of IFAC Symposium on Computer Aided Design, W Lafayette, IN, USA*, (1982), 611–618.

[88] Madala, H. R., "System Identification Tutorials," *Technical Report 1982:042T*, University of Lulea, Lulea, Sweden, (1982).

[89] Madala, H. R., "A New Harmonical Algorithm for Digital Signal Processing," *Proceedings of IEEE Acoustics, Speech, and Signal Processing, San Diego, CA, USA*, (1984), 671–674.

[90] Madala, H. R., "Layered Inductive Learning Algorithms and Their Computational Aspects," in Bourbakis N. G. (ed.) *Applications of Learning and Planning Methods*, (World Scientific, Singapore, 1991), 49–69.

[91] Madala, H. R., "Comparison of Inductive Versus Deductive Learning Networks," *Complex Systems*, **5**, 2, (1991), 239–258.

[92] Madala, H. R., "Simulation Studies of Self Organizing Network Learning," *International Journal of Mini and Microcomputers*, **13**, 2, (1991), 69–76.

[93] McCulloch, W. S. and Pitts, W., "A Logical Calculus of the Ideas Immanent in Nervous Activity," *Bull Math Biophys*, **5**, (1943), 115–133.

[94] Mehra, R. K., "GMDH Reviews and Experience," *Proceedings of IEEE Conference on Decision and Control, New Orleans*, (1977), 29–34.

[95] Minsky, M. and Papert, S., *Perceptrons: an Introduction to Computational Geometry*, (MIT Press, Cambridge 1969).

[96] Newell, A. and Simon, H. A., "Computer Simulation of Human Thinking," *Science*, **134**, (1961), 2011–2017.

[97] Newell, A. and Simon, H. A., *Human Problem Solving*, (Englewood Cliffs, New Jersey: Prentice Hall 1972).

[98] Nguyen, D. H. and Widrow, B., "Neural Networks for Self Learning Control Systems," *IEEE Control Sys Mag April*, (1990), 18–23.

[99] Patarnello, S. and Carnevali, P., "Learning Capabilities of Boolean Networks," in Aleksander I. (ed.), *Neural Computing Architectures—the Design of Brain-Like Machines*, (The MIT Press, Cambridge, MA., 1989), 117–129.

[100] Patarnello, S. and Carnevali, P., "Learning Networks of Neurons with Boolean Logic," *Europhysics Letters*, **4**, 4, (1987), 503–508.

[101] Poggio, T. and Girosi, F., "Networks for Approximation and Learning," *Proceedings IEEE*, **78**, 9, (1990), 1481–1497.

[102] Price, W. C. and Chissick, S. S. (eds.), *The Uncertainty Principle and Foundations of Quantum Mechanics: a Fifty Years' Survey*, (John Wiley Sons, New York, 1977), 572.

[103] Psaltis, D. and Farhat, N., "Optical Information Processing Based on an Associative-Memory Model of Neural Nets with Thresholding and Feedback," *Optics Letters*, **10**, 1985, 98–100.

[104] Rao, C. R., *Linear Statistical Inference and Its Applications*, (John Wiley, NY 1965).

[105] Rosenblatt, F., "The Perceptron—a Probabilistic Model for Information Storage and Organization in the Brain," *Psychological Review*, **65**, 6, (1958) 386–408.

[106] Rosenblatt, F., *Principles of Neurodynamics: Perceptrons and the Theory of Brain Mechanisms*, (Spartan Books 1962).

[107] Rumelhart, D. E., McClelland, J. R. and the PDP Research Group, *Parallel Distributed Processing: Explorations in the Micro Structure of Cognition*, (Vol.1, Cambridge, Massachusetts: MIT Press 1986).

[108] Sawaragi, Y., Soeda, T. and Tamura, H., "Statistical Prediction of Air Pollution Levels Using Nonphysical Models," *Automatica*, **15**, 4, (1979) 453–460.

[109] Scott, D. S. and Hutchison, C. E., *Modeling of Economical Systems*, (Univ of Massachusettes, Boston, 1975), 115.

[110] Shankar, R., *The GMDH*, (Master of Electrical Engineering Thesis, Univ of Delaware, Newark, 1972), 250.

[111] Shannon, C. E., *The Mathematical Theory of Communication*, (Univ of Illinois press, Urbana, 1949), 117.

[112] Shelud'ko, O. I., "GMDH Algorithm with Orthogonalized Complete Description for Synthesis of Models by the Results of a Planned Experiment," *Soviet Automatic Control*, **7**, 5, (1974), 24–33.

[113] Simon, H. A., *Models of Discovery*, (D. Reidel Publishing Co., Dordrecht, Holland, 1977).

[114] Stepashko, V. S., "Optimization and Generalization of Model Sorting Schemes in Algorithms for the Group Method of Data Handling," *Soviet Automatic Control*, **12**, 4, (1979), 28–33.

[115] Stepashko, V. S., "A Combinatorial Algorithm of the Group Method of Data Handling with Optimal Model Scanning Scheme," *Soviet Automatic Control*, **14**, 3, (1981), 24–28.

[116] Stepashko, V. S., "A Finite Selection Procedure for Pruning an Exhaustive Search of Models," *Soviet Automatic Control*, **16**, 4, (1983), 88–93.

[117] Stepashko, V. S., "Noise Immunity of Choice of Model Using the Criterion of Balance of Predictions," *Soviet Automatic Control*, **17**, 5, (1984), 27–36.

[118] Stepashko, V. S., "Selective Properties of the Consistency Criterion of Models," *Soviet Journal of Automation and Information Sciences*, **19**, 2, (1986), 38–46.

[119] Stepashko, V. S. and Kocherga, Yu L., "Classification and Analysis of the Noise Immunity of External Criteria for Model Selection," *Soviet Automatic Control*, **17**, 3, (1984), 36–47.

[120] Stepashko, V. S., "Asymptotic Properties of External Criteria for Model Selection," *Soviet Journal of Automation and Information Sciences*, **21**, 6, (1988), 84–92.

[121] Stepashko, V. S. and Zinchuk, N. A., "Algorithms for Calculating the Locus of Minima for a Criterion of Accuracy of Models," *Soviet Journal of Automation and Information Sciences*, **22**, 1, (1989), 85–90.

[122] Tamura, H. and Kondo, T., "Large spatial Pattern Identification of Air Pollution by Computer Model of Source Receptor and Revised GMDH," *Proceedings IFAC Symposium on Environmental Systems Planning, Design and Control, Kyoto, Japan*, (1977), 167–171.

[123] Tumanov, N. V., "A GMDH Algorithm with Mutually Orthogonal Partial Descriptions for Synthesis of Polynomial Models of Complex Objects," *Soviet Automatic Control*, **11**, 3, (1978), 82–84.

[124] Tou, J. T. and Gonzalez, R. C., *Pattern Recognition Principles*, (Addison-Wesley Publ. Co., Reading, MA, 1974), 377.

[125] van Zyl, J. G., "Experiments in Socioeconomic Forecasting Using Ivakhnenko's Approach," *Appl. Math. Modeling*, **2**, 3, (1978) 49–56.

[126] von Neumann, J., *Theory of Self Reproducing Automata*, (University of Illinois Press, Urbana 1966).

[127] Vysotskiy, V. N., Ivakhnenko, A. G., and Cheberkus, V. I., "Long Term Prediction of Oscillatory Processes by Finding a Harmonic Trend of Optimum Complexity by the Balance-of-Variables Criterion," *Soviet Automatic Control*, **8**, 1 (1975), 18–24.

[128] Vysotskiy, V. N., "Optimum Partitioning of Experimental Data in GMDH Algorithms," *Soviet Automatic Control*, **9**, 3 (1976), 62–65.

[129] Vysotskiy, V. N. and Ihara, J., "Improvement of Noise Immunity of GMDH Selection Criteria by Using Vector Representations and Minimax Forms," *Soviet Automatic Control*, **11**, 3 (1978), 1–8.

[130] Vysotskiy, V. N. and Yunusov, N. I., "Improving the Noise Immunity of a GMDH Algorithm Used for Finding a Harmonic Trend with Nonmultiple Frequencies," *Soviet Automatic Control*, **10**, 5 (1977), 57–60.

[131] Widrow, B. and Hoff, M. E., Jr., "Adaptive Switching Circuits," *Western Electronic Show and Convention Record 4*, Institute of Radio Engineers, (1960), 96–104.

[132] Widrow, B., Winter, R. G. and Baxter, R. A., "Layered Neural Nets for Pattern Recognition," *IEEE Transactions on Acoustics, Speech, and Signal Processing*, **36**, 7, (1988), 1109–1118.

[133] Wiener, N., *Cybernetics: or Control and Communication in the Animal and the Machine*, (The Technology Press and Wiley, 1948; The MIT Press, 1961, 2nd edition).

[134] Yurachkovskiy, Yu P., "Convergence of Multilayer Algorithms of the Group Method of Data Handling," *Soviet Automatic Control*, **14**, 3 (1981), 29–34.

[135] Yurachkovskiy, Yu P. and Groshkov, A. N., "Application of the Canonical Form of External Criteria for Investigating their Properties," *Soviet Automatic Control*, **12**, 3 (1979), 76–80.

[136] Yurachkovskiy, Yu P. and Mamedov, M. I., "Internal Convergence of Two GMDH Algorithms," *Soviet Automatic Control*, **18**, 1, (1985), 96–100.

[137] Yurachkovskiy, Yu P., "Use of Karhunen-Loeve Expansion to Construct a Scalar Convolution of a Vector Criterion," *Soviet Journal of Automation and Information Sciences*, **20**, 1, (1987), 14–22.

[138] Yurachkovskiy, Yu P., "Analytical Construction of Optimal Quadratic Discriminating Criteria," *Soviet Journal of Automation and Information Sciences*, **21**, 1, (1988), 1–10.

Index

100.028 Coefficients :
3747.018 -1788.922 -735.772 -156.845 36.246 252.496 -155.687 241.027
-359.088